W0228109

Serono Symposia, USA
Norwell, Massachusetts

PROCEEDINGS IN THE SERONO SYMPOSIA, USA SERIES

FUNCTION OF SOMATIC CELLS IN THE TESTIS
Edited by Andrzej Bartke

GLYCOPROTEIN HORMONES: Structure, Function, and Clinical Implications
Edited by Joyce W. Lustbader, David Puett, and Raymond W. Ruddon

GROWTH HORMONE II: Basic and Clinical Aspects
Edited by Barry B. Bercu and Richard F. Walker

TROPHOBLAST CELLS: Pathways for Maternal-Embryonic Communication
Edited by Michael J. Soares, Stuart Handwerger, and Frank Talamantes

IN VITRO FERTILIZATION AND EMBRYO TRANSFER IN PRIMATES
Edited by Don P. Wolf, Richard L. Stouffer, and Robert M. Brenner

OVARIAN CELL INTERACTIONS: Genes to Physiology
Edited by Aaron J.W. Hsueh and David W. Schomberg

CELL BIOLOGY AND BIOTECHNOLOGY: Novel Approaches to Increased
Cellular Productivity
Edited by Melvin S. Oka and Randall G. Rupp

PREIMPLANTATION EMBRYO DEVELOPMENT
Edited by Barry D. Bavister

MOLECULAR BASIS OF REPRODUCTIVE ENDOCRINOLOGY
Edited by Peter C.K. Leung, Aaron J.W. Hsueh, and Henry G. Friesen

MODES OF ACTION OF GnRH AND GnRH ANALOGS
Edited by William F. Crowley, Jr., and P. Michael Conn

FOLLICLE STIMULATING HORMONE: Regulation of Secretion and Molecular
Mechanisms of Action
Edited by Mary Hunzicker-Dunn and Neena B. Schwartz

SIGNALING MECHANISMS AND GENE EXPRESSION IN THE OVARY
Edited by Geula Gibori

GROWTH FACTORS IN REPRODUCTION
Edited by David W. Schomberg

UTERINE CONTRACTILITY: Mechanisms of Control
Edited by Robert E. Garfield

NEUROENDOCRINE REGULATION OF REPRODUCTION
Edited by Samuel S.C. Yen and Wylie W. Vale

FERTILIZATION IN MAMMALS
Edited by Barry D. Bavister, Jim Cummins, and Eduardo R.S. Roldan

GAMETE PHYSIOLOGY
Edited by Ricardo H. Asch, Jose P. Balmaceda, and Ian Johnston

GLYCOPROTEIN HORMONES: Structure, Synthesis, and Biologic Function
Edited by William W. Chin and Irving Boime

THE MENOPAUSE: Biological and Clinical Consequences of Ovarian Failure:
Evaluation and Management
Edited by Stanley G. Korenman

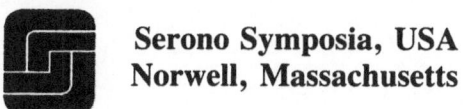

Serono Symposia, USA
Norwell, Massachusetts

Andrzej Bartke   Editor

Function of Somatic Cells in the Testis

With 165 Figures

Springer-Verlag
New York Berlin Heidelberg London Paris
Tokyo Hong Kong Barcelona Budapest

Andrzej Bartke, Ph.D.
Department of Physiology
School of Medicine
Southern Illinois University
Carbondale, IL 62901-6512
USA

Proceedings of the XIIth North American Testis Workshop on Function of Somatic Cells in the Testis, sponsored by Serono Symposia, USA, held April 13 to 16, 1993, in Tampa, Florida.

For information on previous volumes, please contact Serono Symposia, USA.

Library of Congress Cataloging-in-Publication Data
Function of somatic cells in the testis/Andrzej Bartke, editor.
    p. cm.
    "Proceedings of the XIIth North American Testis Workshop on Function of Somatic Cells in the Testis, sponsored by Serono Symposia, USA, held April 13 to 16, 1993, in Tampa, Florida"—T.p. verso.
    At head of title: Serono Symposia, USA, Norwell, Massachusetts.
    Includes bibliographical references and indexes.
    ISBN-13: 978-1-4612-7619-7   e-ISBN-13: 978-1-4612-2638-3
    DOI: 10.1007/978-1-4612-2638-3
    1. Testis—Physiology—Congresses. 2. Somatic cells—Congresses.
I. Bartke, Andrzej. II. Serono Symposia, USA. III. North American Testis Workshop on Function of Somatic Cells in the Testis (1993: Tampa, Fla.)
    [DNLM: 1. Hybrid Cells—physiology—congresses. 2. Testis—physiology—congresses. WJ 830 F979 1993]
    QP255.F86 1994
    612.6'1—dc20
    DNLM/DLC   93-39374

Printed on acid-free paper.

© 1994 Springer-Verlag New York, Inc.
Softcover reprint of the hardcover 1st edition 1994

All rights reserved. This work may not be translated or copied in whole or in part without the written permission of the publisher (Springer-Verlag New York, Inc., 175 Fifth Avenue, New York, NY 10010, USA), except for brief excerpts in connection with reviews or scholarly analysis. Use in connection with any form of information storage and retrieval, electronic adaptation, computer software, or by similar or dissimilar methodology now known or hereafter developed is forbidden.
The use of general descriptive names, trade names, trademarks, etc., in this publication, even if the former are not especially identified, is not to be taken as a sign that such names, as understood by the Trade Marks and Merchandise Marks Act, may accordingly be used freely by anyone.
While the advice and information in this book are believed to be true and accurate at the date of going to press, neither the authors, nor the editors, nor the publisher, nor Serono Symposia, USA, nor Serono Laboratories, Inc., can accept any legal responsibility for any errors or omissions that may be made. The publisher makes no warranty, expressed or implied, with respect to the material contained herein.
Permission to photocopy for internal or personal use, or the internal or personal use of specific clients, is granted by Springer-Verlag New York, Inc., for libraries registered with the Copyright Clearance Center (CCC), provided that the base fee of $5.00 per copy, plus $0.20 per page is paid directly to CCC, 21 Congress Street, Salem, MA 01970, USA. Special requests should be addressed directly to Springer-Verlag New York, Inc., 175 Fifth Avenue, New York, NY 10010, USA.

Production coordinated by Marilyn Morrison and managed by Francine McNeill; manufacturing supervised by Vincent Scelta.
Typeset by Best-set Typesetter Ltd., Hong Kong.

9 8 7 6 5 4 3 2 1

XIIth NORTH AMERICAN TESTIS WORKSHOP ON FUNCTION OF SOMATIC CELLS IN THE TESTIS

Scientific Committee

Andrzej Bartke, Ph.D., Chairman
Southern Illinois University
Carbondale, Illinois

Organizing Secretary

Bruce K. Burnett, Ph.D.
Serono Symposia, USA
100 Longwater Circle
Norwell, Massachusetts

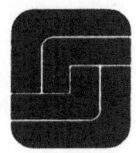

Preface

The XIIth North American Testis Workshop was held in Tampa, Florida, April 13 to 16, 1993. The program consisted of 23 invited lectures and 120 poster presentations. In keeping with the tradition of the Testis Workshop, every attempt was made to emphasize recent developments in the study of endocrine and gametogenic functions of the male gonad. Although the main emphasis, reflected in the title of the proceedings, was on the function and control of somatic cells in the testis, one session was devoted to germ cells and one to the studies of male reproductive development and function. The last session of the workshop was included in the program of the Postgraduate Course of the American Society of Andrology, which held its annual meeting in Tampa immediately following the workshop. This volume consists of 21 chapters prepared by the speakers and 8 chapters contributed by presenters of the posters.

The Program Committee consisted of Drs. Nancy Alexander, C. Wayne Bardin, Andrzej Bartke (Chairman), Kevin J. Catt, Claude Desjardins, Maria Dufau, Michael Griswold, Norman Hecht, Aaron Hsueh, Dolores J. Patanelli, Bernard Robaire, and Richard Sherins. The workshop was generously funded by Serono Symposia, USA, which also agreed to handle registration, local arrangements, and publication of this volume. On behalf of the Program Committee and all participants, I would like to thank Dr. Bruce Burnett of Serono Symposia, USA and his staff for their efforts and support in every aspect of the organization of the workshop. Travel support for young investigators presenting posters was provided by the National Institute of Child Health and Human Development, by Serono Symposia, USA, and by funds generated by Dr. Bernard Robaire for support of the 1991 Testis Workshop.

I would also like to thank Dr. Don Cameron and members of the ASA Local Arrangements Committee, who helped in many aspects of planning and organization; Dr. D. "Pat" Patanelli for her help in organizing a meeting of the Program Committee; all those who served as Session Chairpersons; and especially those who presented their most recent data in lectures and in poster sessions, thus ensuring that the workshop was exciting and up-to-date.

Finally, it seems appropriate to recall that over 20 years have elapsed since Dr. Mortimer Lipsett organized the first Testis Workshop at NIH. This was a one-day meeting held in 1972 that was attended, I believe, by some 30 or 40 people. Since that time very impressive progress has been made, but many problems remain unresolved. Hopefully, the fascination of so many investigators with the function of the male gonad will not only continue to expand the present understanding of the biology of the testis, but also lead to the development of novel methods for the treatment of male infertility and for male contraception.

ANDRZEJ BARTKE

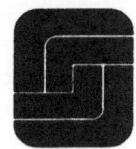

Contents

Part III.   Signaling Molecules and Their Receptors

Part V. Recent Developments in the Studies of Male Reproductive Functions

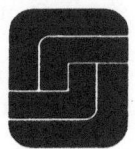

Contributors

ARMANDO G. AMADOR, Department of Obstetrics and Gynecology, Division of Research, Southern Illinois University School of Medicine, Springfield, Illinois, USA.

RICARDO H. ASCH, Department of Obstetrics and Gynecology, Division of Reproductive Endocrinology and Infertility, Center for Reproductive Health, University of California Irvine, Orange, California, USA.

ODILE AVALLET, INSERM, Debrousse Hospital, Lyons, France.

PAUL A. BAIN, Department of Obstetrics and Gynecology, Steroid Research Unit, Women's Hospital, University of Michigan, Ann Arbor, Michigan, USA.

ANDRZEJ BARTKE, Department of Physiology, Southern Illinois University School of Medicine, Carbondale, Illinois, USA.

MATTI BERGENDAHL, Department of Physiology, University of Turku, Turku, Finland.

MARLIES BETKA, Department of Biology, Boston University, Boston, Massachusetts, USA.

CARLA BOITANI, Institute of Histology and General Embryology, La Sapienza University, Rome, Italy.

NOUREDDINE BOUJRAD, Department of Anatomy and Cell Biology, Georgetown University Medical Center, Washington, District of Columbia, USA.

TERRY R. BROWN, Department of Population Dynamics, Division of Reproductive Biology, School of Hygiene and Public Health, Johns Hopkins University, Baltimore, Maryland, USA.

DAVID BUNICK, Department of Veterinary Biosciences, University of Illinois, Urbana, Illinois, USA.

MARÍA BURGOS-TRINIDAD, Department of Obstetrics and Gynecology, Steroid Research Unit, Women's Hospital, University of Michigan, Ann Arbor, Michigan, USA.

GLORIA V. CALLARD, Department of Biology, Boston University, Boston, Massachusetts, USA.

YING-TAI CHANG, Department of Pediatrics/Endocrinology, University of Illinois at Chicago, Chicago, Illinois, USA.

MICHEL CHRÉTIEN, Laboratory of Molecular Neuroendocrinology, Clinical Research Institute of Montreal, University of Montreal, Montreal, Quebec, Canada.

FRANCK CHUZEL, Claude-Bernard University and INSERM, Debrousse Hospital, Lyons, France.

MARCO CONTI, Department of Gynecology and Obstetrics, Division of Reproductive Biology, Stanford Medical Center, Stanford, California, USA.

PAUL S. COOKE, Department of Veterinary Biosciences, University of Illinois, Urbana, Illinois, USA.

ANNE CORLU, INSERM, Pontchaillou Hospital, University of Rennes I, Rennes, France.

CRISTINA D'ALESSANDRIS, Institute of Histology and General Embryology, La Sapienza University, Rome, Italy.

PETER DE BOER, Department of Genetics, Wageningen Agricultural University, Wageningen, The Netherlands.

DIRK G. DE ROOIJ, Department of Cell Biology, Medical School, Utrecht University, Utrecht, The Netherlands.

JENNIFER DORRINGTON, Banting and Best Department of Medical Research, C.H. Best Institute, University of Toronto, Toronto, Ontario, Canada.

E.M. EDDY, Gamete Biology Section, Laboratory of Reproductive and Developmental Toxicology, National Institute of Environmental Health Sciences, National Institutes of Health, Research Triangle Park, North Carolina, USA.

ULRICH EISEL, Unit on Developmental Biology, Laboratory of Cell Biology, National Institute of Mental Health, Bethesda, Maryland, USA.

GAETANO FRAJESE, Department of Internal Medicine and Endocrinology, II University of Rome, Tor Vergata, Rome, Italy.

K.D. FULCHER, Gamete Biology Section, Laboratory of Reproductive and Developmental Toxicology, National Institute of Environmental Health Sciences, National Institutes of Health, Research Triangle Park, North Carolina, USA.

NADINE GÉRARD, Groupe d'Etude de la Reproduction chez le Mâle, INSERM, University of Rennes I, Rennes, France.

PAOLO GHIRRI, Cattedra di Neonatologia, University of Pisa, Pisa, Italy.

SUSHMITA GHOSH, Laboratory of Structural Biology, Department of Physiology, Southern Illinois University School of Medicine, Carbondale, Illinois, USA.

TAMARA GOETZ, Department of Biochemistry and Biophysics, Washington State University, Pullman, Washington, USA.

MICHAEL D. GRISWOLD, Department of Biochemistry and Biophysics, Washington State University, Pullman, Washington, USA.

CHRISTIANE GUGUEN-GUILLOUZO, INSERM, Pontchaillou Hospital, University of Rennes I, Rennes, France.

ANNE-MAARIT HAAVISTO, Department of Physiology, University of Turku, Turku, Finland.

HARRI HAKOVIRTA, Department of Anatomy, Institute of Biomedicine, University of Turku, Turku, Finland.

MATTHEW P. HARDY, The Population Council, Rockefeller University, New York, New York, USA.

LESLIE HECKERT, Department of Biochemistry and Biophysics, Washington State University, Pullman, Washington, USA.

REX A. HESS, Department of Veterinary Biosciences, University of Illinois, Urbana, Illinois, USA.

CHRISTA HEYTING, Department of Genetics, Wageningen Agricultural University, Wageningen, The Netherlands.

AARON J.W. HSUEH, Department of Gynecology and Obstetrics, Division of Reproductive Biology, Stanford University School of Medicine, Stanford, California, USA.

JAMES R. HUDSON, JR., Research Genetics, Inc., Huntsville, Alabama, USA.

ILPO HUHTANIEMI, Department of Physiology, University of Turku, Turku, Finland.

SAVERIA IONA, Institute of Histology and General Embryology, La Sapienza University, Rome, Italy.

ANDRZEJ JAKUBOWIAK, Department of Obstetrics, Gynecology and Reproductive Sciences, University of Texas Medical School, Houston, Texas, USA.

ANDRZEJ JANECKI, Department of Obstetrics, Gynecology and Reproductive Sciences, University of Texas Medical School, Houston, Texas, USA.

BERNARD JÉGOU, Group for the Study of Male Reproduction, INSERM, University of Rennes I, Rennes, France.

L. JENKIN, Department of Anatomy, School of Medical Sciences, University of Bristol, Bristol, UK.

JOAN C. JORGENSEN, School of Veterinary Medicine, University of Wisconsin–Madison, Madison, Wisconsin, USA.

ANTTI KAIPIA, Department of Anatomy, Institute of Biomedicine, University of Turku, Turku, Finland.

HENRI KERCRET, Group for the Study of Male Reproduction, INSERM, University of Rennes I, Rennes, France.

SHAFIQ A. KHAN, Banting and Best Department of Medical Research, C.H. Best Institute, University of Toronto, Toronto, Ontario, Canada.

JOHN D. KIRBY, Department of Urology, Northwestern University Medical School, Chicago, Illinois, USA.

BERNARD KNEIP, INSERM, Pontchaillou Hospital, University of Rennes I, Rennes, France.

LYNNE A. KRUMMEN, Cell Biology, Genentech, Inc., South San Francisco, California, USA.

CHUNG YING KU, Department of Biochemistry and Molecular Biology, University of Texas Medical School at Houston, Houston, Texas, USA.

FERNAND LABRIE, Medical Research Council Group in Molecular Endocrinology, CHUL Research Center and Laval University, Quebec, Canada.

CAROL LINDER, Department of Biochemistry and Biophysics, Washington State University, Pullman, Washington, USA.

QINGXIAN LU, Department of Biochemistry and Molecular Biology, University of Texas Medical School at Houston, Houston, Texas, USA.

VAN LUU-THE, Medical Research Council Group in Molecular Endocrinology, CHUL Research Center and Laval University, Quebec, Canada.

JENNIE P. MATHER, Cell Biology, Genentech, Inc., South San Fancisco, California, USA.

TIINA MATIKAINEN, Department of Physiology, University of Turku, Turku, Finland.

ARTUR MAYERHOFER, Department of Anatomy and Cell Biology, Ulm University, Ulm, Germany.

MAJAMBU MBIKAY, Laboratory of Molecular Neuroendocrinology, Clinical Research Institute of Montreal, University of Montreal, Montreal, Quebec, Canada.

MICHAEL P. McGUINNESS, Department of Anatomy and Cell Biology, Temple University School of Medicine, Philadelphia, Pennsylvania, USA.

CLAUDE J. MIGEON, Pediatric Endocrine Laboratories, Johns Hopkins Hospital, Baltimore, Maryland, USA.

LUCIA MONACO, Institute of Histology and General Embryology, La Sapienza University, Rome, Italy.

ANNA RITA MORENA, Institute of Histology and General Embryology, La Sapienza University, Rome, Italy.

C. MORI, Department of Anatomy, Kyoto University Faculty of Medicine, Sakyo-ku, Kyoto, Japan.

KOICHI MURONO, Department of Population Dynamics, Division of Reproductive Biology, School of Hygiene and Public Health, Johns Hopkins University, Baltimore, Maryland, USA.

H.D. NICHOLSON, Department of Anatomy, School of Medical Sciences, University of Bristol, Bristol, UK.

D.A. O'BRIEN, Laboratories for Reproductive Biology, Departments of Pediatrics and Cell Biology and Anatomy, University of North Carolina at Chapel Hill, Chapel Hill, North Carolina, USA.

JOANNE M. ORTH, Department of Anatomy and Cell Biology, Temple University School of Medicine, Philadelphia, Pennsylvania, USA.

PIRJO PAKARINEN, Department of Physiology, University of Turku, Turku, Finland.

VASSILIOS PAPADOPOULOS, Department of Anatomy and Cell Biology, Georgetown University Medical Center, Washington, District of Columbia, USA.

MARTTI PARVINEN, Department of Anatomy, Institute of Biomedicine, University of Turku, Turku, Finland.

PASQUALE PATRIZIO, Department of Obstetrics and Gynecology, Division of Reproductive Endocrinology and Infertility, University of California Irvine, Orange, California, USA.

ANITA H. PAYNE, Departments of Obstetrics and Gynecology, Biological Chemistry and the Reproductive Sciences Program, Women's Hospital, University of Michigan, Ann Arbor, Michigan, USA.

GEORGES PELLETIER, Medical Research Council Group in Molecular Endocrinology, CHUL Research Center and Laval University, Quebec, Canada.

TARJA-LEENA PENTTILÄ, Department of Anatomy, Institute of Biomedicine, University of Turku, Turku, Finland.

ANTTI PERHEENTUPA, Department of Physiology, University of Turku, Turku, Finland.

HÅKAN PERSSON, Department of Medical Chemistry, Laboratory of Molecular Neurobiology, Karolinska Institute, Stockholm, Sweden.

ANTOINE H.F.M. PETERS, Department of Genetics, Wageningen Agricultural University, Wageningen, The Netherlands.

DONALD W. PFAFF, Laboratory of Neurobiology and Behavior, The Rockefeller University, New York, New York, USA.

MARIE-LAURE RAFFIN-SANSON, Endocrinology Unit, Cochin Hospital, Paris, France.

KAY REYNOLDS, Unit on Developmental Biology, Laboratory of Cell Biology, National Institute of Mental Health, Bethesda, Maryland, USA.

MICHELLE RIDDICK, Unit on Developmental Biology, Laboratory of Cell Biology, National Institute of Mental Health, Bethesda, Mayrland, USA.

MARYVONNE RISSEL, INSERM, Pontchaillou Hospital, University of Rennes I, Rennes, France.

PENELOPE E. ROBERTS, Cell Biology, Genentech, Inc., South San Francisco, California, USA.

LONNIE D. RUSSELL, Laboratory of Structural Biology, Department of Physiology, Southern Illinois University School of Medicine, Carbondale, Illinois, USA.

JOSÉ M. SAEZ, INSERM, Debrousse Hospital, Lyons, France.

BARBARA M. SANBORN, Department of Biochemistry and Molecular Biology, University of Texas Medical School at Houston, Houston, Texas, USA.

PATRICIA SCHERER, Department of Population Dynamics, Division of Reproductive Biology, School of Hygiene and Public Health, Johns Hopkins University, Baltimore, Maryland, USA.

HELENA SCHTEINGART, CONICET, Endocrinology Unit, R. Gutierrez Children's Hospital, Buenos Aires, Argentina.

MARLENE SCHWANZEL-FUKUDA, Laboratory of Neurobiology and Behavior, The Rockefeller University, New York, New York, USA.

NABIL G. SEIDAH, Laboratory of Neuroendocrine Biochemistry, Clinical Research Institute of Montreal, University of Montreal, Montreal, Quebec, Canada.

CLAUDIO SETTE, Department of Gynecology and Obstetrics, Division of Reproductive Biology, Stanford Medical Center, Stanford, California, USA.

JACQUES SIMARD, Medical Research Council Group in Molecular Endocrinology, CHUL Research Center and Laval University, Quebec, Canada.

AMIYA P. SINHA HIKIM, Laboratory of Structural Biology, Department of Physiology, Southern Illinois University School of Medicine, Carbondale, Illinois, USA.

FRANCINE SIROIS, Laboratory of Molecular Neuroendocrinology, Clinical Research Institute of Montreal, University of Montreal, Montreal, Quebec, Canada.

OLOF SÖDER, Pediatric Endocrinology Unit, Karolinska Hospital, Stockholm, Sweden.

MARIO STEFANINI, Institute of Histology and General Embryology, La Sapienza University, Rome, Italy.

ANNA STEINBERGER, Department of Obstetrics, Gynecology and Reproductive Sciences, University of Texas Medical School, Houston, Texas, USA.

VIQAR SYED, Rajput Colony, Rawalpindi, Pakistan.

HAIDY TADROS, Laboratory of Molecular Neuroendocrinology, Clinical Research Institute of Montreal, University of Montreal, Montreal, Quebec, Canada.

JUHA S. TAPANAINEN, Department of Gynecology and Obstetrics, Division of Reproductive Biology, Stanford University School of Medicine, Stanford, California, USA.

KATJA TEERDS, Department of Cell Biology and Histology, Utrecht University Veterinary School, Utrecht, The Netherlands.

JORMA TOPPARI, Department of Anatomy, Institute of Biomedicine, University of Turku, Turku, Finland.

HENK JG VAN DE KANT, Department of Cell Biology, Medical School, Utrecht University, Utrecht, The Netherlands.

PAUL T. VAN DER SAAG, Hubrecht Laboratory, Netherlands Institute for Developmental Biology, Utrecht, The Netherlands.

ANS M.M. VAN PELT, Department of Cell Biology, Medical School, Utrecht University, Utrecht, The Netherlands.

ELENA VICINI, Department of Gynecology and Obstetrics, Division of Reproductive Biology, Stanford Medical Center, Stanford, California, USA.

MICHÈLE VIGIER, INSERM, Debrousse Hospital, Lyons, France.

AIDA WAHAB, Pediatric Endocrinology Unit, Karolinska Hospital, Stockholm, Sweden.

J.E. WELCH, Gamete Biology Section, Laboratory of Reproductive and Developmental Toxicology, National Institute of Environmental Health Sciences, National Institutes of Health, Research Triangle Park, North Carolina, USA.

DAVID E. WILDT, National Zoological Park, Smithsonian Institution, Washington, District of Columbia, USA.

WILLIAM W. WRIGHT, Department of Population Dynamics, Division of Reproductive Biology, School of Hygiene and Public Health, Johns Hopkins University, Baltimore, Maryland, USA.

GERI L. YOUNGBLOOD, Department of Obstetrics and Gynecology, Steroid Research Unit, Women's Hospital, University of Michigan, Ann Arbor, Michigan, USA.

ZHIFENG ZHOU, Department of Population Dynamics, Division of Reproductive Biology, School of Hygiene and Public Health, Johns Hopkins University, Baltimore, Maryland, USA.

ANDREAS ZIMMER, Unit on Developmental Biology, Laboratory of Cell Biology, National Institute of Mental Health, Bethesda, Maryland, USA.

ANNE ZIMMER, Unit on Developmental Biology, Laboratory of Cell Biology, National Institute of Mental Health, Bethesda, Maryland, USA.

1

Endangered Species Spermatozoa: Diversity, Research, and Conservation

DAVID E. WILDT

The global explosion in human population has placed extraordinary pressures on the earth's resources. The result has been a rate of species extinction rivaling that seen during the demise of the dinosaurs. It is a dual-tiered disaster. Not only is the sheer number of species on the planet (biodiversity) being drastically reduced, but the well-being of extant populations is being compromised by relentless exploitation, fragmentation, and pollution of natural ecosystems. The crisis is not going unchallenged, and there are a host of treatises arguing for a wake-up call—that much more attention be given to preserving nature. Perhaps the most eloquent (and alarming) is E.O. Wilson's recent book, *The Diversity of Life*, that documents the impact of global ecosystem destruction, predicting extinction for perhaps 20% of all existing species within the next 30 years (1). In addition to enriching the soil and providing the very air we breathe, ecosystems conceal a wealth of new information on still undeveloped pharmaceuticals, crops, alternative food sources, timber, fibers, petroleum substitutes, and other potential amenities. For these reasons, he convincingly declares that "biological diversity is the key to the maintenance of the world as we know it." Wilson, like many others, makes a plea for developing systematic action plans that will provide stewardship for the earth's resources. To do otherwise or to presume that humankind can thrive in an impoverished biological world is both arrogant and reckless.

The preservation of rare animal and plant species often is entangled in political, economic, and scientific self-interests. Despite these complications, there seems to be real progress on at least two fronts. First, there is a growing realization that preserving native ecosystems and economic development are not necessarily mutually exclusive. Habitat and species can be preserved while fulfilling human needs, a concept known as *sustainable development* (2). Examples include extracting nontimber products

1

from rainforests (i.e., rubber) or highly regulated commercial wildlife hunting in national parks. These types of approaches provide incentive for maintaining high-quality habitat and the species living therein. Second, a growing number of scientists, managers, and politicians are recognizing the real and catastrophic implications of losing biological and genetic diversity. An entire discipline, *conservation biology*, has emerged that seeks to define and understand the many factors that influence the survival of ecosystems, species, subspecies, and even individuals.

In general, there is unanimous consensus that the emphasis should be on maintaining natural ecosystems (in situ conservation), but for some it already is too late. In a sense, many wildlife refuges have become large zoos without fences. Encroachments and destruction have produced small and fragmented populations (similar to that found in captivity) that, if mismanaged further, are vulnerable to loss because of genetic impoverishment. This realization is driving the formation of national and international groups that are cooperating to preserve both habitat and individual species. Zoo scientists play a pivotal role in this new cooperation. Applying computer-based models (developed for managing small captive populations) to predicting wild population viability (3) is one example. Another is the actual reintroduction of zoo-bred offspring into the wild (4–6). In addition to developing new and improved strategies for propagating endangered species (7–10), zoos also contribute through public education/outreach programs that stress conservation themes and even through formal "adoption" of natural reserves (11). But the most important and rapidly evolving function of a zoo is its role as a research center. Of course, all good science explores the unknown, but zoo-oriented research tends to address questions that influence the existence or well-being of *entire species*. Perhaps nothing is more challenging and satisfying in research than understanding and helping to preserve a whole, living thing.

It is not unusual for most scientific professionals to be unaware of the specifics of the biodiversity crisis or recent efforts to address the problems. The issue is real, and for those desiring a "conversion experience," the author recommends studying reference 1. The sequelae involve coming to terms with how science (in this case, reproductive biology) can contribute to conservation. The purpose of this chapter is twofold. First, it is not well understood that high-quality research can be conducted in wildlife species and that findings can be relevant across a range of taxa (including humans). Therefore, one objective is to review some of our laboratory's research on sperm morphology and function in wild felids (cats), virtually all of which are threatened by severe losses in genetic diversity and by extinction. The aim is to demonstrate that novel basic research can be conducted in these unconventional taxa and that findings can be related directly or indirectly to species/population status and, perhaps, to fertility potential. Second, there is an increasing number of organized ex situ captive breeding programs for endangered animals in which species are propagated strictly on the basis of their genes. Formal *Species Survival Plans* (SSPs) allow management of entire regional populations (many zoos together, rather than individual zoos) while dictating which animals may be mated together (12). The theory is that only genetically vigorous offspring are produced. Assisted reproduction techniques, now used routinely for accelerating livestock production and combating human infertility, have profound potential for managing endangered species. This assertion will be supported by reviewing our recent applied studies.

Genetic Diversity and Reproductive/Health Fitness, Especially Sperm Integrity

Direct human actions exert the most pressure on the extinction process, but habitat quality, demographic variables, and unpredictable catastrophes (e.g., disease and climatic change) also influence species and population survival. One other variable—the level of genetic variation within a species or population—also has attracted attention. In theory, as population size constricts, the incidence of incestuous matings increases, producing a homozygous or inbred population that is less healthy and less reproductively efficient. Producers of livestock have understood for years the sinister effects of monomorphic genotypes, a fact that was long ignored by wildlife managers. Both zoos and natural ecosystems provide ideal venues for studying the cause and impact of inbreeding depression. Due to limited space and because most of their specimens are rare, zoos are destined to deal with small populations. Managers of native ecosystems face a similar dilemma because habitat destruction and fragmentation eliminate natural corridors of genetic exchange. The result is a creeping tendency towards genetic homogenization, with species and populations becoming vulnerable to loss. A vortex is formed in which small population size combined with poor management causes a continued loss of fitness, with an eventual spin-off to extinction (Fig. 1.1).

There is evidence that reduced genetic diversity influences the viability of a wildlife population; specifically, its survivability and reproductive health. The first suggestion of this relationship came in 1979 from Ralls et al. (13) who surveyed survival of captive-bred ungulates with known pedigrees. The results (which in retrospect should not have been surprising) were at the time alarming because juvenile mortality was rampant in inbred compared to outbred offspring. The findings immediately motivated managers throughout North American zoos to begin propagating animals according to genotype, and one eventual result was the formation of the North American SSP program (12). Another outcome was increased interest in systematic, molecular assessments of genetic variation in natural and captive populations, especially in the context of simultaneous

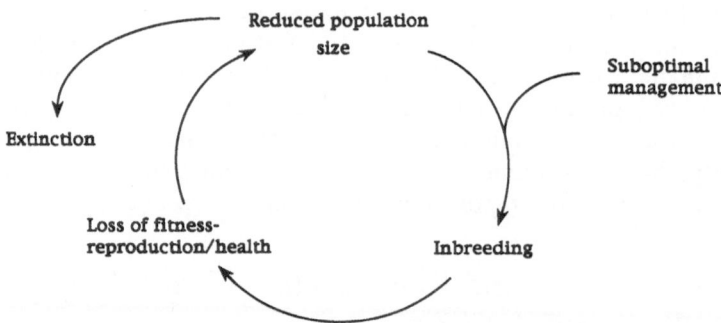

Scenario for Population Demise/Extinction

FIGURE 1.1. One scenario for extinction. Small population size, either with or without poor management, results in incestuous matings and inbreeding depression often expressed by decreased reproductive fitness and increased susceptibility to disease and congenital malformations. The cycle continues with further reductions in population size and more inbreeding, eventually leading to extinction.

measurements of reproductive traits. In this context, we have benefited from a long collaboration with the molecular genetics laboratory of O'Brien and associates at the National Cancer Institute. For more than a decade, concurrent studies of genetic variation and reproductive characteristics have produced compelling evidence that genetic impoverishment is related to poor ejaculate traits in Felidae species.

Our initial findings occurred by serendipity. During a survey of electroejaculate characteristics in a large population of cheetahs (*Acinonyx jubatus*) in southern Africa (14), we collected blood samples, almost as an afterthought. The latter were analyzed for allozyme polymorphisms by O'Brien and colleagues, who made the startling discovery that the cheetah was genetically depauperate, having no more genetic variation than inbred laboratory mice (15–17). The results were especially interesting considering that all male cheetahs were producing extraordinarily high proportions of structurally abnormal spermatozoa (~70% pleiomorphisms/ ejaculate) (14, 18, 19). The findings also generated debate that the lost variability was somehow contributing to the cheetah's historic reluctance to reproduce in captivity and its tendency to succumb to infectious disease (16, 17, 19).

Two general questions became the focus of our research. First, is there a cause-effect relationship between reduced genetic variation and poor seminal quality in felids, especially an increased incidence of malformed spermatozoa? Second, what is the physiological significance and impact of pleiomorphic sperm in felid ejaculates? Specifically, do these structurally deformed cells participate in sperm-oocyte interaction and fertilization?

Addressing these questions has been difficult in the cheetah, in part because all cheetahs (captive and free-ranging) are not only genetically similar, but all produce high proportions of abnormally shaped sperm (14, 18). The lack of an outbred "control" population for study presents obvious limitations. In a recent detailed survey of captive cheetahs ($n = 60$), proven breeders produced comparable ejaculate quality (including proportions of pleiomorphic sperm) to age-matched, unproven males (19). Such findings only reinforce the uncertainty of the purpose or influence of these malformed cells. Is it possible to defend, from a physiological perspective, that pleiomorphic sperm in cheetahs are somehow normal? Based on evidence to follow, this appears not to be the case.

Incidence of Sperm Pleiomorphisms in Felids: A Relationship to Genetic Variability?

During the past 15 years, we have evaluated reproductive traits in more than 1000 individual felids representing 28 of the 37 known species in the family Felidae. Three consistent observations have emerged (7, 20, 21). First, there are species- or population-specific sperm concentration (number) and motility characteristics. Second, the taxon as a whole exhibits a higher incidence of teratospermia than most other mammals. Third, some felid species (or populations) ejaculate mostly structurally normal spermatozoa, whereas others largely produce malformed cells (Fig. 1.2).

These differences in sperm structure may be of genetic origin, but empirical, cause-effect studies have not yet been done. One can imagine the ethical problems of purposely inbreeding endangered species to verify such a hypothesis. Therefore, our investigations have been mostly opportunistic and retrospective, often taking advantage of unique populations that have become (for whatever reason) genetically compromised. Some evidence that felid ejaculate quality is sensitive to genetic influences is founded on brief case studies. Other findings are based on more systematic and longitudinal studies.

Evidence in a Leopard Cat Subspecies

One case report involves *Felis bengalensis euptilura*, a subspecies of leopard cat that has been readily propagated at the Tallinn Zoopark in Estonia (on the Baltic Sea coast). Similar to the common domestic cat in size, the *euptilura* subspecies is indigenous to eastern Russia, Manchuria, and Korea and generally unknown to North American captive breeding programs. Regimented electroejaculation of nine adult *euptilura* revealed total sperm numbers, motility ratings, and pleiomorphisms (15 structural types) (Table 1.1) similar to published values for normospermic domestic cats (14, 22). However, 2 age-matched males failed to conform to the

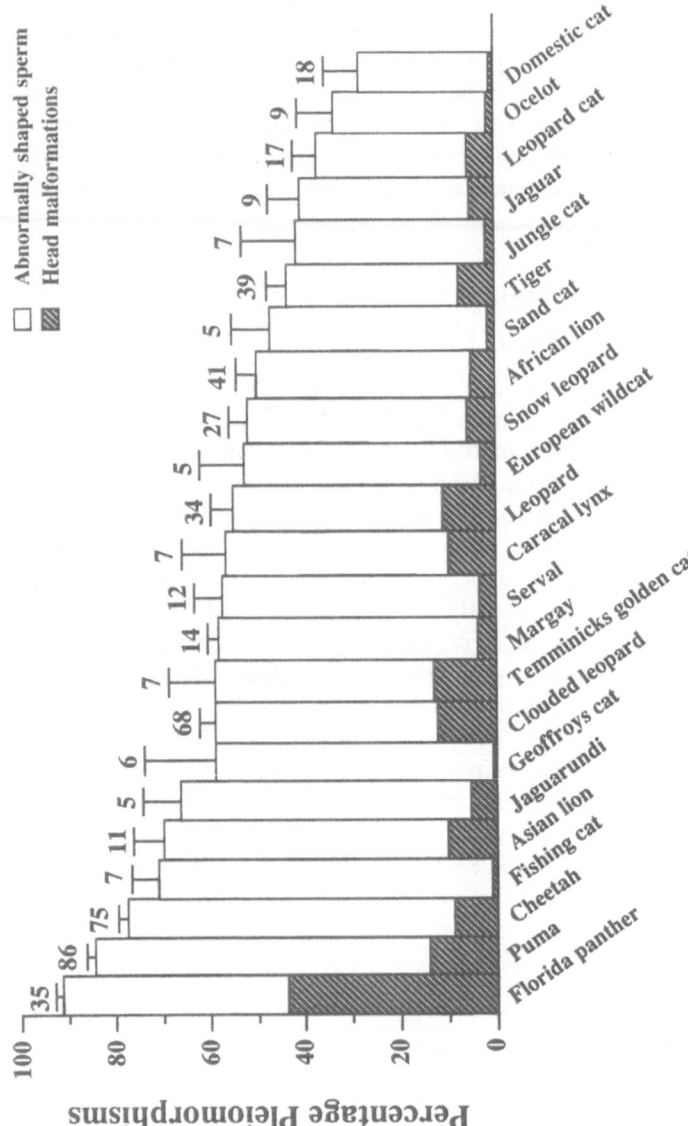

FIGURE 1.2. Proportion of total sperm pleiomorphisms and head defects (hatched area) in 23 species (or subspecies) in the Felidae family. Values are means ± SEM. The Florida panther is presented as a unique subspecies because of the high proportion of spermatozoa with head defects, primarily an acrosomal malformation. Domestic cat data are based on a normospermic population. Number above each bar indicates number of ejaculates.

TABLE 1.1. Electroejaculate traits of leopard cats (*Felis bengalensis euptilura*) known to be outbred ($n = 9$) versus 2 males produced by a sib-to-sib mating.

	Population mean (±SEM)	Male 10442	Male 10443
Total sperm number per ejaculate ($\times 10^6$)	9.9 ± 2.9	2.6	1.0
Sperm motility index*	63.6 ± 4.3	40.0	62.5
Structurally normal sperm (%)	74.4 ± 3.2	48.0	58.5
1. Abnormal acrosome	0.2 ± 0.1	0.0	1.0
2. Detached head	2.8 ± 1.2	3.5	2.5
3. Microcephaly	0.8 ± 0.4	0.5	0.0
4. Macrocephaly	0.3 ± 0.2	1.0	0.0
5. Bicephaly	0.1 ± 0.1	0.0	0.5
6. Bent neck region	0.4 ± 0.1	4.0	1.5
7. Bent midpiece with residual cytoplasmic droplet	6.0 ± 0.6	21.5	21.0
8. Bent midpiece without residual cytoplasmic droplet	1.4 ± 0.6	2.0	3.0
9. Deranged midpiece/sheath	0.5 ± 0.2	2.5	0.0
10. Tightly coiled flagellum	3.3 ± 1.5	11.0	3.5
11. Bent flagellum with residual cytoplasmic droplet	0.1 ± 0.1	0.0	0.0
12. Bent flagellum without residual cytoplasmic droplet	6.9 ± 2.0	4.0	3.0
13. Biflagellate	0.1 ± 0.1	0.0	0.5
14. Proximal cytoplasmic droplet	0.9 ± 0.4	1.0	5.0
15. Distal cytoplasmic droplet	1.8 ± 0.6	1.0	0.0

*Sperm motility index values were based on a calculation that considers overall sperm percent motility and type of forward progressive status, as described in reference 22.

population pattern (Table 1.1), producing only a fraction of the sperm number in the 9 counterparts. Although the sperm motility index was lower only for male 10422, both individuals produced more pleiomorphic sperm (26% and 16%, respectively) than the group mean, mostly an increased incidence of bent-neck and midpiece defects (Table 1.1).

The reason may have been related to the pedigree of the two outlier individuals. Both were siblings born to sibling parents, the sire being one of two brothers housed long-term with a full sister (Fig. 1.3). Because the parents of the sire and dam were unrelated (the male being wild-caught), the implication was that one generation of full-sibling matings may have adversely influenced both sperm numbers and sperm integrity in F1 offspring.

Evidence in Lion Subspecies

In collaboration with O'Brien and associates, we have examined genetic diversity and sperm characteristics in two populations of lions, one free-living in the Serengeti ecosystem in eastern African (*Panthera leo spp.*) and another (*Panther leo persica*, the Asian lion) in an isolated national park in western India. Compared to African lions, Asian lions have

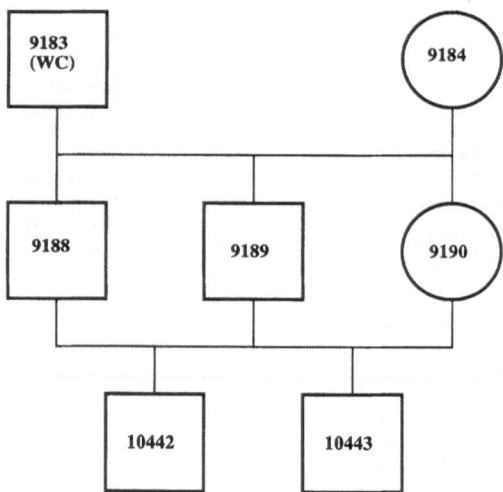

FIGURE 1.3. Lineage of 2 leopard cats (*Felis bengalensis euptilura*) known to produce poorer ejaculate quality than age-matched males managed under the same conditions (Table 1.1). Parents of both males were littermates not separated at sexual maturity. The sire may have been either male 9188 or 9189. The paternal ancestor was a wild-caught (WC) male from Korea.

markedly less heterozygosity based on allozyme polymorphisms and mitochondrial DNA fingerprint variation (23–25). These findings are reasonable because the ~300 extant Asian lions are descendents of ancestors that experienced a severe population contraction (to fewer than 20 individuals) in the first quarter of the twentieth century (26). These lions now strain the limited habitat of the Gir Forest Sanctuary and surrounding ecosystem (~1400 km^2), although a captive breeding program also exists at a nearby zoo.

In an earlier report (26), we documented that sperm number and structural integrity (but not sperm motility) were depressed in the genetically impoverished Asian lion compared to its African counterpart. More recently, we have reevaluated Asian males known to be fertile (e.g., proven breeders) versus those known to be infertile (based on serial couplings with multiple females and failure to produce young). Table 1.2 depicts sperm characteristics for (i) the original African "control" population, (ii) a group of 4 proven-breeder Asian males, and (iii) 2 Asian individuals (Ravi and Deep) known to be sterile. Again, Asian lions as a population tended to produce fewer total sperm of comparable motility, but with more structural deformities. Particularly striking, however, were the differences between the proven and sterile Asian males: Infertile individuals tended to produce many fewer total sperm, an overall poorer sperm motility index, and markedly higher proportions of defective-

TABLE 1.2. Comparison of electroejaculate traits for outbred African lions (*Panthera leo spp.*) free-living in Tanzania ($n = 8$), Asian lions (*P. leo persica*) that were proven breeders ($n = 4$), and 2 individual Asian lions known to be sterile.

	Population mean (±SEM)		Male Ravi	Male Deep
	African lions	Asian lions		
Total sperm number per ejaculate ($\times 10^6$)	323.4 ± 29.9	224.3 ± 55.5	27.6	46.8
Sperm motility index*	80.5 ± 4.2	76.2 ± 4.8	55.0	50.0
Structurally normal sperm (%)	75.2 ± 7.2	51.8 ± 5.4	11.0	11.0
1. Abnormal acrosome	1.1 ± 0.3	4.2 ± 1.5	13.0	1.0
2. Detached head	0.0 ± 0.0	8.5 ± 2.3	16.0	25.0
3. Microcephaly	0.2 ± 0.0	0.3 ± 0.3	3.0	1.0
4. Macrocephaly	0.6 ± 0.2	0.2 ± 0.2	4.0	0.0
5. Bicephaly	0.2 ± 0.0	0.2 ± 0.2	0.0	0.0
6. Bent neck region	0.7 ± 0.1	1.2 ± 0.5	0.0	2.0
7. Bent midpiece with residual cytoplasmic droplet	2.3 ± 0.6	8.2 ± 3.0	18.0	33.0
8. Bent midpiece without residual cytoplasmic droplet	2.1 ± 0.6	2.8 ± 1.0	1.0	4.0
9. Deranged midpiece/sheath	1.9 ± 0.4	1.2 ± 0.9	11.0	2.0
10. Tightly coiled flagellum	2.3 ± 0.5	5.8 ± 1.6	8.0	11.0
11. Bent flagellum with residual cytoplasmic droplet	0.0 ± 0.0	0.0 ± 0.0	0.0	2.0
12. Bent flagellum without residual cytoplasmic droplet	0.9 ± 0.3	0.8 ± 0.2	1.0	4.0
13. Proximal cytoplasmic droplet	7.5 ± 1.1	11.0 ± 4.2	10.0	4.0
14. Distal cytoplasmic droplet	5.0 ± 1.8	1.0 ± 0.4	4.0	0.0

*Sperm motility index values were based on a calculation that considers overall sperm percent motility and type of forward progressive status, as described in reference 22.
Source: African lion data adapted from Wildt, Bush, Goodrowe, et al. (26).

appearing cells (predominantly detached heads, bent midpieces, and tightly coiled flagellum defects) (Table 1.2).

Are these observations relevant to the genetic history of the Asian lion? We contend that the most logical explanation for the significant differences in ejaculate traits is the outbred versus inbred genotypes of the African versus Asian population/subpopulations. The recent appearance of sterile Asian lion males suggests a continued cascade in gene loss, predisposing some males to physiological infertility. It is important to reemphasize that the entire Asian population (both wild and captive) is genetically invariant (23, 24). Both of Ravi's parents were wild-caught (Fig. 1.4), but the ejaculate traits of the sire Sundar (evaluated in 1986; total sperm/ejaculate: 74.6×10^6; sperm motility index: 51.8; structurally abnormal sperm: 71.8%, including 15.4% detached heads, 11.4% sperm with a bent midpiece, and 13.7% sperm with a tightly coiled flagellum) were poorer than proven breeder males listed in Table 1.2. Sundar also was a grandsire of Deep, who, in turn, was an offspring of a sib-to-sib

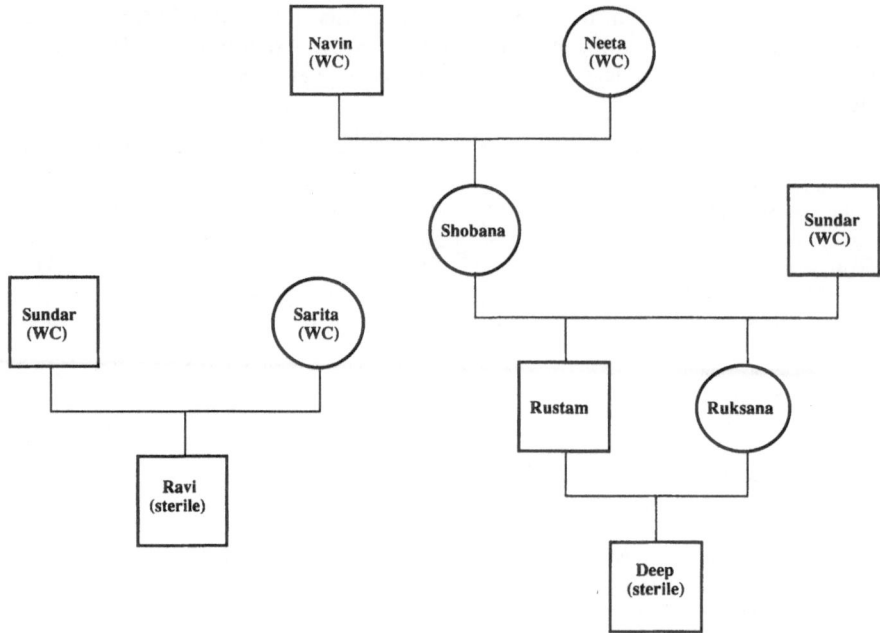

FIGURE 1.4. Lineage of 2 Asiatic lions (*Panthera leo persica*) known to be sterile and producing poor-quality ejaculates (Table 1.2). Sundar, the sire of Ravi and the ancestor of Deep, also was known to produce poor-quality semen, even though wild-caught (WC) (see text).

mating (Fig. 1.4). Therefore, considering that (i) there are quantitative differences in genetic variation and in ejaculate traits between African and Asian lions and (ii) sterile Asian lions tend to be offspring of deliberate inbreeding or sires with poor semen quality, then it seems likely that genotype is influencing reproductive fitness. In short, the genetic vigor of wild Asian lions is being compromised by a restricted habitat that inevitably promotes incestuous matings. The problem is accentuated by suboptimal captive management that is accelerating inbreeding and causing documented sterility.

Evidence in the Puma: Florida Panther Example

The Florida panther (*Felis concolor coryi*) is a subspecies of puma (cougar, mountain lion) on the brink of extinction. Historically, the panther ranged throughout the southeastern United States as far west as eastern Texas and as far north as Tennessee (27). However, human residential and agricultural development has squeezed the extant wild population, totaling 30 individuals, into southern Florida, principally within the Big Cypress Swamp and the adjacent Everglades National Park ecosystems (28, 29). Although the critical status of the Florida panther has been known for 25 years and although legally protected, the population is not self-sustaining, and computer modeling has predicted subspecies extinction within 25 to 40 years (30).

The potential loss of this charismatic, large-sized predator has motivated federal and state authorities to mobilize a massive research effort into the natural history and biology of the subspecies. Original tracking, prey base evaluations, habitat utilization, and natural mortality studies (begun in 1976) were succeeded by more intensive investigations into population genetic structure, medical causes of mortality, habitat contamination, and in our case, reproductive function.

The Florida panther has several unique phenotypic and genotypic traits not found in other puma subspecies throughout North and South America. Panthers have a broad flat frontal region of the skull, a distinct whorl of hair (cowlick) on the mid-dorsal thorax, and a 90° kink in the vertebrate at the tail tip (31). On the basis of allozyme polymorphisms, the frequency of *polymorphic loci* (P) and the average heterozygosity estimate values are 4.9% and 1.8%, respectively (29). By contrast, pumas from other geographic regions have relatively abundant allozyme diversity, including 11 polymorphic loci ($P = 27\%$) and average heterozygosity estimates from 1.8% to 6.7%. Overall genomic diversity also has been quantified using hypervariable minisatellite genetic loci (DNA fingerprinting). Using clones of two feline-specific minisatellite genomic families, O'Brien and associates have shown that the Florida panther expresses 85% less variation than other puma populations (29). The level of genetic variability in authentic Florida panthers is practically analogous to that measured in the homogeneous Asian lion (discussed earlier).

This genetic uniformity is exerting both a reproductive and general health cost. The Florida panther population is reproducing, but at a nonsustaining and poorly efficient level. For example, one adult male has been maintained long-term in captivity with female pumas of western US origin. Despite multiple pairings and observed copulations, no pregnancy has ever been established. We have systematically evaluated seminal and testicular characteristics in Florida panthers and free-living counterparts in four other geographic regions, as well as "generic" individuals in zoos (27). The results (Table 1.3) revealed several important findings. Regardless of site of origin, the puma as a species tended to ejaculate relatively high proportions of pleiomorphic spermatozoa. However, compared to all groups, Florida panthers produced poorer-quality electroejaculates, including ~94% of all sperm being malformed. The most prevalent defect was an acrosomal anomaly observed much less frequently in other felid species or populations. Compared to the other puma subspecies, more Florida panther sperm had this acrosomal malformation (>40%), as well as a defective midpiece that consisted primarily of an abnormal mitochondrial sheath (swollen, partial, or complete aplasia: >5%).

TABLE 1.3. Comparison of electroejaculate traits (mean ± SEM) for Florida panthers compared to free-ranging pumas from Texas, Colorado, and Latin America and captive generic males from North American zoos.

	Florida (n = 40)	Texas (n = 20)	Colorado (n = 8)	Latin America (n = 6)	Captive (n = 66)
Total sperm number per ejaculate ($\times 10^6$)	4.8 ± 2.1^a	$34.4 \pm 13.7^{a,b}$	$40.4 \pm 11.2^{a,b}$	$33.6 \pm 20.5^{a,b}$	62.8 ± 11.6^b
Sperm motility index*	45.6 ± 5.7^a	$56.9 \pm 5.0^{a,b}$	73.8 ± 2.2^b	71.9 ± 7.2^b	58.5 ± 2.6^b
Structurally normal sperm (%)	6.5 ± 0.7^a	14.0 ± 3.5^b	16.3 ± 2.1^b	37.8 ± 3.3^c	16.5 ± 1.9^b
Abnormal acrosomes (%)	41.8 ± 2.0^a	10.1 ± 2.8^b	18.5 ± 2.5^c	$15.4 \pm 4.3^{b,c}$	10.7 ± 1.8^b
Abnormal midpiece†	5.4 ± 0.7^a	1.3 ± 0.3^b	$3.5 \pm 0.7^{a,b}$	1.8 ± 0.8^b	2.2 ± 0.5^b

Note: n = total number of ejaculates. Row values with different superscripts differ ($P < 0.05$).
* Sperm motility index values were based on a calculation that considers overall sperm percent motility and type of forward progressive status, as described in reference 22.
† Primarily consisting of a deformed mitochondrial sheath (swollen, partial, or complete aplasia).
Source: Adapted, in part, from Barone, Roelke, Howard, Anderson, and Wildt (27).

Transmission electron microscopy of the Florida panther spermatozoon revealed extraneous acrosomal material between the acrosome and the sperm plasma membrane, derangements in acrosomal continuity, and vesicular remnants of the Golgi complex in the head region (27). This deformity has many of the same characteristics as the miniacrosome defect in humans that has been attributed to degeneration during spermiogenesis or absorption of the acrosome by the Sertoli cells (32).

Florida panthers also exhibit an unusually high predisposition to *cryptorchidism* (one or both testicles retained in the body cavity). This developmental defect is heritable in other species and is suspected to result from a sex-limited autosomal gene (33, 34). The condition is uncommon in wild felid species and has been observed in only 2 of more than 50 captive males and never in free-living pumas in Texas, Colorado, British Columbia, or Latin America (29). Retrospective analysis has revealed an alarming increase of cryptorchidism in Florida panthers. From 1970 to 1974, only 1 of 6 males (16.7%) expressed the condition, but the incidence rose to 60% from 1980 to 1984, 72.8% from 1985 to 1989, and 83.8% (5 of 6 cubs) from 1990 to 1992 (27, 29). As a result, 90% of all living males are now afflicted, and recently, 2 young males from a consanguineous mating have been discovered to be bilaterally cryptorchid (and, thus, presumed sterile).

Genetic fixation is having other maladaptive consequences in the Florida panther. An atrial septal cardiac defect has been diagnosed recently as causing mortality in two young pumas, and heart murmurs have been discovered in approximately 80% of all new offspring (29). Heaped on these problems have been recent medical surveys indicating high seroprevalence of circulating antibodies to various infectious pathogens, including feline panleukopenia virus, feline calicivirus, puma lentivirus, rabies virus, and feline corona virus. O'Brien and associates have argued previously that one consequence of genetic compromise is increased vulnerability to infectious agents, primarily due to homogenization of genes involved in immune defense (16). This scenario seems to be occurring in the Florida panther.

The overall findings provide a provocative example of the consequences of demographic reduction and genetic depletion on population survival. The results are quite clear: Once the genotype of a rare population becomes homogeneous, reproduction is only one of many physiological traits impaired. Furthermore, despite the best (late) efforts of government policymakers and science, heroic options for reversing the process are unavailable, and the population is at highest risk for extinction. Certainly, the primary threat to the Florida panther as a true subspecies continues to be direct demographic pressures imposed by humans. Continued habitat loss, prey depletion, and even the recent discovery of mercury toxicity in the environment likely are the clearest threats to survival. However, the by-product of genetic uniformity also contributes to what may be a calamitous finale for the Florida panther. Whatever the ultimate result, the subspecies offers a profound illustration of the absolute need for wildlife managers to be concerned about genetic depletion.

Influence of Teratospermia on Gamete Interaction in Felids

Do pleiomorphic spermatozoa ejaculated by some felid species or populations influence sperm-oocyte interaction or fertilization? In reality, it is likely that few of these sperm reach the oviduct, the site of fertilization. Nevertheless, almost a decade ago Mahadevan and Trounson (35) observed an indirect correlation between the incidence of structurally abnormal sperm in the inseminant and ovum penetration in human *in vitro fertilization* (IVF) trials. More recent studies have demonstrated that the binding ability of sperm to the *zona pellucida* (ZP) and the number of sperm within the ZP are diminished in samples from subfertile men that fail to achieve fertilization during IVF (36–38). The consequences of teratospermia in felids appear similar. In IVF trials we consistently observe a much higher incidence of fertilization and cleavage when inseminants contain high proportions of structurally normal spermatozoa (9). For example, embryo cleavage routinely occurs in 60%–80% of all oocytes cocultured with sperm from normospermic domestic cats (39, 40) or tigers (*Panthera tigris*) (41). In contrast, fewer than 20% of conspecific oocytes form cleaved embryos when co-incubated with teratospermic inseminants from the cheetah (42) or puma (43).

We have systematically examined the ability of sperm from different-quality ejaculates to bind, penetrate, and fertilize oocytes in vitro. These studies have been facilitated by comparing two populations of domestic cats that consistently produce normospermic (<40% abnormally shaped sperm) or teratospermic (>60% pleiomorphisms) ejaculates (22, 44, 45). Subjecting these teratospermic ejaculates to swim-up processing increases the percentage of normal sperm forms recovered (22, 45). For example, teratospermic cat ejaculates routinely averaging only 33.3% normal sperm contained 66.5% normal cells after processing, a value no different from normospermic counterparts (70.7%) (45). When these sperm populations were co-incubated with conspecific, salt-stored oocytes, the number of sperm bound to the outer periphery of the ZP was increased by more than 3-fold using normospermic (mean: 26.2) versus teratospermic (8.3) inseminants. Compared to normospermic males, the sperm from terato-spermic cats were less able to penetrate the inner, bilayered ZP or the *perivitelline space* (PVS) (Fig. 1.5a). Sperm from both groups also were capable of fertilizing in vivo-matured cat oocytes, but the incidence of embryo formation (cleavage) was 36% higher for the normospermic group (Fig. 1.5b). Although fertilization was compromised using morphologically normal sperm from teratospermic cats, all resulting embryos grew at a similar pace, and there was no effect of the origin of the sperm donor on the proportion of embryos eventually becoming morulae in vitro (Fig. 1.5b). Because sperm concentration and motility ratings were maintained constant between cat populations, we conclude that there is a fundamental functional deficit, even in normal-appearing sperm from teratospermic domestic cats (44, 45).

We have also examined the fate of pleiomorphic sperm in the terato-spermic domestic cat and cheetah (Fig. 1.6) (9, 45, 46). Both study populations ejaculated more than 70% malformed sperm per inseminant that then were cocultured in vitro with salt-stored, domestic cat oocytes. Few of the pleiomorphic domestic cat sperm (<30%) bound to the ZP compared to >55% of the malformed cheetah sperm. In both species fewer than 20% of the abnormal sperm penetrated the outer layer of the ZP, and virtually none of the pleiomorphic cell types reached the inner ZP or the PVS. Obviously, the felid ZP plays a significant role in filtering structurally defective sperm. More recent data suggest that the ZP may act as more than a simple barrier/filter to pleiomorphic sperm. In an attempt to increase the incidence of oocyte penetration using terato-spermic ejaculates, we mechanically pierced the ZP by micromanipulation, thus creating channels to potentially facilitate sperm entry (47). Terato-spermic domestic cat ejaculates were swim-up processed and then cocul-tured with conspecific oocytes that were untreated or ZP pierced (6 channels/oocyte). Micromanipulation enhanced the proportion of oocytes with sperm in the inner ZP or the PVS without increasing the number of pleiomorphic cells at either site (Fig. 1.7). Thus, ZP piercing appears to compensate, in part, for the functional deficit preventing penetration by even structurally normal sperm in teratospermic ejaculates. Because the

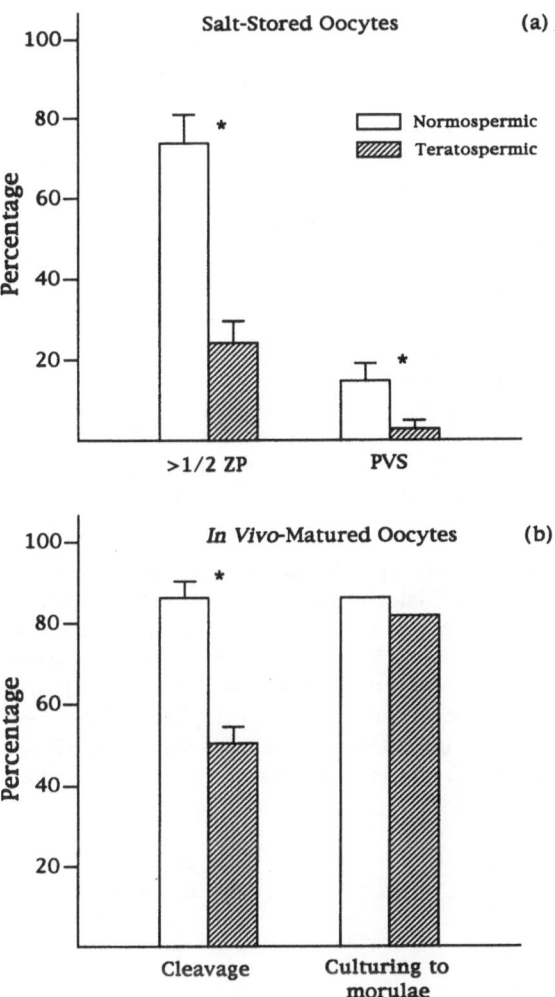

FIGURE 1.5. Penetration in vitro of salt-stored, domestic cat oocytes (*a*) (*n* = 202) or cleavage and embryo development after insemination of in vivo-matured oocytes (*b*) (*n* = 401) using sperm from normo- or teratospermic ejaculates. Categories of zona pellucida (ZP) penetration: >1/2 ZP = % oocytes with sperm in the inner half of the ZP; PVS = % oocytes with sperm in the perivitelline space. The asterisk indicates that means (± SEM) are different (*P* < 0.05). Adapted from Howard, Bush, and Wildt (44) and Howard, Donoghue, Johnston, and Wildt (45).

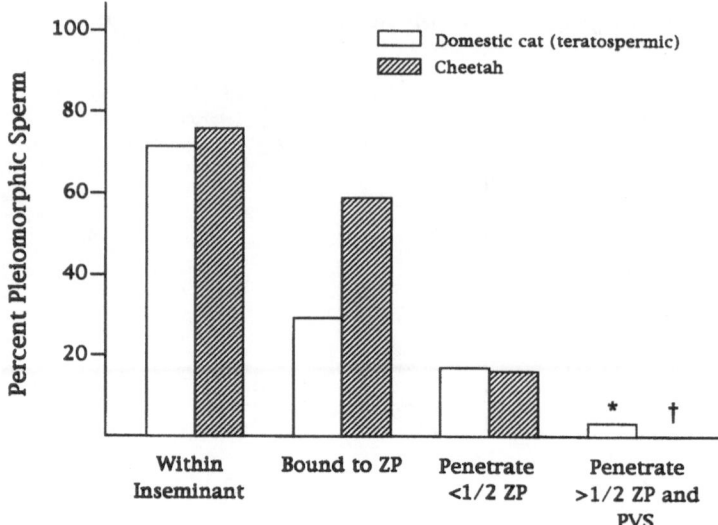

FIGURE 1.6. Illustration of the filtering role of the felid ZP. Teratospermic domestic cat ($n = 6$) and cheetah ($n = 20$) ejaculates were used to inseminate salt-stored domestic cat oocytes. Categories of ZP penetration: <1/2 ZP = % of oocytes with malformed sperm in the outer half of the ZP; >1/2 ZP and PVS = % of oocytes with malformed sperm in the inner half of the ZP or in the PVS. The asterisk indicates only sperm with a residual cytoplasmic droplet entering >1/2 ZP, not the PVS. The dagger indicates a zero value. Adapted from Howard, Donoghue, Johnston, and Wildt (45) and Howard, Barone, Bush, and Wildt (46).

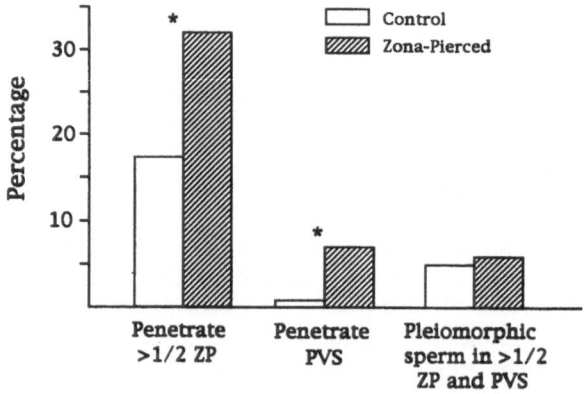

FIGURE 1.7. Impact of no treatment (control) or ZP piercing by micromanipulation on sperm penetration in vitro in the teratospermic domestic cat. Values represent the incidence of penetration of salt-stored oocytes ($n = 122$) and the proportion of structurally abnormal sperm in the inner ZP (>1/2 ZP) or PVS. The asterisk indicates that values are different ($P < 0.05$). Adapted from Roth, Howard, and Wildt (47).

technique does not allow indiscriminate penetration by malformed sperm, the ZP must selectively filter pleiomorphisms by multiple mechanisms (beyond simply acting as a barrier). Alternatively, perhaps these deformed cells are so structurally, biochemically, or metabolically compromised that fertilization will fail regardless of ZP integrity.

Based on these cumulative observations, we are now screening a host of specific factors that might help explain why normal and abnormal sperm from teratospermic males are inefficient at interacting with oocytes. These include sperm ultrastructure, motility/metabolism, ability to achieve the acrosome reaction, and type and concentration of sperm-binding receptors. We believe the findings will have obvious implications to conservation biology. First, there will be a continued affirmation and better understanding about how ejaculate quality relates to reproductive fitness. Second, we anticipate that these efforts will naturally give rise to developing assisted reproduction techniques for circumventing cases of suboptimal fertility. Finally, because men routinely produce high proportions of sperm pleiomorphisms,we also contend that the cat taxon serves as one of the most ethically attractive models available for studying the impact of teratospermia on fertilization as it relates to human reproductive health (48).

Meeting Management and Preservation Challenges Using Assisted Reproduction

The challenges to maintaining biological and genetic diversity can be met using an arsenal of scientific wisdom, logic, and technology. Because the essence of species survival is reproduction, the reproductive physiologist is destined to play an important role. For ex situ (zoo) breeding programs, assisted reproduction could be a powerful management tool for overcoming problems associated with sexual incompatibility, cases of organic infertility, and aged or underrepresented founders unable to contribute to species preservation. Especially important would be the ability to meet various SSP breeding mandates with fewer complications. (For example, transporting frozen semen, rather than living animals, would be less stressful and costly.)

Reproductive biology can also contribute to preserving species in native habitats. Perhaps the most profound impact could be achieved by developing systematic *genetic resource banks*, repositories containing germ plasm, blood products, tissues, and DNA (Table 1.4) (8, 10). Organized sampling and cryopreservation of, for example, spermatozoa from selected, free-living individuals would permit snapshot storage of existing diversity. The effect would be to provide insurance against future human-induced or natural catastrophes capable of reducing genetic diversity further or eliminating entire species. Because of habitat fragmentation

TABLE 1.4. Benefits derived from using assisted reproductive technologies (artificial insemination, IVF, and embryo transfer) in combination with the systematic banking (cryopreservation) of germ plasm/embryos from endangered species.

- Would help resolve the problem of too little space to preserve too many species, subspecies, populations, and individuals
- Would offer unique opportunities for circumventing management problems associated with a limited gene pool, sexual incompatibility, organic infertility, and aged, underrepresented founders
- Would provide a high level of protective insurance from catastrophes (natural disasters and disease epidemics)
- Would allow the interactive movement of biological material between living populations, thus maximizing genetic vigor

and the resulting isolation of populations, these germ plasm reserves could also be used for transferring genetic vigor. We recently described the logistical possibilities of interchanging genes by periodically capturing and artificially inseminating females from one isolated population using sperm from another (8). Each animal is producing surplus germ plasm, and given that this material is on reserve, there is no longer need to supplement zoo populations with animals from the wild. In essence, reproductive technologies could assist in keeping wild populations healthy and numerous. The most important by-product would be the sheer presence of a wild population that in turn justifies and promotes habitat protection.

The advantages of genetic resource banks go beyond supporting in situ populations. A natural benefit would be allowing the interactive movement of biological materials between living populations, especially transporting wild germ plasm to invigorate captive populations that have become genetically stagnant. Finally, sperm (and eventually embryo) repositories could help resolve the major crisis now facing zoos—lack of space. Of the many thousands of taxa deserving attention, biologists generally agree that zoos have sufficient space for conserving fewer than 1000 species of mammals, birds, and reptiles. Some of this problem could be alleviated by maintaining portions of needed genetic diversity in liquid nitrogen, thereby reducing the number of living animals required and freeing space for other species at high risk.

It is possible to use assisted reproductive technology and fresh or frozen-thawed germ plasm to generate offspring in rare wildlife species (reviewed in 9, 10). Table 1.5 lists 7 species in which young have been produced after artificial insemination or IVF/embryo transfer in our laboratory. Fresh sperm were used for artificial insemination in most cases, but it is noteworthy that offspring have been born to 3 endangered species (leopard cat, black-footed ferret, and Eld's deer) using thawed spermatozoa and to 2 species (cheetah and puma) known to be terato-

TABLE 1.5. Recent examples of producing offspring from rare or endangered species using artificial insemination (AI) or IVF followed by embryo transfer (ET).

Species	Technique	Fresh or frozen-thawed germ plasm	No. of pregnancies	No. of offspring
Leopard cat				
(*Felis bengalensis*)	AI	Fresh and frozen	2	3
Cheetah				
(*Acinonyx jubatus*)	AI	Fresh	3	7
Tiger				
(*Panthera tigris altaica*)	AI	Fresh	1	1
(*Panthera tigris tigris*)	IVF/ET	Fresh	1	3
Puma				
(*Felis concolor stanleyana*)	AI	Fresh	1	1
Clouded leopard				
(*Neofelis nebulosa*)	AI	Fresh	1	2
Black-footed ferret				
(*Mustela nigripes*)	AI	Fresh and frozen	3	7
Eld's deer				
(*Cervus eldi*)	AI	Frozen	9	10

Source: Adapted from Wildt, Monfort, Donoghue, Johnston, and Howard (9); Donoghue, Johnston, Seal, et al. (41); Howard and Doherty (50); Howard (51); Howard, Donoghue, Goodrowe, et al. (52); Donoghue, Johnston, Armstrong Simmons, and Wildt (53); and Monfort, Asher, Wildt, et al. (54).

spermic. We have emphasized in earlier reports (48, 49) that the rate of these successes is directly proportional to the amount of fundamental (basic) reproductive knowledge available for each species of interest. It is now well established that species-specific idiosyncrasies often prevent domestic livestock or laboratory animal technologies from being easily adapted to wild counterparts (9, 55). However, given a sound database, all evidence suggests that basic reproductive knowledge can be translated into living young. Our ability to produce multiple pregnancies in some species (Table 1.5) supports the assertion that these technical approaches hold exciting promise for conservation.

Summary and Concluding Perspective

Our overall challenge is not in conforming to the idea that more biodiversity is good (one would be hard-pressed to find disagreement on this point), but in meeting the staggering problem of identifying, prioritizing, and then preserving the earth's most valuable specimens. What is the role of the reproductive biologist? One high priority is generating the reproductive life history database for individual species in crisis, information critical for allowing successful natural or assisted reproduction to proceed.

Another seems to be in developing formal genetic resource banks useful for insuring existing diversity and managing more species than currently is possible in ex situ captive breeding programs. Certainly, because of continued fragmentation of in situ habitats, we also foresee infusions of genetic vigor into isolated wild populations via assisted reproductive technology as a viable option. Therefore, two additional contributions of reproductive biologists to conservation include (i) developing strategies to maximize genetic health while improving management efficiency and (ii) identifying when losses of genetic diversity appear to be influencing reproductive performance.

Although it has been difficult to define unequivocally a direct relationship between the level of genetic variation and fertility, we have provided evidence for compromised physiology and health in felid species, subspecies, and individuals known to be at least partially genetically impoverished. Such observations have led to a series of detailed studies on the impact of a specific unique characteristic (sperm pleiomorphisms) on sperm-oocyte interaction and fertilization in vitro. These studies produced evidence that both structurally abnormal and normal sperm from teratospermic ejaculates are compromised in fertilization ability.

Although of basic interest to gamete biologists, such results may be more important when considered in the context of conservation as a whole and the need for more research and improved management. For example, any evidence illustrating the potentially adverse effects of genetic loss certainly can be incentive for managers to manage better. However, more important is the utility of these types of findings for compelling policymakers to begin seriously supporting conservation biology research. In reality, the few reproductive biologists lured into this field are absolutely fascinated with the potential, but simultaneously panicked and impeded by the lack of any formal support mechanism. Amazingly, there is no National Institutes of Health for Endangered Ecosystems or Species, no government-based resource for considering competitive applications to fund conservation-oriented basic and applied research. No one can argue the critical importance of continued research funding to enhance the human condition; obviously, it should remain at the apex of the priority list. But considering that so much of the earth's biological and genetic diversity is approaching extirpation, should not this area of research at least be *on* a priority list?

In summary, there are three elements essential to allowing reproductive researchers to contribute to real conservation: (i) access to species in crisis, (ii) working models of how reproductive physiology can be used to generate genetically valuable young from rare species, and (iii) resources. Two of these elements now are available. First, increased interinstitutional collaboration among conservationists and the production of action plans that guide individual species preservation have become the norm. Second, examples are available demonstrating how high-quality basic research can be translated (with assisted reproduction) to produce endangered young. Therefore, the only missing element is the resources needed to fund this new brand of research. Hopefully, this perspective will provide a greater understanding for the scientific community of the potential for preserving global biological and genetic diversity, as well as the dilemmas to be faced. For those scientists or their colleagues involved with policymaking, it is to be hoped that these words will provoke action.

Acknowledgments. The data and ideas discussed in various forms in this manuscript were generated in collaboration with Drs. JoGayle Howard, Terri Roth, Mark Barone, William Rall, and Mitchell Bush of the National Zoological Park, Stephen O'Brien of the National Cancer Institute, Melody Roelke of the Florida Game and Fresh Water Fish Commission, Victor Shille of the University of Florida, and Lyndsay Phillips of the Chicago Zoological Park to whom the author is greatly indebted. The author also thanks the numerous collaborators affiliated with the field projects at the Tallinn Zoopark (Estonia), the Gir Forest Sanctuary and Sakkarbaug Zoo (India), and the many zoological institutions that allowed generous access to animals and available information; Thomas Wood and Jennifer Buff for summarizing data and preparing figures; and Drs. JoGayle Howard, Terri Roth, William Swanson, and Patricia Schmidt for reviewing an early version of the manuscript. Projects were supported by grants from the National Institutes of Health (HD-23853; RR-00045), the U.S. Fish and Wildlife Service, the Ralston Purina Big Cat Survival Fund administered through the Conservation Endowment Fund of the American Association of Zoological Parks and Aquariums, British Airways, Friends of the National Zoo, and NOAHS Center.

References

1. Wilson EO. The diversity of life. Cambridge: Harvard University Press, 1992.
2. Munro DA, Holdgate MW, eds. Caring for the earth: a strategy for sustainable living. Gland, Switzerland: International Union for Conservation of Nature and Natural Resources, 1991.
3. Lacy RC, Kreeger TJ. VORTEX manual. Apple Valley: Captive Breeding Specialist Group, Species Survival Commission, International Union for Conservation of Nature and Natural Resources, 1992.
4. Smith B. The Arabian oryx comes back to its native desert. Smithsonian 1980;11:130-1.
5. Kleiman DB. Reintroduction of captive mammals for conservation. Bioscience 1989;39:152-61.
6. Banks V. The red wolf gets a second chance to live by its wits. Smithsonian 1988;18:100-8.
7. Wildt DE. Potential applications of IVF technology for species conservation. In: Bavister BD, Cummins J, Roldan ERS, eds. Fertilization in mammals. Norwell, MA: Serono Symposia, USA, 1990;349-64.

8. Wildt DE. Genetic resource banks for conserving wildlife species: justification, examples and becoming organized on a global basis. Anim Reprod Sci 1992;28:247–57.

9. Wildt DE, Monfort SL, Donoghue AM, Johnston LA, Howard JG. Embryogenesis in conservation biology—or, how to make an endangered species embryo. Theriogenology 1992;37:161–84.

10. Wildt DE, Seal US, Rall WF. Genetic resource banks and reproductive technology. In: Cloud JG, Thorgaard GH, eds. Genetic conservation of salmonid fishes. New York: Plenum, 1993.

11. Tilson RL. Preserving critical habitat: the Minnesota Zoo's adopt-a-park program. Annu conf Am Assoc Zoo Prk Aquar, 1991:386–90.

12. Hutchins M, Wiese RJ. Beyond genetic and demographic management: the future of the Species Survival Plan and related AAZPA conservation efforts. Zoo Biol 1991;10:285–92.

13. Ralls K, Brugger K, Ballou J. Inbreeding and juvenile mortality in small populations of ungulates. Science 1979;206:1101–3.

14. Wildt DE, Bush M, Howard JG, et al. Unique seminal quality in the South African cheetah and a comparative evaluation in the domestic cat. Biol Reprod 1983;29:1019–25.

15. O'Brien SJ, Wildt DE, Goldman D, Merril CR, Bush M. The cheetah is depauperate in genetic variation. Science 1983;221:459–62.

16. O'Brien SJ, Roelke ME, Marker L, et al. Genetic basis for species vulnerability in the cheetah. Science 1985;227:1428–34.

17. O'Brien SJ, Wildt DE, Bush M. The cheetah in genetic peril. Sci Am 1986;254:68–76.

18. Wildt DE, O'Brien SJ, Howard JG, et al. Similarity in ejaculate-endocrine characteristics in captive versus free-ranging cheetahs of two subspecies. Biol Reprod 1987;36:351–60.

19. Wildt DE, Brown JL, Bush M, et al. Reproductive status of cheetahs (*Acinonyx jubatus*) in North American zoos: the benefits of physiological surveys for strategic planning. Zoo Biol 1993.

20. Howard JG, Bush M, Hall LL, Wildt DE. Morphological abnormalities in spermatozoa of 28 species of nondomestic felids. Proc X int cong Anim Reprod Artif Insem, 1984;2:57–9.

21. Wildt DE, Phillips LG, Simmons LG, et al. A comparative evaluation of ejaculate and hormonal characteristics of the captive male cheetah, tiger, leopard and puma. Biol Reprod 1988;38:245–55.

22. Howard JG, Brown JL, Bush M, Wildt DE. Teratospermic and normospermic domestic cats: ejaculate traits, pituitary-gonadal hormones and improvement of spermatozoal motility and morphology after swim-up processing. J Androl 1990;11:204–15.

23. O'Brien SJ, Martenson JS, Packer C, et al. Biochemical genetic variation in geographic isolates of African and Asiatic lions. Natl Geogr Res 1987;3: 114–24.

24. Gilbert DA, Packer C, Pusey AE, Stephens JC, O'Brien SJ. Analytical DNA fingerprinting in lions: parentage, genetic diversity and kinship. J Hered 1991;82:378–86.

25. Gilbert DA, Reid YA, Gail MH, et al. Application of DNA fingerprints for cell-line individualization. Am J Hum Genet 1990;47:499–514.

26. Wildt DE, Bush M, Goodrowe KL, et al. Reproductive and genetic consequences of founding isolated lion populations. Nature 1987;329:328–31.

27. Barone MA, Roelke ME, Howard JG, Anderson AE, Wildt DE. Reproductive characteristics of male Florida panthers: comparative studies from Florida, Texas, Colorado, Latin America and North American zoos. J Mamm 1993.

28. Roelke ME. Florida panther biomedical investigation final performance report, E-1 II-E-6. Gainesville: Florida Game and Freshwater Fish Commission, 1990.

29. Roelke ME, Martenson JS, O'Brien SJ. Consequences of demographic reduction and genetic depletion in the endangered Florida panther (submitted).

30. Seal US, Lacy RC. Florida panther (*Felis concolor coryi*) viability analysis and species survival plan. United States Fish and Wildlife Service Cooperative Agreement #14-16-0004-90-902, 1989.

31. Goldman EA. Classification of the races of the puma. In: Young SP, Goldman EA, eds. The puma: mysterious American cat. Washington: American Wildlife Institute, 1946:175–302.

32. Baccetti D, Burrini AG, Collodel G, Piomboni P, Renieri T. A "miniacrosome" sperm defect causing infertility in two brothers. J Androl 1991;12:104–11.

33. Thomas WP, Howard MH Jr. Cryptorchidism and related defects in dogs: epidemiologic comparisons with man. Teratology 1975;12:51–6.

34. Rothschild MF, Christian LL, Blanchard W. Evidence for multigene control of cryptorchidism in swine. J Hered 1988;79:313–4.

35. Mahadevan MM, Trounson AO. The influence of seminal characteristics on the success of human in vitro fertilization. Fertil Steril 1984;42:400–5.

36. Burkman LJ, Coddington CC, Franken DR, Kruger TF, Rosenwaks Z, Hodgen GD. The hemizona assay (HZA): development of a diagnostic test for the binding of human spermatozoa to the human hemizona pellucida to predict fertilization potential. Fertil Steril 1988;49:688–97.

37. Oehninger S, Coddington CC, Scott R, et al. Hemizona assay: assessment of sperm dysfunction and prediction of in vitro fertilization outcome. Fertil Steril 1989;51:665–70.

38. Coddington CC, Franken DR, Burkman LJ, Oosthuizen WT, Kruger T, Hodgen GD. Functional aspects of human sperm binding to the zona pellucida using the hemizona assay. J Androl 1991;12:1–8.

39. Johnston LA, Donoghue AM, O'Brien SJ, Wildt DE. Culture medium and protein supplementation influence in vitro fertilization and embryo development in the domestic cat. J Exp Zool 1991;257:350–9.

40. Johnston LA, Donoghue AM, O'Brien SJ, Wildt DE. Influence of temperature and gas atmosphere on in vitro fertilization and embryo development in the domestic cat. J Reprod Fertil 1991;92:377–82.

41. Donoghue AM, Johnston LA, Seal US, et al. In vitro fertilization and embryo development in vitro and in vivo in the tiger (*Panthera tigris*). Biol Reprod 1990;43:733–47.

42. Donoghue AM, Howard JG, Byers AP, et al. Correlation of sperm viability with gamete interaction and fertilization in vitro in the cheetah (*Acinonyx jubatus*). Biol Reprod 1992;46:1047–56.

43. Miller AM, Roelke ME, Goodrowe KL, Howard JG, Wildt DE. Oocyte recovery, maturation and fertilization in vitro in the puma (*Felis concolor*). J Reprod Fertil 1990;88:249–58.
44. Howard JG, Bush M, Wildt DE. Teratospermia in domestic cats compromises penetration of zona-free hamster ova and cat zona pellucida. J Androl 1991;12:36–45.
45. Howard JG, Donoghue AM, Johnston LA, Wildt DE. Zona pellucida filtration of structurally abnormal spermatozoa and reduced fertilization in teratospermic cats. Biol Reprod 1993.
46. Howard JG, Barone MA, Bush M, Wildt DE. A heterologous salt-stored zonae pellucidae assay for assessing sperm capacitation and the impact of teratospermia in the cheetah (*Acinonyx jubatus*). J Androl 1991;(suppl):101.
47. Roth TL, Howard JG, Wildt DE. Zona pellucida-piercing enhances zona penetration by spermatozoa of normospermic and teratospermic domestic cats. Theriogenology 1993;39:299.
48. Wildt DE. Fertilization in cats. In: Dunbar BS, O'Rand M, eds. A comparative overview of mammalian fertilization. New York: Plenum, 1991:299–328.
49. Wildt DE, Schiewe MC, Schmidt PM, et al. Developing animal model systems for embryo technologies in rare and endangered wildlife. Theriogenology 1986;25:33–51.
50. Howard JG, Doherty J. First endangered cats produced by frozen semen and artificial insemination. Am Assoc Zoo Prk Aquar Commun, May, 1992:12.
51. Howard JG. Successful laparoscopic artificial insemination in the leopard cat. Am Assoc Zoo Prk Aquar Commun, June, 1991:15.
52. Howard JG, Donoghue AM, Goodrowe KL, et al. Successful induction of ovarian activity and laparoscopic intrauterine insemination in the cheetah (*Acinonyx jubatus*). J Zoo Wildl Med 1992;23:288–300.
53. Donoghue AM, Johnston LA, Armstrong DL, Simmons LG, Wildt DE. Birth of a Siberian tiger cub (*Panthera tigris altaica*) following laparoscopic intrauterine insemination. J Zoo Wildl Med 1993.
54. Monfort SL, Asher GW, Wildt DE, et al. Successful intrauterine insemination of Eld's deer (*Cervus eldi thamin*) with frozen-thawed spermatozoa. J Reprod Fert 1993.
55. Schiewe MC, Bush M, Phillips LG, Citino S, Wildt DE. Comparative aspects of estrous synchronization, ovulation induction and embryo cryopreservation in the scimitar-horned oryx, bongo, eland and greater kudu. J Exp Zool 1991;58:75–88.

Part I

Functional Significance of Morphological Relationships

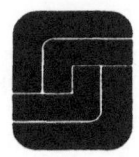

2

Stage-Related Functions of Sertoli Cells: Lessons from Lower Vertebrates

GLORIA V. CALLARD, MARLIES BETKA, AND JOAN C. JORGENSEN

Primarily from studies in mammals, it is evident that testicular functions have complex control mechanisms that include several categories of endocrine, paracrine, and autocrine mediators and involve crosstalk among germinal elements and multiple somatic cell types (e.g., Sertoli cells, Leydig cells, peritubular myoid cells, and fixed and transient leucocytes). As the only somatic cell element intimately in contact with germ cells, Sertoli cells have a pivotal role. What has hampered efforts to define functions of Sertoli cells during spermatogenic development is the organizational complexity of the mammalian testis. By contrast, the testis of certain lower vertebrates is much simpler in organization and has advantageous anatomic or functional characteristics that facilitate observation and experimental manipulation stage by stage. As part of a program of research examining the functional interdependence of spermatogenesis and steroidogenesis, we have carried out many studies using the spiny dogfish shark (*Squalus acanthias*) and have developed methods for studying spermatogenesis and its regulation using intact germinal units in vitro. Reviewed in this chapter are published and unpublished results from this laboratory.

Evolutionary History

To appreciate how novel information obtained from sharks can support and extend concepts derived from the mammalian prototype, it is useful at the outset to consider conserved versus variable features of spermatogenesis in vertebrates. A survey of widely separated taxa reveals striking similarities in the basic processes by which male germ cells are produced, including (i) the major events of spermatogenesis; (ii) germ cell stages and cytological features; (iii) synchronous development of differentiating germ cells as syncitially connected clones; (iv) organization of succeeding generations in strict spatial and temporal order; (v) a central role for Sertoli cells, variously termed *supporting, nutrient, companion, sustentacular, nurse,* or *cyst* cells; and (vi) a steroid-rich environment (reviewed in 1–5). Compared to conserved features, differences are relatively superficial; for example, in the final form of the mature spermatozoon, the number of spermatogonial divisions, the timing between stages, the primary source of intratesticular steroids (Sertoli cells vs. Leydig cells), and the organization of the testis.

When contemporary vertebrates are arranged in their order of branching from mainline evolution, a trend is seen from a renewable population of Sertoli cells associated with a single germ cell generation toward an increasingly versatile "immortal" Sertoli cell population that has the ability to nurture new germ cell generations before earlier generations are complete. As a trade-off for increased structural and functional complexity, Sertoli cells come to share their role with additional somatic elements residing outside the germinal compartment (e.g., Leydig cells). These trends correlate well with the acquisition of permanent seminiferous tubules for housing the primary germinal units (germ cell clones plus associated Sertoli cells) and with cytologic and functional differentiation of the interstitium, including peritubular cells and Leydig cells. The latter, situated in close proximity to the vascular supply, evolve as the primary source of steroids exported to the circulation, usurping the earlier role of Sertoli cells in providing steroids to the germinal compartment. At the same time, germ cell clones housed entirely within the tubule lose their original spherical configuration, becoming more irregular in form and "open" to the tubular environment and to germinal clones in more or less advanced stages of development. Figure 2.1 details similarities and differences in sharks and mammals, with a view to emphasizing cross-species relatedness and clarifying terminology.

Shark Testicular Organization

A detailed history and comprehensive bibliography of information pertaining to the reproductive physiology and biochemistry of male elasmobranch fishes (sharks, skates, and rays) can be found in earlier reviews (1–3, 6). The organization of the testis in male sharks—typified by the spiny dogfish, but also including the European spotted dogfish (*Scyliorhinus*) and several Japanese dogfish species (*Mustelus ssp.*)—has several main technical advantages.

First, the basic germinal unit is a closed, spherical spermatocyst (lobule, follicle, or ampulla) comprising a single germ cell clone and a second clonal population of stage-synchronized Sertoli cells. The number of

Sertoli cells and germ cells per cyst is strictly dependent on the stage of spermatogenesis and, for any one species, is predictable. For example, in the spiny dogfish, cyst formation begins with a single stem cell of each type (7). Initially, mitoses are synchronized, maintaining a germ cell-somatic cell ratio of 1:1 through 9 divisions; however, Sertoli cells cease dividing prior to the final 4 spermatogonial mitoses. Upon entry into meiosis, therefore, the germ cell-Sertoli cell ratio is 16:1 and, at this stage, a unit (spermatoblast) comprising a single Sertoli cell and its 16–member germ cell cohort becomes readily apparent. Following two meiotic divisions and spermiogenesis, there are 64 germ cells per Sertoli cell and approximately 424 Sertoli cells (27,136 spermatozoa) per cyst. Sertoli cells degenerate or are lost into the semen after sperm release. Thus, in species with a *cystic* mode of spermatogenesis, Sertoli cells, like germ cells, undergo a single cycle of proliferation, differentiation, and loss.

A second advantageous feature of sharks is that cysts are situated at the terminae of collecting ducts and embedded directly in a connective tissue matrix. The absence of a tubule homolog correlates well with the absence of typical peritubular myoid cells and Leydig cells, although Leydig-like elements, which are small, sparse, and undifferentiated through all developmental stages, may be evolutionary forerunners of the latter (8, 9). Accordingly, somatic cell functions are embodied primarily or exclusively in Sertoli cells (9, 10).

Third, cysts are arranged in maturational order in a simple linear fashion across the diameter of the testis from germinal ridge to efferent ducts, resulting in a topographic separation of developmental stages and a readily visible zonation (Fig. 2.2). New cysts are formed continuously through the seasonal period of reproductive activity in adults. They originate at fixed germinal sites on the dorsolateral aspect of the testis and move steadily away from the germinal zone as they grow and mature, followed closely by successively younger cysts. As they approach the mesoventral aspect of the testis, located just beneath the epigonal organ (a lymphomyeloid tissue in sharks, which have no bone marrow), they contain the most advanced germ cell stages. When spermatogenesis is complete, an opening is affected into the attached terminal branch of the collecting ducts, and the spermatozoa exit the testis via the efferent ducts. Each column of migrating cysts progresses at the same rate away from the germinal zone; therefore, in cross-section, cysts in the same stage of differentiation are recognizable as roughly concentric rows around a germinal hilus. Even without the aid of a dissecting microscope, several zones of maturation are readily distinguishable on the basis of opacity, color, and position relative to the germinal zone and epigonal tissue. The fractional area of each cross-section occupied by the different zones varies seasonally in species with annual cycles, presumably a reflection of the number of surviving cysts in a given stage and the duration of that stage.

I. GERM CELLS

Elasmobranchs and Mammals

A. Stages of development are primordial germ cells, gonocytes, spermatogonia, spermatocytes, spermatids, spermatozoa.

B. Major events are stem cell renewal, commitment to a differentiative pathway, mitotic proliferation, meiosis, spermiogenesis, spermiation.

C. Cytological changes during development are similar, with the exception of spermiogenesis and the final form of the mature spermatozoon, which has species-specific characteristics.

D. Germ cells develop as isogenetic clones; all cells of a clone are derived from a single spermatogonium, remain connected by intercellular bridges, and are synchronized in development.

E. Succeeding germ cell generations are related in strict spatial and temporal order.

Elasmobranchs

F-1. A single germ cell clone, together with a second clonal population of Sertoli cells, forms spherical unit (spermatocyst) bounded by an acellular basal lamina.

Mammals

F-2. A germ cell clone is flattened with irregular borders and no boundary membrane but is associated with a cohort of Sertoli cells.

II. SERTOLI CELLS

Elasmobranchs and Mammals

A. Somatic cells termed follicle, cyst, nurse or Sertoli cells have an intimate anatomic arrangement with germ cells and mediate between the soma and germ cell clones in all respects.

B. Structural and functional changes are coordinated with germ cell development

Mammals

C-2. Sertoli cell proliferation is generally limited to early development.

D-2. A Sertoli cell of adult testis simultaneously nurtures 4 or 5 germ cell generations and displays base to lumen differentiation in relation to stage.

E-2. Sertoli cells possess a limited number of steroidogenic activities but are not believed to synthesize steroids de novo.

Elasmobranchs

C-1. Sertoli cells undergo cycles of proliferation, development and degeneration in the adult testis.

D-1. A given Sertoli cell is associated with a single germ cell generation and displays stage-dependent characteristics

E-1. Sertoli cells are the primary steroidogenic elements of the testis.

III. SECONDARY GERMINAL COMPARTMENT

Mammals

A-2. Seminiferous tubules are contiguous with the intratesticular collecting ducts and encompass germ cell clones and associated Sertoli elements

Elasmobranchs

A-1. None present; spermatocysts are attached to the terminae of collecting ducts and embedded directly in testicular stroma.

IV. LEYDIG CELLS

Mammals

A-2. Leydig cells, located in the interstices of tubules are the primary steroid secretory elements of the testis and vary structurally and functionally in relation to stage.

Elasmobranchs

A-1. Definitive Leydig cells are absent through all spermatogenic stages but small, undifferentiated Leydig-like precursor elements are seen.

FIGURE 2.1. Comparison of testicular organization and spermatogenesis in sharks and mammals: a summary of conserved vs. variable features. After Callard (2).

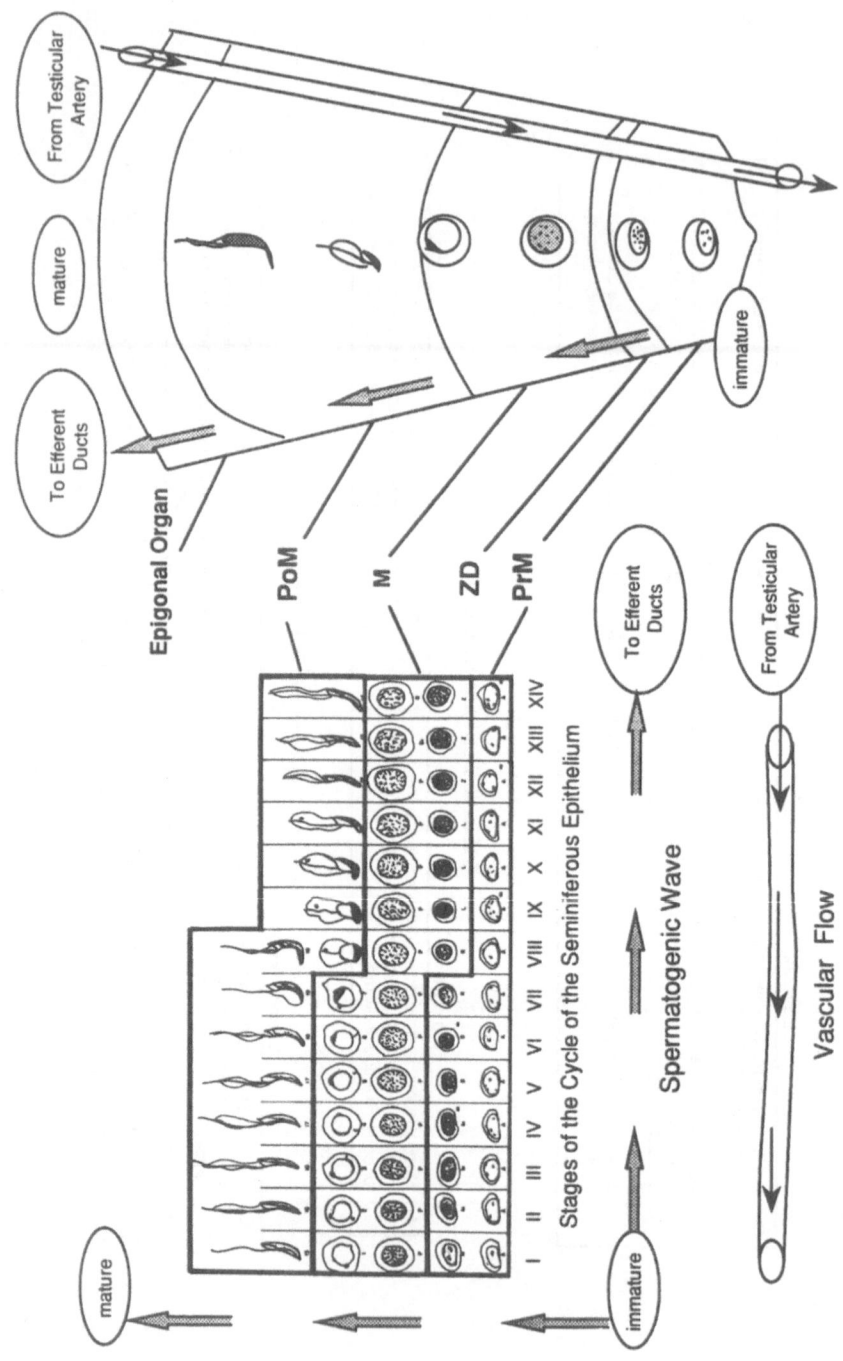

Although mammalian germ cells do not actually migrate across the diameter of the testis, as in the shark, a maturational wave radiates along the length of each tubule away from the rete (Fig. 2.2). The spermatogenic wave, together with the base-to-lumen migration of succeeding germ cell generations at a single point on the tubule, constitutes the mammalian counterpart of the spatial order seen in sharks. Pilsworth and Setchell (4) have likened the mammalian spermatogenic wave to a string of open-ended, contiguous cysts, a hypothetical model that closely resembles the actual condition in the shark testis. Although it is technically difficult to ascertain the exact number of individual cells in a syncitial germ cell clone after the early spermatogonial stages in mammals, the theoretical number in the rat, based on electron microscopic observation of intercellular bridges in serial sections, has been estimated as 512 in the spermatid stage (11), plus correspondingly fewer numbers in each less mature clone, giving an estimated total of 1000–1500 per Sertoli cell. Given that the number of mature spermatids per Sertoli cell is 10 (12), the number of Sertoli cells per spermatocyst homolog in rats can be estimated at 50. When elasmobranchs and mammals are compared, therefore, it is evident that evolution has materially decreased the size of the germinal unit concomitant with a decreased nurturing efficiency, but at the same time has increased the versatility of individual Sertoli cells by enabling simultaneous nurturing of successive generations (1, 2).

Stage-Related Characteristics In Vivo

Cytologically and functionally, the mammalian Sertoli cell reflects germ cell order in its base-to-lumen differentiation and in variations corresponding to the stages of the seminiferous epithelial cycle (13). In sharks, developmental synchrony of Sertoli cells with a single germ cell gen-

◄───

FIGURE 2.2. Organization of stages of germ cell development in mammalian vs. shark testis and countercurrent relation with blood flow. In the mammalian testis (left), succeeding germ cell generations are arranged from base to lumen (germ cell associations) and in descending maturational order from the rete along the length of the tubule (stages of the cycle of the seminferous epithelium, I–XIV in the rat), constituting the spermatogenic wave. On the right is a cross-section of shark testis showing the simple linear order of germ cell stages from germinal ridge across the diameter of the testis to efferent ducts. Corresponding regions in mammalian vs. shark testis are delimited: premieotic (PrM), meiotic (M), and postmeiotic (PoM). Also shown in shark testis is the zone of degeneration (ZD) and epigonal organ, a lymphomyeloid tissue that encapsulates the testis at the mature pole. Note the parallel-antiparallel arrangement of vascular flow (thin arrows) in relation to maturational order (shaded arrows) in both mammals, as described in reference 49, and sharks, as described in reference 15.

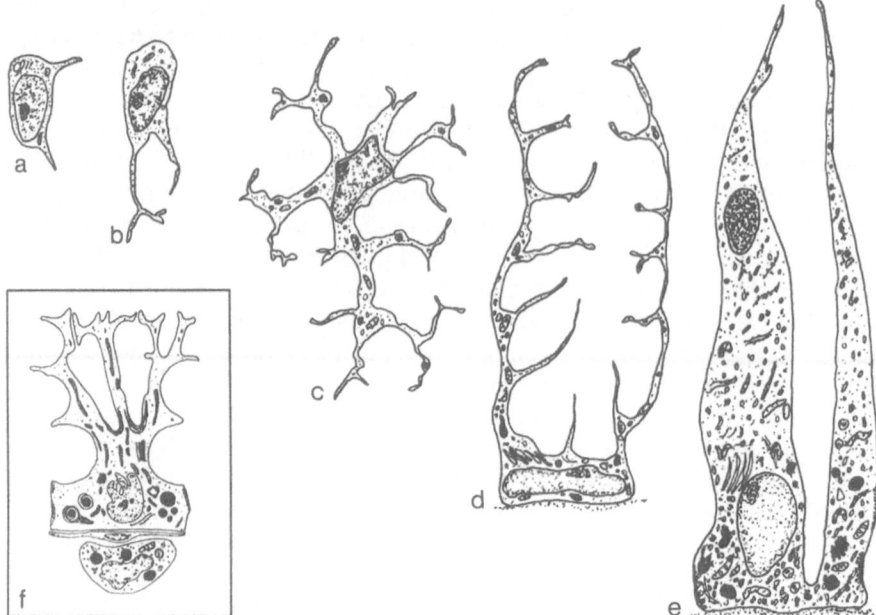

FIGURE 2.3. Diagram of Sertoli cells in the spiny dogfish shark (*Squalus acanthias*) vs. mammalian testis. Shark Sertoli cells (*a–e*) are associated with a single clonal population of germ cells throughout development and display corresponding stage-related morphological change during (*a*) stem cell, (*b*) early spermatogonial, (*c*) late spermatogonial, (*d*) spermatocyte, and (*e*) mature spermatid stages of development. Leydig cells and myoid cells are absent or undifferentiated through all stages. The mammalian Sertoli cell (*f*) is differentiated from base to lumen, reflecting its association simultaneously with 4 or 5 germ cell generations; peritubular myoid cells and Leydig cells are present outside the germinal compartment. After Holstein (7) and Fawcett (13).

eration facilitates observation of stage-related change and has prompted numerous investigators to study their morphological differentiation (reviewed in 1, 2) (Fig. 2.3). Using electron microscopy, we have focused especially on maturation-related increases in organelles associated with steroidogenesis (8, 10). Also, using tissues from defined regions (stages) of shark testis for direct biochemical analysis, we have documented several indicators of steroid biosynthesis, metabolism, and action (14–19). Generally, tissues were grossly divided into *premeiotic* (PrM) (stem cells, spermatogonia, and preleptotene spermatocytes), *meiotic* (M) (spermatocytes and round spermatids), and *postmeiotic* (PoM) (elongating and mature spermatids) stages. However, it is possible to study more precisely defined substages of these three major divisions, and several

schemes have been proposed (7, 20, 21). Results from this laboratory are
reviewed below and summarized in Figure 2.4.

Morphology of Sertoli Cells

Salient changes during spermatogenesis in shark Sertoli cells include a
change in position of the nucleus from an adluminal to a basal position
within the cyst during late spermatogonial stages; dramatic shape changes
due to extensive development of cytoplasmic processes that engulf proliferating germ cells during mitotic and meiotic stages; and a marked
increase in cytoplasmic volume that first becomes apparent during
spermiogenesis as the germ cells condense (7, 10, 22–25) (Fig. 2.3). At
the electron microscopic level, there is a remarkable increase in lipid
droplets, mitochondria with tubulovesicular cristae, and smooth endoplasmic reticulum (7, 10). The latter increases so greatly in abundance
that at the end of spermiogenesis, it fills the Sertoli cell as a mass of
tubules. Only in the squirrel, and then only in regions immediately
adjacent to mature spermatids, has such a vast system of smooth membranes been described.

In shark Sertoli cells, these cytologic changes during successive spermatogenic stages correspond exactly to the increase in microsomal
enzymes that regulate steroidogenesis (Fig. 2.4; see below) and with
histochemical studies showing a strong positive reaction for *3β-hydroxy-
steroid dehydrogenase* (3βHSD) in Sertoli cell cytoplasm and the lumens
of cysts with mature spermatids (26, 27). The 3βHSD has also been
visualized in semen, a finding consistent with the presence of pinched-off
Sertoli cell remnants (cytoplasts) (28), steroidogenic activity (29), and
high endogenous steroid concentrations (30) in the seminal fluid of the
same species. These data, together with biochemical and cell culture
assays (see below), support the conclusion that Sertoli cells are the
primary steroidogenic element of shark testis.

Like their mammalian counterparts, shark Sertoli cells give the cytologic
appearance of protein synthesis and secretion, with increasing activity
during spermatogenesis (Fig. 2.3). This is marked by the development of
a prominent Golgi, an increase in rough endoplasmic reticulum, and
the accumulation of secretory products in its cisternae (23, 31). Also,
electron-dense granules are observed at the apical pole of the fully mature
Sertoli cell and seem to be secreted into the lumen (31). At the end of
spermiogenesis, a large, proteinaceous structure appears, the *problematic
body* (23, 31, 32). The latter is present in the cytoplasm at the apices of
Sertoli cells, and when the apical cytoplasm is sloughed at spermiation,
the problematic body is carried away with the spermatozoa into the efferent ducts and is readily observed in semen collected from the ampulla
d. deferens (28). Cytochemical techniques show that the proteins of the
problematic body are rich in lysine, cysteine, and tryptophan (31), but

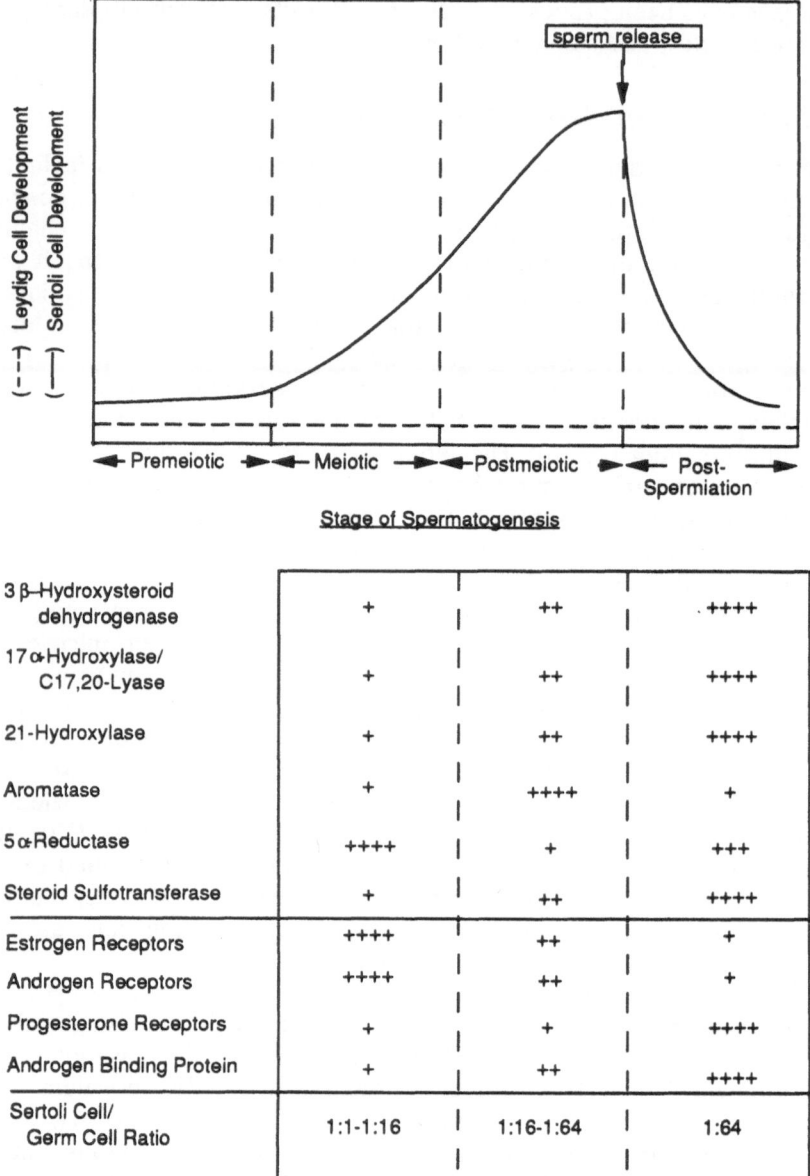

FIGURE 2.4. Summary of stage-related structural and functional characteristics in the testis of the spiny dogfish shark (*Squalus acanthias*) during spermatogenesis. Sertoli cell growth and differentiation are keyed to spermatogenesis, whereas Leydig-like cells are small, sparse, and undifferentiated through all stages. Several indicators of steroid synthesis and action were measured in staged tissues: the stage having maximal levels was set at (++++); lower levels in other stages are shown accordingly (+++ to +). For actual units of measure, see text for original references. Also shown are Sertoli cell/germ cell ratios by stage. After Callard (2, 64).

whether it is a mechanism for sequestering and degrading proteins formed during spermiogenesis or is somehow involved in the transport and delivery of proteins to the efferent ducts for maintenance of spermatozoa is a matter of speculation.

Several other fine-structural features of mature Sertoli cells are worthy of note. While the apical surface of the nucleus remains smooth, that facing the basal lamina develops fingerlike projections, and intranucleolar granules and lamellar bodies appear in association with the nucleolus (31). These resemble structures of unknown significance in mammalian Sertoli cells. Also, gap junctions are seen between adjacent Sertoli cells in the basal regions where the plasmalemmas of the two cells appear to be fused and seem to be reinforced by microfilaments and by parallel arrays of tubules of the smooth reticulum (32). Exactly when gap junctions are formed and whether junctional complexes prevent penetration of macromolecular markers into the spermatocyst lumen have not been systematically studied in sharks; however, an extracellular marker (Evans blue dye) that is homogeneously distributed in immature testicular regions does not penetrate cysts in mature testicular regions (15).

Steroid Synthesis and Metabolism

Radiolabeled tracer analysis of homogenates and subfractions, together with biochemical and radioimmunoassay analysis of organic extracts and secreted products, has definitively established that steroidogenic pathways leading from cholesterol to biologically active C-21, C-19, and C-18 steroids are identical in the testis of sharks and mammals. However, some notable differences in the quantity of certain metabolites are seen (reviewed in 1, 2). For example, *11-deoxycorticosterone* (DOC), *17α-hydroxyDOC* (Reichstein's Substance S), and a variety of other progestins hydroxylated at C-17, C-20β, or C-21 are relatively abundant products of shark testis and are also found in high amounts in semen (33, 34). The functional significance, if any, is not known, although hydroxylated progestins, acting on the outer cell membrane, are involved in final oocyte maturation in teleost fish and may play a role in spermiation in males. Also, conjugating pathways leading to steroid glucuronides and sulfates are exceedingly active in shark testis, such that a 1-h, 1-pass perfusion of the testis in situ with [³H]androgen or [³H]estrogen results in >90% recovered from testicular cytosol as polar metabolites (15). This mechanism of steroid inactivation may have utility in regulating ligand binding in organs like the testis where steroids are synthesized in close proximity to receptors.

Using radiolabeled tracer methodology and microsomal subfractions derived from staged testicular tissues, we observed that 3βHSD, the enzyme catalyzing conversion of 5 precursors to the 4,3-ketone A ring of hormonally active steroids, increases progressively during germ cell maturation (5-fold, PoM > M > PrM) (Fig. 2.4) (16). Maturation-related increases in microsomal 3βHSD are also obtained when microsomes from isolated spermatocysts or Sertoli cells are assayed (16, 35). Moreover, comparison of specific enzyme activities in whole testis versus spermatocysts or Sertoli cells indicates that virtually 100% of testicular 3βHSD can be accounted for by that associated with Sertoli cells. Thus, biochemical results correspond exactly to the development of smooth endoplasmic reticulum seen in Sertoli cells by electron microscopy and support the view that this cell type is the exclusive, or primary, site of steroid biosynthesis. Other enzymes that display maturation-related increases in activity, although with different fold changes, are 17α-hydroxylase/C-17,20-lyase, the key enzyme in androgen biosynthesis (2- to 12-fold); 21-hydroxylase, an enzyme that utilizes progesterone to form DOC (3- to 20-fold); and androgen/estrogen sulfotransferase (2-fold) (14, 15, 34).

A distinctly different stage-related distribution pattern is seen with aromatase, the rate-limiting enzyme in androgen to estrogen conversion, which is 2- to 7-fold higher where germ cells are undergoing meiosis than in less mature or more mature regions (14). By contrast, 5α-reductase, which regulates *testosterone* (T) to *5α-dihydrotestosterone* (DHT) transformation, differs from all other patterns (PrM > PoM ≫ M), although in contrast to the prostate and other androgen targets in mammals, the necessity of this transformation for spermatogenesis has not been established (16).

It is important to note here that the timing of maximal androgen production and low aromatization in shark testis (mature spermatid stage) agrees with observations in rats showing that stage VIII of the seminiferous cycle, a time approximately coincident with the release of the most advanced germ cell generation, is associated with Leydig cells significantly larger than in other stages, highest tubular androgen levels, and the presence of an aromatase inhibitor (36). Taken together, these data support the view that biologically active versus inactive steroid ligands are developmentally regulated and result in quantitatively and qualitatively unique steroidal microenvironments at each germ cell stage.

Steroid Receptors and Binding Protein

The strongest indication that steroid regulation of gene expression has been widely conserved as a mechanism of control in vertebrate testis is the presence in several nonmammalian species of steroid receptors that resemble those of mammals in their important physicochemical characteristics, despite minor species differences in ligand specificity and chromatin binding characteristics (14, 17, 18, 37). Furthermore, studies using staged shark tissues have been useful in providing clues to the steroid-sensitive control points during spermatogenesis (Fig. 2.4). Both *androgen receptors* (ARs) and *estrogen receptors* (ERs) are highest in regions with

stem cells and spermatogonia and are virtually nondetectable in regions
with mature germ cells (14, 17). Additionally, the highest percentage of
occupied ERs and the highest number of ER-specific chromatin binding
sites are present in PrM stages (14, 18). Virtually all testicular ARs are
occupied, as determined by exchange assays (17).

These distributions support the hypothesis that endogenous androgen
and estrogen may cooperate in regulating gene expression prior to meiosis
and are consistent with observations showing that hypophysectomy most
profoundly alters the transition from spermatogonia to spermatocytes in
the shark (27, 38, 39). Also, effects of hypophysectomy are mimicked
early each year when a *zone of degeneration* (ZD) (Fig. 2.2) appears at
the junction of the spermatogonial-spermatocyte transition, an event that
may reflect the normal seasonal waning of gonadotropin secretion and
consequent steroidogenesis.

Recently, we identified a *progesterone receptor* (PR) in shark testis
(17). The PRs clearly differ from the ARs in their zonal distribution
(PoM $\gg$ M = PrM) (Fig. 2.4) and steroid binding characteristics, but are
unable to distinguish between *progesterone* (P) and DOC as ligands.
Although cultured spermatocysts produce both P and DOC de novo (34,
40), when [^{3}H]P is added to cultured spermatocysts, >90% of radioactivity recovered from nuclear subfractions is unchanged P, and despite
active conversion to [^{3}H]DOC, little or no nuclear-associated DOC is
detectable (Piferrer and Callard, unpublished observations).

Patterned distributions of steroid receptors in shark testis are consistent
with available data in rodents. Although ERs and PRs have not been
measured in relation to the stage of the seminiferous cycle in the rat, AR
binding activity is 50% higher in cell nuclei from stages IX–XII and
XIII–I when compared to other stages (41). This has been interpreted as
an effect of androgen on meiotic divisions in stage XIV, but does not
exclude an influence on earlier germ cell generations that are present in
the same segments. Significantly, several bouts of mitoses occur in stages
having the highest nuclear AR activity. Attempts to pinpoint androgen-
sensitive steps in vivo using hypophysectomy, gonadotropin, or steroid
replacement to manipulate testicular steroid levels have revealed that
hormones do not influence the duration of the cycle, but somehow increase the efficiency of the process (42). Most profoundly affected by
androgen deprivation are early meiotic stages of development, specifically
the transition from preleptotene to pachytene spermatocytes. In androgen-
resistant (AR-defective) mice, spermatogenesis is arrested in the first
meiotic divisions (43), buttressing the idea that androgen has actions early
in spermatogenesis, although the possibility of effects at later stages
cannot be ruled out.

Although we have not localized steroid receptors in shark testis to
Sertoli cells specifically, the absence of other differentiated somatic
elements, together with data from mammals showing that Sertoli cells

and/or Sertoli cell lines have receptors for androgen, estrogen, and glucocorticoids (44, 45), suggests that Sertoli cells are the steroid target in
sharks. Androgens and glucocorticoids positively or negatively regulate a
number of Sertoli cell functions in rats, but Leydig cells and peritubular
cells are steroid targets as well (45, 46). The most compelling evidence
that androgen exerts control over spermatogenesis via one or more
somatic cell types comes from studies in which germ cells from AR-
deficient mice developed normally in tubules of wild-type chimeras (43).

In addition to conventional steroid receptors that are detected in both
nuclear and cytosolic extracts and that adhere to DNA-cellulose affinity
columns, a nonreceptor steroid binding protein has been characterized in
cytosolic subfractions of shark testis (19, 47). In its physicochemical
properties (high affinity, broad specificity, apparent molecular weight,
isoelectric point, and dimeric structure), it resembles *androgen binding
protein* (ABP) of mammalian testis and is presumed to be the shark
counterpart. When its intratesticular distribution is examined, ABP concentration increases progressively through spermatogenic development,
corresponding exactly to the hypertrophy and differentiation of Sertoli
cells and maximal androgen biosynthetic potential (Fig. 2.4), data that
are consistent with the protein secretory appearance of Sertoli cells at
mature stages and the known Sertoli cell origin and androgen dependence
of ABP in mammals. Interestingly, dissected rat tubular segments secrete
ABP maximally at stage VIII, just prior to release of the most advanced
cohort of germ cells (36). Although secretion of ABP by shark Sertoli
cells has not been established, this protein may serve as a steroid reservoir
in the testis or during long-term sperm storage in the excurrent ducts and
may thus account for high endogenous steroid levels in shark semen (30).

Intratesticular Vascular Communication Pathway

Analysis of the enzyme and receptor distribution patterns described
above reveals a paradox. With the exception of PRs, which are maximally
concentrated in testicular regions where P-synthesizing potential is
maximal, maximal androgen and estrogen biosynthesis (PoM and M
stages, respectively) is spatially (temporally) separated from regions in
which ARs and ERs are concentrated (PrM stages) (Figs. 2.2 and 2.4).

To investigate whether steroids synthesized in more advanced stages
of development have access to corresponding receptors in less mature
stages, we perfused the testis in situ with Evans blue dye via the genital
artery and observed that blood enters the testis via the capsule immediately beneath the epigonal organ: The pattern of flow is PoM → M
→ PrM (Fig. 2.2) (15, 48). Since the PoM region of shark testis has an
abundance of ABP and steroid sulfotransferase, both of which compete
with receptors for active hormone, we tested whether [³H]T entering via

this mature region has access to nuclear binding sites in PrM regions (48). Radiolabeled steroids identified as *17β-estradiol* (E_2), DHT, and T were concentrated in cell nuclei in PrM, but not in other regions. These results verify localization of ER and AR in early spermatogenic stages and underline the importance of in situ aromatization and 5α reduction under simulated in vivo conditions.

In contrast to nuclear-associated (receptor-bound) steroids, radio-labeled steroids recovered from whole homogenates, which include free metabolites and both nonreceptor-bound and receptor-bound fractions, show a different distribution profile, with the majority of perfused radio-labeled T recovered as ether-insoluble conjugates from PoM regions immediately adjacent to the site of blood entry into the testis and forming a gradient that corresponds to the flow pattern (15). Perfusion with [^{3}H]E_2 gives a distinctly different zonal distribution than that obtained with [^{3}H]T. Perfused E_2 is extensively metabolized to estrone sulfate and is not concentrated in any one region, nor are significant amounts bound to cell nuclei in PrM regions where ERs are localized. We infer from these results that steroid conjugating pathways may be a mechanism for screening out hormonal signals entering the testis from the general circulation. This would be essential in a situation where steroids synthesized in situ are acting as parahormonal regulators.

Data from sharks are consistent with the idea that steroids are part of an intratesticular signaling system that operates within (autocrine) and between (paracrine) stages of development. Is this between-stage communication pathway in shark testis unique, or is there a comparable communication pathway in mammals? As shown diagrammatically in Figure 2.2, the different stages comprising the spermatogenic wave in the rat tubule are separated by only a few millimeters and, hence, can readily communicate by simple diffusion of chemical messengers. However, capillaries are arranged longitudinally, and blood entering the testis proper via the region of the efferent ducts flows parallel to the spermatogenic wave in decreasing order (mature → immature) along the length of the tubule (49). Additionally, although androgen levels are maximal at stage VIII of the rat seminiferous cycle (36), AR occupancy peaks in stages IX–I (41), implying the route mature → immature, as we have observed in shark testis.

Culture of Intact Germinal Units Versus Isolated Sertoli Cells

From the foregoing, it is reasonable to postulate that androgen, estrogen, and progestin, each acting at selected stages of development and preferentially affecting certain genes and physiological processes, together constitute a concerted mechanism of spermatogenic control. Although correlational data are useful for constructing hypotheses, we were motivated to develop an in vitro spermatogenesis system to test our hypothesis directly. Since steroids are believed to affect spermatogenesis indirectly via receptors in somatic elements that in turn are regulated by germinal elements, an absolute requirement for an in vitro approach is that germ cells and Sertoli cells be in their normal physiological context.

Accordingly, techniques were developed for harvesting and culturing shark spermatocysts from defined stages (35, 40, 50). Stanley (24) was the first to describe the phase contrast appearance of living spermatocysts from trypsin-treated tissues of the European spotted dogfish. More recently, phase contrast properties of spermatocysts in different stages of spermiogenesis were described in the same species (21). An expanded description of the transillumination characteristics of spermatocysts corresponding to regions IV–XVII as defined by Mellinger (20) have been documented (51).

In our in vitro system, phase contrast and light microscopy confirm the stage and cellular composition of cysts and show that they retain their closed, spherical configuration for at least 15 days in culture (50). Interestingly, isolated rodent Sertoli cells allowed to reaggregate in Matrigel resemble cultured shark cysts in form, and occasional spermatogonia trapped within the aggregates progress into meiosis (52). Although the developmental advance of shark spermatocysts in vitro has not yet been studied cytologically, *[³H]thymidine* ([³H]Tdr) incorporation indicates that DNA synthesis in culture is as predicted from known events in vivo (PrM ≫ M = PoM) and is maintained quantitatively and qualitatively for at least 7 days in nutrient-supplemented media (50). In autoradiograms, germ cells in spermatogonial and preleptotene spermatocyte stages are labeled all-or-none in a given cyst, whereas cysts in later stages are unlabeled, indicating that low incorporation rates in M-stage and PoM-stage cultures are probably due to DNA repair (53).

We also compared intact spermatocysts to isolated Sertoli cells from corresponding stages (35, 50). As predicted from observations in situ, cultured Sertoli cells display stage-related differences in cell size and shape, nuclear size and shape, and the abundance of lipid droplets and other cellular inclusions. Sertoli cells from PrM and M stages begin attaching immediately upon seeding and eventually form a confluent monolayer remarkably similar to cultured Sertoli cells from immature rat testis. By contrast, Sertoli cells from PoM stages remain unattached with decreasing viability during the culture period, a result consistent with their in vivo status as terminally differentiated cells destined for degeneration. One day after seeding, DNA synthesis by Sertoli cells corresponds to the pattern obtained with intact cysts (PrM ≫ M = PoM). This result is predictable since mitoses are known to be synchronized in germ cells and Sertoli cells during the first 9 spermatogonial divisions in the spiny dogfish in vivo (7). In marked contrast to freshly isolated Sertoli cells, however, 7 days in culture effects a 40- to 200-fold increase in DNA synthesis and alters the stage-related pattern to PrM = M > PoM, an indication that Sertoli cells dedifferentiate in vitro. One interpretation is that germ cells and/or the three-dimensional cyst conformation normally serve as a brake on Sertoli cell proliferation (50). Whatever the explanation, these data reinforce the conclusion that intact germinal units are preferable to isolated Sertoli cells for regulatory studies. Reviewed below are data showing that cultured spermatocysts continue to express at least some normal growth and differentiated traits and respond selectively to added regulators (e.g., steroids and growth factors), manipulation of second messengers (e.g., cAMP), and sublethal toxicant exposure.

Positive and Negative Growth Control

Factors and mechanisms that regulate spermatogonial proliferation and their rate of entry into meiosis are largely unknown. Since PrM stages of spermatogenesis have the highest AR and ER concentrations, we were prompted first to investigate growth regulatory effects of steroids using [^{3}H]Tdr incorporation as an end point. Initial experiments show that E_2, T, DHT, and P have no consistent effect on DNA synthesis when added to basal medium at various doses and times after seeding. However, IGF-I (15 ng/mL) doubles DNA synthesis of PrM cysts, and this response is mimicked by insulin at higher doses (10 µg/mL) or by 10% fetal bovine serum, while shark serum, purified shark relaxin (a member of the insulinlike peptide family), transferrin, *epidermal growth factor* (EGF), *platelet-derived growth factor* (PDGF), and *transforming growth factor β* (TGFβ) are ineffective (53).

None of these substances alters low rates of DNA synthesis in M- and PoM-stage cysts; however, IGF-I stimulates a dose-dependent increase in DNA synthesis by isolated Sertoli cell monolayers, indicating that all or part of IGF-I's actions on the germ cells in the intact cyst are indirect. In addition to known additives, a <30-kd fraction of spent media from PrM-stage Sertoli cells stimulates DNA synthesis in PrM spermatocysts, whereas that from M-stage Sertoli cells inhibits DNA synthesis (53). Although SDS-PAGE analysis of the same media reveals distinct stage-related protein banding patterns, it has not been determined whether growth regulatory bioactivity corresponds to any of the bands. It may be relevant here that seasonal arrest of spermatogenic activity and removal of the gonadotropin-containing pituitary lobe in sharks cause regression of the testis and reduce [^{3}H]Tdr incorporation in vivo after intratesticular injection, but only in fish collected in summer (April to September) during the period of spermatogenic activity (38, 39).

Early reports of a spermatogonial chalone associated with spermatid stages in rodent testis (54) prompted us to investigate the effects of more mature spermatocysts on DNA synthesis by PrM cysts (55, 56). Coculture of PrM cysts in a two-chamber assembly with M- or PoM-stage spermatocysts decreases DNA synthesis (PoM > M), and inhibitory activity is even greater in cocultures with epigonal organ, a lymphomyeloid tissue encapsulating the testis adjacent to the PoM region and sharing a common vascular pathway: epigonal → PoM → M → PrM (Fig. 2.2) (15).

Growth inhibitory bioactivity (chalone) is detectable in spent media from cultured epigonal fragments, cytosolic subfractions of epigonal tissues, and white blood cells, but not red blood cells, plasma, or muscle. The effects of the endogenous chalone are dose dependent, have a short response latency (<3 h), are reversible (<16 h after washout), and counteract the stimulatory effects of insulin on DNA synthesis in PrM cysts. Epigonal organ is a rich source of this chalone. Initial characterization studies show the epigonal chalone is water and acid soluble, heat stable, trypsin and pronase insensitive, and <10 kd in molecular size, but inhibition cannot be mimicked by known steroids, prostaglandins, or growth factors tested. These results suggest that a growth inhibitory signal secreted by lymphomyeloid tissue directly into the blood entering the testis, or by immune cells derived therefrom, may regulate the number of germ cell clones that enter the cell cycle and may also account for the orderly diametric arrangement of germ cell stages in this animal model.

Programmed Cell Death

As an alternative to positive and negative regulation of mitotic and meiotic proliferation, it is possible that the size of the PrM cyst population may be determined by controlling the number of cysts surviving from one developmental stage to the next. In shark testis, the seasonal appearance of a localized *zone of degeneration* (ZD) between the last generation of spermatogonia and the first appearance of spermatocytes, which is mimicked by hypophysectomy (27, 38, 39), prompted us to examine the possibility that programmed cell death is a normal regulatory mechanism during spermatogenesis (Jorgensen and Callard, unpublished observations).

When DNA is extracted from staged tissues or spermatocysts and size-fractionated on agarose gels, *ladders* characteristic of nucleosomal oligomers are seen (Fig. 2.5). The proportion of DNA fragments, relative to high molecular weight DNA, is greater from tissues in ZD than in PrM tissues. This suggests that nucleosomal cleavage, an early biochemical marker of cysts programmed to die, occurs rarely, if at all, after cysts enter meiosis. Programmed cell death continues on schedule in PrM cysts maintained for 7 days in vitro, as indicated by time-dependent accumulation of fragmentation products. However, ladders are never seen in M

1   2 3   4 5   6 7   8 9   10 11 12 13

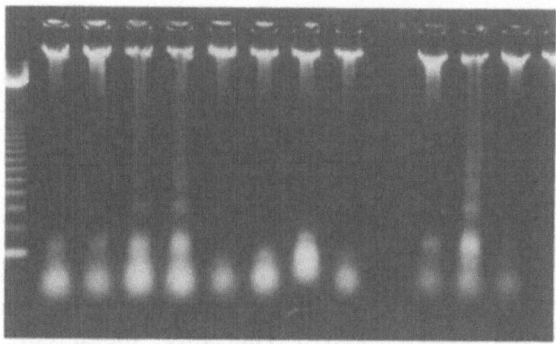

FIGURE 2.5. Electrophoretic analysis of DNA cleavage products extracted from spermatocysts from PrM (lanes 2–3), ZD (lanes 4–5), M (lanes 6–7), and PoM (lanes 8–9) regions of the testis and from dissected testicular tissues from the same four regions (lanes 10–13, respectively). A standard 123-bp DNA ladder is shown at left.

or PoM cultures despite random DNA fragmentation indicative of cell lysis.

Reexamination of published and unpublished light micrographs of shark testis reveals germ cells with characteristics of apoptosis. These appeared to be affected all-or-none in a given cyst, although Sertoli cells are normal in appearance. Germ cells with the appearance of apoptosis have been recognized in rodent testis as well (57). How cell death is programmed in shark testis and whether steroids are involved are questions of intense research interest in this laboratory.

Differentiated Functions

Steroid Synthesis

Based on tracer analysis of steroidogenic enzyme activities showing stage-related distributions in whole testis, spermatocysts, and isolated Sertoli cells, it is reasonable to predict that cultured cysts in PrM, M, and PoM stages would differentially utilize endogenous substrates to synthesize and secrete radioimmunoassayable steroids in vitro. Data show that P, T, and smaller amounts of E_2 are secreted by cultured cysts, although the addition of 25-hydroxycholesterol, a cell-permeant form of cholesterol, is essential to maximize steroid output, presumably by substituting for lipoprotein-mediated cholesterol uptake in vivo (40). These studies provide direct evidence that the complete array of steroidogenic enzymes for the pathway from cholesterol to estrogen is present within the primary germinal unit.

Also, yields of E_2 when T is supplied as substrate are as predicted from the distribution of aromatase activity (M $\gg$ PrM = PoM). Paradoxically, however, T- and P-production do not agree with the stage-related pattern of enzyme activities, and although manipulation of second messengers by the addition of dibutyryl cAMP, 3-isobutyl-1-methylxanthine, forskolin, and other regulators are effective in increasing or decreasing steroid output, these agents do not entirely eliminate the discrepancy between tracer analysis and radioimmunoassays. Further analysis reveals that steroids are actively converted to polar metabolites, and this process is stage related (PoM > M > PrM). Therefore, measurement of radio-immunoassayable free steroids in a static culture system underestimates true steroidogenic potential, especially in PoM stages, and in future studies it may be necessary to develop a perifusion system that more closely simulates vascular perfusion of the testis in vivo. The pattern of in vitro steroid secretion and responses to cAMP by cysts isolated from the spotted dogfish were reported in an earlier study, but in some respects differ from our data (58).

Protein Synthesis, Secretion, and Phosphorylation

Protein composition, synthesis, and secretion are frequently used as indicators of differentiation in developmental systems and, with respect to cultured spermatocysts, are readily quantifiable end points for assessing culture conditions and responses to regulators. Patterns of synthesized proteins, whether or not they are individually identified, can be used as markers for the progression of cysts through spermatogenesis in vitro and as reagents for isolation and characterization of intrinsically programmed versus regulated gene expression. Accordingly, we have begun to characterize proteins associated with or secreted by spermatocysts at specific stages (Figs. 2.5 to 2.7).

Striking differences are revealed when proteins in PrM, M, and PoM stages are compared quantitatively or qualitatively (59 and Betka and Callard, unpublished observations). Although many proteins are common to all stages, 10–20 proteins are obviously stage dependent or stage specific, even after one-dimensional SDS-PAGE. Secreted proteins are more distinctly stage related than those retained by cysts, and some bands are similar in electrophoretic mobility to identified products of rodent Sertoli cells.

As shown in Figure 2.6, patterns of [³H]methionine-labeled secreted proteins differ in PrM- versus PoM-stage cultures, but secretion is virtually undetectable in M-stage fluorograms, a consistent finding with different amino acid precursors. The latter may reflect formation of a

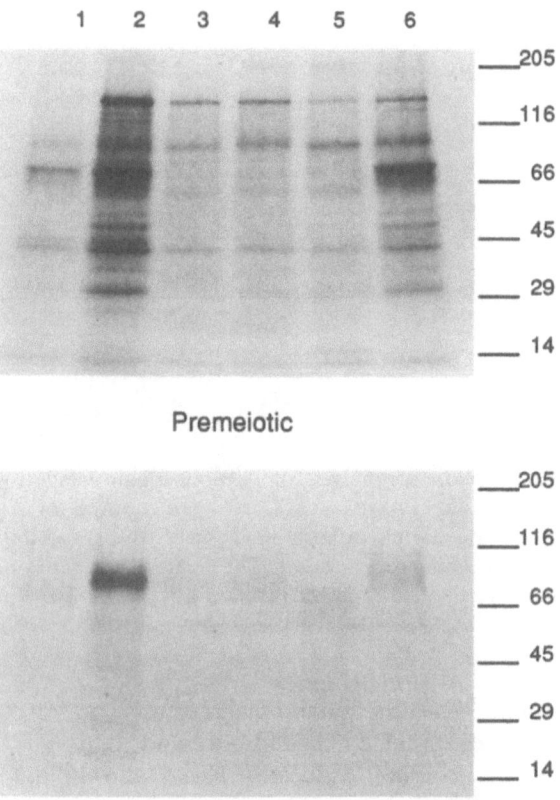

Premeiotic

Postmeiotic

FIGURE 2.6. Electrophoretic analysis of [^{35}S]methionine-labeled proteins secreted by spermatocysts in PrM vs. PoM stages of development in response to various regulators: lane 1 (retinoic acid: 1 μM); lane 2 (T: 1 μM); lane 3 (P: 1 μM); lane 4 (25-hydroxycholesterol: 60 μM); lane 5 (insulin: 10 μg/mL); and lane 6 (serum-free basal media). Cysts (1 or 1.5 mg protein/well for PrM vs. PoM stages, respectively) were cultured for 3 days in basal media with/without additives and labeled (10 μCi/well) for the final 24 h. Equal volumes of processed media were analyzed. No labeled proteins were visible on fluorograms of M-stage cysts in the same experiments (not shown). Molecular weight standards (kd) are on right.

blood-testis barrier during meiosis because when total protein synthesis (cysts plus media) is measured in the same cultures, M and PoM cysts are equivalent (Betka and Callard, unpublished observations). Figure 2.6 also shows that added T markedly increases all secreted proteins and selectively increases specific bands in both PrM and PoM stages; for example, a heterogeneous band that resembles rodent SGP-1 in size (68–70 kd). By contrast, insulin, 25-hydroxycholesterol, and P decrease

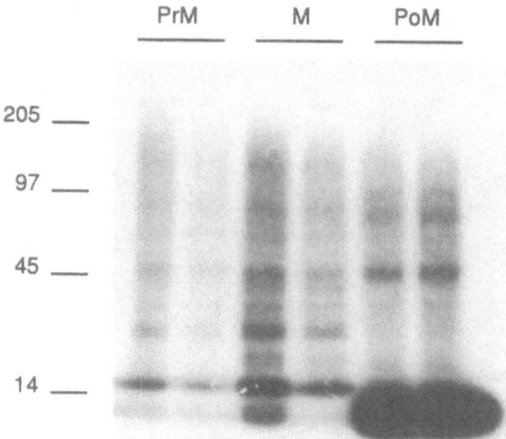

FIGURE 2.7. Electrophoretic analysis of [^{32}P]phosphate-labeled proteins in spermatocysts in PrM, M, and PoM stages. Cysts were cultured for 28 h in basal media and labeled with [^{32}P]orthophosphoric acid (40 μCi/well) for the final 4 h. Equal quantities of cyst protein were loaded. Duplicate lanes represent different cultures. Molecular weight standards (kd) are on left.

the overall protein secretory activity of PrM cysts and selectively obliterate certain bands; for example, 53- and 70-kd bands. Retinoic acid suppresses a different set of proteins, especially 32-kd and 100-kd bands in PrM stages (Fig. 2.6). Which bands are germ cell derived versus Sertoli cell derived remains to be investigated.

Figure 2.7 shows distinct stage-related phosphorylation patterns of spermatocyst proteins. Note especially a heavily phosphorylated band in PoM cysts at <14 kd. In an earlier study Gusse and Chevaillier (21) used 1–4 individual staged spermatocysts obtained by manual dissection from the shark testis and microelectrophoretic analysis to show the progressive replacement during spermiogenesis of basic nuclear proteins (somatic histones)—which are present in immature spermatids—by intermediate basic proteins at the beginning of nuclear elongation and, in turn, the appearance of spermatid-specific protamines during nuclear spiralization and final maturation of mature spermatids. These protein changes in germ cell nuclei have been correlated with changes in chromatin structure during spermiogenesis. More recently, the same authors reported the amino acid sequence of a spermatid-specific intermediate protein isolated from nuclei of spermatid-enriched testicular zones (60). This protein (SI) of dogfish, which is only the second transition protein to be sequenced, is characterized by di-, mono- and nonphosphorylated forms and is similar in size (11 kd) to the heavily phosphorylated protein seen in PoM cysts (Fig. 2.7).

Toxicology

Although the testis is known to be adversely affected by low-dose toxicant exposure, there is a need for animal studies to elucidate mechanism and to identify valid cellular and molecular markers that predict reproductive and developmental consequences. *Cadmium* (Cd) is one of the most toxic trace metals known and exerts at least some of its actions via affecting gene expression. Since the testis is exquisitely sensitive to Cd toxicity, we investigated the utility of spermatocyst cultures for investigating the effects of Cd on protein synthesis.

As shown in Figure 2.8, two [³H]cysteine-rich proteins are labeled in M and PoM cysts, but not in PrM stages. The 43-kd band may be the shark counterpart of *SPARC* (secreted protein, acidic, rich in cysteine), a Ca^{++} binding protein associated with remodeling events in rodent testis (61). The <14-kd band is a doublet and although not well defined in the gel shown, corresponds to *metallothioneins* (Mt) after isolation from shark testicular cytosols by a standard protocol (62). Significantly, Mt synthesis is restricted to M and PoM stages, but the addition of Cd increases expression in PoM stages only. A recent study reported high constitutive levels of Mt protein and mRNA in rodent testis and enrichment in

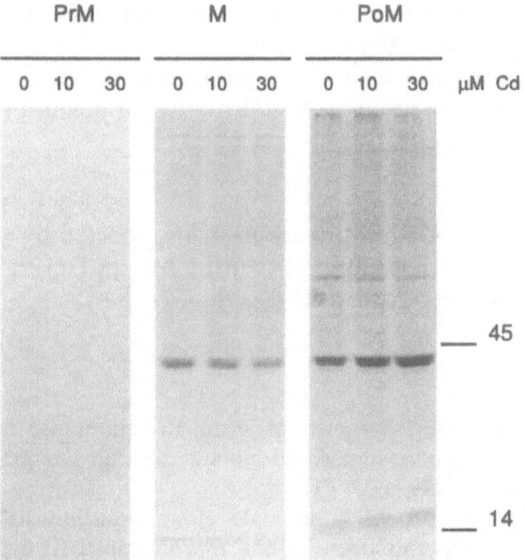

FIGURE 2.8. Electrophoretic analysis of [³⁵S]cysteine-labeled proteins extracted from spermatocysts in PrM, M, and PoM stages. Cysts were cultured for 1 day in medium containing [³⁵S]cysteine (5 µCi/well), without (0) or with cadmium (Cd: 10 or 30 µM). Equal amounts of radioactivity were loaded. Molecular weight standards (kd) are on right.

pachytene germ cells relative to Sertoli cells or other germ cell stages, but in vivo Cd treatment did not increase Mt levels (63). Whether the discrepancy with our data can be explained by species differences or by in vivo versus in vitro Cd sensitivity requires further study.

Conclusions and Future Perspectives

From the foregoing, it is evident that the shark testis model has considerable potential for studying spermatogenesis stage by stage and even more importantly, for investigating the role of Sertoli cells and the control of spermatogenesis under defined conditions in vitro. To date, it has not been determined which steps in spermatogenesis are intrinsically programmed, which are steroid dependent, what specific steroids are involved, which genes or gene networks are positively or negatively regulated, how physiological processes are differentially affected, or what the consequences are for development; nor do we know how events in germinal and somatic elements are coordinated. Obviously, there are drawbacks to using an unconventional animal model (seasonal availability, a dearth of background information, and difficulties in laboratory maintenance and experimental manipulation). However, recombinant DNA technology should facilitate the isolation and characterization of shark genes and their products, thus obviating problems of species specificity that have plagued investigators using nonmammals heretofore. In conclusion, the advantages of testicular organization in sharks recommend their continued use for obtaining new information of general relevance to male reproductive biology.

Acknowledgments. Original research was supported by a grant from NIH (HD-16715) and carried out in part at the Mount Desert Island Biological Laboratory, Salsbury Cove, ME 04672.

References

1. Callard GV. Spermatogenesis. In: Pang P, Schreibman M, eds. Vertebrate endocrinology: fundamentals and biomedical implications; 4(A). New York: Academic Press, 1991:303–41.
2. Callard GV. Reproduction in male elasmobranch fishes. In: Kinne KH, ed. Oogenesis, spermatogenesis and reproduction; 10. Basel: Karger, 1991: 104–54.
3. Roosen-Runge EC. Spermatogenesis in animals. Cambridge University Press, 1977:1–214.
4. Pilsworth LM, Setchell BP. Spermatogenic and endocrine functions of the testis of invertebrate and vertebrate animals. In: Burger H, de Kretser D, eds. The testis. New York: Raven Press, 1981:9–38.

5. Setchell BP, Pilsworth LM. The functions of the testes of vertebrate and invertebrate animals. In: Burger H, de Kretser D, eds. The testis. New York: Raven Press, 1989:1–66.
6. Dodd JM, Sumpter JP. Fishes. In: Lamming GE, ed. Marshall's physiology of reproduction; vol 1: reproductive cycles of vertebrates. New York: Churchill Livingstone, 1984:1–126.
7. Holstein AF. Zur frage der lokalen steuerung der spermatogenese beim dornhai (*Squalus acanthias L.*). Z Zellforsch 1969;93:265–81.
8. Pudney J, Callard GV. Identification of Leydig-like cells in the interstitium of the shark testis (*Squalus acanthias*). Tissue Cell 1984;18:375–82.
9. Pudney J. Comparative cytology of the non-mammalian Sertoli cell. In: Russell LD, Griswold MD, eds. The sertoli cell. Clearwater, FL: Cache River Press, 1993:611–58.
10. Pudney J, Callard GV. Development of the agranular endoplasmic reticulum in the Sertoli cell of the shark *Squalus acanthias* during spermatogenesis. Anat Rec 1984;209:311–21.
11. Dym M, Fawcett DW. Further observations on the numbers of spermatogonia, spermatocytes and spermatids connected by intercellular bridges in the mammalian testis. Biol Reprod 1971;4:195–215.
12. Russell LD, Peterson RN. Determination of the elongate spermatid-Sertoli cell ratio in various mammals. J Reprod Fertil 1984;70:635–41.
13. Fawcett DW. Ultrastructure and function of the Sertoli cell. In: Hamilton D, Greep R, eds. Handbook of physiology, endocrinology, vol 5: male reproductive system. Washington, DC: American Physiological Society, 1975: 21–56.
14. Callard GV, Pudney JA, Mak P, Canick JA. Stage-dependent changes in steroidogenic enzymes and estrogen receptors during spermatogenesis in the testis of the dogfish *Squalus acanthias*. Endocrinology 1985;177:1328–35.
15. Cuevas ME, Miller W, Callard GV. Sulfoconjugation of steroids and the vascular pathway of communication in dogfish testis. J Exp Zool 1992;264: 119–29.
16. Cuevas ME, Collins K, Callard GV. Stage-related changes in steroid converting enzyme activities in *Squalus* testis: synthesis of biologically active metabolites via 3β-hydroxysteroid dehydrogenase/isomerase(3βHSD/isomerase) and 5α-reductase. Steroids 1993;58:87–94.
17. Cuevas ME, Callard GV. Androgen and progesterone receptors in shark (*Squalus*) testis: characteristics and stage-related distrubution. Endocrinology 1992;130(4):2173–82.
18. Ruh MF, Singh RH, Mak P, Callard GV. Tissue and species specificity of nuclear acceptor sites for the estrogen receptor of *Squalus* testis. Endocrinology 1986;118:811–8.
19. Mak P, Callard GV. A novel steroid binding protein in the testis of the dogfish *Squalus acanthias*. Gen Comp Endocrinol 1987;68:104–12.
20. Mellinger J. Stades de la spermatogenese chez *Scyliorhinus caniculus (L.)*: description, donnees histochimiques, variations normales et experimentales. Z Zellforsch 1965;67:653–73.
21. Gusse M, Chevaillier P. Ultrastructural and chemical study of chromatin during spermiogenesis of fish *Schyliorhinus caniculus*. Cytobiologie 1978;16: 421–43.

22. Collenot G. Apparition et evolution de l'activite endocrine du testicule de *Scyliorhinus canicula L.* (elasmobranche). Ann Embryol Morphol 1970;2: 461–77.

23. Collenot G, Damas D. Etude ultrastructurale de la cellule de Sertoli au cours de la spermiogenese chez *Scyliorhinus canicula L.* Cah Biol Mar 1980;21: 209–19.

24. Stanley HP. The structure and development of the seminiferous follicle in *Scyliorhinus caniculus* and *Torpedo marmorata* (elasmobranchii). Z Zellforsch 1966;75:453–68.

25. Stephan MP. L'evolution de la cellule de Sertoli des selaciens apres la spermatogenese. C R Soc Biol (Paris) 1902;54:775–6.

26. Collenot G, Ozon R. Mise en evidence biochimique et histochimique d'une 5, 3β-hydroxysteroide dehydrogenase dans le testicule de *Scyliorhinus canicula L.* Bull Soc Zool Fr 1964;26:40–2.

27. Simpson TH, Wardle CS. A seasonal cycle in the testis of the spurdog, *Squalus acanthias*, and the sites of 3β-hydroxysteroid dehydrogenase activity. J Mar Biol Assoc UK 1967;47:699–808.

28. Pudney J, Callard GV. Sertoli cell cytoplasts in the semen of the spiny dogfish (*Squalus acanthias*). Tissue Cell 1986;18:375–82.

29. Simpson TH, Wright RS, Renfrew J. Steroid biosynthesis in the semen of dogfish (*Squalus acanthias*). J Endocrinol 1964;31:11–20.

30. Simpson TH, Wright RS, Gottfried H. Steroids in the semen of dogfish (*Squalus acanthias*). J Endocrinol 1963;26:489–98.

31. Collenot G, Damas D. Mise in evidence de la nature proteique de corps enigmatiques presents dans le testicule de *Scyliorhinus canicula L.* (elasmobranche). Cah Biol Mar 1975;16:39–46.

32. Moyne G, Collenot G. Unusual nucleolar fine structure in the Sertoli cells of the dogfish *Scyliorhinus canicula (L.).* Biol Cell 1982;44:239–48.

33. Simpson TH, Wright RS, Hunt SV. Steroid biosynthesis in the testis of the dogfish (*Squalus acanthias*). J Endocrinol 1964;31:29–38.

34. Barry TP, Thomas P, Callard GV. Stage-related production of 21-hydroxylated progestins by the dogfish (*Squalus acanthias*) testis. J Exp Zool 1993.

35. DuBois W, Mak P, Callard GV. Sertoli cell functions during spermatogenesis: the shark testis model. Fish Physiol Biochem 1989;7:221–7.

36. Parvinen M. Regulation of the seminiferous epithelium. Endocr Rev 1982;3: 404–17.

37. Callard GV, Mak P. Exclusive nuclear localization of estrogen receptors in *Squalus* testis. Proc Natl Acad Sci USA 1985;32:1336–40.

38. Dobson S, Dodd M. Endocrine control of the testis in the dogfish *Schyliorhinus canicula L.*, I. Effects of partial hypophysectomy on gravimetric, hormonal and biochemical aspects of testis function. Gen Comp Endocrinol 1977;32:41–52.

39. Dobson S, Dodd M. Endocrine control of the testis in the dogfish *Scyliorhinus canicula L.*, II. Histological and ultrastructural changes in the testis after partial hypophysectomy (ventral lobectomy). Gen Comp Endocrinol 1977;32:53–71.

40. Cuevas ME, Callard GV. In vitro steroid secretion by staged spermatocysts (Sertoli/germ cell units) of dogish (*Squalus acanthias*) testis. Gen Comp Endocrinol 1992;88:151–65.

41. Isomaa V, Parvinen M, Janne OA, Bardin CW. Nuclear androgen receptors in different stages of the seminiferous epithelial cycle and the interstitial tissue of rat testis. Endocrinology 1985;116:132–7.
42. Vernon RG, Go VLW, Fritz IB. Hormonal requirement of the different cycles of seminiferous epithelium during reinitiation of spermatogenesis in long-term hypophysectomized rats. J Reprod Fertil 1975;42:77–90.
43. Lyon MF, Glenister PH, Lamoreux ML. Normal spermatozoa from androgen resistant germ cells of chimaeric mice and the role of androgen in spermatogenesis. Nature 1975;258:620–2.
44. Nakhla AM, Mather JP, Janne OA, Bardin CW. Estrogen and androgen receptors in Sertoli, Leydig, myoid and epithelial cells: effect of time in culture and cell density. Endocrinology 1984;115:121–8.
45. Levy FO, Ree AH, Eikvar L, Govindan MV, Jahn T, Hansson V. Glucocorticoid receptors and glucocorticoid effects in rat Sertoli cells. Endocrinology 1988;124:430–6.
46. Ritzen EM, Hansson V, French FS. The Sertoli cell. In: Burger H, de Kretser D, eds. The testis. New York: Raven Press, 1989:269–302.
47. Mak P, Callard GV. A novel steroid binding protein in the testis of the dogfish *Squalus acanthias*. Gen Comp Endocrinol 1987;68:104–12.
48. Callard GV, Mak P, DuBois W, Cuevas M. Regulation of spermatogenesis: the shark testis model. J Exp Zool 1989;(suppl 2):353–64.
49. Setchell BP. The mammalian testis; Ithaca, NY: Cornell University Press, 1978;61–5.
50. DuBois W, Callard GV. Culture of intact Sertoli/germ cell units and isolated Sertoli cells from *Squalus* testis, I. Evidence of stage-related function in vitro. J Exp Zool 1991;258:359–72.
51. Sourdaine P, Jegou B. Dissociation and identification of intact seminiferous lobules from the testis of the dogfish (*Scyliorhinus canicula*). Cell Tissue Res 1989;255:199–207.
52. Hadley MA, Byers SW, Suarez-Quian CA, Kleinman HK, Dym M. Extracellular matrix regulates Sertoli cell differentiation, testicular cord formation and germ cell development in vitro. J Cell Biol 1985;101–2:1511–22.
53. DuBois W, Callard GV. Culture of intact Sertoli/germ cell units and isolated Sertoli cells from *Squalus* testis, II. Stimulatory effects of IGF-I and other factors on DNA synthesis. J Exp Zool 1993.
54. Clermont Y, Mauger A. Existence of a spermatogonial chalone in the rat testis. Cell Tissue Kinet 1974;7:165–72.
55. Piferrer F, Redding M, DuBois W, Callard GV. Stage-specific stimulatory and inhibitory regulation of the spermatogenic progression: studies in *Squalus acanthias*. Fish Physiol Biochem 1993.
56. Piferrer F, Callard GV. Regulation of DNA synthesis during premeiotic stages of spermatogenesis by a growth inhibitory factor from the immune system. Proc Endocr Soc, June 9–12, 1993, Las Vegas, NV.
57. Allan DJ, Harmon BV, Kerr JFR. Cell death in spermatogenesis. In: Potten CS, ed. Perspectives on mammalian cell death. UK: Oxford University Press, 1987:229–58.
58. Sourdaine P, Jegou B. Dissociation and identification of intact seminiferous lobules from the testis of the dogfish (*Scyliorhinus canicula L.*). Cell Tissue Res 1988;255:199–207.

59. Betka M, Callard GV. Stage-related and regulated changes in protein synthesis during spermatogenesis in vitro: the shark (*Squalus acanthias*) testis model. Bull MDIBL 1993;32:113–4.
60. Chauviere M, Martinage A, Briand G, Sautiere P, Chevaillier P. Nuclear basic protein transition during sperm differentiation: amino acid sequence of a spermatid-specific protein from the dogfish *Scyliorhinus caniculus*. Eur J Biochem 1987;169:105–11.
61. Vernon RB, Sage H. The calcium-binding protein SPARC is secreted by Leydig and Sertoli cells of the adult mouse testis. Biol Reprod 1989;40:1329–40.
62. Otsuka F, Koizumi S, Kimura M, Ohsawa M. Silver staining for carboxymethylated metallothioneins in polyacrylamide gels. Anal Biochem 1988;168:184–92.
63. De SK, Enders GC, Andrews GK. High levels of metallothionein messenger RNAs in male germ cells of the adult mouse. Mol Endocrinol 1991;5:628–36.
64. Callard GV. The autocrine and paracrine role of steroids during spermatogenesis: studies in *Squalus acanthias* and *Necturus maculosus*. J Exp Zool 1992;261:132–42.

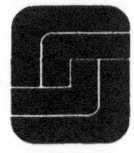

3

Structure-Function Relationships in Somatic Cells of the Testis and Accessory Reproductive Glands

LONNIE D. RUSSELL, AMIYA P. SINHA HIKIM, SUSHMITA GHOSH, AND ANDRZEJ BARTKE

Somatic cells of the male are key to the normal functioning of the male reproductive system. Their physiology has been extensively measured using endocrine and endocrine receptor assays, but until recently, little has been learned about the structural manifestations of cell function and the correlation between structure and function. The somatic cells of the testis serve as appropriate models to study the behavior of active and inactive cells of all kinds, given that virtually all cells in a complex organism have active and inactive states and function in response to stimuli. Some cells develop from inactive precursors to show cyclic functionality in sexually mature animals. Others develop from relatively inactive cells to active cells that once active, function continuously in mature animals.

In this chapter we summarize published and submitted work (1–14) that details morphometric analysis primarily of Leydig cells and Sertoli cells and to some degree, testis macrophages, blood vessels, and myoid and urethral gland cells in conjunction with endocrine determinations to answer several basic questions about the structure of these cells and the relationship of structure and function. Furthermore, questions about the similarities between hypophysectomized models and seasonal breeding models, as well as species differences, are addressed. Since there are literally thousands of measurements/determinations that have been made in the aforecited studies, this chapter focuses on a summary of the findings in relation to broad questions that are posed. There are many findings of a specific nature that are not mentioned in this chapter due to space limitations. The reader is encouraged to seek out specific publications when details are sought.

What Models Are Utilized to Study Active and Inactive Cells?

Three models were used to understand somatic cell function in the male.

Photoperiod Model

The first model employs a seasonal breeder, the golden hamster. This species undergoes periods where the testis is stimulated by increased secretion of gonadotropins and prolactin (recrudescence occurring spontaneously or in response to long-day light stimulation) followed by periods in which gonadotropin levels and prolactin are low and the testis is regressed (in a short-day environment). These changes, normally brought about in a natural environment, can be produced in the laboratory by altering the lighting conditions for a 12- to 13-week period to simulate a short-day or a long-day environment (1, 15, 16). Thus, the photoperiod-induced gonadal regression and recrudescence represent physiological states in which the testis and also reproductive behavior undergo a phase of inactivity followed by a phase of activity (Fig. 3.1).

In the photoperiod model, sacrifice intervals were also selected after placement of animals in long days and in short days that represented short-term responses with the purpose of studying the transitional periods

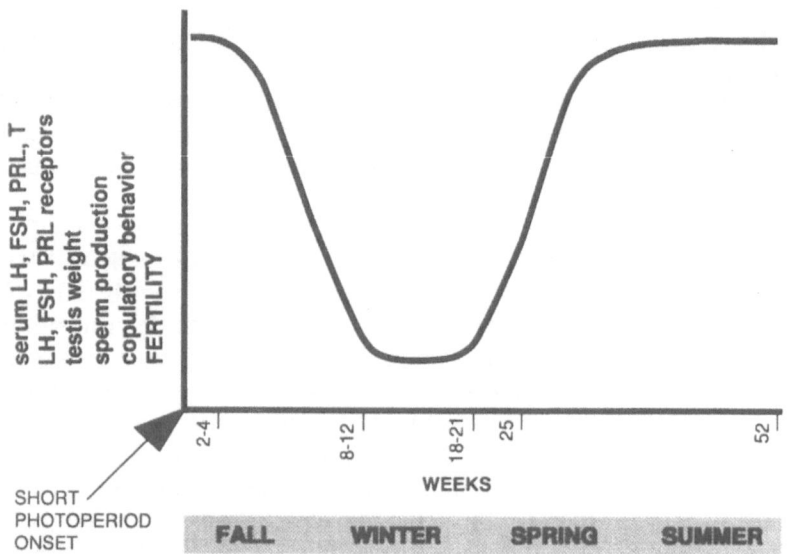

FIGURE 3.1. Reproductive parameters plotted against season in a seasonally breeding species such as the golden hamster. Reprinted with permission from Sinha Hikim, Bartke, and Russell (13).

in cells that are on their way to becoming active or becoming inactive. In selecting animals for a determination of the short-term effects of changing photoperiod, the focus was on obtaining a short-term morphological pattern of response (17). The short-term pattern of responses was characterized either, in the case of regression, by the initial appearance of a cohort of degenerating germ cells among a much larger population of viable germ cells, or in the case of recrudescence, by the development of a slightly larger population of viable germ cells than was present in the regressed testis. In other words, the initial signs of inactivity and activity, respectively, were perceptible in animals selected for these studies.

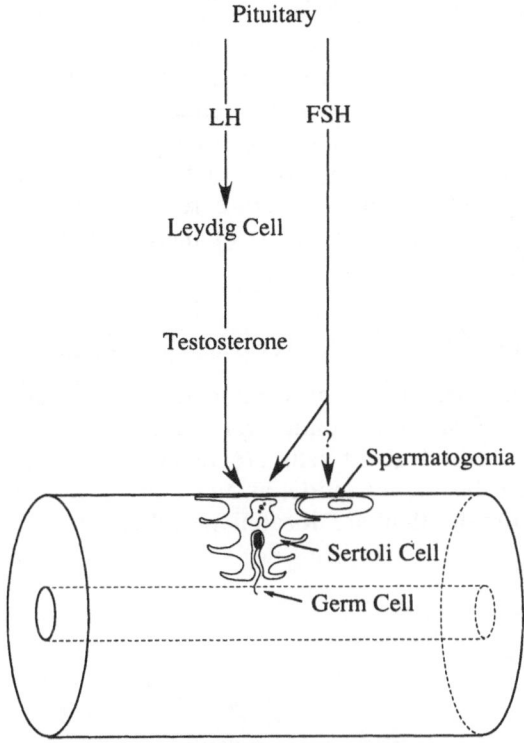

FIGURE 3.2. A very basic representation of the hormonal control of testis function by gonadotropins. Luteinizing hormone (LH) is released by the pituitary and binds to receptors on Leydig cells that in turn stimulate a cascade of events leading to testosterone (T) release. The T stimulates Sertoli cells by binding to receptors on their nuclei. Follicle stimulating hormone (FSH) binds to spermatogonia (likely) and has receptors on the Sertoli cell plasma membrane via a second-messenger system. Thus, both FSH and T work to support spermatogenesis by activities mediated by the Sertoli cell. Not shown in this diagram is the well-documented role of prolactin or lack of prolactin in mediating the effects of photoperiod-induced recrudescence or regression, respectively, in the hamster. Reprinted with permission from Russell, Ettlin, Sinha Hikim, and Clegg (17).

Hypophysectomy Model

Second, the classical hypophysectomy model was utilized to virtually eliminate gonadotropin secretion, specifically LH, whose receptors are on the Leydig cell, and also FSH, whose receptors are on the Sertoli cell (and most likely spermatogonia) (Fig. 3.2). Prolactin, which is considered to be important in the seasonal breeding model (15, 16) is also eliminated by hypophysectomy. To induce complete testicular regression, rats were hypophysectomized for 28 days and hamsters for 20 days. These time periods provided the maximal regressive features judged by morphology.

Animals that were hypophysectomized were also studied in short-term experiments to study the early degenerative responses of the testes, as defined above. In all studies of either a long-term or a short-term nature, the functional parameter was considered to be the status of the germ cell population or the level of the major endocrine product of the Leydig cell, testosterone. Quantitation of germ cells was undertaken in conjunction with morphometric analysis to show by objective means that the germ cell population had either regressed (hypophysectomy model and photoperiod-induced regression) or increased (photoperiod-induced recrudescence).

Testosterone-Dependent, Accessory Gland Model

In a third model, urethral gland cells in testis-intact and castrated mice were examined. Castrated animals were given either testosterone or oil vehicle. Since urethral gland cells are testosterone dependent (7) (Fig. 3.3), we utilized this mouse model as an example of hormone-dependent accessory sex organs that are located peripheral to the testis in which

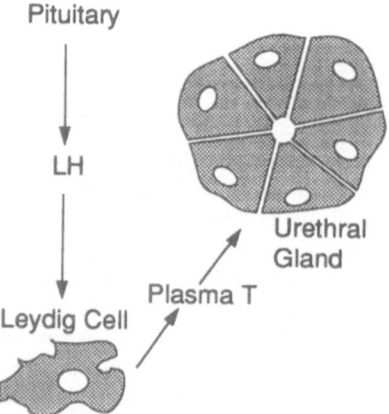

FIGURE 3.3. Dependence of urethral gland cells in the mouse on testosterone.

active (stimulated) and inactive (unstimulated) states of cellular activity can be studied.

Morphometric Techniques

Morphometry is the technique for determining the three-dimensional characteristics (volume) from objects that appear two-dimensional in microscope sections and/or is the technique for determining the two-dimensional characteristics (surface areas) from objects that show one dimension in microscope sections. Stereological principles are employed in making these determinations (17–19). Morphometry was conducted at both the light and electron microscope levels using accepted techniques to obtain volume and surface area parameters relating to somatic and germ cells of the testis. Although the techniques utilized are laborious, considerable information can be obtained from such methodology. Volume and surface area were expressed as volume densities or surface densities and as absolute volumes or surface areas. A *volume density* or *surface density* is an expression of volume or surface, respectively, per unit area. In this chapter we do not report both absolute and relative data, but use the data form best suited to illustrate the answer to the question being asked.

All Sertoli cell morphometry in the current study was performed in tubules at stage VII of spermatogenesis. Data are expressed on a per-cell basis. Note that proportions in the pie charts presented below may differ from published data due to the expression of cell constituents on a per-cell basis and not per cell cytoplasm, as was previously published.

What Are the Morphological Characteristics of Active and Inactive Male Somatic Cells?

A comparison of active and inactive hamster cells was undertaken to determine the extremes in cellular activity.

Leydig Cells: Photoperiod Model

In the seasonal breeding hamster model, the number of Leydig cells declined significantly (by 29%) after photoperiod-related gonadal regression. Active hamster Leydig cells were approximately 3 times larger than inactive Leydig cells. The change in cell size after photoperiod-related regression was due primarily to cytoplasmic reduction (74%) and to a lesser degree, nuclear size decrease (50%).

The total volumetric proportion (density) of the constituents in active and inactive Leydig cells remained approximately similar for the

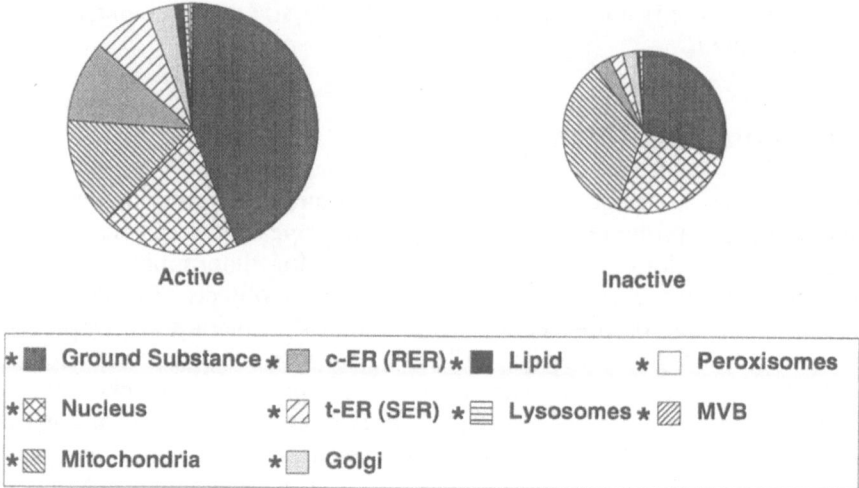

FIGURE 3.4. Pie charts comparing active and inactive Leydig cells in the hamster photoperiod model. In this and subsequent pie charts, the size of the pie is proportional to the size of the cell. Pie-shaped wedges are shown for the various components of the cell and occupy a space proportional to the space occupied by the constituent within the cell. Significantly different parameters in active vs. inactive cells are indicated by an asterisk in the legend.

two groups, although the regressed cell was slightly more packed with organelles (Fig. 3.4). As the volume of the cell decreased, a significant volumetric loss in all Leydig cell subcellular components was recorded except for heterochromatin. The volume loss for subcellular components, except heterochromatin, ranged from about 50% to 90% (Fig. 3.5). If organelles were analyzed individually, there was a lesser percentage loss of mitochondria as compared with other organelles.

Declines in the volumes and surface areas of most organelles and cell components ranged from approximately 50% to 90% (Fig. 3.6). All cell components showed significant differences in surface area. In terms of surface area, the nucleus showed less percentage loss of surface than most other cellular constituents.

Summary

Leydig cells responded to a decrease in their state of physiologic activity by a slight decrease in numbers and a modest decrease in nuclear size, but large decreases in virtually all other cell components. Although cell components decreased in volume, they did so in proportion to the reduction in cell volume.

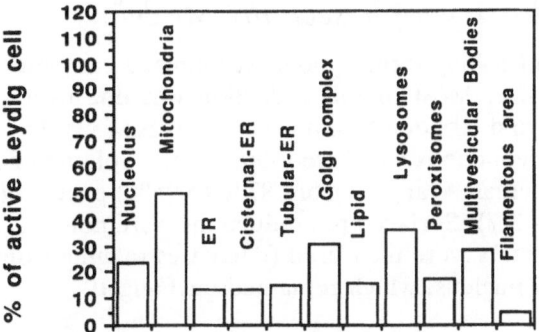

FIGURE 3.5. Volumes of Leydig cells and their constituents in the photoperiod model shown as a percentage of the active cell. All values are significantly different. Reprinted with permission from Sinha Hikim, Amador, Klemcke, Bartke, and Russell (12), © The Endocrine Society, 1989.

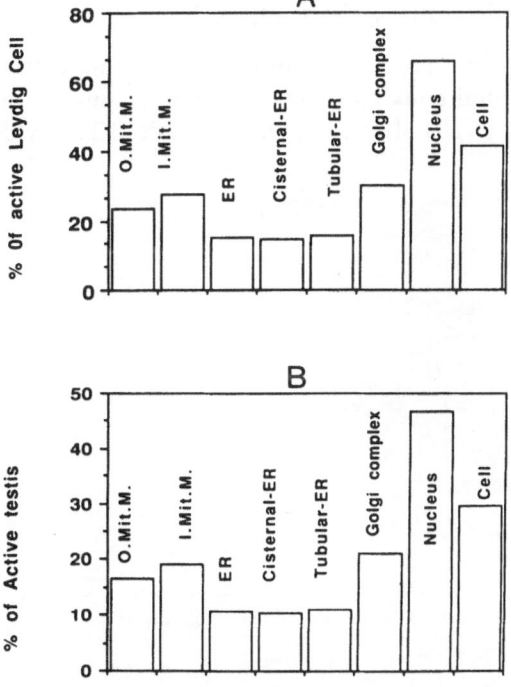

FIGURE 3.6. Surface areas of Leydig cells and their constituents in the photoperiod model shown as a percentage of the active Leydig cell (A) and active testis (B). All values are significantly different. Reprinted with permission from Sinha Hikim, Amador, Klemcke, Bartke, and Russell (12), © The Endocrine Society, 1989.

Leydig Cells: Hypophysectomy Model

Leydig cells in a long-term hypophysectomized rat were only 12.7% of their former size. Most of this reduction was due to a cytoplasmic decrease (83%) and less due to a nuclear decrease (52%).

Significant reductions were recorded for the volume of all cytoplasmic organelles (decreases ranging from 87% to 95%) except heterochromatin and lipid (Fig. 3.7). Surface area reductions were noted in the cell (76%), and all parameters were measured (decreases ranging from 86% to 96%) except for the nucleus, which remained unchanged.

Summary

Leydig cells from hypophysectomized rats thus showed dramatic declines in virtually all volume and surface area parameters. Cytoplasmic reductions are greater than nuclear reductions, although organelles remain in approximately the same proportions within the cytoplasm of active and inactive cells.

Sertoli Cells: Photoperiod Model

There was no difference in numbers of Sertoli cells after complete photoperiod-related gonadal regression. The volume of individual Sertoli cells decreased by 72% after complete regression. The cytoplasmic volume

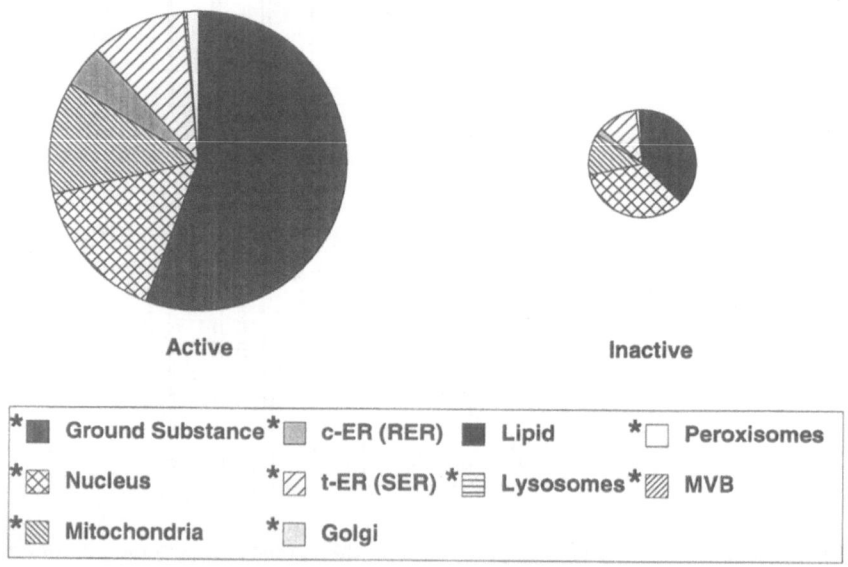

FIGURE 3.7. Pie charts comparing active and inactive Leydig cells in the hypophysectomy model.

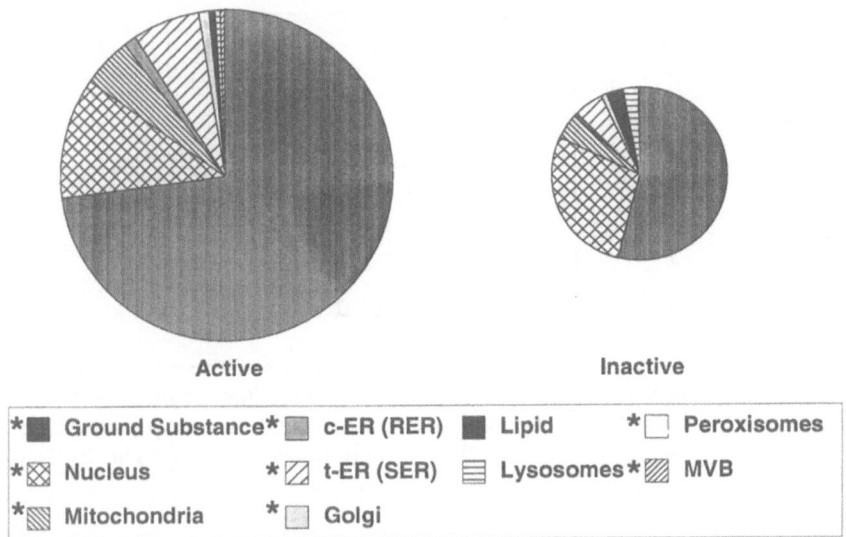

FIGURE 3.8. Pie charts comparing active and inactive Sertoli cells in the hamster photoperiod model. The size of the pie is proportional to the size of the cell. Pie-shaped wedges are depicted for the various components of the cell. Significantly different parameters are indicated by the asterisk in the legend.

decrease (88%) was the factor primarily responsible for the decrease in cell size, although the nuclear volume decreased somewhat (34%). Virtually all cytoplasmic constituents showed significant volume (Figs. 3.8 and 3.9) and surface area (Fig. 3.10) reductions (ranging from 52% to 91%) except for the volumes of lipid and lysosomes, the two constituents that remained unchanged.

Sertoli Cells: Hypophysectomy Model, Rat and Hamster

The response of hamster and rat Sertoli cells to hypophysectomy was to decrease their cell volume by 82% and 56%, respectively, and their surface area by 89% and 62%, respectively. Nuclear decreases were less responsible for overall cell size decreases than were cytoplasmic decreases. Cell components were decreased in the hamster (decreases ranging from 83% to 91%) except for lysosomes, which increased 1.3-fold (Fig. 3.11A). In the rat the volume of all cell components was decreased (decreases ranging from 60% to 69%) except lipid, which was increased 5-fold, and lysosomes, which were unchanged (Fig. 3.11B). The surface areas of all subcellular organelles measured decreased significantly (decreases ranging from 33% to 76%).

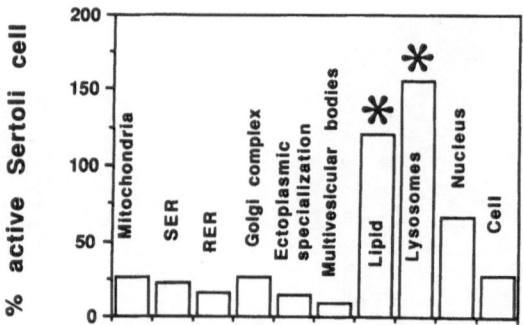

FIGURE 3.9. Bar graph showing Sertoli cell and cell constituent volumes in the photoperiod model. All values are expressed as a percentage of those in the active cell. Asterisks indicate significant differences. Reprinted with permission from Sinha Hikim, Amador, Klemcke, Bartke, and Russell (11), © The Endocrine Society, 1989.

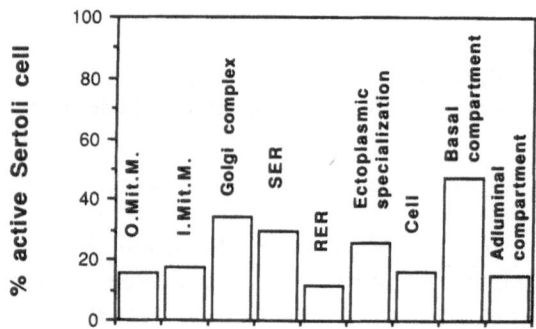

FIGURE 3.10. Bar graph showing Sertoli cell surface areas in the photoperiod model. All values are expressed as a percentage of those in the active cell. Asterisks indicate significant differences. Reprinted with permission from Amador, Klemcke, Sinha Hikim, Bartke, and Russell (11), © The Endocrine Society, 1989.

Summary

Hypophysectomy in the hamster and rat produces dramatic declines in the volume and surface areas of the cell and its components. Lipid and lysosomes are the exceptions in that they remain the same or are increased. Although most subcellular components of a cell decrease in proportion to the volume of the cell, the nucleus does not, and it is the primary factor responsible for the increased crowding noted in the re-gressed cell.

A

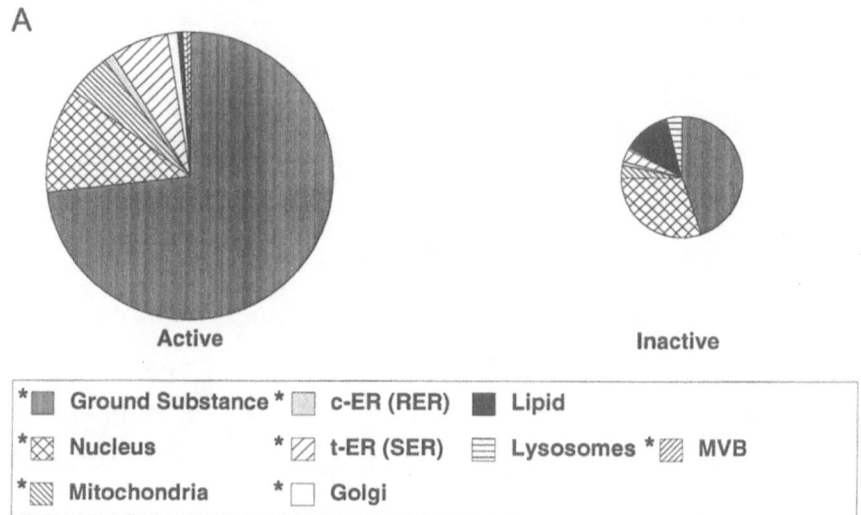

B

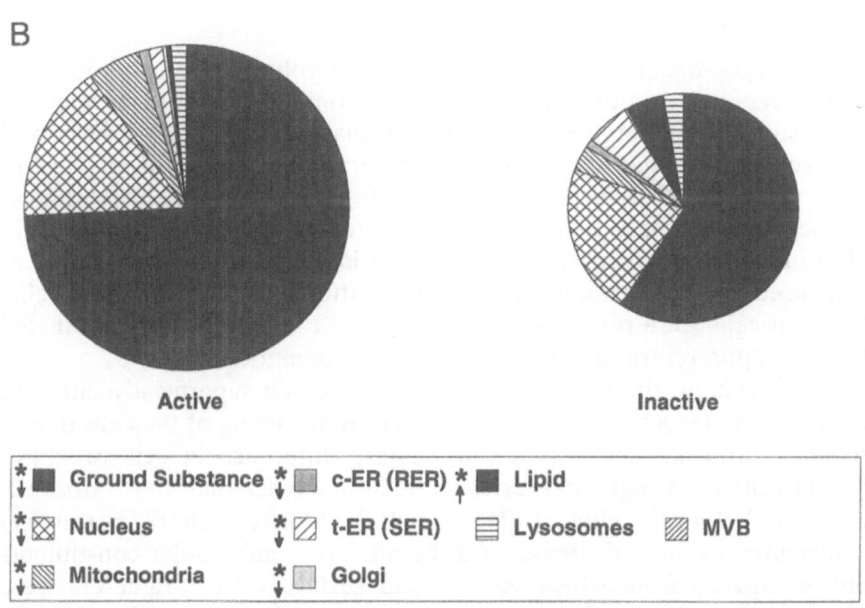

FIGURE 3.11. Pie chart showing the response of Sertoli cells to hypophysectomy in the hamster (20 days) (*A*) and rat (*B*).

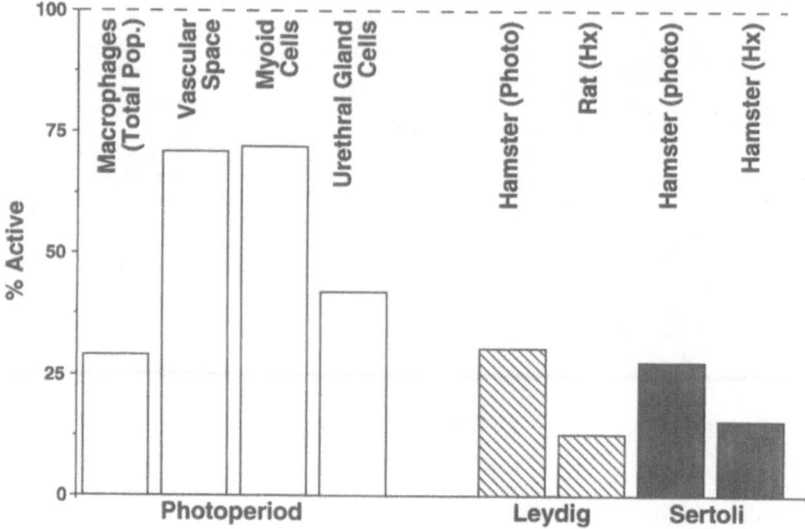

FIGURE 3.12. Cell size in a variety of cell types expressed as a percentage of the size of an active cell. For comparison, Leydig and Sertoli cell responses are shown to the right. The vascular space was measured and not vasculature cell volume. In terms of macrophages, only the total volume of macrophages was measured, leaving in doubt whether or not the number or size of macrophages was altered. (Hx = hypophysectomy model; photo = photoperiod model.)

Other Somatic Cells

Macrophages underwent a 69% loss in total volume after photoperiod-related regression, although it was not determined whether this was due to a change in cell numbers or cytoplasmic mass (Fig. 3.12). It is suspected that cell number had changed since there are no reports of changes in macrophage cell size during transitions between active and inactive states.

The vascular space in photoperiod-related regression decreased 29% after induction of regression as a result of loss in vascular cells. This was evidenced by the remaining basal lamina after regression of these cells. Vascular cells were restored, as evidenced by the presence of endothelial budding upon returning animals to a long photoperiod (6).

Myoid cells in the photoperiod model appeared superficially different in active and inactive testes. This resulted from folding of the cells due to tubular shrinkage and was not a primary difference in cell structure. Myoid cells from regressed testes were 28% smaller than those from active testes. Both the nucleus (24% smaller) and cytoplasm (37% smaller) contributed to these decreases. Of the numerous subcellular constituents and organelles from which volume and surface area parameters were obtained, only the numbers of pinocytotic vesicles of myoid cells were reduced in gonadally inactive animals.

Cell and nuclear size of cells in the urethral glands.[a]

Mice	Pelvic urethral glands		Sinus urethral glands	
	Cell	Nucleus	Cell	Nucleus
Sham castrates	1899.8 ± 326.5	184.0 ± 23.7	2793.2 ± 548.4	189.5 ± 18.0
Oil-treated castrates	469.7* ± 40.5	106.7* ± 14.0	438.0* ± 83.3	100.5* ± 9.5
Testosterone-treated castrates	1805.0 ± 238.9	202.4 ± 20.0	2601.8 ± 339.5	193.0 ± 6.0

[a]μm^3 + SEM.
*Significantly different at the 95% significance level.

FIGURE 3.13. Urethral gland size in intact, castrated, and testosterone-maintained castrated mice. Reprinted with permission from Parr, Ren, Russell, Prins, and Parr (7).

The urethral gland cells of castrated mice showed a 58% reduction in size compared with the cells from intact animals. A nuclear decrease in size accounted for only 25% of the overall decrease. Exogenous testosterone was capable of fully maintaining the size of urethral gland cells in castrated animals (Fig. 3.13).

Summary

Most somatic cells of the testis and those outside of the reproductive tract respond dramatically to lack of hormonal stimulation by demonstrating large reductions in size and in the volumes and surface areas of their constituents. Hamster myoid cells and the vascular cells are somewhat atypical of the other cells described in that although they likely possess androgen receptors, they respond minimally. Their role as typical hormone-responsive cells, demonstrating responses like either the Leydig or Sertoli cells, must be questioned. Testicular macrophages may possess FSH receptors (20) and thus may be responsible to this endocrine signal. The results of this study indicate that there are many elements in the testis responding to one degree or another to loss of homonal stimulation. Future studies should determine if all cell types in the testis are regulated by one hormone (e.g., LH-stimulated testosterone) through a cascade of effects or if the factors of many hormones come into play to independently produce a smaller testis with changes throughout.

How Do Leydig and Sertoli Cells Compare in Their Response When the Same Model (Photoperiod) Is Used to Induce Regression?

Leydig cells and Sertoli cells were compared in the photoperiod model (Fig. 3.14). Both cell types show dramatic reductions in their constituents.

In general, the Sertoli cell responded to photoperiod-induced testicular regression in a manner similar to Leydig cells. While most organelle

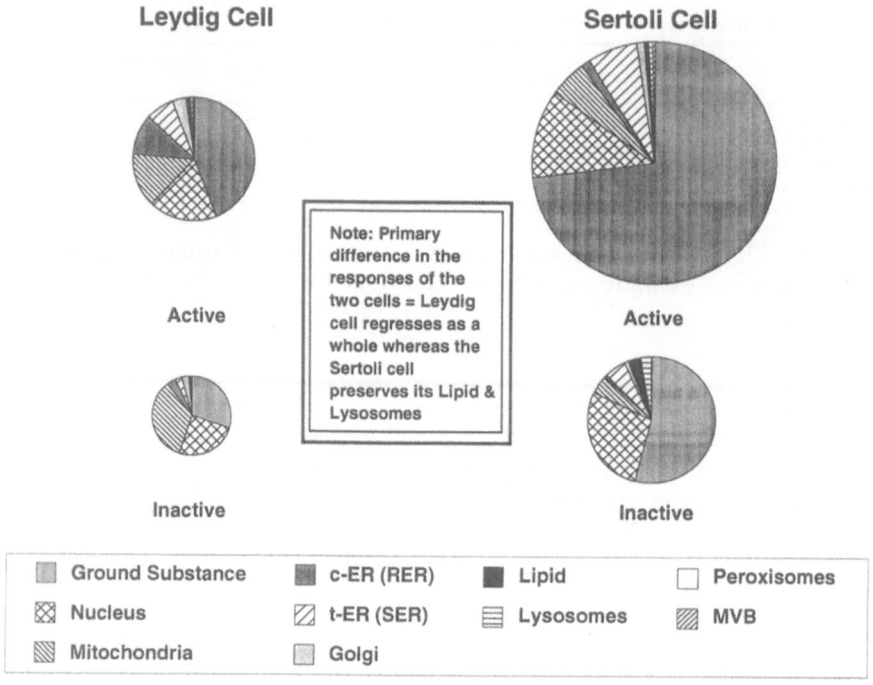

FIGURE 3.14. Comparison of Leydig cells and Sertoli cells in the hamster photo-period model.

volume and surface area parameters decreased in proportion to the cyto-plasmic decrease, the nucleus did not. Consequently, the regressed cell appeared more crowded. There is, however, one major difference in the response of the two cell types. While all components of the Leydig cell are markedly regressed, the Sertoli cell maintains its lysosomal and lipid constituents. Apparently, it preserved its subcellular lysosomal apparatus to continue to phagocytose degenerating germ cells that were produced. It is generally thought that during phagocytosis, the lipidic material from these cells is stored in lipid droplets within the Sertoli cell. Thus, the Sertoli cell maintained one essential functional feature even in what might be considered an inactive state.

Is a Hypophysectomy Model Similar to the Seasonal Breeding Model When the Two Conditions Are Compared in the Same Species?

In the hamster, the photoperiod-related gonadal regression model was compared to the hypophysectomy model. Maximally regressed states in

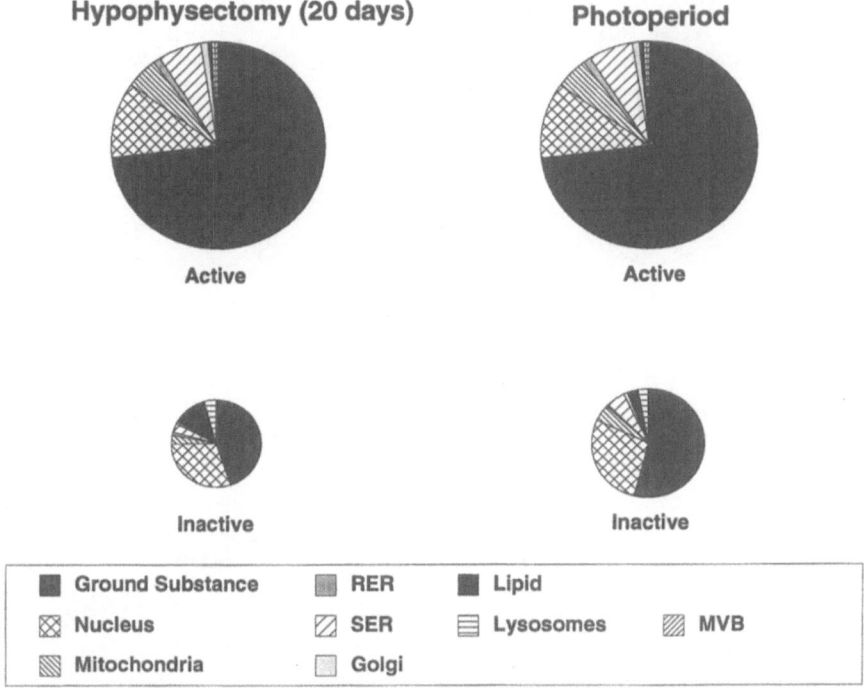

FIGURE 3.15. A comparison of Sertoli cell volumetric parameters in the hypophysectomy model and the seasonal breeding model, both models utilizing the hamster.

both models were compared to the active state. Only changes after regression were compared since there is currently no ideal model to completely reverse the effects of hypophysectomy. Figure 3.15 portrays the Sertoli cell in both conditions in terms of cell size and the volumes of its constituents. Surface area parameters are not shown. Hypophysectomy resulted in a qualitatively similar, but slightly more severe, pattern of regressive changes in Sertoli cells than after photoperiod-induced gonadal regression.

Given that the same type of response is obtained from a qualitative standpoint, the results suggest that both hypophysectomy- and photoperiod-induced regression are caused by similar mechanisms in the hamster. Both have in common depressed gonadotropin secretion that appears to be the likely cause. However, the two models differ slightly in the degree of regression, probably the result of more severe gonadotropin depression found after hypophysectomy.

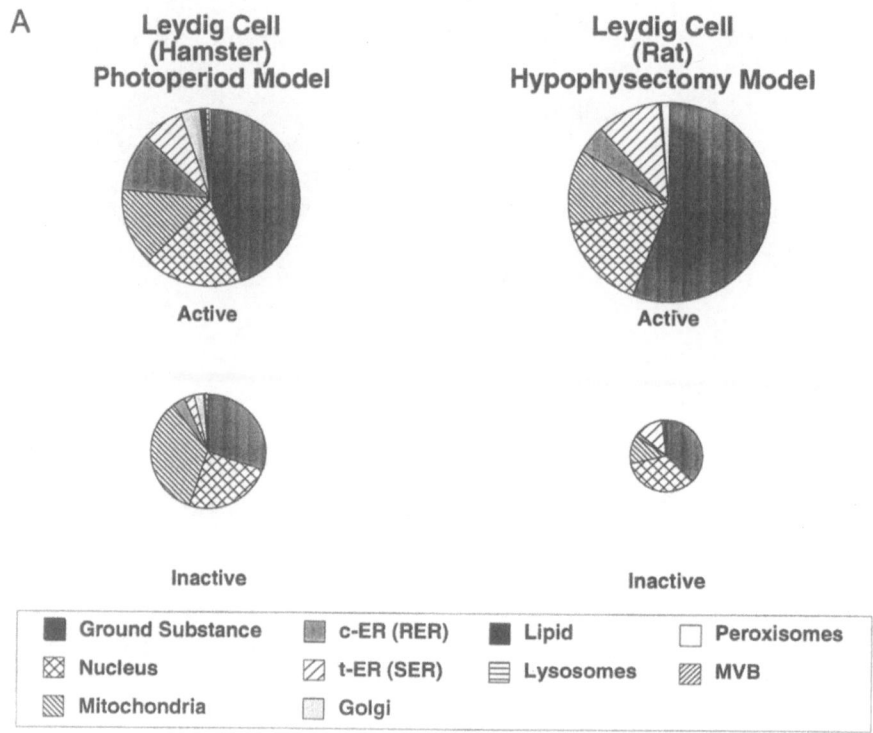

FIGURE 3.16. A cross-species comparison of seasonal breeding in the hamster with hypophysectomy in the rat: Leydig cell (*A*) and Sertoli cell (*B*). Significantly different values are indicated by an asterisk.

Are the Regressive Changes Seen after Photoperiod-Related Seasonal Regression in One Species (Hamster) Similar to Hypophysectomy in Another Species (Rat)?

Leydig Cell

In general, a comparison between photoperiod-related changes and hypophysectomy (Fig 3.16A) revealed that the percentage decreases for any particular parameter of the Leydig cell were greater after hypophysectomy in the rat than those due to photoperiod-induced regression in the hamster, suggesting the importance of a relatively greater loss of gonadotropin secretion posthypophysectomy.

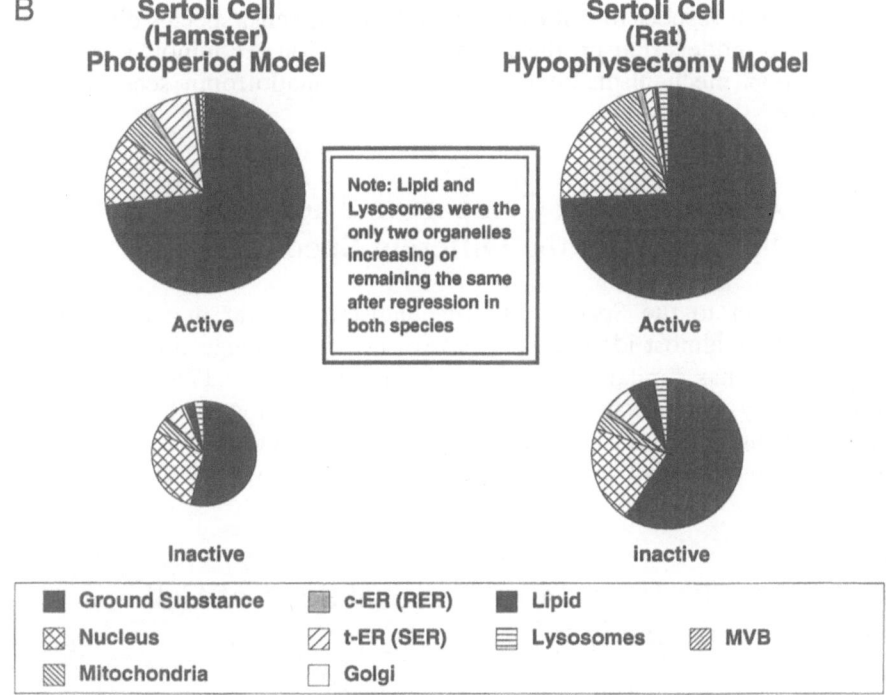

FIGURE 3.16. *Continued*

Sertoli Cell

Comparison of hypophysectomy and photoperiod-related testicular regression in the Sertoli cell also shows identical patterns of response (Fig. 3.16B). This is evident in terms of cell size and proportion of organelles.

Summary

The Leydig cell responded similarly in both photoperiod-induced regression and hypophysectomy in a cross-species comparison. With respect to the Sertoli cell, the hypophysectomy model is generally similar to the seasonal regression model in that lipid and lysosomes were the subcellular components that did not show a pattern similar to other organelles; that is, they did not respond by decreasing their volume. There was one difference in the two models: After hypophysectomy lipid increased rather than remained unchanged, as was seen after photoperiod-related regression (Fig. 3.16B).

Thus, with respect to both Leydig and Sertoli cells, species differences in the two models that produce gonadal regression are minor, suggesting that similar mechanisms related to loss of gonadotropin secretion are operative.

How Do Sertoli Cells Compare in Their Response to Hypophysectomy in the Different Species?

With respect to the Sertoli cell, the hamster and rat hypophysectomy models show almost identical patterns of response, suggesting that hypophysectomy has similar effects in both species (Figs. 3.17A and 3.17B). Lipid and lysosomes show patterns not evident in other cell types. Note, however, in Figure 3.17B that the magnitude of effect is much greater in the hamster in all areas except for lysosomes, where the pattern is reversed.

What Are the Early Morphological Signs of Cell Activity and Inactivity?

Response of the Germ Cell Population

Study of short-term hypophysectomized hamsters, short-term hypophysectomized rats, and hamsters undergoing the early phases of photoperiod-induced regression of the testis show a morphological pattern of response in the testis that is qualitatively similar. The pattern of germ cell degeneration is the same as that described by Russell and Clermont (21), Bartlett et al. (22), and Ghosh et al. (2) in the rat and is primarily a stage VII–VIII response whereby specific germ cell types (preleptotene spermatocytes, pachytene spermatocytes, and step 7 and step 19 spermatids) degenerate. As always, the short-term pattern of response is one in which a few degenerating cells are seen among a much larger population of viable cells (Table 3.1).

In the hamster photoperiod-induced testicular recrudescence model, there was a noticeable repopulation of basal compartment germ cells (Table 3.2). The mode of repopulation of these cells—that is, which cells were now dividing/surviving to cause an increase in the numbers of basal compartment cells—has not been determined.

Leydig and Sertoli Cells: Short-Term Regression in the Photoperiod Model

The response of somatic cells was determined in animals sacrificed after short-term exposure to a short (inhibitory) photoperiod, and significant

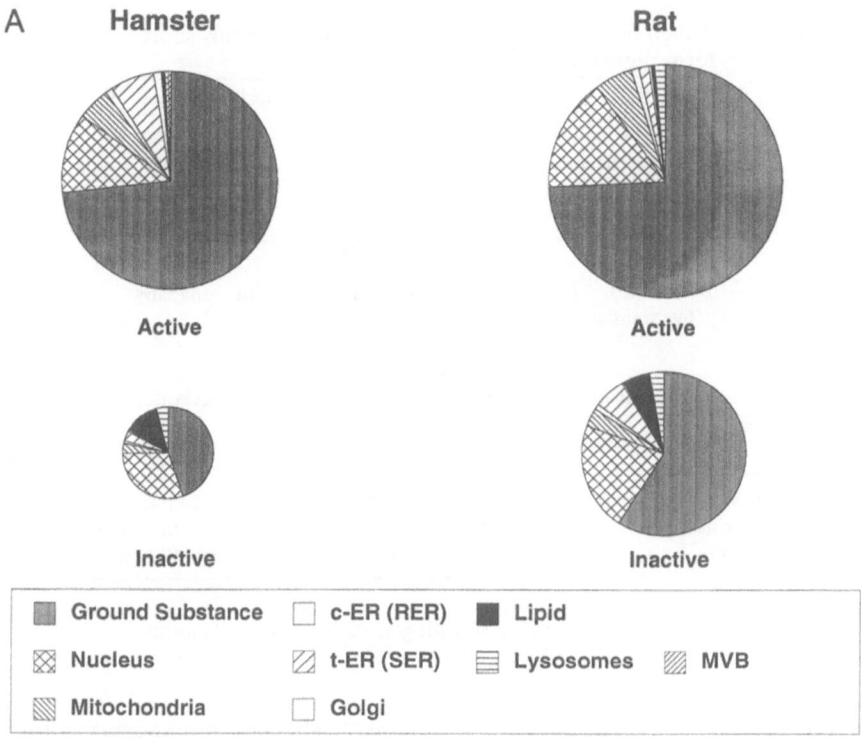

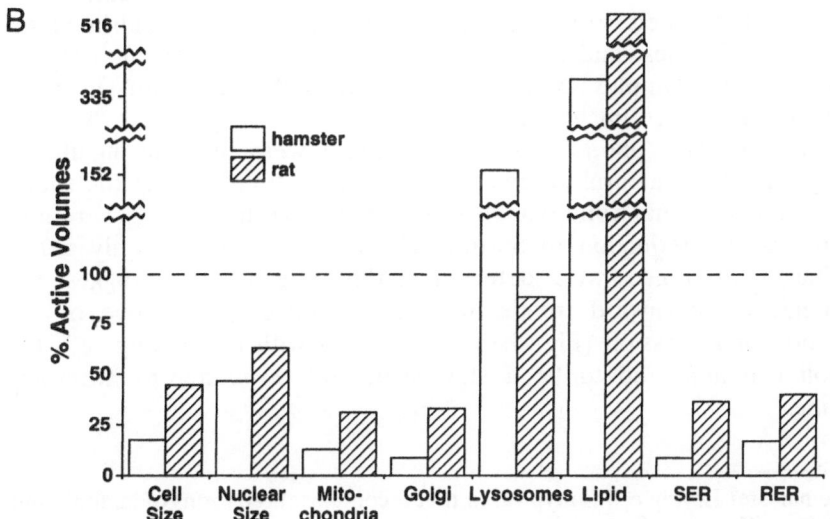

FIGURE 3.17. Pie chart (*A*) showing a comparison of the hamster and rat Sertoli cell in response to hypophysectomy and bar graph (*B*) showing a cross-species comparison of the effect of hypophysectomy in rat and hamster Sertoli cells.

TABLE 3.1. Sertoli cell-germ cell ratios during short-term regression of the hamster testis.

Cell types	Active ($n = 5$)	Early regressed ($n = 5$)
Preleptotene spermatocytes	2.27 ± 0.12*	1.73 ± 0.19[†]
Pachytene spermatocytes	2.46 ± 0.09	1.88 ± 0.22[†]
Round spermatids	8.17 ± 0.56	5.87 ± 0.71[†]

* Mean ± SE.
[†] $P < 0.05$. (P-value refers to comparison with gonadally active animals.)

TABLE 3.2. Sertoli cell-germ cell ratios during short-term recrudescence of the hamster testis.

Cell types	Regressed ($n = 5$)	Early recrudesced ($n = 5$)
Preleptotene spermatocytes	0.96 ± 0.07*	1.45 ± 0.15[†]
Pachytene spermatocytes	0.20 ± 0.07	0.35 ± 0.03
Round spermatids	0.04 ± 0.02	0.04 ± 0.01

* Mean ± SE.
[†] $P < 0.02$. (P-value refers to comparison with gonadally regressed hamsters.)

differences were determined. In the hamster short-term-exposure photoperiod model where regression was induced, there were significant decreases in cell volume and in the volume and surface area parameters in organelles associated with steroid biosynthesis (inner mitochondrial membrane, peroxisomes, and tubular [smooth] endoplasmic reticulum*) and also the Golgi complex. Volume and surface area parameters in other organelles remained unchanged (Fig. 3.18).

In the hamster photoperiod model, there were minimal significant changes in the structural parameters of the Sertoli cell during the short-term regression that simultaneously resulted in germ cell degeneration. There was no reduction in Sertoli cell volume. Although only a few significant differences were detected at this time, the data suggest the existence of downward trends in most parameters, indicative of the initiation of regression (Fig. 3.18). Compared with the Leydig cell, the Sertoli cell appeared to be a slow structural responder to hormonal declines.

---

* The hamster Leydig cell is atypical in that it contains three forms of endoplasmic reticulum. There are flattened cisternae of ribosome-populated reticulum, as well as flattened cisternae of ribosome-free reticulum. These two forms are apparently interconnected (13). There is also a minor component, an anastomosing tubular system of endoplasmic reticulum typical of smooth endoplasmic reticulum in Leydig cells of many species.

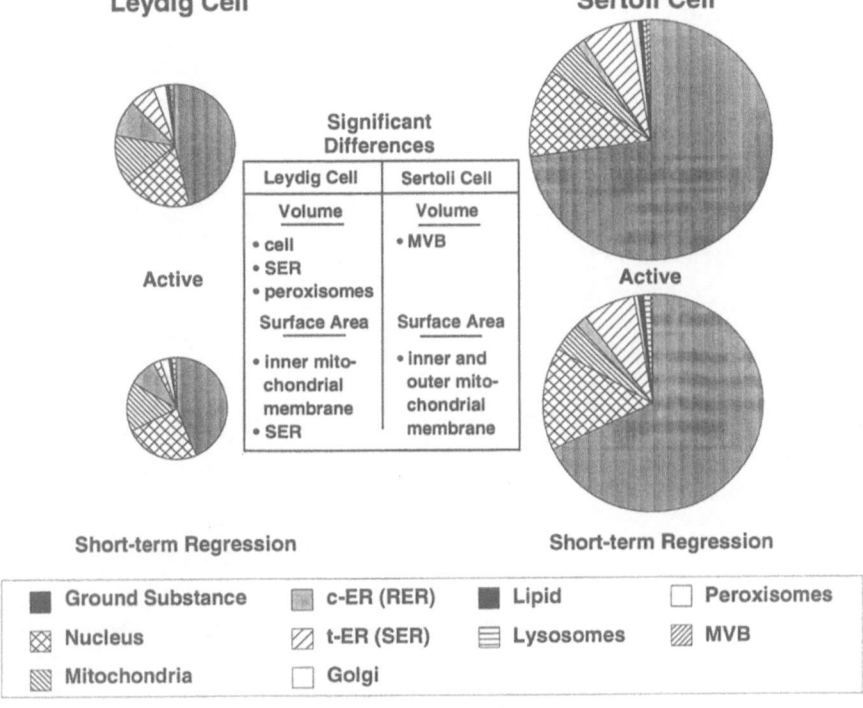

The following legend appears within the figure:

Leydig Cell	Sertoli Cell
Volume	Volume
• cell	• MVB
• SER	
• peroxisomes	
Surface Area	Surface Area
• inner mito-chondrial membrane	• inner and outer mito-chondrial membrane
• SER	

Significant Differences

Leydig Cell — Active — Short-term Regression

Sertoli Cell — Active — Short-term Regression

Ground Substance	c-ER (RER)	Lipid	Peroxisomes
Nucleus	t-ER (SER)	Lysosomes	MVB
Mitochondria	Golgi		

FIGURE 3.18. Comparison of hamster Leydig and Sertoli cells in short-term experiments that resulted in regression of the testis.

Leydig and Sertoli Cells: Short-Term Recrudescence in the Photoperiod Model

With respect to the Sertoli cell during early (short-term) recrudescence, the cytoplasm volume and the surface areas of the rough and smooth endoplasmic reticulum and mitochondrial membranes were significantly increased. During early recrudescence there was a significant increase in the volume of individual Leydig cells. Increases were also noted in nuclear, nucleolar, and cytoplasmic volumes and the surface areas of cisternal (rough) endoplasmic reticulum and mitochondrial membranes (Fig. 3.19).

Sertoli Cells: Hypophysectomy Model

After short-term hypophysectomy, the rat Sertoli cell showed few significant declines in volume or surface area parameters. Notable were declines in the cell surface area and secondary lysosomes (endocytosed germ cells) (Fig. 3.20).

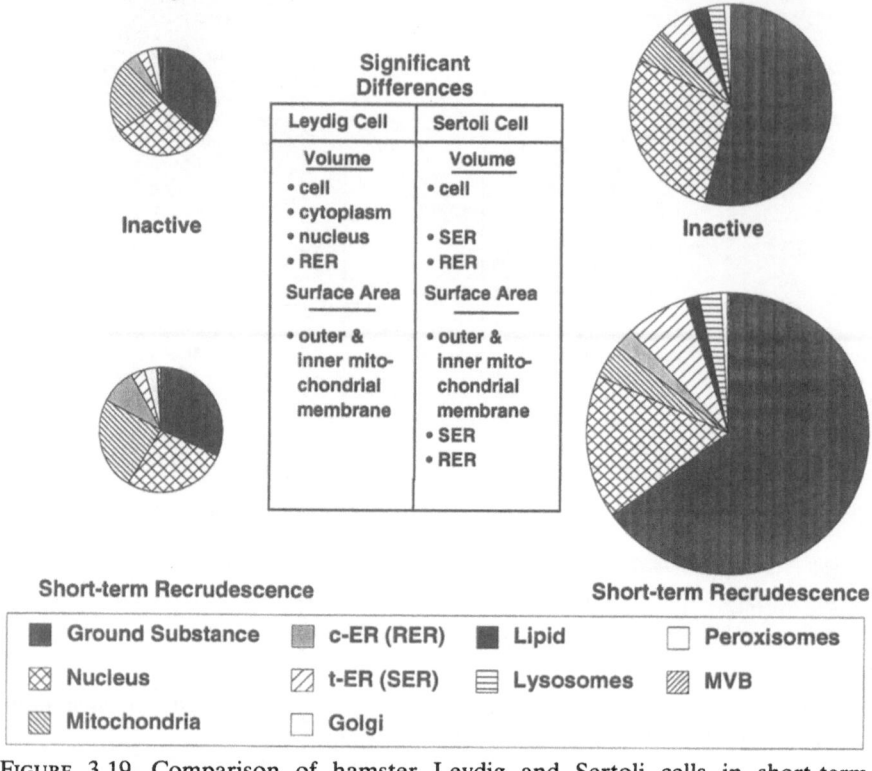

FIGURE 3.19. Comparison of hamster Leydig and Sertoli cells in short-term experiments that resulted in recrudescence of the testis.

Urethral Gland Cells

Within 2–4 h of placing males with female mice, there is an increase in the nuclear size of urethral gland cells as compared with urethral gland cell nuclei from isolated males (unpublished observations).

Summary

Overall, these results from short-term sacrifice intervals indicate that Leydig cells first lose organelles associated with steroid biosynthetic processes during regressive conditions (Table 3.3). However, to restore these organelles in conditions of recrudescence, there must first be a reestablishment of the synthetic machinery of the cell that includes parameters reflective of the function of the nucleus, mitochondria, and rough endoplasmic reticulum.

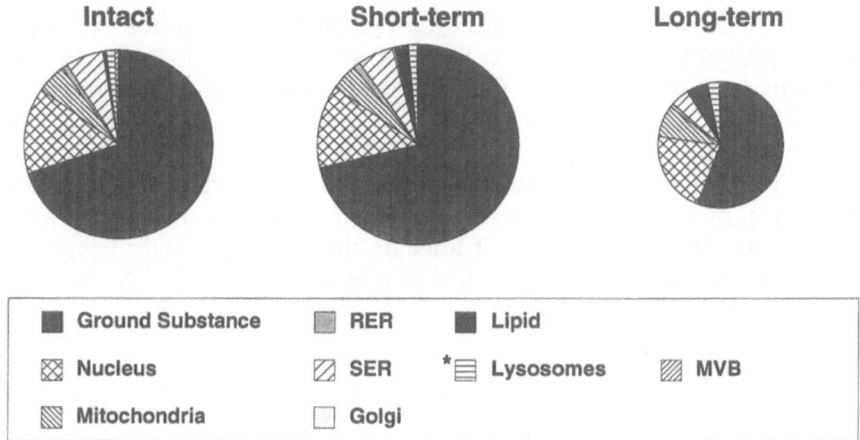

FIGURE 3.20. Short-term effects of hypophysectomy in the rat compared with pituitary-intact animals and animals undergoing long-term hypophysectomy. Significant differences are noted by an asterisk.

TABLE 3.3. Significant changes relating to Leydig and Sertoli cells noted in short-term experiments.

Leydig cells				Sertoli cells			
Early regression		Early recrudescence		Early regression		Early recrudescence	
Volume	Surface area	Volume	Surface area	Volume	Surface area	Volume	Surface area
Cell	Inner	Cell	Outer and inner	MVB	Outer and inner	Cell	Cell
SER	mitochon-	Cytoplasm	mitochon-		mitochon-	Nucleus	Nucleus
Peroxisomes	drial	Nucleus	drial		drial	SER	Inner and outer
	membrane	RER	membrane		membrane	RER	mitochon-
	SER		RER				drial
							membranes
							SER
							RER

A lack of Sertoli cell response after short-term hypophysectomy or short-term photoperiod regression is puzzling in the light of functional (germ cell) declines within the epithelium. While there is no hard and fast rule, it has been generally assumed that cell structure is reflective of cell function. This was not manifested in the present study at the sacrifice interval chosen. Possibly examining more animals would have allowed detection of more subtle changes. Also, since the Sertoli cell is primarily responsive to testosterone, it is possible that its morphological response is delayed in comparison with that of the Leydig cell because the Sertoli response would be secondary to functional alterations in the Leydig cell.

After short-term recrudescence, the response of the Sertoli cell more closely resembles the Leydig cell in terms of the cell's ability to first regain its synthetic machinery (endoplasmic reticulum and Golgi) prior to reestablishing its former complement of cell organelles, volume, and surface area.

Urethral gland cells responded in a matter of hours to a mating stimulus. Placement of males with females is known to acutely increase plasma testosterone levels (23). Thus, it appears that a morphological response can be extremely rapid using nuclear size as a parameter.

Hormone Determinations

Photoperiod Model: Hamster

Plasma *luteinizing hormone* (LH), *follicle stimulating hormone* (FSH) *testosterone* (T), and testis tissue T after photoperiod-induced testicular regresssion are significantly lower compared with the active testis. LH binding sites were higher in gonadally active animals than in animals having undergone regression. However, the density of binding sites expressed as sites per μm^2 of Leydig cell plasma membrane in the two states of gonadal activity were not different. This indicates that receptors decrease, but since the Leydig cell surface area is also decreasing, there remains a constant number of sites per unit area on the Leydig cell.

FSH receptors expressed per mg protein were significantly higher in the regressed compared with the active testis. This is most likely due to the loss of germ cells and the concentration of Sertoli cells in the testis since the number of receptors per cell and per testis was significantly lower in the regressed testis. When the comparison was made of the number of receptors/μm^2 of basal compartment Sertoli cell membrane, there were no differences in the concentration between the two groups. This indicates that the Sertoli cell loses FSH receptors, but maintains them in the same concentration on the basal compartment membrane, where receptors are known to be located. As mentioned above, the same results were obtained with regard to LH binding sites in the Leydig cells.

After short-term placement of gonadally active male hamsters in a short-day environment, there was no significant depression of LH or T, although FSH is significantly depressed. The concentration (fmol/mg protein) of FSH receptors was elevated significantly, but the content (fmol/testis) remained unchanged. Thus, the early hormonal changes resulting in regression appear to be related to FSH and its receptors.

After short-term placement of gonadally inactive hamsters in a long-day environment, plasma LH, FSH, and plasma T were elevated significantly. Neither the content nor the concentration of FSH receptors had changed.

Hypophysectomy Model: Hamster

Plasma LH, FSH and T declined significantly after long-term hypophysectomy as compared with the pituitary-intact, gonadally active animal. FSH receptor content (fmol/testis) decreased significantly after long-term hypophysectomy, although the concentration (fmol/mg protein) remained the same. In a short-term hypophysectomized animal, the levels of LH, FSH, and T also declined significantly as compared with the gonadally active animal.

Hypophysectomy Model: Rat

Plasma LH, FSH, T, and tissue T declined significantly after long-term hypophysectomy as compared with the pituitary-intact males. FSH receptors expressed as fmol/mg protein remained unchanged, but decreased when expressed per testis, per unit area of Sertoli cell, or per unit area of the basal compartment plasma membrane of the Sertoli cell. These data suggest a much greater loss of FSH receptors than was noted for the hamster photoperiod model. The concentration of LH receptors increased when expressed as fmol/g testis and decreased when expressed as content (fmol/testis) or number per cell, but remained unchanged when expressed as concentration per unit area of Leydig cell membrane.

After short-term hypophysectomy both plasma and tissue T decreased significantly. There was a significant decrease in the content of LH receptors (fmol/testis) and the number of receptors per cell, but the number of receptors as expressed per μm^2 of Leydig cell plasma membrane did not change. Like the photoperiod model, the Leydig cell appeared to lose receptors, but maintained their concentration as the Leydig cell surface area decreased. FSH receptors declined only after short-term hypophysectomy when the data were expressed per unit area of Setoli cell basal compartment plasma membrane.

Is There a Correlation of Endocrine Parameters with Active and Inactive Cells?

Leydig Cell: Photoperiod Model

In the hamster photoperiod model, the Leydig cell structural parameters correlated strongly and positively with LH levels (r-values between 0.68 and 0.99) and testicular (r-values between 0.72 and 0.96) and plasma concentrations of T (r-values between 0.56 and 0.97). These data are in agreement with results obtained in the rat implanted with T plus estradiol (24, 25) in which suppression of gonadotropins with steroids leads to testicular regression. In this model steroid production is strongly correlated with the surface area of cytoplasmic organelles involved in steroid biosynthesis (26). However, our data using gonadally active and inactive animals show that virtually all cell parameters—not only those involved in steroid production—correlated with hormonal profiles, suggesting that in these contrasting conditions there were no specific correlations of structural parameters with hormonal levels.

However, when results obtained from the short-term photoperiod-induced regression model were added to those from the gonadally active and inactive groups, the calculations revealed specific correlation of LH and T, particularly with structural components of steroid-secreting organelles. This supports the concept of early elimination of organelles related to the major function attributed to this cell. Certainly, these are the most abundant organelles and would be those preferentially eliminated before the cell lost its synthetic capacity. In this model there was also a correlation of FSH levels with Leydig cell structure, suggesting either paracrine interactions from FSH-stimulated cells within the tubule or simply a similarity in LH and FSH patterns during gonadal regression.

Sertoli Cell: Photoperiod Model

When correlations were based solely on active and inactive testes using the photoperiod hamster model, Sertoli cell structural parameters correlated strongly and positively with FSH (r-values between 0.57 and 0.76) and testicular (r-values between 0.75 and 0.91) and plasma (r-values between 0.63 and 0.86) concentrations of T. When short-term responses were included in the correlations, there were few positive and significant correlations. Only rough endoplasmic reticulum correlated with structural parameters and plasma FSH and plasma T.

Myoid Cell

In active and inactive states of the photoperiod hamster model, myoid cell volume and volumes of pinocytotic vesicles within the myoid cell were weakly correlated with plasma T. No other subcellular parameters of the myoid cell showed correlations.

Leydig Cells: Rat Hypophysectomy Model

In the rat hypophysectomy model, correlations of volumes and surface areas of regressing Leydig cells with serum and tissue T-levels were positive and significant with the exception of nuclear components, lipid, and lysosomes. There was also a significant and positive correlation of the number of LH receptors per cell and most surface area and morphological parameters.

Sertoli Cells: Rat Hypophysectomy Model

With respect to the Sertoli cell in three states of activity (active, short-term hypophysectomy, and long-term hypophysectomy), there were few positive correlations between structural parameters and FSH, plasma, and testicular T.

Sertoli Cells: Hamster Hypophysectomy Model

In the hamster hypophysectomy model, using active, short-term hypophysectomized and long-term hypophysectomized animals, there were selective weak correlations of morphological parameters of the Sertoli cell with plasma FSH and plasma T, although no particular patterns could be ascertained.

Summary

To summarize, when correlations are made between gonadally active and inactive animals in either the hypophysectomy model or the photoperiod-induced gonadal regression model, correlations between the stimulus and response of the Leydig cell are positive and significant. There is a positive relationship between the structural components of steroid-producing organelles and these hormones. When correlations are made between gonadally active animals, inactive animals, and animals in transitional states of testicular activity in either the hypophysectomy model or the photoperiod-induced gonadal regression model, the Sertoli cell shows a poor correlation of structure and function. This appears to be due to the lack of response of Sertoli cells in short-term conditions after significant hormonal declines have occurred since correlations made without the addition of the short-term group are highly positive and significant.

Overall Conclusion

This is the first study to address the structural manifestation of functionally active and inactive somatic cells of the testis. At the extremes of cellular activity, virtually all somatic cells show marked differences in volume and surface area parameters. Leydig cells, Sertoli cells, and urethral gland cells show marked differences in active and inactive states, whereas myoid cells show only minimal differences.

The photoperiod-induced testicular regression model and the hypophysectomy model in the hamster show nearly identical qualitative responses, indicating that structural changes observed in these two models have the same cause, namely, reduction of gonadotropin and prolactin secretion. Similarly, there is a cross-species similarity between the hamster photoperiod-induced testicular regression model and the rat hypophysectomy model.

While all Leydig cell structural components are diminished in volume and surface area in inactive states, short-term regressive changes in Leydig cells point to the steroid-producing organelles as being preferentially susceptible. In recrudescing conditions, nuclear and endoplasmic reticulum increases are the sentinels of activity. These particular changes indicate that the Leydig cell must first rebuild the structural components that are the machinery to re-install the steroid synthetic capacity of the cell.

Most Sertoli cell structural components in the short-term experiments (hypophysectomy and photoperiod-induced regression) remain unchanged. Thus, germ cell degeneration may take place without statistically significant changes in the structural parameters of the Sertoli cell. However, in recrudescing conditions the Sertoli cell behaves similarly to the Leydig cell by first bringing back structural components that are necessary to synthesize other cellular components.

In general, there is a strong correlation between the hormone that stimulates the Leydig cell (LH), LH receptors per cell, and the major steroid product (T) of the Leydig cell. However, the Sertoli cell structural components show little correlation with the levels of its stimuli (FSH and T). This is due primarily to the lack of change in the structure of the Sertoli cell in short-term experiments in which hormonal levels have declined.

Acknowledgments. The primary author acknowledges those individuals who have contributed to the data presented. These include Drs. A.G. Amador, V. Chandrashekar, T.J. Corbin, L.R. França, P. Grasso, H.G. Klemcke, M. Kurohmaru, A. Mayerhofer, L.E. Reichert, Hong Ping Ren, E.L. Parr, M.B. Parr, and G.S. Prins.

References

1. Bartke A, Sinha Hikim AP, Russell LD. Sertoli cell structure and function in seasonally breeding mammals. In: Russell LD, Griswold MD, eds. The Sertoli cell. Clearwater, FL: Cache River Press, 1993:349–64.
2. Ghosh S, Sinha Hikim AP, Russell LD. Further observations on stage-related effects seen after short-term hypophysectomy in the rat. Tissue Cell 1991; 23:613–30.
3. Ghosh S, Bartke A, Grasso P, Reichert LE, Russell LD. Structural manifestations of the rat Sertoli cell to hypophysectomy: a correlative morphometric and endocrine study. Endocrinology 1992;131:485–97.
4. Ghosh S, Bartke A, Grasso P, Reichert LJ, Russell LD. The structural response of the hamster Sertoli cell to hypophysectomy: a correlative morphometric and endocrine study. Anat Rec 1992;234:513–29.

5. Kurohmaru M, Sinha Hikim AP, Mayerhofer A, Bartke A, Russell LD. Golden hamster myoid cells during active and inactive states of spermatogenesis: correlation of testosterone levels with structure. Am J Anat 1990; 188:319–27.
6. Mayerhofer A, Sinha Hikim AP, Bartke A, Russell LD. Changes in the testicular microvasculature during photoperiod-related seasonal transition from reproductive quiescence to reproductive activity in the adult golden hamster. Anat Rec 1989;224:495–507.
7. Parr MB, Ren HP, Russell LD, Prins GS, Parr EL. Urethral glands of the male mouse contain secretory component and immunoglobulin A and are targets of testosterone. Biol Reprod 1992;47:1031–9.
8. Russell LD, Chandrashekar V, Bartke A, Sinha Hikim AP. The hamster Sertoli cell in early testicular regression and early recrudescence (submitted).
9. Russell LD, Corbin TJ, Ren HP, Amador AG, Bartke A, Ghosh S. Structural changes in rat Leydig cells post hypophysectomy: a morphometric and endocrine study. Endocrinology 1992;131:498–508.
10. Sinha Hikim AP, Chandrashekar V, Bartke A, Russell LD. Sentinels of Leydig cell structure and function in the golden hamsters during early testicular regression and recrudescence. Int J Androl 1993.
11. Sinha Hikim AP, Amador AG, Klemcke HG, Bartke A, Russell LD. Correlative morphology and endocrinology of Sertoli cells in hamster testes in active and inactive states of spermatogenesis. Endocrinology 1989;125:1829–43.
12. Sinha Hikim AP, Amador AG, Klemcke H, Bartke A, Russell LD. Structural/function relationships in active and inactive hamster Leydig cell: a correlative morphometric and endocrine study. Endocrinology 1989;125:1844–56.
13. Sinha Hikim AP, Bartke A, Russell LD. The seasonal breeding hamster as a model to study structure-function relationships in the testis. Tissue Cell 1988;20:63–78.
14. Sinha Hikim AP, Bartke A, Russell LD. Morphometric studies on hamster testes in gonadally active and inactive states: light microscope findings. Biol Reprod 1988;39:1225–37.
15. Gaston S, Menaker M. Photoperiodic control of hamster testis. Science 1967;158:925–8.
16. Bartke A. Male hamster reproductive endocrinology. In: Siegel HI, ed. The hamster. New York: Plenum, 1985:73–98.
17. Russell LD, Ettlin RA, Sinha Hikim AP, Clegg ED. Histological and histopathological evaluation of the testis. Clearwater, FL: Cache River Press, 1990:210–66.
18. Weibel ER, Bolender RP. Stereological techniques for electron microscopic morphometry. In: Hayat MA, ed. Principles and techniques of electron microscopy; vol 3. New York: Van Nostrand-Reinhold, 1973:237–96.
19. Bozzola JJ, Russell LD. Quantitative electron microscopy. In: Electron microscopy: principles and techniques for biologists. Boston: Jones and Bartlett, 1992:287–304.
20. Yee JB, Hutson JC. Testicular macrophages: isolation, characterization and hormonal responsiveness. Biol Reprod 1983;29:1319–26.
21. Russell LD, Clermont Y. Degeneration of germ cells in normal, hypophysectomized and hormone treated hypophysectomized rats. Anat Rec 1977; 187:347–66.

22. Bartlett JM, Kerr JB, Sharpe RM. The effect of selective destruction and regeneration of rat Leydig cells on the intratesticular distribution of testosterone and morphology of the seminiferous epithelium. J Androl 1986;7:240–53.
23. Macrides F, Bartke A, Dalterio S. Strange females increase plasma testosterone levels in male mice. Science 1975;189:1104–6.
24. Ewing LL, Gorski RA, Sbordorne RJ, Tyler JV, Desjardins C, Robaire C. Testosterone-estradiol filled polydimethylsiloxane subdermal implants: effect on fertility and masculine sexual and aggressive behavior of male rats. Biol Reprod 1979;21:765–72.
25. Keeney DS, Mendis-Handagama SMLC, Zirkin BR, Ewing LL. Effect of long term deprivation of luteinizing hormone on Leydig cell volume, Leydig cell number, and steroidogenic capacity of the rat testis. Endocrinology 1988;123:2906–15.
26. Ewing LL, Wing T-Y, Cochran RC, Kromann N, Zirkin BR. Effect of luteinizing hormone on Leydig cell structure and testosterone secretion. Endocrinology 1983;112:1763–9.

4

Local Growth Factor Effects During Spermatogenesis

Martti Parvinen, Harri Hakovirta, Antti Kaipia, Tarja-Leena Penttilä, Jorma Toppari, Viqar Syed, Bernard Jégou, Håkan Persson, Aida Wahab, and Olof Söder

In the mammalian seminiferous epithelium, the cell associations follow each other in a wavelike fashion. The cell associations are characterized by constant cell composition and are identified usually by changing morphology of the acrosome system and of the nucleus in the early spermatids. Cells at DNA synthesis—that is, the six different classes of spermatogonia and preleptotene spermatocytes—are also accurately localized in defined stages of the cycle of the seminiferous epithelium.

Growth factors have been suggested to play an important role in the differentiation of the cells of the seminiferous epithelium. The first factor characterized was the seminiferous growth factor (1); since then, several other growth factors have been found in the testis (2, 3). However, very little information is available on local growth factor actions in the seminiferous epithelium, particularly about their role in the interactions between Sertoli cells and spermatogenic cells. We have developed new methods for analysis of stage-specific DNA synthesis in the seminiferous epithelium and the action of different growth factors separately on DNA synthesis of type A, intermediate, and type B spermatogonia and on preleptotene spermatocytes that synthesize DNA specifically for meiosis.

Methods

Transillumination-assisted microdissection of rat seminiferous tubules was used for separation of seminiferous tubule segments from defined stages of the cycle of the seminiferous epithelium. The segments were selected to represent various phases of the DNA synthesis of spermatogonia and spermatocytes in the following way (Fig. 4.1): Stage I (end of the weak spot zone) contains type A4 spermatogonia that differentiate through 2

DURATION OF THE CYCLE [DAYS]

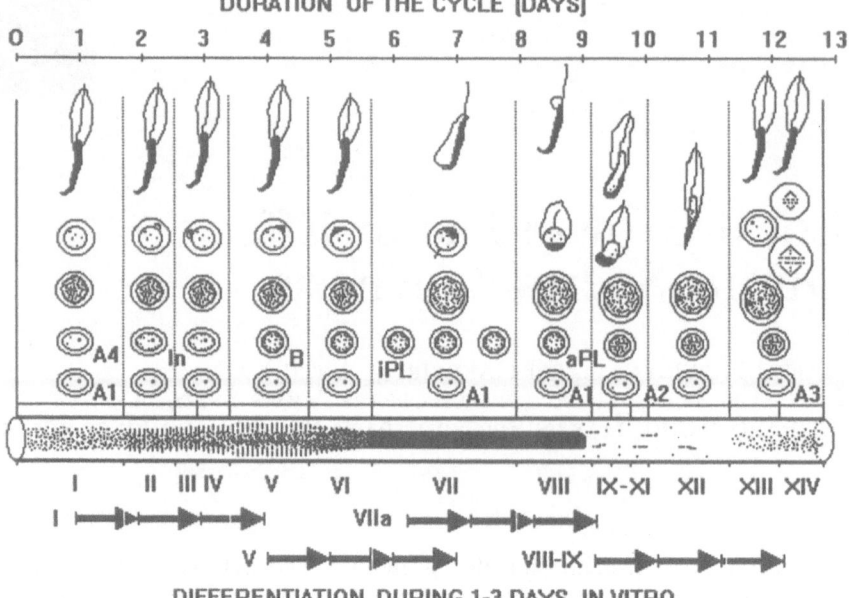

DIFFERENTIATION DURING 1-3 DAYS IN VITRO

FIGURE 4.1. Map of rat spermatogenesis, modified from reference 34, in relation to the transillumination pattern of freshly isolated unstained seminiferous tubules (schematic tracing according to reference 35). The length of each of the 14 stages is adjusted to correspond to the duration of the cycle, as described in reference 36. The arrows indicate the rate of differentiation in vitro, as described in reference 4. A1–A4, In, and B represent type A, intermediate, and type B spermatogonia, respectively. (iPL = resting preleptotene spermatocytes at stage VII; aPL = preleptotene spermatocytes in meiotic DNA synthesis at stage VIII.)

mitoses during 3 days to produce intermediate and type B spermatogonia; in stage V, type B spermatogonia differentiate to inactive preleptotene spermatocytes; in stage VIIa, inactive preleptotene spermatocytes differentiate to active preleptotene spermatocytes and early zygotene; and in stages VIII–IX, active preleptotene spermatocytes enter the prophase of meiosis (leptotene and zygotene) and type A1 spermatogonia differentiate through type A2 to A3. The selection of the seminiferous tubule segments by this protocol covers virtually all stages of spermatogonial differentiation and the specific effects of growth factors can be investigated in vitro (4).

Interleukin-1

Testes of rat and human have been shown to secrete large amounts of interleukin-1-like activity (5) that originates from the Sertoli cells (6). In

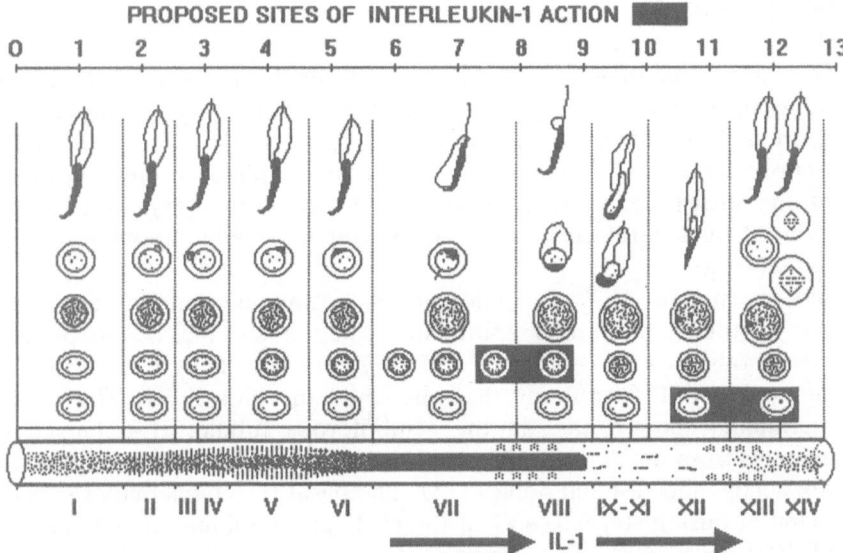

FIGURE 4.2. Proposed sites of action of IL-1α in rat spermatogenesis. Significant stimulations of DNA synthesis in vitro were found in stage VIIa and VIII–IX tubule segments cultured for 48 h. The suggested target cells are preleptotene spermatocytes and type A2–A3 spermatogonia, respectively, as described in reference 4.

different stages of rat seminiferous epithelium, there exists a correlation between total DNA synthesis and *interleukin-1* (IL-1) activity (7). Stage VII of the cycle shows background values of both parameters. These observations led to a hypothesis that IL-1 might be a spermatogonial growth factor.

To test this, staged seminiferous tubule segments were incubated in vitro with concentrations of 0.1 and 0.5 ng/mL of human recombinant IL-1α. Significant stimulation of DNA synthesis was found in stage VIIa and VIII–IX tubules after 48 h of culture at 34°C. This was interpreted to indicate promotion of DNA synthesis of preleptotene spermatocytes and of types A2 and A3 spermatogonia (4) (Fig. 4.2). The results suggest that IL-1 is both a mitotic and meiotic growth factor during spermatogenesis; this view is also supported by earlier in vivo studies in hypophysectomized rats (8). More recently, we have analyzed the IL-1 bioactivity in 1-mm seminiferous tubule segments staged by the improved transillumination technique (9) and found that the activity sharply drops at the beginning of stage VI and equally sharply raises again at the beginning of stage VIII (Söder et al., unpublished observation). This further suggests that IL-1 regulates the DNA synthesis at a very accurate cellular level in the seminiferous epithelium.

Interleukin-6

The cytokine *interleukin-6* (IL-6) is produced by several cell types, including fibroblasts, monocytes, endothelial cells, and T cells, and it is an important regulator of the host defense response (10). Among the major functions of IL-6 is the stimulation of acute phase protein synthesis and the growth and differentiation of lymphocytes (11). IL-6 is also produced by several endocrine organs, but knowledge about testes has been lacking.

We investigated IL-6 production and its hormonal regulation during the cycle of the seminiferous epithelium using a specific IL-6-dependent mouse-mouse hybrid cell line, 7TDI (12). The specificity was verified by a monoclonal IL-6 antibody that abolished the bioactivity. The culture experiments showed that in the seminiferous tubule, IL-6 bioactivity originates from the Sertoli cells, and it is stimulated by lipopolysaccharide, latex beads, and residual bodies (13). The basal IL-6 bioactivity showed a maximum during stages II–VI of the cycle and a minimum at stages VII and VIII (Fig. 4.3). The increase during stages IX–XII is obviously stimulated at least in part by residual bodies that are released from the maturing spermatids at stage VIII of the cycle.

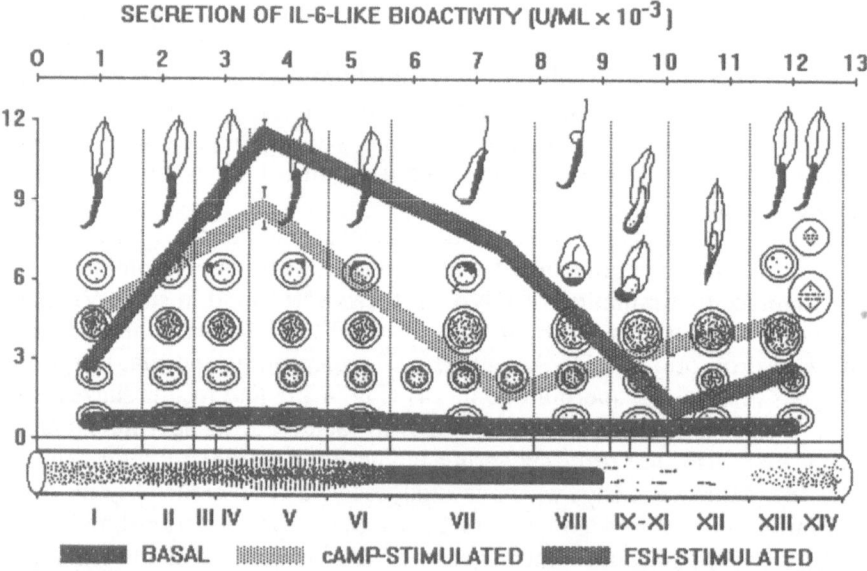

FIGURE 4.3. Secretion of IL-6-like bioactivity related to the cycle of the seminiferous epithelium. The tubule segments from stages II–VI, VII–VIII, IX–XII, and XIII–I were analyzed and stimulated by FSH and cAMP, as described in reference 13.

Follicle stimulating hormone (FSH) had a strong stimulatory action on IL-6 bioactivity, with the relative stimulation being particularly high at stages VII–VIII. Stimulation by dbcAMP at stages II–VI and XIII–I was comparable to that with FSH, but showed a clear difference at stages VII–VIII and IX–XII (Fig. 4.3). This suggests that FSH-induced effects on IL-6 production are not only mediated through cAMP, but via another pathway(s) yet to be determined. The function of IL-6 in the testis is unknown, but in concert with IL-1 it may play a role in the paracrine regulation of spermatogenesis.

Nerve Growth Factor

Besides nervous tissue, *β-nerve growth factor* (NGFβ) and its mRNA are expressed in spermatocytes and early spermatids of rat and mouse (14). Its low-affinity receptor is expressed in Sertoli cells under negative control of testosterone (15). The strictly stage-specific (VIIcd and VIII) expression of this receptor, together with the expression of tyrosine kinase receptor trk, an essential component of the high-affinity *NGF receptor* (NGF-R), led to a hypothesis that the specific function of NGF might reside in that region of the seminiferous epithelium. The stage-specific DNA synthesis was analyzed in vitro, and a stimulation was found in stages VIII–IX after 24 h and in stage VIIa after 48 and 72 h in culture, suggesting a specific action on meiotic DNA synthesis (16) (Fig. 4.4).

The action of NGF on preleptotene spermatocytes is apparently mediated by another Sertoli cell factor that is still unknown. It may be a *survival factor* of Sertoli cells since a beneficial effect of NGF for cultured human seminiferous tubule segments has been demonstrated (17). Furthermore, NGF has been demonstrated to stimulate androgen binding protein mRNA expression in the rat seminiferous tubules, probably by prolonging the duration of the most active stages (VII and VIII) of the cycle (18).

Insulin-Like Growth Factors

Insulin-like growth factors (IGFs) are peptide mitogens that are structurally homologous to proinsulin (19). IGF-I and IGF-II are synthesized in multiple tissues, including the testis (20) where they have been suggested to have a paracrine role (21). Receptors for IGF-I have been demonstrated in meiotic cells (22). We investigated the effects of IGF-I, *truncated IGF-I* (trIGF-I), IGF-II, human *growth hormone* (GH), and insulin on stage-specific DNA synthesis of rat spermatogenesis in vitro (23).

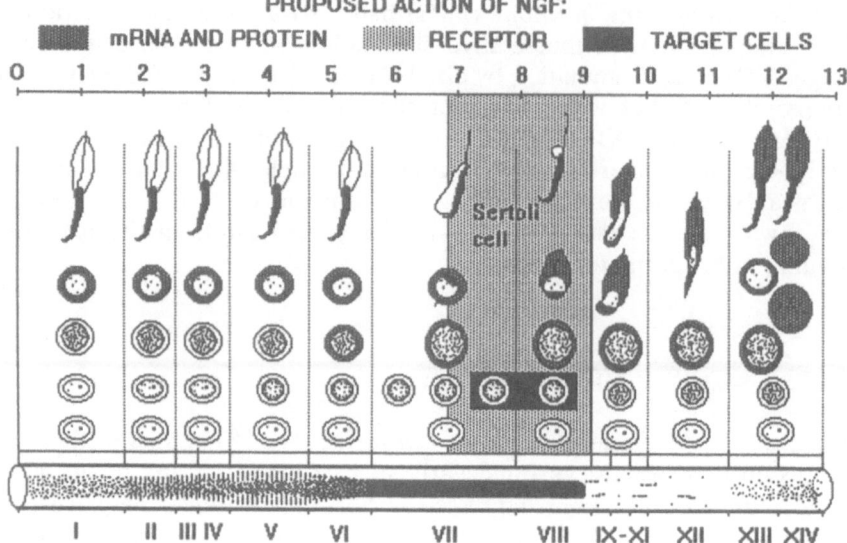

FIGURE 4.4. Summary of findings about the NGF action in the rat seminiferous epithelium. It is produced by spermatogenic cells, has stage-specific receptors in Sertoli cells, and promotes meiotic DNA synthesis of preleptotene spermatocytes, as decribed in reference 16.

In physiological temperature (34°C), IGF-I, trIGF-I, and IGF-II stimulated DNA synthesis of originally stage VIII–IX seminiferous tubules, wherefrom spermatogenesis progressed to stages XIII–XIV during 3 days. During this time the type A1 spermatogonia differentiate through type A2 to A3, and the preleptotene spermatocytes stop their DNA synthesis as they differentiate through leptotene to zygotene stage of the prophase of meiosis. It is proposed that type A3 spermatogonia are target cells for IGF-I and IGF-II action (Fig. 4.5).

Another site where trIGF-I and IGF-II had an effect was stage V that differentiated to stage VII during 3 days in vitro. The proposed target cells are type B spermatogonia that differentiate to inactive preleptotene spermatocytes during that time. A slight but significant dose-dependent effect was obtained with GH at stage I, where the type A4 spermatogonia differentiated to intermediate-type spermatogonia at stages II and IV of the cycle during 2 days (Fig. 4.5). At 37°C, largely similar effects were found, except that only insulin had effects in this temperature, whereas GH showed no effects at 37°C. It was concluded from these results that IGFs and IGF-related peptides are predominantly spermatogonial growth factors that act at the beginning and at the end of spermatogonial differentiation, but that cells at the onset of meiosis were not stimulated.

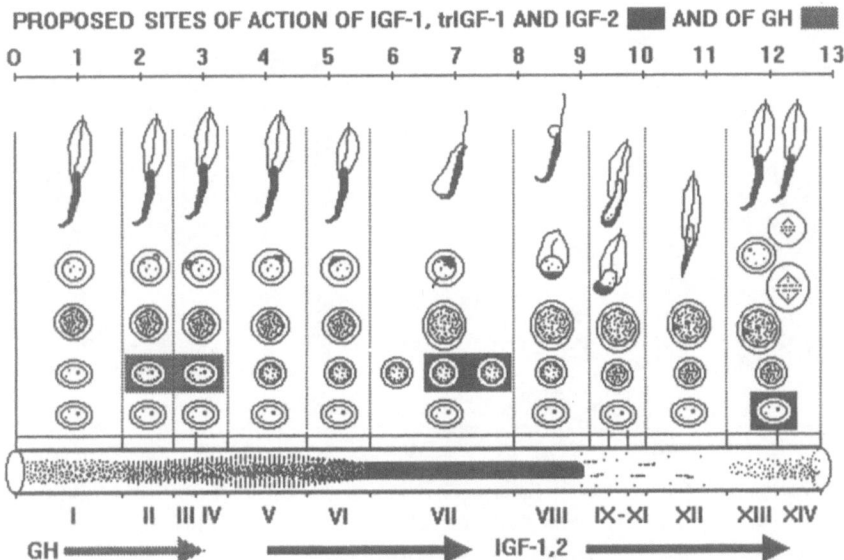

FIGURE 4.5. Similar action of IGF-I and IGF-II on originally stage V and VIII–IX tubule segments. After 3 days the proposed target cells have differentiated to inactive preleptotene and type A3 spermatogonia, suggesting an activation site during the beginning and end of spermatogonial differentiation, as described in reference 23.

Inhibin and Activin

Inhibin, a member of the TGFβ-family, consists of two dissimilar subunits: an α-chain combined to either βA or βB (24). Activin is either a homodimer of inhibin βA subunits or a βA-βB heterodimer. Activin and inhibin have opposite effects on pituitary FSH secretion, and they were suggested to have opposite paracrine actions on spermatogonial proliferation (25, 26).

Follistatins also inhibit FSH release from the pituitary and modulate inhibin and activin actions as specific binding proteins (27). A high-affinity activin A receptor mRNA has been localized in pachytene spermatocytes and round spermatids (28, 29), and germ cells are capable of binding activin A and inhibin A molecules (30). In the seminiferous epithelium, inhibin α-subunit and βB subunit mRNAs are expressed in a stage-specific manner, suggesting a local paracrine action (31). We used in situ hybridization for more accurate localization of mRNAs for inhibin βA and βB subunits, follistatin, and activin A receptor and activin receptor IIB2 to gain further insight into their role in the function of the seminiferous epithelium.

mRNA EXPRESSIONS:

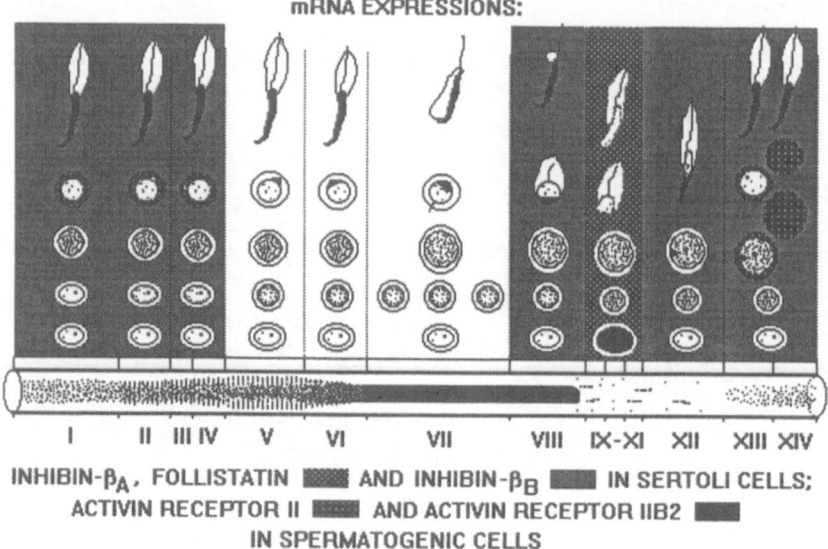

INHIBIN-β_A, FOLLISTATIN ▨▨▨ AND INHIBIN-β_B ▨▨▨ IN SERTOLI CELLS;
ACTIVIN RECEPTOR II ▨▨▨ AND ACTIVIN RECEPTOR IIB2 ▨▨▨
IN SPERMATOGENIC CELLS

FIGURE 4.6. Summary of the results of in situ hybridizations of mRNAs related to the proposed action of activin and inhibin in the seminiferous epithelium, as described in references 28 and 33.

Inhibin βA mRNA was expressed in a highly stage-specific manner (Fig. 4.6). It was first detected in Sertoli cells at stage VIII of the cycle. The onset was coincident with spermiation, and the expression continued through stages IX–XI of the cycle. A stage-specific expression pattern was first detected in 3-week-old postnatal rats, where the hybridization was found most abundantly around leptotene spermatocytes in Sertoli cell cytoplasm. Follistatin mRNA showed an identical expression pattern (28).

Inhibin βB mRNA started to increase in the Sertoli cells at stage VIII of the cycle, reached a maximum at stage XIII, and declined during stages IV–V (Fig. 4.6). A high level of mRNA expression was found in 2-week-old rats, and it showed a decline thereafter. The selective absence of spermatogonia or spermatocytes 7 and 17 days after irradiation with 3 Gy of X rays did not have any effect on stage-specific expression of mRNAs of inhibin βA, βB, or follistatin.

Activin A receptor mRNA was first detected in diplotene and diakinetic primary spermatocytes immediately prior to the meiotic divisions. The expression continued through the meiotically dividing cells at stage XIV and in step 1–4 spermatids in both adult and pubertal rats. A low level of hybridization could be observed over Sertoli cells at all stages of the cycle. The Northern hybridization of poly(A)$^+$ RNA revealed a 2.4-kb band of inhibin βA mRNA, as well as a predominant 2.6-kb transcript and minor 1.2- and 4.6-kb transcripts of follistatin. The activin receptor mRNA showed 3.0- and 6.0-kb transcripts, whereas inhibin βB cRNA probe detected 3.5- and 4.2-kb transcripts.

Another activin receptor, ActR-IIB, was recently cloned (32). This gene encodes potentially 4 isoforms of activin receptor, 2 of which (ActR-IIB1 and ActR-IIB2) have a higher affinity for activin A than the previously cloned activin A receptor. We analyzed the expression of the activin receptor IIB2 in rat testis by in situ hybridization (33) and found the strongest expression in stages IX–XI of the cycle in basal parts of the epithelium, where it showed a clusterlike distribution both in Sertoli cells and in type A1 and A2 spermatogonia. Since the first mitoses of spermatogenesis occur in type A1 and A2 spermatogonia, it is possible that activin regulates the onset of spermatogenesis, and through the other receptor ActRII, also participates in the regulation of events associated with meiotic divisions and early spermiogenesis.

Another interesting hypothesis is that under the influence of activin, the spermatogenic cells secrete another factor(s) that influences the spermatogonial proliferation. Such a mechanism seems likely, with the NGF that is produced by spermatogenic cells, has receptors in Sertoli cells, and acts on cells active in meiotic DNA synthesis through a factor(s) that is unknown at the present time.

Conclusions and Future Prospects

Androgens and FSH are still considered the most important factors in the regulation of spermatogenesis. Sertoli cells contain receptors for these hormones, and their effects on developing germ cells are apparently mediated by several growth factors. There are examples of regulation of growth factor secretion by FSH (IL-6) and also of down-regulation of growth factor receptor by androgens (NGF). The available data suggest that mitotic and meiotic DNA syntheses are differentially regulated by growth factors. NGF is a meiotic growth factor, whereas IGF-I and IGF-II act predominantly on the mitotic phase of DNA synthesis. IL-1 has effects on both types of DNA synthesis. All of these growth factors have their effects on a limited area of the seminiferous epithelium. This suggests that each of the six mitoses of spermatogonia and the premeiotic DNA synthesis are regulated by different growth factors and perhaps by a combination of factors. This will be the topic of future research.

References

1. Feig LA, Bellvé AR, Horbach-Erickson NH, Klagsbrun M. Sertoli cells contain a mitogenic polypeptide. Proc Natl Acad Sci USA 1980;77:4774–8.
2. Bellvé AR, Zheng W. Growth factors as autocrine and paracrine modulators of male gonadal functions. J Reprod Fertil 1989;85:771–93.

3. Skinner MK. Cell-cell interactions in the testis. Endocr Rev 1991;12:45–77.
4. Parvinen M, Söder O, Mali P, Fröysa B, Ritzén EM. In vitro stimulation of stage-specific deoxyribonucleic acid synthesis in rat seminiferous tubule segments by interleukin-1α. Endocrinology 1991;129:1614–20.
5. Khan S, Söder O, Syed V, Gustafsson K, Lindh M, Ritzén EM. The rat testis produces large amounts of an interleukin-1-like factor. Int J Androl 1987;10: 494–503.
6. Gérard N, Syed V, Bardin CW, Genetet N, Jégou B. Sertoli cells are the site of interleukin-1α synthesis in rat testis. Mol Cell Endocrinol 1991;82:R13–6.
7. Söder O, Syed V, Callard GV, et al. Production and secretion of an interleukin-1-like factor is stage-dependent and correlates with spermatogonial DNA synthesis in the rat seminiferous epithelium. Int J Androl 1991;14: 223–31.
8. Pöllänen P, Söder O, Parvinen M. Interleukin-1α stimulation of spermatogonial proliferation in vivo. Reprod Fertil Dev 1989;1:85–7.
9. Kangasniemi M, Kaipia A, Mali P, Toppari J, Huhtaniemi I, Parvinen M. Modulation of basal and FSH-stimulated cyclic AMP production in rat seminiferous tubules staged by an improved transillumination technique. Anat Rec 1990;227:62–76.
10. van Snick J. Interleukin-6: an overview. Annu Rev Immunol 1990;8:253–78.
11. Lotz M, Jirik F, Kabouridis P, et al. B cell stimulatory factor 2/interleukin-6 is a costimulant for human thymocytes and T-lymphocytes. J Exp Med 1988;167:1253–8.
12. van Snick J, Cayphas S, Vink A, et al. Purification and NH$_2$-terminal amino acid sequence of a T cell derived lymphokine with growth factor activity for β-cell hybridomas. Proc Natl Acad Sci USA 1986;83:9679–83.
13. Syed V, Gérard N, Kaipia A, Bardin CW, Parvinen M, Jégou B. Identification, ontogeny and regulation of an interleukin-6 like (IL-6) factor in the rat testis. Endocrinology (in press).
14. Ayer-LeLièvre C, Olson L, Ebendal T, Hallböök F, Persson H. Nerve growth factor mRNA and protein in the testis and epididymis of mouse and rat. Proc Natl Acad Sci USA 1988;85:2628–32.
15. Persson H, Ayer-LeLièvre C, Söder O, et al. Expression of β-nerve growth factor receptor mRNA in Sertoli cells downregulated by testosterone. Science 1990;247:704–7.
16. Parvinen M, Pelto-Huikko M, Söder O, et al. Expression of β-nerve growth factor and its receptor in rat seminiferous epithelium: specific function at the onset of meiosis. J Cell Biol 1992;117:629–41.
17. Seidl K, Holstein AF. Organ culture of human seminiferous tubules: a useful tool to study the role of nerve growth factor in the testis. Cell Tissue Res 1990;261:539–47.
18. Lönnerberg P, Söder O, Parvinen M, Ritzén EM, Persson H. β-nerve growth factor influences the expression of androgen binding protein messenger ribonucleic acid in the rat testis. Biol Reprod 1992;47:381–8.
19. Rinderknecht E, Humbel RE. The amino acid sequence of human insulin-like growth factor-I and its structural homology with proinsulin. J Biol Chem 1978;253:2769–76.
20. Hall K, Ritzén EM, Johnsonbaugh RE, Parvinen M. Pubertal rise of immunoreactive somatomedin and its eventual source. In: Spencer EM, ed.

Insulin-like growth factors/somatomedins: basic biochemistry, biology, clinical importance. New York: Walter de Gruyter, 1983:611–4.

21. Borland K, Mita M, Oppenheimer CL, et al. The actions of insulin-like growth factors I and II on cultured Sertoli cells. Endocrinology 1984;114: 240–6.

22. Tres LL, Smith EP, van Wyk JJ, Kierszenbaum AL. Immunoreactive sites and accumulation of somatomedin-C in rat Sertoli-spermatogenic cell co-cultures. Exp Cell Res 1986;162:33–50.

23. Söder O, Bang P, Wahab A, Parvinen M. Insulin-like growth factors selectively stimulate spermatogonial but not meiotic, deoxyribonucleic acid synthesis during rat spermatogenesis. Endocrinology 1992;131:2344–50.

24. Mason AJ, Hayflick JS, Ling N, et al. Complementary DNA sequences of ovarian follicular fluid inhibin show precursor structure and homology with transforming growth factor-β. Nature 1985;318:659–63.

25. Mather JP, Attie KM, Woodruff TK, Rice GC, Phillips DM. Activin stimulates spermatogonial proliferation in germ-Sertoli cell cocultures from immature rat testis. Endocrinology 1990;127:3206–14.

26. van Dissel-Emiliani FMF, Grootenhuis AJ, de Jong F, de Rooij DG. Inhibin reduces spermatogonial numbers in testes of adult mice and chinese hamsters. Endocrinology 1989;125:1898–903.

27. Nakamura T, Takio K, Eto Y, Shibai H, Titani K, Sugino H. Activin-binding protein from rat ovary is follistatin. Science 1990;247:836–8.

28. Kaipia A, Penttilä TL, Shimasaki S, Ling N, Parvinen M, Toppari J. Expression of inhibin βA and βB, follistatin and activin-A receptor messenger ribonucleic acids in the rat seminiferous epithelium. Endocrinology 1992; 131:2703–10.

29. de Winter JP, Themmen APN, Hoogerbrugge JW, Klaij IA, Grootegoed JA, de Jong FH. Activin receptor mRNA expression in rat testicular cell types. Mol Cell Endocrinol 1992;83:R1–8.

30. Woodruff TK, Borree J, Attie KM, Cox ET, Rice GC, Mather JP. Stage-specific binding of inhibin and activin to subpopulations of rat germ cells. Endocrinology 1992;130:871–81.

31. Bhasin S, Krummen L, Morelos BS, et al. Stage-dependent expression of inhibin alpha and beta B subunits during the cycle of the rat seminiferous epithelium. Endocrinology 1989;124:987–91.

32. Attisano L, Wrana JL, Cheifetz S, Massague J. Novel activin receptors: distinct genes and alternative splicing generate a repertoire of serine/threonine kinase receptors. Cell 1992;68:97–108.

33. Kaipia A, Parvinen M, Toppari J. Localization of activin receptor (ActR-IIB2) mRNA in the rat seminiferous epithelium. Endocrinology 1993;132: 477–9.

34. Perey B, Clermont Y, Leblond CP. The wave of the seminiferous epithelium in the rat. Am J Anat 1961;108:47–77.

35. Parvinen M, Vanha-Perttula T. Identification and enzyme quantitation of the stages of the seminiferous epithelial wave in the rat. Anat Rec 1972;174: 435–50.

36. Clermont Y, Harvey SC. Duration of the cycle of the seminiferous epithelium of normal, hypophysectomized and hypophysectomized-hormone treated albino rats. Endocrinology 1965;76:80–9.

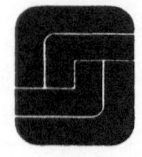

5

Stage-Specific Expression of the Cathepsin L Gene by Rat Sertoli Cells

WILLIAM W. WRIGHT

The seminiferous epithelium, the tissue in which spermatogenesis occurs, is unique in that it contains multiple generations of cells in a single developmental lineage that replicate and mature in a synchronous and coordinated manner. These cells are the developing male gametes. This coordinated development of spermatogenic cells, first described in detail by Leblond and Clermont (1), insures that specific types of germ cells are associated with one another, defining in any cross section of a seminiferous tubule one of the stages of the cycle of the seminiferous epithelium. In the rat, 14 stages have been defined (1). This synchronous development of spermatogonia, spermatocytes, and one or two generations of spermatids, illustrated in Figure 5.1, raises the question of whether there is a central organizer in the testis to insure that germ cells progress through spermatogenesis according to a specific developmental schedule. If such an organizer extrinsic to the germ cells exists, it is likely that it is the only somatic cell within the seminiferous epithelium, the Sertoli cell.

Since its discovery in 1893, the Sertoli cell has been an important focus for studies of the regulation of spermatogenesis (2). Sertoli cells envelop and adhere to the germ cells, express receptors for FSH and testosterone, hormones that are required for spermatogenesis, and secrete a variety of micronutrient carrier proteins, growth factors, and proteases (2–5). If a specific Sertoli cell product influences a particular event in spermatogenesis, it follows either that this product would be synthesized at a limited number of stages of the cycle or that the ability of germ cells to respond to the Sertoli cell product would be restricted to specific stages of the cycle. Thus, the analysis of stage-specific germ cell-Sertoli cell interactions holds promise for determining the mechanism by which these cells interact and the consequences of these interactions to spermatogenesis.

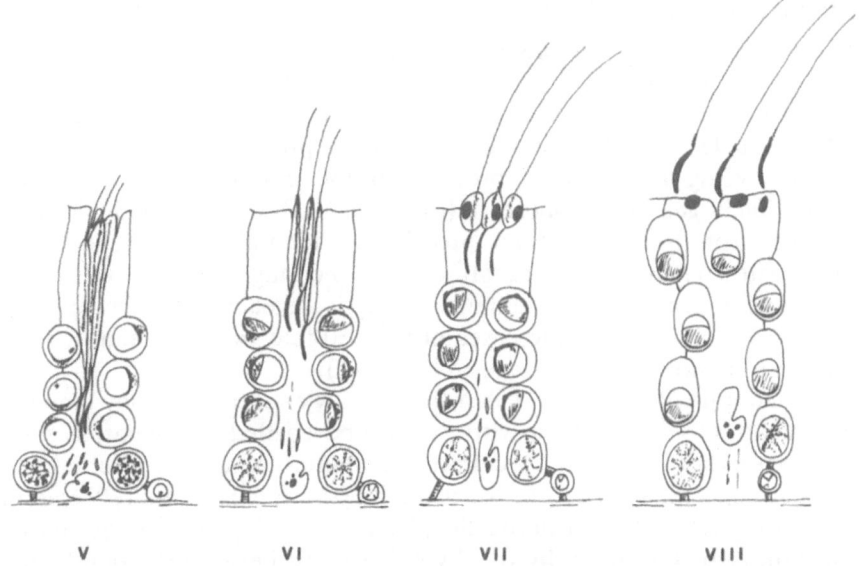

V   VI   VII   VIII

FIGURE 5.1. Stages V–VIII of the cycle of the seminiferous epithelium of the rat. This figure illustrates the simultaneous development of multiple generations of germ cells around a single Sertoli cell. The figure depicts the following types of germ cells in a stage V tubule: type B spermatogonia, pachytene spermatocytes, step 5 spermatids and step 17 (compacted) spermatids. As the tubule progresses from stage V to stage VIII of the cycle, all these germ cells undergo simultaneous and coordinated maturation. Type B spermatogonia divide at the end of stage VI and give rise to preleptotene spermatocytes. From stage V to VIII, there is a significant increase in the size of pachytene spermatocytes. Simultaneously, round spermatids complete deposition of the acrosome, and at stage VIII their nuclei migrate to one end of the cell. Finally, the compacted spermatids, which are located in deep infoldings of the Sertoli cell at stage V, move to the apex of the Sertoli cell by stage VII and are released into the lumen of the seminiferous tubule at the end of stage VIII.

For the past 10 years, our laboratory has investigated germ cell-Sertoli cell interactions by studying one protein, initially called *cyclic protein-2* (CP-2), that is synthesized and secreted in significant amounts only at stages VI and VII of the cycle of the seminiferous epithelium of the rat (6–9). This chapter summarizes our rationale for studying this protein, provides current knowledge on its stage-specific synthesis and secretion, documents the stage-specific expression of the gene that encodes this protein, and reviews our demonstration that CP-2 is the proenzyme form of the cysteine protease cathepsin L. Finally, we discuss the potential functions of this protein in the seminiferous epithelium.

Discovery of Cyclic Protein-2 and Analysis of the Stage-Specific Synthesis and Secretion of the Protein

Cyclic protein was initially identified by culturing mature rat seminiferous tubules at defined stages of the cycle in the presence of ^{35}S-methionine and analyzing radiolabeled proteins in the culture medium by two-dimensional gel electrophoresis (6). This analysis identified 15 stage-specific proteins; the second protein we identified, CP-2, exhibited the greatest changes with progression of the stages of the cycle. Two-dimensional scanning densitometric analysis revealed that the apparent secretion of this protein increased at least 30-fold from its nadir at stages XII–II to its maximum at stages VI–VII (6).

As subsequent studies revealed that CP-2 was secreted in low amounts by immature Sertoli cells in vitro, we proposed that the apparent stage-specific secretion of CP-2 by cultured seminiferous tubules resulted from the stage-specific synthesis of the protein by Sertoli cells and reflected the effect of germ cells at a particular phase of development on Sertoli cell function (10). Prompted by this hypothesis, we began an in-depth study of this protein with the intent of elucidating basic aspects of germ cell-Sertoli cell interactions and the consequences of these interactions to germ cell development. To pursue these studies, CP-2 was first purified to biochemical homogeneity from 220 mL of rat seminiferous tubule fluid, and the purified protein was used to generate a monospecific, polyclonal antiserum (7).

The availability of an antiserum to CP-2 allowed us to test directly the hypothesis that the apparent stage-specific secretion in vitro of CP-2 at stages VI and VII of the cycle resulted from the stage-specific synthesis of the protein and not from the stage-specific differences in its degradation or uptake within the cultured segments of tubule (7). Thus, seminiferous tubules at stages II, VI, VIIa,b, VIII, and XII of the cycle were cultured for 1 or 17 h with ^{35}S-methionine and protein immunoprecipitated from homogenates of the tubules (1-h sample) or from medium (17-h sample).

The results (Figs. 5.2A and 5.2B) demonstrated similar stage-specific patterns for the synthesis and secretion of CP-2. However, additional studies demonstrated that >17 h was required for CP-2 to be quantitatively exported out of the 2-mm segments of seminiferous tubule and into the culture medium (7). This suggested that the protein was quantitatively secreted by the Sertoli cells, but then reversibly bound to the plasma membranes of Sertoli or germ cells. It is noteworthy, as discussed in a later section of this chapter, that receptors for CP-2 have been identified on both Sertoli cells and germ cells (11).

A. Synthesis of CP-2/Cathepsin L

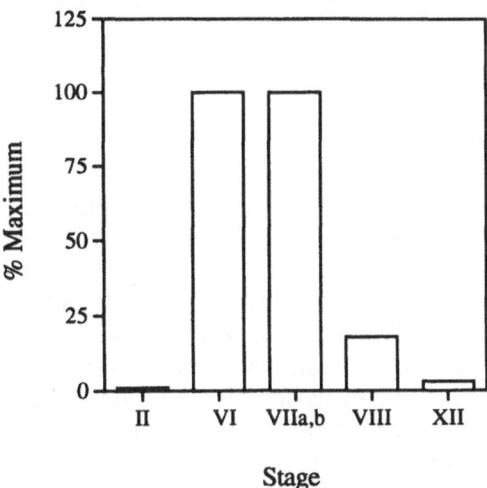

B. Secretion of CP-2/Cathepsin L

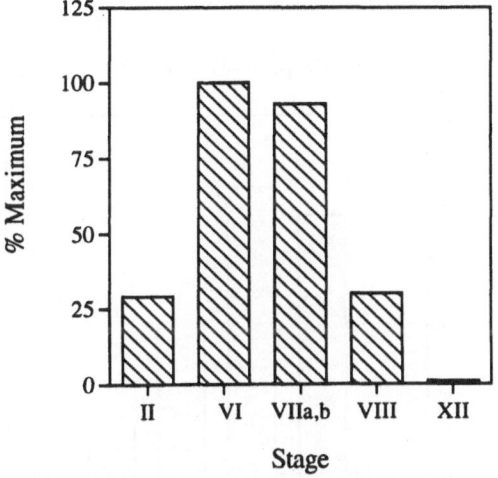

FIGURE 5.2. Stage-specific synthesis (*A*) and secretion (*B*) of CP-2/cathepsin L. Synthesis was measured by incubating 5-cm seminiferous tubules at stages V, VI, VIIa,b, VIII, and XII of the cycle for 1 h with ^{35}S-methionine, immunoprecipitating the protein from tubule homogenates, isolating the immunoprecipitated protein by SDS gel electrophoresis, and quantifying radioactivity with X-ray film and scanning densitometry. Data are expressed as the percent at the stage exhibiting maximum synthesis. Secretion was measured by incubating 5 cm of tubules at specific stages for 17 h and measuring the amount of radiolabeled CP-2/cathepsin L in the medium. Data are expressed as the percent at the stage exhibiting maximum secretion.

Cloning of a cDNA for CP-2 mRNA and Demonstration That This Protein Is the Proenzyme Form of Cathepsin L

To determine whether stage-specific synthesis of CP-2 resulted from the stage-specific expression of the gene encoding this protein and to explore the biochemical identity of CP-2, we first used our antiserum to screen a rat testis expression library to identify cDNAs that encode CP-2 mRNA (9). A 1.8-kb cDNA was cloned and used to analyze steady state levels of CP-2 mRNA at specific stages of the cycle. This analysis demonstrated a stage-specific pattern of expression of the transcript that was similar to what was observed for the synthesis and secretion of the protein (Fig. 5.3); CP-2 mRNA was barely detectable at stages II and XII, but was abundant at stages VI and VIIa,b (9). We therefore concluded that the stage-specific synthesis and secretion of this protein resulted from stage-specific changes in the steady state levels of CP-2 mRNA.

Sequence analysis of this cDNA revealed an open reading frame that encoded 247 amino acids and a long 3′ untranslated sequence (9).

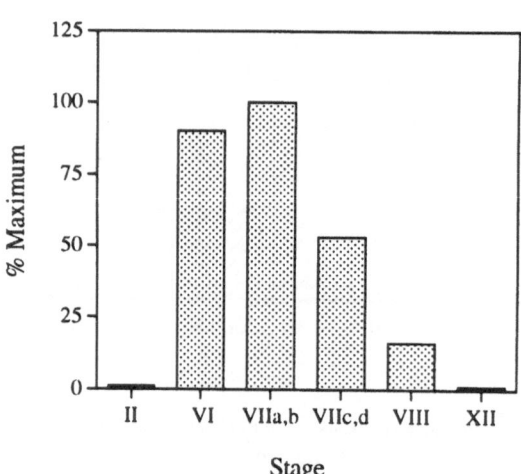

CP-2/Cathepsin L mRNA

FIGURE 5.3. Stage-specific expression of CP-2/cathepsin L mRNA. Two micrograms of total RNA from seminiferous tubules at stages II, VI, VIIa,b, VIIc,d, VIII, and XII of the cycle were fractionated on agarose gels, transferred to nitrocellulose, probed with ^{32}P− CP-2/cathepsin L cDNA, and exposed to X-ray film. Radioactivity was quantified by scanning densitometry of the X-ray film, and data are expressed as percent at the stage exhibiting maximum CP-2/cathepsin L mRNA content.

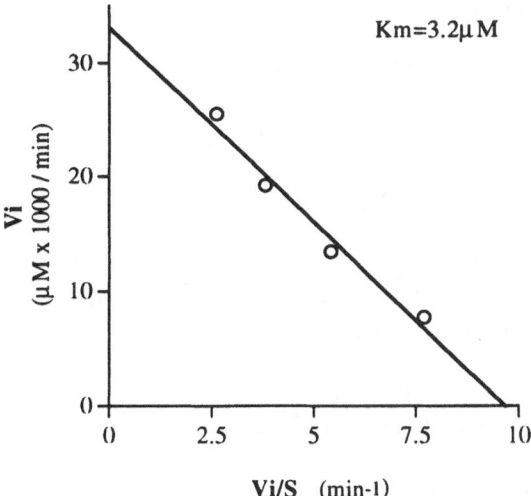

FIGURE 5.4. Eadie-Hofstee analysis of the affinity of CP-2/cathepsin L for the fluorescent substrate Z-phe-arg amino methyl coumarin. Highly purified CP-2/cathepsin L was activated by preincubation for 80 min at 37°C in 100 mM sodium formate and 1 mM EDTA, pH = 3.0. Z-phe-arg amino methyl coumarin was then titrated against the activated protein, and the mixture was incubated at 30°C for 2–8 min. The initial velocities of the formation of the fluorescent product, amino methyl coumarin, were calculated for each substrate concentration and K_m of activated CP-2/cathepsin L for the substrate determined by Eadie-Hofstee analysis.

Comparison of this sequence with all sequences in Genbank revealed that CP-2 was identical to the proenzyme form of the cysteine protease cathepsin L. This identification of CP-2 as procathepsin L was verified by immunoprecipitation of CP-2 with antiprocathepsin L and by the demonstration of the proteolytic activity of the purified protein (Fig. 5.4).

This identification of CP-2 as procathepsin L was of interest because a significant amount was known about the regulation and function of this protein in other biological settings. Procathepsin L is secreted by a number of transformed cell lines, and in some cases the level of cathepsin L enzyme activity in the cell correlates with the metastatic potential of the cells (12–14). Procathepsin L is also secreted by the endometrium of the cat uterus, as well as the murine placenta (15, 16). Finally, procathepsin L has been implicated in bone resorption (17). All of these examples involve instances of tissue reorganization and raise the question of whether procathepsin L had a similar function in the seminiferous epithelium.

Potential Functions of CP-2/Cathepsin L in the Seminiferous Epithelium

Our discussion of the potential functions of CP-2/cathepsin L in the seminiferous epithelium must address the following questions: (i) Where is CP-2/cathepsin L localized in the seminiferous epithelium; (ii) what are the possible biological consequences of this localization; and (iii) how can a protease with an ability to degrade a large number of proteins have a specific and limited biological function in this tissue?

Immunocytochemical studies by Zabludoff et al. (8) demonstrated that in the rat CP-2/cathepsin L was detectable only in stage V–VIII seminiferous tubules and at stages V–VII was concentrated around the heads of the compacted spermatids. As noted in Figure 5.1, from stage V to stage VII, the compacted spermatids move from deep crypts, formed by infoldings of Sertoli cell plasma membrane, to the apical surface of the Sertoli cell. The colocalization of CP-2/cathepsin L with the spermatids at stage V to the end of stage VII led us to suggest that CP-2 was targeted at the compacted spermatids and facilitated the movement of these spermatids toward the apical surface of the epithelium, a process that is a prerequisite for spermiation (8, 9).

The hypothesis that CP-2/cathepsin L facilitates the translocation of compacted spermatids within the seminiferous epithelium should be placed in the context of recent observations on adhesion of compacted spermatids to Sertoli cells and other proposed mechanisms for the movement of these spermatids within the epithelium. Adhesion of Sertoli cells to compacted spermatids occurs primarily in the region of the head of the developing spermatid and appears to be mediated by β1 integrins that are located in this area and persist up to the end of stage VII (18).

This site of adhesion is associated with a specific structure within the Sertoli cell, the ectoplasmic specialization, that resides immediately beneath the Sertoli cell plasma membrane and consists of a sequence of hexagonally packed actin filaments adjacent to a fenestrated endoplasmic reticulum (19). From stages V to VII, the ectoplasmic specialization moves toward the apical pole of the Sertoli cells along with the compacted spermatids; the structure is lost at the end of stage VII and replaced by tubulobulbar complexes of the Sertoli cell (20).

Vogel has proposed that the movement of the compacted spermatids from stage V to stage VII is due to the action of mechanoenzymes that slide the ectoplasmic specialization along microtubules in the Sertoli cell (19, 21). This hypothesis was recently challenged by the observation that the only two mechanoenzymes known to cause movement along microtubules, cytoplasmic dynein and kinesin, are not associated with the ectoplasmic specialization of stage V–VII tubules (22, 23). In contrast, both mechanoeyzymes are found in substantial amounts in the ectoplasmic specializations at stages IX–XIV, the stages when the bundles of compacted spermatids form and move into the crypts formed by Sertoli cells (23). Thus, it seems unlikely that kinesin or cytoplasmic dyneins cause the movement of compacted spermatids from deep crypts to the apex of the Sertoli cell.

Therefore, this movement at stages V–VII must be the result of complex biochemical and cellular processes and may involve the function of the secreted proenzyme CP-2/cathepsin L. Based on the correlation between the movement of the compacted spermatids at stages V–VII, the stage-specific secretion and distribution of CP-2/cathepsin L, and the loss of the ectoplasmic specialization and adhesion molecules at the end of stage VII, we suggest that this proenzyme has one or more of the following potential functions in the seminiferous epithelium: (i) to degrade the adhesion molecules that tether the spermatids to the Sertoli cells, (ii) to modify the plasma membrane of Sertoli cells or spermatids in a way that promotes movement of spermatids, and (iii) to cleave cell surface molecules that in turn stimulate second-messenger systems in a manner analogous to what has been reported for the thrombin receptor (24). Insight into these possibilities requires localization of CP-2/cathepsin L in the seminiferous epithelium at the ultrastructural level, experiments that are currently under way in our laboratory.

A second question arises in considering the function of this protein in the seminiferous epithelium: How can a protease with an ability to cleave a number of proteins produce a specific and limited effect within the seminiferous epithelium? A number of lines of evidence, summarized in Figure 5.5, suggest that a limited and specific function is possible.

First, as discussed above, CP-2/cathepsin L is synthesized at a few stages of the cycle and apparently targeted to a specific subcellular domain, the space between the compacted spermatids and Sertoli cells. Second, cell surface mannose 6-phosphate receptors have been identified on Sertoli cells, spermatocytes, and spermatids (11). While the distribution of such receptors within the tissue is currently unknown, if they are localized in the region surrounding the compacted spermatids, these receptors hold promise for concentrating the protease around the compacted spermatids.

Additionally, two inhibitors of cathepsin L are synthesized in the seminiferous epithelium, α2 macroglobulin and cystatin C (25, 26). Thus, any activated CP-2/cathepsin L in the lumen of the seminiferous epithelium will have its proteolyic activity neutralized. The potential importance of these inhibitors is suggested by the recent demonstration by Tsuruta and Griswold that cystatin C mRNA is expressed in a stage-specific manner that is the opposite for CP-2/cathepsin L (27). Thus, cystatin C mRNA levels are low at stages when CP-2/cathepsin L mRNA levels are high, and vice versa.

These results predict that at stages VI–VII, CP-2/cathepsin L that is secreted into the microenvironment separating the compacted spermatids and Sertoli cells will not have its enzyme activity inhibited. However,

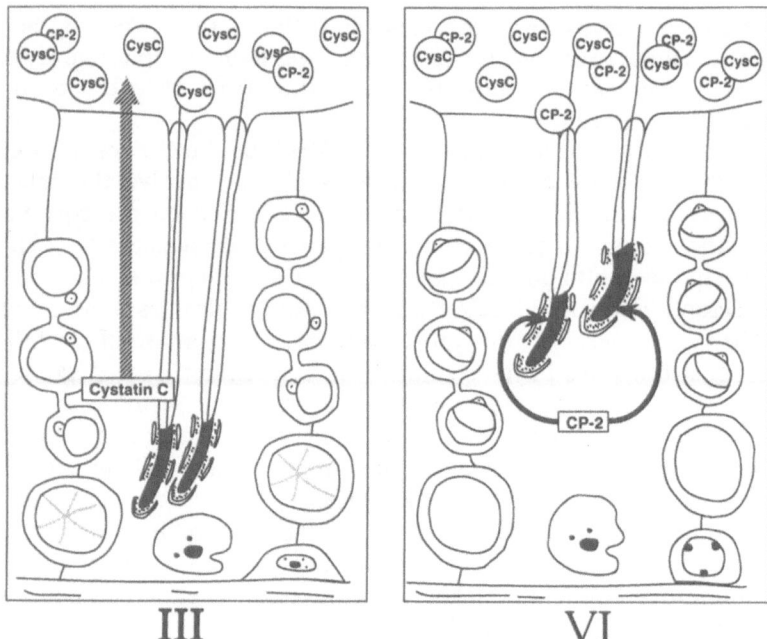

FIGURE 5.5. Proposal to explain how CP-2/cathepsin L may have a specific and limited function in the seminiferous epithelium. At most stages of the cycle (stage III is shown here as an example), the cysteine protease inhibitor, cystatin C (CysC), is synthesized by Sertoli cells and secreted into the lumen of the seminiferous tubule. However, at stages VI and VII, CP-2/cathepsin L (CP-2) is secreted into the microenvironment separating the Sertoli cells and compacted spermatids. Since cystatin C synthesis is minimal at these stages, CP-2/cathepsin L activity will not be inhibited in this microenvironment. However, once the proenzyme enters the fluid of the tubule, it is bound by the cystatin C in this fluid, and enzyme activity is inhibited. Thus, a specific and limited biological effect of CP-2/cathepsin L is possible because it is secreted into a microenvironment in the tubule at stages VI and VII and because its activity is inhibited by cystatin C in the luminal fluid.

once the protein enters the lumen of the seminiferous tubule, it will be bound by one of two inhibitors, either cystatin C or α2 macroglobulin. In summary, CP-2/cathepsin L may have a specific effect in the seminiferous epithelium that involves targeting of the protein to a specific domain of the seminiferous epithelium, concentration of the protein in that domain by cell surface receptors, and, finally, inhibition of the proteolytic activity of activated proenzyme in the lumen of the seminiferous tubule by two protease inhibitors.

Acknowledgments. The research in the author's laboratory is supported by NIH Grants HD-17989 and AG-08321 and Population Center Grant 2-P30 HD-6308.

References

1. Leblond CP, Clermont Y. Definition of the stages of the cycle of the seminiferous epithelium in the rat. Ann NY Acad Sci 1952;55:548–73.
2. Bardin CW, Cheng CY, Musto NA, Gunsalus GL. The Sertoli cell. In: Knobil E, Neil JD, eds. The physiology of reproduction; vol 1. New York: Raven Press, 1988:933–74.
3. Griswold MD. Protein secretions of Sertoli cells. Int Rev Cytol 1988;110: 134–56.
4. Skinner MK. Secretion of growth factors and other regulatory factors. In: Russell LD, Griswold MD, eds. The Sertoli cell. Clearwater, FL: Cache River Press, 1993:237–48.
5. Fritz IB, Tung PS, Ailenberg M. Proteases and antiproteases in the seminiferous epithelium. In: Russell LD, Griswold MD, eds. The Sertoli cell. Clearwater, FL: Cache River Press, 1993:217–36.
6. Wright WW, Parvinen M, Musto NA, et al. Identification of stage-specific proteins synthesized by rat seminiferous tubules. Biol Reprod 1983;29: 257–70.
7. Wright WW. Germ cell-Sertoli cell interactions: analysis of the biosynthesis and secretion of cyclic protein-2. Dev Biol 1988;130:45–56.
8. Zabludoff SD, Erickson-Lawrence M, Wright WW. Sertoli cells, proximal convoluted tubules in the kidney and neurons in the brain contain cyclic protein-2. Biol Reprod 1990;43:15–24.
9. Erickson-Lawrence M, Zabludoff SD, Wright WW. Cyclic protein-2, a secretory product of rat Sertoli cells, is the proenzyme form of cathepsin L. Mol Endocrinol 1991;5:1789–98.
10. Wright WW, Luzarraga ML. Isolation of cyclic protein-2 from rat seminiferous tubule fluid and Sertoli cell culture medium. Biol Reprod 1986;35: 761–72.
11. O'Brien DA, Gabel CA, Rockett DL, Eddy EM. Receptor-mediated endocytosis and differential synthesis of mannose 6-phosphate receptors in isolated spermatogenic and Sertoli cells. Endocrinology 1989;125:2973–84.
12. Sloane BF, Rozhin J, Moin K, Ziegler G, Fong D, Muschel RJ. Cysteine endopeptidases and their inhibitors in malignant progression of rat embryo fibroblasts. Biol Chem Hoppe Seyler 1992;373:589–94.
13. Brodt P, Reich R, Moroz LA, Chambers AF. Differences in the repertoires of basement membrane degrading enzymes in two carcinoma sublines with distinct patterns of site-selective metastasis. Biochim Biophys Acta 1992;1139: 77–83.
14. Doherty PJ, Hua L, Liau G, et al. Malignant transformation and tumor promoter treatment increase levels of a transcript for a secreted glycoprotein. Mol Cell Biol 1985;5:466–73.
15. Jaffe RC, Donnelly KM, Mavogianis PA, Verhage HG. Molecular cloning and characterization of a progesterone-dependent cat endometrial secretory

protein complementary deoxyribonucleic acid. Mol Endocrinol 1989;3: 1807–14.

16. Hamilton RT, Bruns KA, Delgado MA, et al. Developmental expression of cathepsin L and cRas in the mouse placenta. Mol Reprod Dev 1991;30: 285–92.

17. Krieger NS, Sukhatame VA, Bushinsky DA. Conditioned medium from rat oncogene-transformed NIH 3T3 cells induces bone resorption in vitro. J Bone Miner Res 1990;5:159–64.

18. Palombi F, Salanova M, Tarone G, Farini D, Stefanini M. Distribution of β1 integrin subunit in rat seminiferous epithelium. Biol Reprod 1992;47: 1173–82.

19. Vogel AW, Pfeiffer DC, Redenbach DM, Brove BD. Sertoli cell cyto-skeleton. In: Russell LD, Griswold MD, eds. The Sertoli cell. Clearwater FL: Cache River Press, 1993:40–86.

20. Russell LD. Spermatid-Sertoli tubulobulbar complexes as devices for elimi-nation of cytoplasm from the head region of late spermatids of the rat. Anat Rec 1979;194:233–46.

21. Redenbach DM, Boelkelheide K, Vogel AW. Binding between mammalian spermatid-ectoplasmic specialization complexes and microtubules. Eur J Cell Biol 1992;59:433–48.

22. McIntosh JR, Porter ME. Enzymes for microtubule-dependent motility. J Biol Chem 1989;264:6001–4.

23. Hall ES, Eveleth J, Jiang C, Redenbach DM, Boekelheide K. Distribution of the microtubule-dependent motors, cytoplasmic dynein and kinesin, in rat testis. Biol Reprod 1992;46:817–28.

24. Vu T-KH, Hung DT, Wheaton VI, Coughlin SR. Molecular cloning of a functional thrombin receptor reveals a novel proteolytic mechanism of receptor activation. Cell 1991;64:1057–68.

25. Cheng CY, Grima J, Stahler MS, Gugleilmotti A, Silvestrini B, Bardin CW. Sertoli cell synthesizes and secretes a protease inhibitor α_2-macroglobulin. Biochemistry 1990;29:1063–8.

26. Esnard A, Esnard F, Guillou F, Gauthier F. Production of the cysteine proteinase inhibitor cystatin C by rat Sertoli cells. FEBS Lett 1992;300:131–5.

27. Tsuruta KJ, Griswold MD. Sertoli cells and germ cells synthesize cystatin C. Biol Reprod 1991;44(suppl 1):101.

Part II

Functions of Somatic Cells in the Testis

6

Characterization, Regulation, and Tissue-Specific Expression of Steroidogenic Enzymes

ANITA H. PAYNE, GERI L. YOUNGBLOOD, MARÍA BURGOS-TRINIDAD, AND PAUL A. BAIN

Biosynthesis of gonadal steroid hormones from cholesterol requires the action of five enzymes (Fig. 6.1). The initial step in this pathway is the conversion of the C27 steroid, cholesterol, to the $C21-\Delta^5-3\beta$-hydroxysteroid, pregnenolone, that is catalyzed by the cytochrome P450 enzyme, cholesterol side-chain cleavage ($P450_{scc}$). This enzyme is located in the inner mitochondrial membrane. Pregnenolone diffuses across the mitochondrial membrane and is further metabolized by enzymes associated with the smooth endoplasmic reticulum.

In the mouse pregnenolone is first converted to the $C21-\Delta^4$-3-ketosteroid, progesterone, by the action of *3β-hydroxysteroid dehydrogenase/Δ^5-Δ^4-isomerase* (3βHSD). The next reaction is catalyzed by the cytochrome P450 enzyme, *17α-hydroxylase/C17-20 lyase* ($P450_{c17}$). This enzyme is a single protein that brings about two reactions; first, the hydroxylation of progesterone that is then followed by cleavage of the C17−20 bond to yield the C19 steroid, androstenedione, the immediate precursor of testosterone. The final reaction in the biosynthesis of testosterone is the reduction of the 17-ketone of androstenedione by *17-ketosteroid reductase* (17KSR). Androstenedione or testosterone is converted to the estrogen, estrone or estradiol, respectively, by the action of the cytochrome *P450 enzyme, aromatase* ($P450_{arom}$). This chapter discusses the chromosomal mapping and tissue-specific expression of mouse genes encoding steroidogenic enzymes, as well as the transcriptional regulation of *Cyp17*, the gene that encodes $P450_{c17}$.

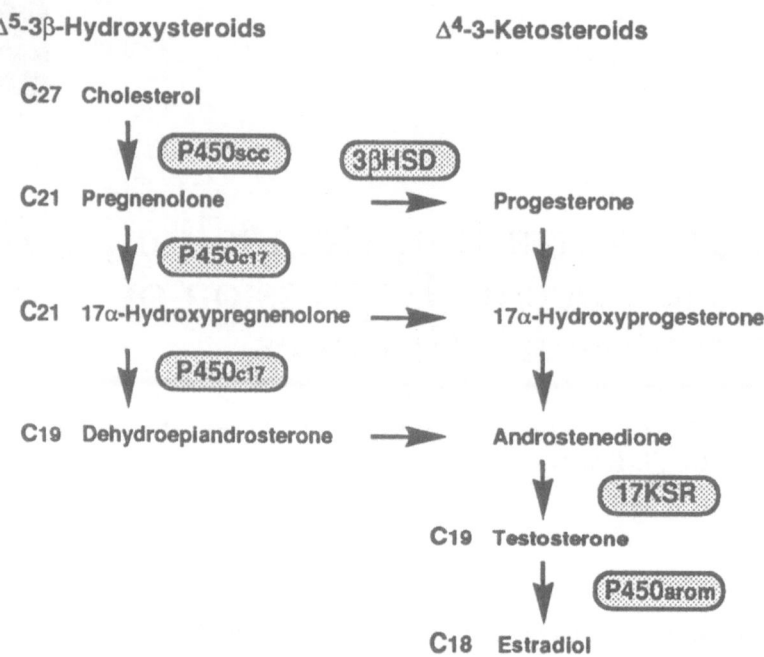

FIGURE 6.1. Steroid hormone biosynthesis from cholesterol in gonads. ($P450_{scc}$ = cytochrome P450 cholesterol side-chain cleavage; 3βHSD = 3β-hydroxysteroid dehydrogenase/Δ^5-Δ^4-isomerase; $P450_{c17}$ = cytochrome P450 17α-hydroxylase/C17-20 lyase; 17KSR = 17-ketosteroid reductase; $P450_{arom}$ = cytochrome P450 aromatase.)

Mouse Genes Encoding Steroidogenic Enzymes

Chromosomal Mapping

Table 6.1 lists mouse genes encoding steroidogenic enzymes whose chromosomal location has been determined in our laboratory. *Cyp11a* and *Cyp19* encode $P450_{scc}$ and $P450_{arom}$, respectively (1). Using *restriction fragment length variations* (RFLV) and recombinant strains of mice, we demonstrated that these two genes are found closely linked on mouse chromosome 9 (Fig. 6.2) (2). It was found that *Cyp11a* and *Cyp19* are most closely linked to another gene, *Cyp1a-1*, encoding a cytochrome P450 enzyme, P1450, that catalyzes aryl hydrocarbon hydroxylase (3). The structural genes for *CYP1A1* (4), *CYP11A1* (5), and *CYP19* (6) in the human genome have been mapped to chromosome 15 in a region that is conserved from mouse chromosome 9.

The structural gene encoding $P450_{c17}$ (*Cyp17*) was localized utilizing an interspecific testcross to mouse chromosome 19, distal to *Got-1* (7).

TABLE 6.1. Mouse genes encoding steroidogenic enzymes.

Gene	Enzyme	Tissue of expression
Cyp11a	P450$_{scc}$	Gonads, adrenal glands
Cyp17a	P450$_{c17}$	Gonads
Cyp19	P450$_{arom}$	Gonads, brain
Hsd3b-1	3βHSD I	Gonads, adrenal glands
Hsd3b-2	3βHSD II	Kidney (liver)
Hsd3b-3	3βHSD III	Liver (kidney)
Hsd3b-4	3βHSD IV	Kidney

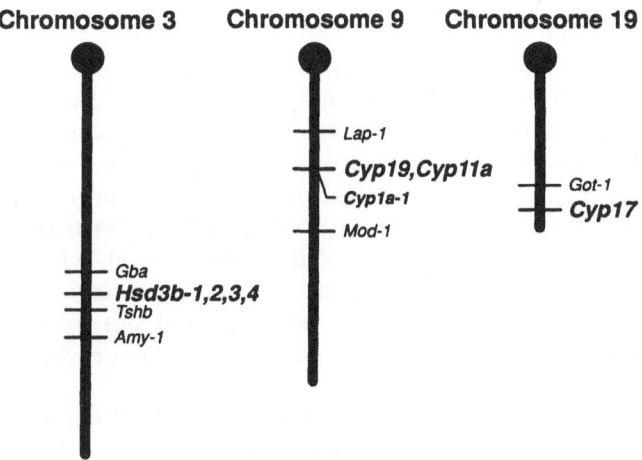

FIGURE 6.2. Illustration of chromosomal location of mouse genes that encode steroidogenic enzymes in relation to previously mapped loci on mouse chromosomes 3, 9, and 19. Hsd3b-1,2,3,4 encode 3βHSD I, II, III, and IV; Cyp19 encodes P450$_{arom}$; Cyp11a encodes P450$_{scc}$; and Cyp17 encodes P450$_{c17}$.

Recently, Fan et al. (8) localized the human structural gene for *CYP17* to chromosome 10q24.3 close to human *GOT1*. In humans, however, *CYP17* was found just proximal to *GOT1*, suggesting that an inversion took place during the karyotic evolution of human chromosome 10 (8).

As shown in Table 6.1 and Figure 6.2, we have identified four distinct genes in the mouse (*Hsd3b-1,2,3,4*) that encode different isoforms of the enzyme 3βHSD. We recently mapped the chromosomal location of these four genes by linkage analysis using gene-specific probes derived from the 3′ untranslated regions of the 3βHSD cDNA clones (9). The four *Hsd3b* genes were found to be closely linked within a segment of chromosome 3 between *Tshb* and *Gba*. This segment of mouse chromosome 3 shows conservation of gene order and physical distance with the centromeric region of human chromosome 1 (10). To date, two human *HSDB3* genes have been identified (11, 12). Human *HSDB3* has been mapped to 1p13 by in situ hybridization (13). Our results suggest that the human *HSDB3* genes will be found closely linked on the short arm of human chromosome 1 proximal to *TSHB*. More recently, using pulse field gel electrophoresis of mouse genomic DNA and specific probes representing each of the 3βHSD cDNA clones, we demonstrated that the four *Hsd3b* genes are found within a 400-kb segment, suggesting that they arose by tandem duplication (Bain, Payne, unpublished data).

Tissue-Specific Expression

The tissue-specific expression of the mouse steroidogenic enzymes is listed in Table 6.1. In a recent study we analyzed the promoter region of the mouse *Cyp11a* gene (13). Fragments containing different lengths, 5′ of the coding region, were subcloned into vectors containing the *chloramphenicol acetyl transferase* (CAT) reporter gene and were transiently transfected into mouse Y-1 adrenocortical cells or into mouse MA-10 Leydig cells. It had been reported previously that constructs containing 1500 bp of the 5′ flanking region of the *Cyp11a* gene when transfected into Y-1 adrenal cells exhibited increased basal expression of a growth hormone reporter gene. The same construct did not exhibit increased expression over that observed from a minimal promoter when transfected into the MA-10 Leydig cells (14).

This finding suggested that there may be additional sequences that are required for the expression of *Cyp11a* in Leydig cells. In Y-1 adrenal cells, sequences between −423 bp and −1200 bp are essential for basal expression (Fig. 6.3). In Leydig cells, very little expression of the *Cyp11a* gene constructs was observed in plasmids containing up to −2500 bp 5′ of the coding region. As shown in Figure 6.3, a marked increase in expression was observed in MA-10 Leydig cells with a construct that contained up to 5000 bp 5′ of the coding region. Transfection into MA-10 Leydig cells of a construct that had sequences between −198 bp (minimal promoter) and −2500 bp deleted exhibited an increase in basal expression similar to cells that had been transfected with the −5000 bp CAT construct. These results demonstrate that sequences between −2500 bp and −5000 bp of the coding region of the *Cyp11a* gene are essential for expression in Leydig cells, but not for expression in adrenal cells (Fig. 6.3). The data suggest that there may be cell-specific factors that regulate the expression in Leydig cells that differ from factors necessary for expression in adrenal cells.

In the mouse *Cyp17*, which encodes $P450_{c17}$, is only expressed in gonads and not in adrenal glands (15). *Cyp19*, which encodes $P450_{arom}$, is expressed in ovaries and testes as well as the brain (16). No expression was detected in other tissues of the mouse, such as placenta, adipose tissue, or liver (16).

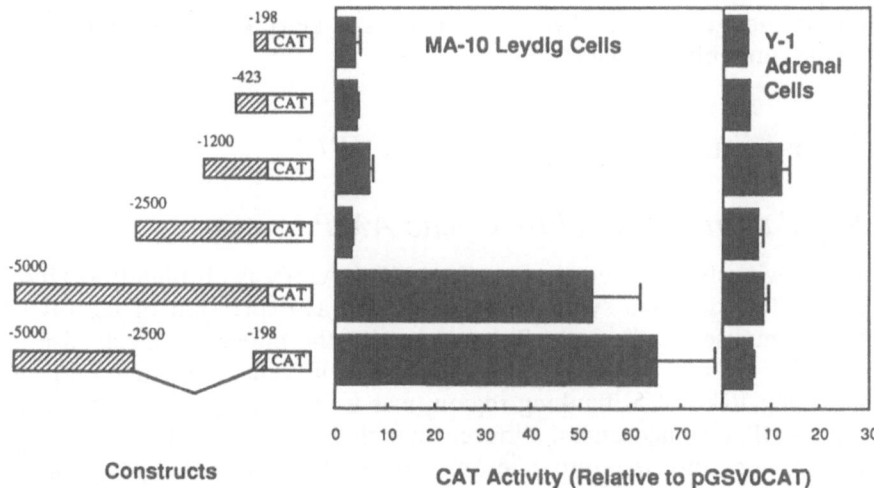

Constructs

CAT Activity (Relative to pGSV0CAT)

FIGURE 6.3. Identification of Leydig cell-specific sequences for the expression of mouse *Cyp11a*. Different-sized fragments 5' of the *Cyp11a* coding region were subcloned into the multiple cloning site of a promoterless plasmid (pGSVOCAT). The *Cyp11a* constructs are shown on the left-hand side of the figure. Ten micrograms of the indicated plasmid and 5 μg of SV2β-gal were transfected transiently into MA-10 Leydig tumor cells or Y-1 adrenocortical tumor cells. Cells were harvested 36 h following transfection, and CAT activity was determined in cell extracts by measuring the amount of [³H]acetylated chloramphenicol produced during 2 h for MA-10 cells or 1 h for Y-1 cells. Transfection efficiency was corrected for β-galactosidase activity. CAT activity is expressed relative to the promoterless pGSVOCAT plasmid. Reprinted with permission from Payne, Youngblood, Sha, Burgos-Trinidad, and Hammond (13).

As shown in Table 6.1, there are four distinct mouse *Hsdb3* genes that are expressed in a tissue-specific manner. Ribonuclease protection analysis using total RNA from various mouse tissues demonstrated that 3βHSD I is expressed only in the classical steroidogenic tissues, the gonads and adrenal glands of both sexes (17). 3βHSD II and III are expressed in kidney and liver, with greater expression of 3βHSD II in kidney than in liver (17). 3βHSD IV is exclusively expressed in the kidney of both male and female mice (Clarke, Bain, Payne, unpublished data). 3βHSD I and 3βHSD III proteins transiently expressed in nonsteroidogenic COS-1 cells can catalyze the conversion of the Δ^5-3β-hydroxysteroids, pregnenolone and dehydroepiandrosterone, to the Δ^4-3-ketosteroids, progesterone and androstenedione, respectively (18), and thus exhibit the enzyme activity attributed to this class of steroidogenic enzymes. 3βHSD IV protein, although 77% homologous to 3βHSD I, exhibits only 3-ketosteroid reductase activity and does not have the capacity to dehydrogenate and isomerize Δ^5-3β-hydroxysteroids to Δ^4-3-ketosteroids (Clarke, Bain, Payne, unpublished data).

Transcriptional Regulation of *CYP17*

Regulation of Cyp17 *by Cyclic AMP*

We previously reported that *cyclic AMP* (cAMP) is absolutely essential for P450$_{c17}$ de novo synthesis (19), as well as expression of its mRNA (20) in mouse Leydig cells. To characterize the molecular mechanism of cAMP regulation of *Cyp17* expression in Leydig cells, the structural gene, including 10 kb of 5' flanking region, was isolated (21). To test the gene for cAMP responsiveness, different-length fragments of the 5' flanking region were inserted into a promoterless pGem vector containing the CAT reporter gene (pGSVOCAT$_t$) (21). The resulting constructs were transiently transfected into MA-10 tumor Leydig cells and after 24 h were treated for 12 h with 500 μM cAMP.

Among the different constructs examined, the greatest cAMP-induced increase in *Cyp17*-CAT expression was observed with the construct that contained -1021 bp of the 5' flanking region (21). To identify further the sequences in the -1021 bp fragment that are essential for cAMP induction, deletions of this region were made. Transfection of these constructs into MA-10 cells and treatment with cAMP identified a *cAMP-responsive region* (CRR) between -346 and -245 bp (Fig. 6.4). Additional evidence that this 101-bp CRR confers cAMP responsiveness to the *Cyp17* promoter was obtained by inserting the CRR upstream of a heterologous promoter (thymidine kinase) and demonstrating that this region can confer cAMP responsiveness to the heterologous promoter (21). The -1021 bp fragment was sequenced, and no functional consensus *cAMP response element* (CRE) with the sequence TGACGTCA was identified in this fragment. Alignment of the mouse *Cyp17* sequence to the human (22) and bovine (23) *CYP17* genes starting at the transcription initiation sites through the reported CRRs showed that the positions of the CRRs within the three genes do not appear to align. Computer analysis of the CRRs per se indicates no apparent homology between the CRRs of these three *CYP17* genes (21).

The absence of a consensus CRE in the CRR of the mouse *Cyp17* gene suggests that cAMP induction of mouse *CYP17* does not involve the *CRE binding protein(s)* (CREB). CREB or CREB-like proteins are ubiquitous proteins found in most cells. The cAMP induction of genes containing CRE sequences exhibits rapid alterations in the rate of transcription (24). This increase in rate of transcription appears to be insensitive to cycloheximide and seems to involve rapid modification of a preexisting protein

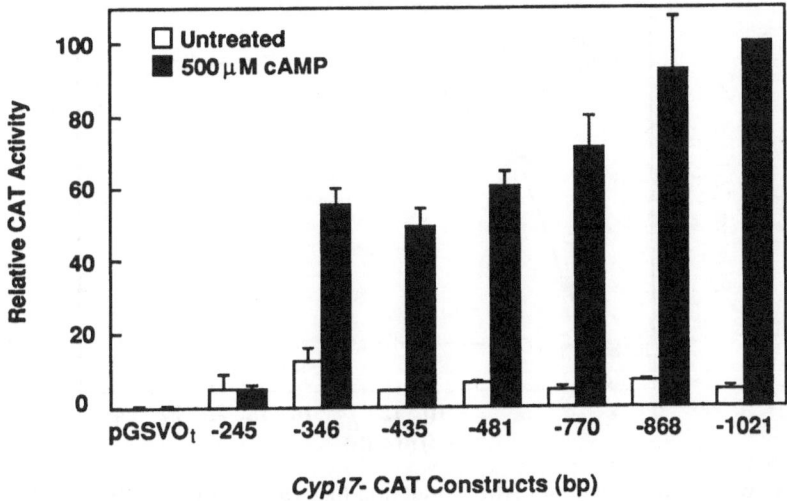

FIGURE 6.4. Identification of sequences responsible for cAMP induction in the 5′ flanking region of the mouse *Cyp17*. MA-10 cells were transfected transiently in duplicate with the *Cyp17*-CAT constructs indicated. Cells were treated for 12 h in the absence or presence of 500 μM 8-bromo-cAMP. The relative expression of −1021-bp-CAT was arbitrarily set at 100 for each experiment. The relative CAT activity represents the mean ± SE for 3 (pGSVOt, −245, −770, and −1021) or 2 (−346, −435, −481, and −868) independent experiments. Transfection efficiency was corrected for β-galactosidase activity. Reprinted with permission from Youngblood and Payne (21).

to a more active form (25). Induction by cAMP of the −1021 bp *Cyp17*-CAT construct in MA-10 Leydig cells produced no increase in CAT expression after 2 h of treatment with cAMP, with a maximal increase observed after 12 h of treatment (Fig. 6.5) (21). This finding, plus the demonstration that cAMP induction of $P450_{c17}$ mRNA in mouse Leydig cells requires protein synthesis (20), suggested that cAMP induction of *Cyp17*-CAT expression is dependent on a newly synthesized protein or proteins.

To determine whether cAMP treatment of MA-10 cells induced a protein that binds to the CRR of *Cyp17*, gel mobility shift assays were performed using nuclear extracts from cAMP-treated and nontreated MA-10 cells, as well as COS-1 cells (a monkey kidney cell line). These nuclear extracts were incubated with the ^{32}P-labeled CRR (−346/−245 bp of *Cyp17*) or a 30-bp ^{32}P-labeled oligonucleotide containing a consensus CRE (H₂B-CRE) (26), and the bound and free DNA were separated by polyacrylamide gel electrophoresis. A DNA-protein complex was formed with the labeled CRR, as shown in Figure 6.6. A DNA-protein complex was also formed with nuclear extracts from nontreated MA-10 cells;

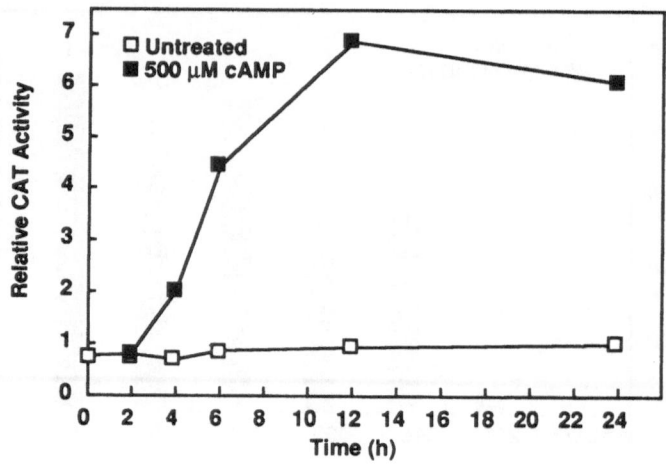

FIGURE 6.5. Time course of cAMP induction of *Cyp17*-CAT expression in MA-10 cells. Cells were transfected transiently with the −1021 bp *Cyp17*-CAT construct. Cells were untreated or treated with 500 μM 8-bromo-cAMP.

however, the intensity of the DNA-protein band from nuclear extracts of cAMP-treated cells was markedly enhanced (Fig. 6.5). There also appears to be a nuclear protein that is induced in COS-1 cells that binds to the CRR. This DNA-protein complex appears to have a somewhat different mobility than the nuclear extract from MA-10 cells. Incubation of a ^{32}P-labeled 30-bp oligonucleotide containing a consensus CRE with nuclear extracts from cAMP- and non-cAMP-treated MA-10 cells results in different mobility bands than those observed with the labeled CRR, and very little difference in the intensity of the shifted bands was observed between nuclear extracts from cAMP-treated and nontreated MA-10 cells. The two shifted bands observed with the CRE-containing oligonucleotide are believed to represent binding of CREB monomer (higher-mobility band) and CREB dimer (lower-mobility band).

To determine specificity of binding of the cAMP-induced protein to the CRR, competition by different nonlabeled fragments representing distinct segments of the CRR were performed (Fig. 6.7). Competition with 10 ng of unlabeled CRR (−346/−245) completely prevented binding to the labeled CRR. Competition with the 5′ fragment of the CRR containing sequences −346/−297 was also highly effective in preventing binding to the labeled CRR, but required somewhat larger amounts of nonlabeled competitor for total prevention of binding. In sharp contrast, no competition was observed at 100 ng of competitor with an oligonucleotide containing sequences −310/−281. Competition with the 3′ fragment of the CRR containing sequences −296/−245 also effectively competed with

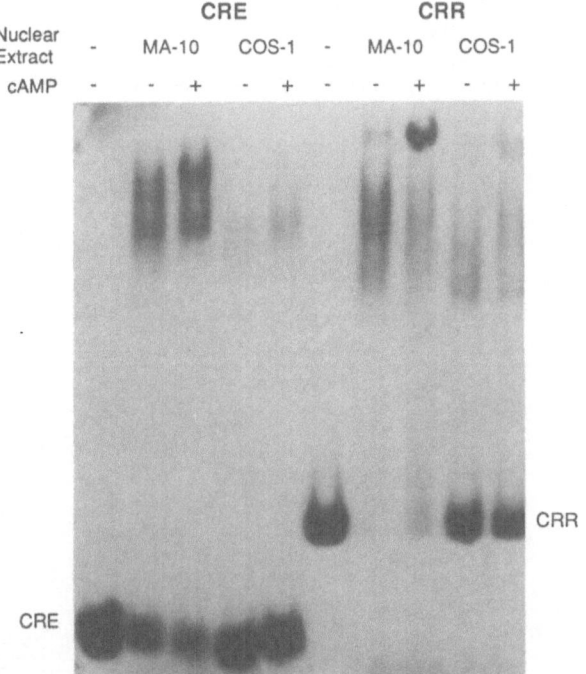

FIGURE 6.6. Binding of nuclear proteins from untreated or cAMP-treated MA-10 Leydig or COS-1 kidney cells to the −346/−245 *Cyp17* fragment (CRR) or H2B-CRE (CRE). The ^{32}P-end-labeled CRE and CRR were incubated with nuclear proteins (5 µg) isolated from MA-10 Leydig cells or COS-1 kidney cells that had been untreated or treated for 12 h with 500 µM 8-bromo-cAMP. The resulting protein-DNA complexes were separated on a 4% polyacrylamide gel containing 2.5% glycerol.

the labeled CRR, but to a somewhat lesser degree than the 5' segment containing the sequences −346/−297. Competition with unlabeled CRE was totally ineffective as a competitor for nuclear protein binding to the CRR.

These results demonstrate that the binding of proteins in the nuclear extract from cAMP-treated MA-10 cells to the CRR is specific and suggest that there may be two distinct binding sites that mediate cAMP responsiveness in *CYP17*-CAT reporter constructs containing the −346/−245 sequences. Although no competition was observed with an oligonucleotide containing sequences between −310 and −281, it cannot be ruled out at present that nucleotides in this region do not contribute to the binding of the induced nuclear protein(s).

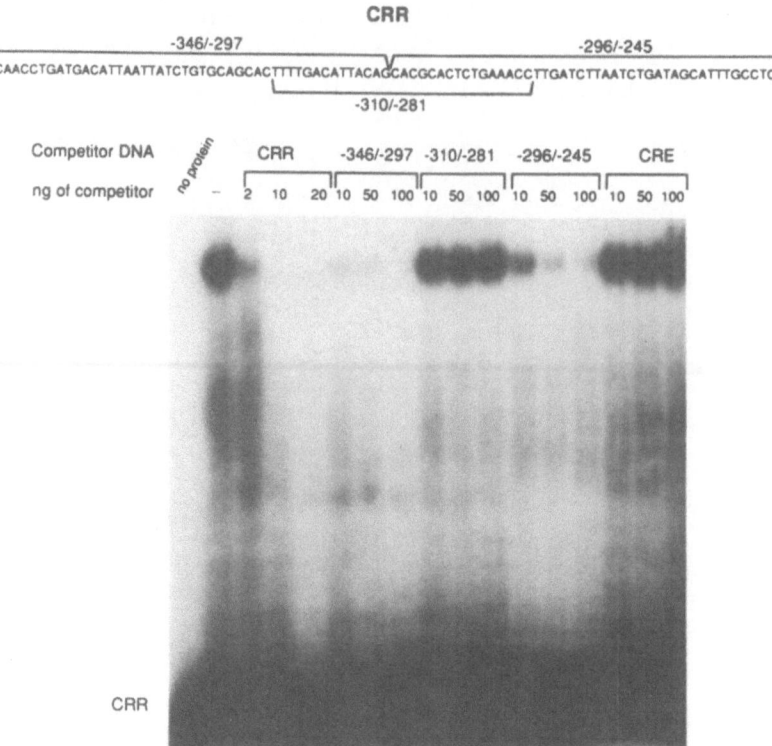

FIGURE 6.7. Competition of ³²P-labeled CRR binding to MA-10 cell nuclear extracts by different CRR fragments or CRE. The ³²P-end-labeled CRR was incubated with nuclear extracts from cAMP-treated MA-10 cells as described in Fig. 6.6 in the presence of increasing amounts of unlabeled CRR fragments as indicated or a 30-bp oligonucleotide containing CRE (CRE). The sequence of the CRR is shown above the gel, with the sequences used as competitor DNA fragments indicated.

Cell-Specific Expression of Cyp17

To examine the cell specificity of expression of *Cyp17*, the −1021-bp *Cyp17*-CAT construct was transfected into Y-1 mouse adrenal cells, another steroidogenic cell line, and into a nonsteroidogenic cell line, CV-1 cells, a transformed monkey kidney cell line. Expression of CAT activity was determined in cells not treated or treated with 500 µM cAMP for 12 h. Figure 6.8 shows that constructs containing different-length fragments of the *Cyp17* promoter were expressed in Y-1 mouse adrenal cell lines, with the highest basal and cAMP-induced expression observed with the construct containing the −1021-bp fragment of *Cyp17*. Neither basal nor cAMP-induced expression was observed when the −1021-bp

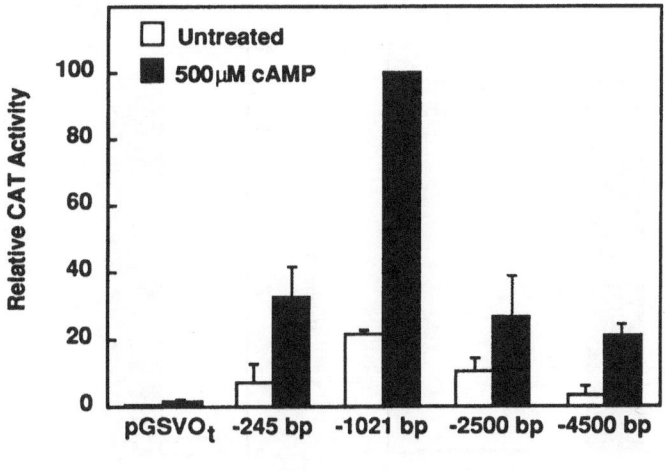

FIGURE 6.8. Expression of *Cyp17*-CAT constructs in Y-1 mouse adrenal cells. Cells were transfected transiently in duplicate with the constructs indicated. Cells were treated in the absence or presence of 500 μM 8-bromo-cAMP for 12 h. The relative expression of −1021-bp CAT was arbitrarily set as 100 for each experiment. The relative CAT activity represents the average for 2 independent experiments, mean ± SE.

Cyp17-CAT construct was transfected into the nonsteroidogenic CV-1 cell line (Fig. 6.9). To demonstrate that the nonexpression of the *Cyp17*-CAT construct in CV-1 cells was specific, another CAT construct containing the CRE of the rat *corticotropin-releasing hormone* (−238-bp CRH) (27) was transfected into CV-1 cells. The data presented in Figure 6.9A demonstrate that the −238-bp CRH-CAT construct was expressed in CV-1 cells, and expression was increased when cells were treated with cAMP. Figure 6.9B shows expression of both the −238-bp CRH-CAT and the −1021-bp *Cyp17*-CAT constructs in MA-10 cells.

Androgen Receptor-Mediated Repression of cAMP-Induced Expression of Cyp17

Previous studies from our laboratory demonstrated that testosterone or the androgen agonist mibolerone repressed cAMP induction of $P450_{c17}$ protein synthesis (28) and mRNA (20) in mouse Leydig cells. The repression by testosterone of cAMP-induced $P450_{c17}$ protein synthesis was prevented by the androgen antagonist hydroxyflutamide (28). These data indicated that endogenous testosterone produced when mouse Leydig cells are treated with cAMP can regulate the amount of $P450_{c17}$ pro-

120   A.H. Payne et al.

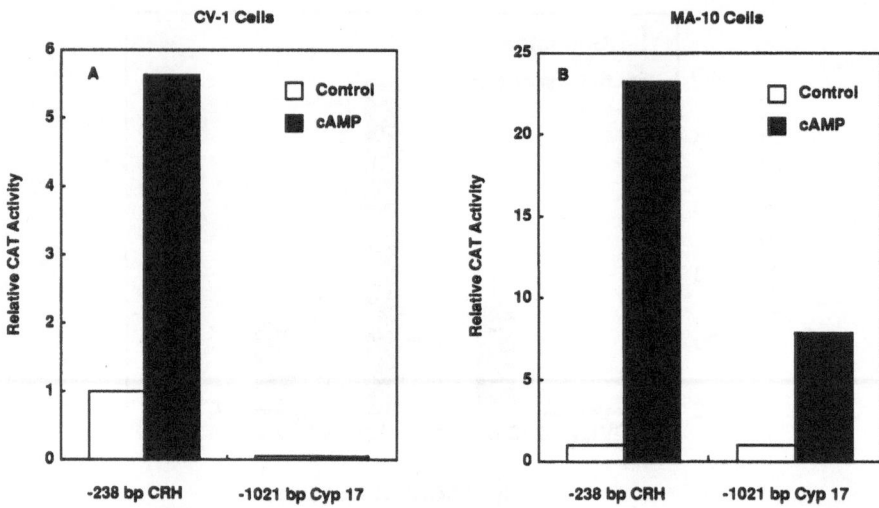

FIGURE 6.9. Cell-specific expression of *Cyp17*. CV-1 (*A*) cells or MA-10 (*B*) cells were transfected with SV2β-gal (4 μg) and either the −238-bp rat corticotropin-releasing hormone (CRH)-CAT construct (5 μg) or the −1021-bp *Cyp17*-CAT construct (5 μg), as described earlier. Cells were untreated or treated for 12 h with 500 μM 8-bromo-cAMP.

tein synthesis and mRNA levels by an androgen receptor-mediated mechanism.

To examine whether the androgen receptor-mediated repression of cAMP-induced $P450_{c17}$ protein synthesis and mRNA levels occurred at the level of transcription, experiments were carried out in MA-10 Leydig cells transfected with the −1021-bp *Cyp17*-CAT construct. MA-10 cells do not express endogenous $P450_{c17}$ mRNA (13) and thus produce progesterone rather than testosterone when stimulated with human chorionic gonadotropin (29) or with cAMP (30). MA-10 cells also do not express a functional androgen receptor (13). To examine the effect of androgens on cAMP induction of *Cyp17*-CAT, MA-10 cells were cotransfected with the −1021-bp *Cyp17*-CAT construct (5 μg) and with increasing amounts of a *mouse androgen receptor* (mAR) expression plasmid (31). MA-10 cells were treated for 12 h with the androgen *dihydrotestosterone* (DHT), cAMP, or cAMP plus DHT. Figure 6.10 illustrates that increasing amounts of mAR resulted in a dose-dependent repression of cAMP-induced *Cyp17*-CAT expression in the absence of DHT. The addition of DHT resulted in a further decrease in CAT activity. DHT treatment alone had little effect on *Cyp17*-CAT expression.

We examined whether the dose-dependent repression by the mAR in the absence of added DHT was due to the large amounts of progesterone produced during cAMP treatment of the MA-10 cells. It has been reported

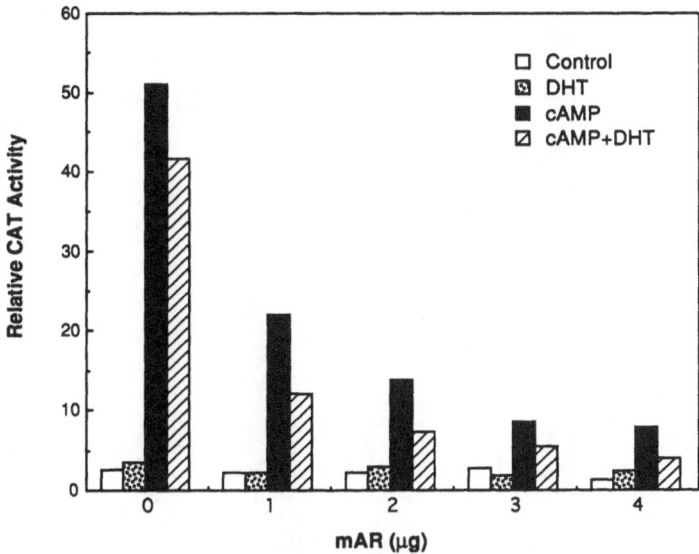

FIGURE 6.10. Effect of mAR on the cAMP-induced expression of *Cyp17*. MA-10 cells were cotransfected by the calcium phosphate precipitation method with a −1021-bp *Cyp17*-CAT construct (5 μg), SV2β-gal (4 μg), and increasing amounts of mAR expression plasmid as indicated. All cultures were corrected to the same total amount of DNA using the parent mAR expression vector lacking the mAR sequences. After 24 h cells were untreated or treated for 12 h with 0.1 μM DHT, 500 μM 8-bromo-cAMP, or cAMP + DHT. CAT activity was determined, and the results are expressed relative to β-galactosidase activity.

that progesterone in high concentrations can elicit the same response as that observed with lower concentrations of DHT from the human androgen receptor (32). A 12-h treatment of MA-10 cells with cAMP resulted in an average accumulation of 5.6 ± 0.8 μM progesterone. The addition of 100 μM aminoglutethimide, together with 500 μM cAMP, reduced the accumulation of progesterone to nonstimulated values, 0.016 ± 0.005 μM and 0.015 ± 0.005 μM, respectively.

The prevention of cAMP-induced progesterone production by the addition of aminoglutethimide during cAMP treatment of MA-10 cells does not abolish the dose-dependent repression observed with increasing amounts of the androgen receptor expression vector in the absence of added DHT (Fig. 6.11). Cotransfection with a *rat androgen receptor* (rAR) expression vector (a gift from Drs. Frank S. French and Elizabeth M. Wilson) (33) at concentrations as low as 0.5 μg markedly repressed cAMP induction of −1021-bp *Cyp17*-CAT expression in the absence of added DHT (Fig. 6.11). In all instances, the addition of 0.1 μM DHT further repressed cAMP induction of the *Cyp17* construct. One microgram

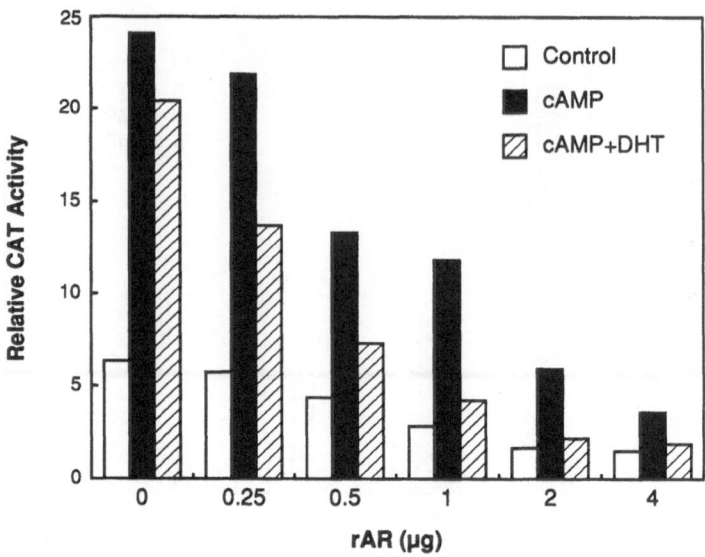

FIGURE 6.11. Effect of rAR on cAMP-induced expression of *Cyp17* in the presence of aminoglutethimide. MA-10 cells were cotransfected as described in Fig. 6.9 with a −1021-bp *Cyp17*-CAT construct, SV2β-gal (4 μg), and increasing amounts of rAR expression plasmid. All cultures were corrected for the same amount of total DNA using the parent rAR expression vector lacking the rAR sequences. All cultures were treated with 100 μM aminoglutethimide, AG (control), AG + 500 μM 8-bromo-cAMP, or AG + cAMP + DHT.

or greater amounts of cotransfected rAR plus added DHT reduced CAT expression to levels observed in the absence of cAMP treatment. In the absence of cotransfected mAR or rAR, the addition of DHT resulted in minimal repression of cAMP-induced expression of −1021-bp *Cyp17*-CAT. The data presented indicate that the androgen-mediated repression of $P450_{c17}$ protein synthesis and mRNA levels observed in mouse Leydig cells reflects androgen receptor-mediated repression of cAMP-induced transcription of mouse *Cyp17*. Future studies will examine the mechanism by which the androgen receptor represses cAMP induction of *Cyp17* expression.

Summary

This chapter reviews the chromosomal location and the tissue- and cell-specific expression of mouse steroidogenic enzymes. Data are presented that identify a CRR at −346/−245 bp relative to the transcription start site of the mouse *CYP17* gene. This CRR does not contain a consensus

CRE, TGACGTCA, suggesting that cAMP induction of *Cyp17* expression does not involve CREB or CREB-like proteins that are found in most cells. We provide evidence for the cAMP induction of a protein or proteins that bind specifically to the CRR and are different from proteins binding to a consensus CRE. This finding is consistent with the hypothesis that cAMP-induced expression of *Cyp17* in Leydig cells is mediated by a newly synthesized nuclear protein(s). *Cyp17* expression appears to occur only in steroidogenic cells, suggesting that there are specific transcription factors in steroidogenic cells that are not present in nonsteroidogenic cells. Data also are presented that show that repression of cAMP induction of *Cyp17* expression by androgens is mediated by the androgen receptor and occurs at the transcriptional level.

Acknowledgments. The authors gratefully acknowledge Dr. Hiroaki Nitta for assistance with some of the illustrations. We also thank Mr. Medardo Maroto for technical assistance and Ms. Rita Lemorie for typing the manuscript. These studies were supported by NIH Grants HD-08358 and HD-17916 to A.H. Payne. M. Burgos-Trinidad is a recipient of NRSA, F32 HD-07672. P.A. Bain was supported in part by NIH Training Grant HD-07048.

References

1. Nelson DR, Kamataki T, Waxman DJ, et al. The P450 superfamily: update on new sequences, gene mapping, accession numbers, early trivial names of enzymes, and nomenclature. DNA Cell Biol 1993;12:1–51.
2. Youngblood GL, Nesbitt MN, Payne AH. The structural genes encoding P450scc and P450arom are closely linked on mouse chromosome 9. Endocrinology 1989;125:2784–6.
3. Hildebrand CE, Gonzales FJ, Kozak CA, Nebert DW. Regional linkage analysis of the dioxin-inducible P-450 gene family on mouse chromosome 9. Biochem Biophys Res Commun 1985;130:396–406.
4. Hildebrand CE, Gonzales FJ, McBride OW, Nebert DW. Assignment of the human 2,3,7,8-tetrachlorodibenzo-p-dioxin-inducible cytochrome P1-450 gene to chromosome 15. Nucleic Acids Res 1985;13:2009–16.
5. Sparkes RS, Klisa KI, Miller WL. Regional mapping of genes encoding human steroidogenic enzymes: P450scc to 15q23–q25; and P450c17 to 10q24–q25. DNA Cell Biol 1991;10:359–65.
6. Chen S, Besman MJ, Sparkes RS, et al. Human aromatase: cDNA cloning, Southern blot analysis, and assignment of the gene to chromosome 15. DNA 1988;7:27–38.
7. Youngblood GL, Sartorius C, Taylor BA, Payne AH. Isolation, characterization and chromosomal mapping of mouse P45017α-hydroxylase/C17-20 lyase. Genomics 1991;10:270–5.
8. Fan Y-S, Sasi R, Winter JSD, Waterman MR, Lin CC. Localization of the human *CYP17* gene (cytochrome P45017α) to 10q24.3 by fluorescence in situ

hybridization and simultaneous chromosome banding. Genomics 1992;14: 1110–1.

9. Bain PA, Meisler MH, Taylor BA, Payne AH. The genes encoding gonadal and nongonadal forms of 3β-hydroxysteroid dehydrogenase/Δ^5-Δ^4 isomerase are closely linked on mouse chromosome 3. Genomics 1993;16:219–23.

10. Kingsmore SF, Moseley WS, Watson ML, Sabina RL, Holmes EW, Seldin MF. Long-range restriction site mapping of a syntenic segment conserved between human chromosome 1 and mouse chromosome 3. Genomics 1990; 7:75–83.

11. Lachance Y, Luu-The V, Labrie C, et al. Characterization of human 3-β-hydroxysteroid dehydrogenase/Δ^5-Δ^4-isomerase gene and its expression in mammalian cells. J Biol Chem 1990;265:20469–75.

12. Lachance Y, Luu-The V, Verreault H, et al. Structure of the human type II 3β-hydroxysteroid dehydrogenase/Δ^5-Δ^4 isomerase (3βHSD) gene: adrenal and gonadal specificity. DNA Cell Biol 1991;10:701–11.

13. Payne AH, Youngblood GL, Sha L, Burgos-Trinidad M, Hammond SH. Hormonal regulation of steroidogenic enzyme gene expression in Leydig cells. J Steroid Biochem 1992;43:895–906.

14. Rice DA, Kirkman MS, Aitken LD, Mouw AR, Schimmer BP, Parker KL. Analysis of the promoter region of the gene encoding mouse cholesterol side-chain cleavage enzyme. J Biol Chem 1990;265:11713–20.

15. Perkins LM, Payne AH. Quantification of P450scc, P45017α and iron sulfur protein reductase in Leydig cells and adrenals of inbred strains of mice. Endocrinology 1988;123:2675–82.

16. Harada N, Yamada K. Ontogeny of aromatase messenger ribonucleic acid in mouse brain: fluorometrical quantitation by polymerase chain reaction. Endocrinology 1992;131:2306–12.

17. Bain PA, Yoo M, Clarke T, Hammond SH, Payne AH. Multiple forms of mouse 3β-hydroxysteroid dehydrogenase/Δ^5-Δ^4 isomerase and differential expression in gonads, adrenal glands, liver, and kidneys of both sexes. Proc Natl Acad Sci USA 1991;88:8870–4.

18. Clarke TR, Bain PA, Sha L, Payne AH. Enzyme characteristics of two distinct forms of mouse 3β-hydroxysteroid dehydrogenase/Δ^5-Δ^4-isomerase complementary deoxyribonucleic acids expressed in COS-1 cells. Endocrinology 1993;132:1971–6.

19. Anakwe OO, Payne AH. Noncoordinate regulation of de novo synthesis of cytochrome P-450 cholesterol side-chain cleavage and cytochrome P-45017α-hydroxylase/C17-20 lyase in mouse Leydig cell cultures: relation to steroid production. Mol Endocrinol 1987;1:595–603.

20. Payne AH, Sha L. Multiple mechanisms for regulation of 3β-hydroxysteroid dehydrogenase/Δ^5-Δ^4-isomerase 17α-hydroxylase/C17-20 lyase cytochrome P450, and cholesterol side-chain cleavage cytochrome P450 messenger ribonucleic acid levels in primary cultures of mouse Leydig cells. Endocrinology 1991;129:1429–35.

21. Youngblood GL, Payne AH. Isolation and characterization of the mouse P45017α-hydroxylase/C17-20-lyase gene (*Cyp17*): transcriptional regulation of the gene by cyclic adenosine 3′,5′-monophosphate in MA-10 Leydig cells. Mol Endocrinol 1992;6:927–34.

22. Picado-Leonard J, Miller WL. Cloning and sequencing of the human gene for P450c17 (steroid 17α-hydroxylase/17,20 lyase) similarity with the gene for P450c21. DNA 1987;6:439–48.
23. Bhasker RC, Adler BS, Dee A, et al. Structural characterization of the bovine *CYP17* (17α-hydroxylase) gene. Arch Biochem Biophys 1989;271: 479–87.
24. Roesler WJ, Vandenbark GR, Hanson RW. Cyclic AMP and the induction of eukaryotic gene transcription. J Biol Chem 1988;263:9063–6.
25. Gonzales GA, Menzel P, Leonard J, Fischer WH, Montminy MR. Characterization of motifs which are critical for activity of the cyclic AMP-responsive transcription factor CREB. Mol Cell Biol 1991;11:1306–12.
26. Andrisani O, Dixon JE. Identification and purification of a novel 120-kDa protein that recognizes the cAMP-responsive element. J Biol Chem 1990; 265:3212–8.
27. Seasholtz AF, Thompson RC, Douglas JO. Identification of a cyclic adenosine monophosphate-responsive element in the rat corticotropin-releasing hormone gene. Mol Endocrinol 1988;2:1311–9.
28. Hales DB, Sha L, Payne AH. Testosterone inhibits cAMP-induced de novo synthesis of Leydig cell cytochrome P-45017α by an androgen receptor-mediated mechanism. J Biol Chem 1987;262:11200–6.
29. Ascoli M. Characterization of several clonal lines of cultured Leydig tumor cells: gonadotropin receptors and steroidogenic responses. Endocrinology 1981;108:88–95.
30. Hales DB, Sha L, Payne AH. Glucocorticoid and cyclic adenosine 3'5'-monophosphate-mediated induction of cholesterol side-chain cleavage cytochrome P450 (P450scc) in MA-10 tumor Leydig cells: increases in mRNA are cycloheximide sensitive. Endocrinology 1990;126:2800–8.
31. He WW, Fischer LM, Sun S, et al. Molecular cloning of androgen receptors from divergent species with a polymerase chain reaction technique: complete cDNA sequence of the mouse androgen receptor and isolation of androgen receptor cDNA probes from dog, guinea pig and clawed frog. Biochem Biophys Res Commun 1990;171:697–704.
32. Kemppainen JA, Lane MV, Sar M, Wilson EM. Androgen receptor phosphorylation, turnover, nuclear transport and transcriptional activation: specificity for steroids and antihormones. J Biol Chem 1992;267:968–74.
33. Tan J, Joseph DR, Quarmby VE, et al. The rat androgen receptor: primary structure, autoregulation of its messenger ribonucleic acid and immunocytochemical localization of the receptor protein. Mol Endocrinol 1988;2:1276–85.

7

Molecular Genetics, Structure-Function Relationships, and Tissue-Specific Expression and Regulation of the 3βHSD Gene Family

FERNAND LABRIE, JACQUES SIMARD, VAN LUU-THE, and GEORGES PELLETIER

Despite its essential role in the biosynthesis of all classes of hormonal steroids, structure of the *3β-hydroxysteroid dehydrogenase/Δ^5-Δ^4-isomerase* (3βHSD) gene family was only recently elucidated (1). This enzyme, as illustrated in Figure 7.1, is required for the biosynthesis of progesterone, glucocorticoids, and mineralocorticoids, as well as androgens and estrogens. 3βHSD is found not only in classical steroidogenic tissues—namely, the placenta, adrenal cortex, ovary, and testis—but also in several peripheral tissues, including the skin, adipose tissue, breast, lung, endometrium, prostate, liver, kidney, epididymis, and brain (1). The widespread distribution of 3βHSD expression indicates that this enzyme is likely to play an important role in the intracrine (2) formation of sex steroids in peripheral target tissues. Such a high level of extragonadal formation of sex steroids is especially important in the human and some other primates whose adrenals secrete large amounts of the precursor sex steroids *dehydroepiandrosterone* (DHEA) and especially DHEA sulfate (2–5).

It is remarkable that humans, in addition to possessing a highly sophisticated endocrine system, have largely vested sex steroid formation in peripheral tissues. In fact, while the ovaries and testes are the exclusive sources of androgens and estrogens in the lower mammals, the situation is very different in higher primates, where active sex steroids are totally or in large part synthesized locally. This providies autonomous control to target tissues that are thus able to adjust the formation and metabolism of sex steroids to local requirements. The situation of a high secretion rate of adrenal precursor sex steroids in men and women is thus completely

126

different from current animal models used in the laboratory, namely, rat, mouse, guinea pig, and all others (except monkeys) where the secretion of sex steroids takes place exclusively in the gonads (2–5). In these animals no significant amounts of androgens and estrogens are left after castration.

Primates are thus unique in having adrenals that secrete large amounts of the precursor steroids DHEA and especially DHEA-S that are converted into *androstenedione* (Δ^4-dione) and then into potent androgens and estrogens in peripheral tissues, depending on the predominance of androgen- or estrogen-synthesizing enzymes in each tissue. Since local formation of androgens and estrogens is likely to play a major role in both normal and tumoral hormone-sensitive tissues, an important proportion of our research program has recently been devoted to this exciting and therapeutically promising area.

We and other groups have cloned cDNAs encoding human (6–8), rat (9–13), macaque (14), bovine (15), and mouse (16) 3βHSD and deduced the amino acid sequences of the corresponding proteins. Furthermore, the structure of two highly homologous human 3βHSD genes has been elucidated (17–19). Moreover, the localization and ontogeny of 3βHSD in the human adrenal, testis, ovary, and placenta, as well as in the same tissues in a rodent, have been studied (20–24).

Structure of Human 3βHSD Isoenzymes and Their Genes

Following purification of 3βHSD from human placenta and development of antibodies against the enzyme in rabbits (25, 26), we have isolated and characterized a first cDNA type (6) and its corresponding gene (17). These sequences have been confirmed by Lorence et al. (7, 19). Northern blot analysis of RNA from human adrenals, gonads, placenta, and mammary gland using human 3βHSD cDNA as a probe identified a single 1.7-kb mRNA species (7). However, the detection of multiple unexpected DNA fragments by Southern blot analysis of human genomic DNA and the heterogeneous clinical picture in classical 3βHSD-deficient patients (27, 28) suggested the presence of multiple 3βHSDs in the human. We thus screened a human adrenal λgt22A cDNA library with human placental cDNA clone hp3βHSD63. The second 3βHSD cDNA type was isolated and sequenced and was chronologically designated as human *type II 3βHSD* (3βHSD II) (8). The structure of the corresponding human 3βHSD II gene has been also elucidated (18) (Fig. 7.2). The human 3βHSD genes corresponding to the human cDNAs type I and II contain 4 exons and 3 introns within a total length of 7.8 kbp (Fig. 7.2). These genes were located by in situ hybridization at the p11–p13 region of chromosome 1 (29).

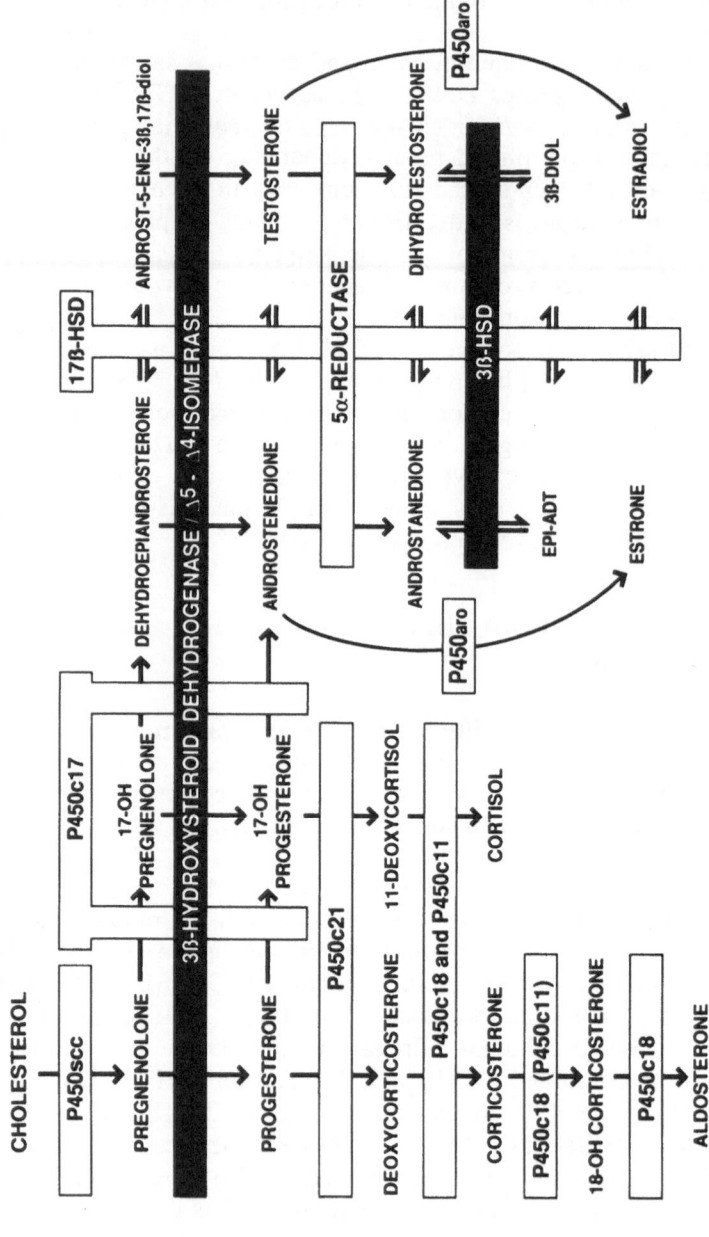

More recently, we also elucidated the structure of three human 3βHSD pseudogenes (30). Using specific cRNA probes, we have observed that the 3βHSD II mRNA population is the almost exclusive species detectable in the human adrenal gland, testis, and ovary. It was possible, however, after longer exposure times, to detect the presence of 3βHSD I-protected mRNA fragments in testis and ovary mRNA, thus indicating the presence of very low levels of 3βHSD I mRNA in the testis and ovary. However, even with overexposed autoradiographs, it was not possible to detect 3βHSD I mRNA in total RNA from either human adrenal or 3βHSD II RNA in human placenta (8). Furthermore, the 3βHSD I mRNA population corresponds to the sole detectable species in human placenta and skin under the experimental conditions used.

This enzyme family is well conserved throughout the course of evolution. An unrooted phylogenetic tree for the mammalian 3βHSD genes based on their coding sequences is illustrated in Figure 7.3. The human 3βHSD II differs from the macaque ovarian 3βHSD by only 13 residues, while 23 different residues distinguish the human 3βHSD II and 3βHSD I proteins. The human 3βHSD II and the macaque 3βHSD genes may have evolved from a common ancestral gene, while the human 3βHSD I gene may have evolved from another member of this gene family that has diverged after a duplication that took place earlier in evolution. The divergence of the liver-specific member (rat type III) of the rat 3βHSD family from other rodent 3βHSD genes is in agreement with its peculiar activity (31). In fact, the rat type III protein that shares 80% homology with the two other rat isoenzymes (9) possesses a unique substrate

◄―――

FIGURE 7.1. Schematic representation of the major mammalian steroidogenic pathways. All P450 are cytochrome enzymes. P450$_{scc}$ is the enzyme catalyzing 20α-hydroxylation, 22-hydroxylation, and hydrolysis of the C20-C22 carbon bond, which is termed cholesterol side-chain cleavage cytochrome P450 and is encoded by the CYP11A1 gene. The 3β-hydroxysteroid dehydrogenase/Δ^5-Δ^4 isomerases are non-P450 enzymes that consist of at least 2 functional type I and type II 3βHSD isoenzymes encoded by 2 homologous genes. P450$_{c21}$ is an enzyme mediating 21-hydroxylation that is encoded by the CYP21B gene. P450$_{c17}$ is an enzyme catalyzing both 17α-hydroxylase and 17-20-lyase activities that is encoded by the CYP17 gene. P450$_{11\beta}$ is an enzyme encoded by the CYP11B1 gene that exhibits 11β-hydroxylase activity exclusively. P450$_{c18}$ is an enzyme that mediates 11β-hydroxylation and the further reactions involved in the biosynthesis of aldosterone. It is encoded by the CYP11B2 gene that is expressed exclusively in the adrenal glomerulosa. 17βHSD, also termed 17KSR, is a non-P450 enzyme. Only the 17βHSD gene encoding the estradiol 17βHSD enzyme has been characterized. The 5α-reductase activity is catalyzed by at least 2 different isoenzymes encoded by the SRD5A1 and SRD5A2 genes. P450$_{arom}$ (P450$_{aro}$) is an enzyme that mediates the aromatization of Δ^4-3-keto C19 steroids into estrogens; it is encoded by the CYP19 gene.

Human Type I 3β-Hydroxysteroid dehydrogenase / Δ5-Δ4 isomerase

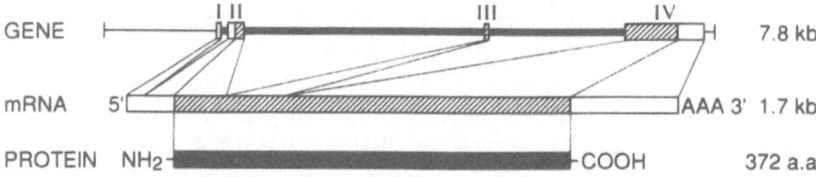

Human Type II 3β-Hydroxysteroid dehydrogenase / Δ5-Δ4 isomerase

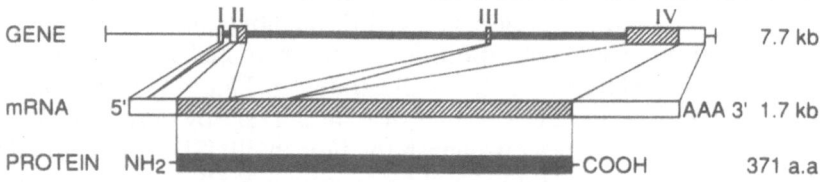

FIGURE 7.2. Structure of human 3βHSD I and 3βHSD II genes, mRNA species, and the corresponding proteins. Exons are represented by boxes in which hatched lines demarcate the coding regions, while open boxes represent the noncoding regions. Introns are represented by black bold lines.

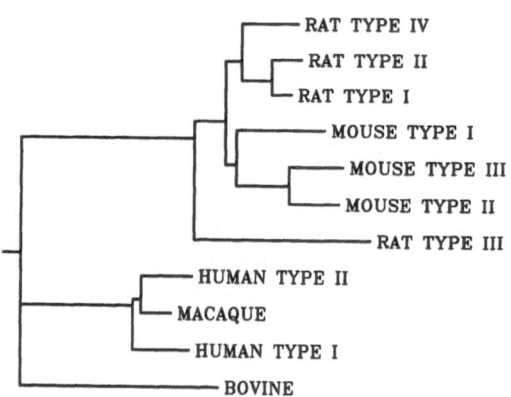

FIGURE 7.3. Unrooted phylogenetic tree for mammalian 3βHSD genes. Generation of the tree was based on the alignment of nucleotide sequences corresponding to the coding regions determined by the programs DNAPARS and FITCH in PHYLIP (Phylogeny Inference Package, version 3.4), generously provided by Dr. J. Felsentein, Dept. of Genetics, University of Washington, Seattle, USA.

specificity almost exclusively limited to the reduction of 3-keto-saturated steroids, such as *dihydrotestosterone* (DHT), Δ^4-dione, and dihydroprogesterone, into their corresponding 3β-hydroxysteroids using NADPH as preferred cofactor (31).

Enzymatic Characteristics of the Expressed Human 3βHSD I and 3βHSD II Isoenzymes

Transfection of HeLa cells with either human 3βHSD I or 3βHSD II cDNA inserts resulted in the production of a single 42-kd protein that cross-reacts with polyclonal antibodies raised against purified human placental 3βHSD and that comigrates with purified human placental 3βHSD (25, 26). We have recently provided evidence for distinct dehydrogenase and isomerase sites within a single 3βHSD protein (32). As revealed by transient expression in heterologous cells, the type I isoenzyme possesses a $3\beta HSD/\Delta^5\text{-}\Delta^4$ isomerase activity higher than type II, with respective K_m values of 0.24 and 1.2 μM for *pregnenolone* (PREG); 0.18 and 1.6 μM for DHEA; and 0.26 and 2.7 μM for DHT. The V_{max} of both types is, however, equivalent (8). These data also show that the affinity of the human 3βHSD II protein is similar for the three substrates tested. Similar 3βHSD activity for the three substrates is also found for the type I protein. The higher K_m value of 3βHSD II, mainly expressed in steroidogenic tissues, could be related to the higher levels of endogenous substrates present in these classical steroidogenic tissues. On the other hand, the approximately 10-fold-higher affinity of 3βHSD I that is preferentially expressed in peripheral intracrine tissues, such as the skin, could greatly facilitate steroid formation from the relatively low concentrations of substrates usually present in these tissues.

Localization and Ontogeny of 3βHSD in Human Testis

In the human male fetus, Leydig cells secrete testosterone that in turn promotes male differentiation of the Wolffian ducts, urogenital sinus, and external genitalia. The production of testosterone by the testis and the concentration of plasma testosterone in the fetus start to rise at the end of the second month of gestation and, shortly thereafter, attain high values that are maintained until late gestation, when they decrease (33). At the time of birth, plasma testosterone levels are very low; shortly afterwards, they begin to rise and remain elevated for approximately 3 months before falling to low levels by the age of 1 year (34–36). The plasma concentration of testosterone remains low until the onset of puberty, when the concentration increases again to reach adult levels by the age of 17 (37, 38).

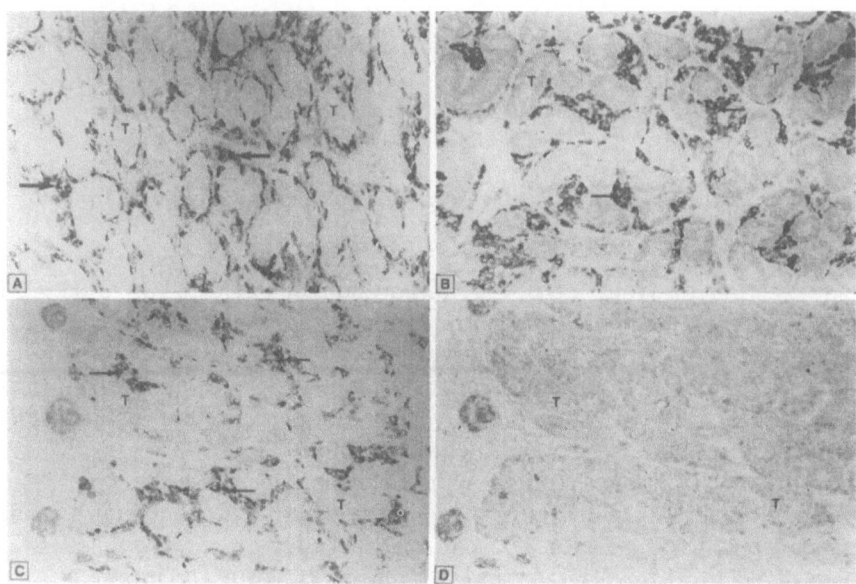

FIGURE 7.4. Sections through human fetal testes immunostained for 3βHSD (×200). *A:* 22 weeks of gestation. Immunostaining can be observed in the cytoplasm of interstitial cells (arrows). The tubules (T) are unstained. *B:* 28 weeks of gestation. Strong immunostaining is present in interstitial cells (arrows) that appear more prominent than those observed at 22 weeks of gestation. (T = tubules.) *C:* 31 weeks of gestation. Heterogeneous intensity of immunostaining of interstitial cells (arrows) can be observed. (T = tubules.) *D:* Shown is a control section adjacent to that shown in part C. Immunoabsorption with excess of human 3βHSD (10^{-6} M) completely prevented labeling. (T = tubules.) Reprinted with permission from Dupont, Luu-The, Labrie, and Pelletier (22).

In order to correlate possible changes in 3βHSD with the well-known variations in testosterone production during development, we have localized 3βHSD by immunocytochemistry during different fetal and postnatal periods of development in the human testis. In the testis of 22-week-old fetuses, immunostaining was observed exclusively in the cytoplasm of interstitial cells, whereas the seminiferous tubular elements remained completely unreactive (Fig. 7.4A) (22). At 28 weeks of fetal life, strong immunolabeling could be detected in the cytoplasm of interstitial cells (Fig. 7.4B). The seminiferous tubules remained unreactive to 3βHSD antibodies. Interestingly, the interstitial cells appeared much larger than those observed at 22 weeks of gestation. In fetuses of 31 weeks (Fig. 7.4C), immunostaining was also observed exclusively in the cytoplasm of interstitial cells. It was of interest to note heterogeneity in labeling; approximately half the interstitial cells were weakly labeled. In

8-month-old infants and during childhood, before puberty, no immuno-staining for 3βHSD could be observed in the testis.

With puberty, dramatic changes were observed in testicular immuno-staining. As observed in sections through a testis of a 15-year-old boy, the seminiferous tubules appeared well developed, and the interstitial cells contained abundant immunoreactive material. The same intensity of im-munostaining was also observed in adult testes (22). No staining could be obtained when preimmunized rabbit serum or the primary antiserum immunoabsorbed with excess of purified 3βHSD was used (Fig. 7.4D).

The present data clearly show that 3βHSD could be detected in Leydig cells in the fetus and that immunolabeling was absent at 8 months after birth, but became detectable at the time of puberty. These immunocyto-chemical results agree well with previous data from Baillie et al. (39), who used enzyme histochemistry to detect 3βHSD activity in testes of human fetuses from 8 to 22 weeks of gestation. Our results, which indicate that 3βHSD is present in fetal testis during the second and third trimester of gestation (22, 28, and 31 weeks), agree well with previous results demonstrating that in the human embryo, plasma testosterone levels are high during the second and third trimesters of gestation and decrease during the last weeks of gestation (33).

In summary, we have demonstrated variation in the testicular content of 3βHSD during pre- and postnatal development in the human. Since the pattern observed in 3βHSD content is similar to that observed for androgen production, it can be suggested that activation of 3βHSD by trophic hormones plays an important role in androgen production during fetal life and postnatal development with all the known consequences on sexual development and growth in the male.

Molecular Basis of Congenital Adrenal Hyperplasia Due to 3βHSD Deficiency

Congenital adrenal hyperplasia (CAH) is the most frequent cause of ambiguous genitalia and adrenal insufficiency in newborn infants (28, 40). CAH consists of a group of syndromes caused by specific enzymatic deficiencies in the adrenal cortex affecting the pathway of cortisol bio-synthesis, thus leading to decreased circulating cortisol levels and a com-pensatory hypersecretion of adrenocorticotropin by the anterior pituitary gland with consequent hyperplasia of the adrenal cortex. Although any of the five enzymatic steps involved in cortisol biosynthesis from cholesterol may be impaired in CAH, 21-hydroxylase deficiency accounts for 90% of the cases, while 11β-hydroxylase, 3βHSD, and 17α-hydroxylase deficien-cies account for the majority of the remaining cases, a side-chain cleavage defect being rare (28, 40).

In contrast to 21-hydroxylase and 11β-hydroxylase deficiencies that are adrenal-specific defects, 3βHSD deficiency impairs steroidogenesis in both the adrenals and the gonads, thus resulting in decreased secretion of not only cortisol and aldosterone, but also of progesterone, androgens, and estrogens by these tissues (27, 28, 40–42). Newborns affected by classical 3βHSD deficiency thus exhibit signs and symptoms of glucocorticoid and mineralocorticoid deficiencies that may be fatal if not diagnosed and treated early, especially in the severe salt-wasting form. Moreover, male newborns exhibit pseudohermaphroditism with incomplete masculinization of the external genitalia due to an impairment of androgen biosynthesis in the testis during fetal development. In contrast, affected females exhibit normal sexual differentiation or partial virilization (27, 28, 43–47).

We have recently elucidated the molecular basis of 3βHSD deficiency in 6 patients presenting the classical form of the disease (48, 49). We have thus analyzed the structure of the highly homologous I and II 3βHSD genes in 4 male pseudohermaphrodite 3βHSD-deficient patients from unrelated families, as well as in the 2 related classical 3βHSD-deficient females showing severe salt wasting, in order to elucidate the molecular basis of classical 3βHSD deficiency since these patients exhibited various degrees of severity of salt loss. We determined the nucleotide sequence of DNA fragments generated by selective polymerase chain reaction amplification that spanned the 4 exons, the exon-intron boundaries as well as the 5′ flanking region of each of the two 3βHSD genes in the 3 male patients. The 5 point mutations characterized were all detected in the 3βHSD II gene, which is the gene predominantly expressed in the adrenals and gonads, while no mutation was detected in the 3βHSD I gene, which is predominantly expressed in the placenta and peripheral tissues.

A homozygous nonsense W171X mutation leading to a predicted truncated protein of 169 amino acids (excluding the first Met) instead of the normal 371 residues of the 3βHSD II protein was detected in the 2 classical 3βHSD-deficient females showing severe salt wasting with a marked blockade of ovarian and adrenal steroidogenesis from the 2 related families previously described by Zachmann et al. (43–45, 48). Furthermore, elucidation of the molecular basis of the classical 3βHSD deficiency in a male patient previously described by Parks et al. (46) reveals that this index case is a compound heterozygote, being heterozygous for the W171X mutation while bearing the same frameshift mutation 186/insC/187 in his other 3βHSD II allele. The 2 other male patients suffering from severe salt-losing 3βHSD deficiency are also compound heterozygotes, one bearing the frameshift mutation 186/insC/187 and the missense mutation Y253N, while the other bears the nonsense mutation W171X and the missense mutation E142K (49).

The influence of the detected missense mutations on enzymatic activity was assessed by in vitro expression analysis of mutant recombinant enzymes generated by site-directed mutagenesis in heterologous mammalian cells. The recombinant mutant 3βHSD II enzymes carrying Y253N or E142K substitutions exhibit no detectable activity. On the other hand, the non-salt-losing patient is homozygous for the missense mutation A245P. This mutation decreases 3βHSD activity by approximately 90% (49). The present findings provide a molecular explanation for the enzymatic heterogeneity responsible for the severe salt-losing form down to the clinically inapparent form of salt loss in classical 3βHSD deficiency revealed exclusively by impaired development of secondary male sex organs. The impairment of steroid formation in both the adrenal and gonadal tissues of patients suffering from classical 3βHSD deficiency thus results from mutation(s) in the 3βHSD II gene causing various levels of impairment of enzymatic activity and, consequently, varying clinical severity of the disease. On the other hand, the finding of a normal 3βHSD I gene provides the basis for the well-recognized, intact, peripheral intracrine steroidogenesis in these patients.

In addition to providing a molecular explanation for the clinical and biological observations of classical 3βHSD deficiency with or without the symptoms of salt loss, the present findings provide the first information on the structure-function relationships of the 3βHSD isoenzyme family by demonstrating the crucial role of residues Glu^{142}, Tyr^{253}, and Ala^{245} for 3βHSD catalytic activity. The importance of the Glu^{142} and Tyr^{253} amino acid residues for 3βHSD activity is well supported by the observation that these two amino acids are conserved in all 3βHSD isoenzymes characterized thus far in the rat, mouse, bovine, macaque, and human (1), as well as in two other members of the 3βHSD superfamily; namely, the bacterial *Nocardia* cholesterol dehydrogenase (50) and the Vaccinia virus open reading frame SalF7L (50) encoding a protein possessing $3βHSD/\Delta^5$-Δ^4 isomerase activity.

Since Ala^{245} is not conserved in the 3βHSD superfamily, it is conceivable that the A245P mutation generating Pro^{245} causes a turn in the polypeptide chain, generating a structural change in the protein that could be responsible for the decrease in activity of this mutant 3βHSD II enzyme. The finding of no detectable 3βHSD activity in cell homogenates in the absence of glycerol (49)—while significant enzymatic activity could be measured in intact transfected cells, as well as in cell homogenates prepared in the presence of glycerol, a known stabilizing agent—could suggest that this mutation weakens the association of the mutant A245P enzyme with proper intracellular membranes. In fact, the absence of salt wasting in this patient could well be explained by the weak, but measurable 3βHSD activity that may permit the formation of mineralocorticoids in sufficient quantities to avoid severe salt wasting, in analogy with some non-salt-losing CAH cases secondary to 21-hydroxylase deficiency (40).

The expected severe inhibition of testosterone biosynthesis by the fetal testis affected by these point mutations provides an explanation for the incomplete masculinization of the external genitalia seen in these 4 male patients. In fact, in a normal male fetus, testicular 3βHSD activity starts to rise at the end of the second month of gestation and is responsible for the production of androgens that are required before 12 weeks of gestation for urethral-fold fusion and penile development. Testicular testosterone secretion then remains elevated during the second and third trimesters of gestation (22, 51, 52).

Ovarian and adrenal fetal 3βHSD activities, which also depend on 3βHSD II gene expression, begin to rise in the midsecond trimester of gestation (23, 51, 52). As frequently seen, the inhibition of 3βHSD II in the adrenals and ovaries was not accompanied by a noticeable alteration of the differentiation of the external genitalia of females patients, as indicated by the absence of ambiguity of external genitalia in the 2 female patients described by Zachmann et al. (43–45).

An elevated ratio of $\Delta^5:\Delta^4$ steroids is considered to be the best biological parameter for the diagnosis of 3βHSD deficiency. The increase in the levels of some Δ^4 steroids and their precursors is best explained by normal peripheral 3βHSD activity exerting its catalytic activity on the elevated levels of substrates for the 3βHSD I enzyme resulting from deficiency of 3βHSD II in the adrenal and the gonads. Although very low levels of 3βHSD I mRNA could be detected in normal gonads by the sensitive ribonuclease protection assay (8), Δ^4 steroids could also originate, to an unknown extent, from gonadal 3βHSD I activity, which possesses a relative 4.5- to 5.9-fold higher activity than that of the 3βHSD II isoenzyme and which could be stimulated following an increase of gonadotropin secretion resulting from low circulating androgen levels. While leading to a better understanding of the phenotypic heterogeneity of severe and mild forms of this autosomal recessive disorder, further characterization of the molecular basis of CAH due to 3βHSD deficiency should also provide important information concerning the structure-function relationships of the 3βHSD superfamily.

Structure and Tissue-Specific Expression of Rat 3βHSD I and 3βHSD II mRNA Species

The nucleotide sequence of four types of rat 3βHSD cDNAs and their corresponding deduced amino acid sequences have recently become available (9, 10, 13). In fact, the sequences of both type I and II cDNAs have an open reading frame of 1119 nucleotides with only 33-bp differences between the two sequences. These sequences have been independently confirmed (11, 12). The predicted rat 3βHSD I and 3βHSD II proteins expressed in the adrenals, gonads, kidneys, placenta, and adipose tissue share 93.8% homology. On the other hand, the liver-specific type III protein sharing 80% homology with the rat type I and II isoenzymes does not display the expected classical 3βHSD activity, but is rather a

3-ketosteroid reductase (3KSR) using NADPH as cofactor. This enzyme catalyzes the conversion of such 3-keto-saturated steroids as DHT into its 3β-hydroxy metabolite (9, 31).

More recently, we have characterized a fourth type of cDNA encoding a novel rat 3βHSD isoenzyme chronologically designated type IV. The structure of the corresponding full-length cDNA was established using the Perkin-Elmer-Cetus RNA PCR kit and poly(A)$^+$ RNA from rat ovary, skin, and day 13 placenta. The type IV isoenzyme shares 90.9%, 87.9%, and 78.8% sequence homology with the rat type I, II, and III proteins, respectively (Table 7.1). The rat type IV is the exclusive 3βHSD mRNA species detectable in the skin and is the predominant mRNA species present in the placenta during the second week of gestation.

Enzymatic Characteristics of Expressed Members of the Rat 3βHSD Family

In order to further characterize the functional difference between the 3βHSD types, transient expression of 3βHSD I and 3βHSD II cDNAs was first performed in HeLa cells. This study demonstrates that the 3βHSD I protein has a relative specificity 64-fold higher than the type II protein using PREG as substrate. The K_m values of the type I and type II enzymes were calculated at 0.74 and 14.3 μM, respectively, thus showing that their different relative specificity largely results from a different affinity for substrates.

More recently, we have demonstrated that the activity of rat type IV is similar to that of rat type I. Since the change of 4 amino acid residues in type II could prevent the formation of a putative *membrane-spanning domain* (MSD) predicted between amino acid residues 75 and 91, chimeric cDNAs containing either type I MSD in type II (II + MSD) or an absence of this MSD in type I (I − MSD) were constructed and transiently expressed (53). The addition of MSD in 3βHSD II markedly increased the affinity, leading to K_m values similar to those found in wild-type 3βHSD I, namely, 0.36 μM for PREG. In fact, the II + MSD chimera encodes a protein having a relative specificity for PREG of 58%, that of native-type 3βHSD I. On the other hand, removal of MSD in the type I protein (I − MSD chimera) decreased the relative specificity of the 3βHSD I protein for PREG to only 0.37%, with a K_m value of 11.7 μM, thus strongly indicating the functional importance of this putative MSD that is predicted in wild-type rat types I, III, and IV, as well as in macaque and human 3βHSD proteins.

Transient expression of the cDNAs in human HeLa cells reveals that the 3βHSD III protein does not display oxidative activity for the classical substrates of 3βHSD, in contrast to the 3βHSD I isoenzyme (31). However, in the presence of NADH, the type III isoenzyme, in common

TABLE 7.1. Percentage of homology between the deduced amino acid sequences and the corresponding nucleotide sequences of human type I and II, macaque, bovine, rat type I, II, III, and IV, and mouse type I, II, and III 3βHSD proteins.

	Protein homology/Nucleotide homology (coding region)										
	Human I	Human II	Macaque	Bovine	Rat I	Rat II	Rat III	Rat IV	Mouse I	Mouse II	Mouse III
Human I		93.5	93.8	78.8	72.3	71.5	68.3	73.2	71.5	71.7	71.0
Human II	93.5		96.2	78.2	72.0	71.0	66.7	73.4	70.7	71.7	70.2
Macaque	94.4	95.2		79.6	73.1	72.0	68.0	75.0	71.5	72.5	71.2
Bovine	81.5	81.0	82.3		74.2	74.2	69.1	76.1	72.8	74.3	73.7
Rat I	78.3	77.8	79.4	77.7		93.8	80.4	90.9	87.9	83.4	82.0
Rat II	78.3	77.5	78.8	77.2	97.1		80.1	87.9	86.3	83.4	81.5
Rat III	75.3	75.4	76.6	75.3	85.7	85.6		78.8	76.6	77.4	75.5
Rat IV	78.1	78.0	79.3	77.0	94.3	93.1	84.7		86.8	84.9	86.8
Mouse I	77.4	76.3	78.1	75.8	91.2	91.0	83.0	89.7		85.3	83.3
Mouse II*	77.4	77.2	79.0	76.1	89.6	89.7	83.5	90.2	89.6		90.6
Mouse III	76.8	76.5	78.1	75.3	88.8	88.8	81.7	89.6	89.3	94.1	

* The data are obtained from the partial sequence of mouse 3βHSD II.

with the type I isoform, converts *5α-androstane-3,17-dione* (A-dione) and DHT into the corresponding 3β-hydroxysteroids. In fact, the type I and type III isoenzymes have the same affinity for DHT, with K_m values of 5.05 and 6.16 μM, respectively. When NADPH is used as cofactor, the affinity of the type III isoform for DHT becomes higher than that of the type I isoform, with K_m values of 0.12 and 1.18 μM, respectively. The type III isoform is thus a 3KSR using NADPH as preferred cofactor and is responsible for the conversion of such 3-keto-saturated steroids as DHT and A-dione into less active steroids (31).

Androgenic 17βHSD Activity of Rat 3βHSD I

Somewhat unexpectedly, we have observed that homogenate obtained from cells transfected with the pCMV 3βHSD I in the presence of the oxidative form of the cofactor—that is, NAD^+—converts DHT into a highly polar metabolite—namely, A-dione—thus demonstrating that the expressed enzyme possesses 17βHSD activity (54). The K_m value for the transformation of DHT into 3β-diol (4.02 ± 0.67 μM) by the expressed 3βHSD I is in the same range as that obtained with the human 3βHSD II (K_m = 2.7 μM) (8), but higher than the K_m obtained with the human placental 3βHSD I. When NAD^+ is used as cofactor, the androgenic 17βHSD activity of rat 3βHSD I shows a K_m of 7.97 ± 2.18 μM (54). As measured in HeLa cell homogenate, the V_{max} of 3βHSD activity of expressed 3βHSD I is much higher than that of 17βHSD activity (24.9 ± 1.18 vs. 1.67 ± 0.13 nmol/min/mg) (54).

However, when the same enzymatic assays are performed in unbroken JEG-3 human choriocarcinoma cells transfected with the rat 3βHSD I, 17βHSD activity is clearly predominant over 3βHSD activity. In fact, cells transfected with rat 3βHSD I metabolize DHT exclusively into Δ^4-dione, while no 3β-diol formation can be detected over basal values obtained in cells transfected with the pCMV plasmid alone (Fig. 7.5) (54). The predominance of the "secondary" 17βHSD activity of rat 3βHSD I expressed in intact mammalian cells in culture strongly suggests the physiological importance of this activity in modulating the intracellular concentration of the active androgen DHT in androgen target tissues.

The observation that the affinity of rat 3βHSD I for conversion of DHT into A-dione is similar to that for its conversion into 3β-diol in homogenates from HeLa cells expressing rat type I isoenzyme also argues for the intrinsic androgenic 17βHSD activity playing an important role in the degradation of the potent androgen DHT. The low levels of bioavailable NADH in intact cells could then force the conversion of DHT into A-dione in spite of a 15-fold lower V_{max} value for this reaction compared to that observed for the conversion of DHT into 3β-diol in homogenates from HeLa cells expressing rat 3βHSD I. However, such a high V_{max} value for the primary 3βHSD activity can well explain the present findings

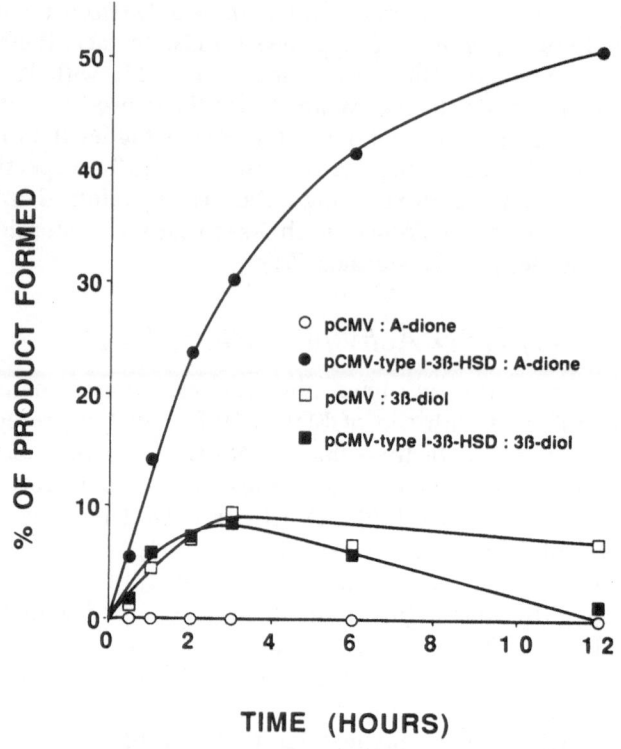

FIGURE 7.5. Time course of androgenic 17βHSD activity of expressed rat 3βHSD I in JEG-3 cells. Five hundred thousand cells were plated in 60-mm wells and transfected with either 5-μg pCMV or 5-μg pCMV 3βHSD I plasmid. Forty-eight hours later, 100 nM [³H]DHT was added into the culture medium. Incubation was stopped at the indicated time intervals, and steroids were determined by HPLC analysis. Reprinted with permission from de Launoit, Simard, Durocher, and Labrie (54), © The Endocrine Society, 1992.

of a rapid conversion of 3β-diol into DHT in the absence of detectable formation of epi-ADT in both HeLa and JEG-3 cells expressing the recombinant rat 3βHSD I.

It is also of interest to mention that the affinity of the androgenic 17β-HSD activity of rat 3βHSD I is higher or similar to that determined using the purified rat testicular (55) or porcine testicular (56) 17βHSD, both having a K_m value of 40 μM for Δ^4-androstenedione. The intrinsic 17βHSD activity of rat 3βHSD I appears to be specific to 5α-androstane 17β-ol steroids since the other standard substrates of 17βHSD—that is, DHEA; androst-5-ene-3β,17β-diol; testosterone; Δ^4-androstenedione; 17β-estradiol; and estrone—were not transformed into their respective metabolites in homogenates of transfected HeLa cells (54) as well as intact HeLa cells expressing this recombinant enzyme (de Launoit, Labrie, Simard, unpublished observations). This suggests that such secondary activity is specifically responsible for controlling the bioavailability of the active androgen DHT.

Ontogeny of 3βHSD in the Rat Testis

In most mammals, including rats, Leydig cells undergo two distinct periods of development, the first one occuring during fetal life and the second one taking place at puberty (57–59), thus leading to two populations of Leydig cells. The fetal and neonatal Leydig cell population, showing 3βHSD activity from 17 days onward, is localized largely in the interstitial space between the seminiferous tubules, while the postnatal Leydig cell population, showing 3βHSD activity from postnatal day 16 onward, surrounds the individual seminiferous tubules (58).

In order to obtain more information about the age-specific expression of 3βHSD in the testis, we have localized this enzyme by immunocytochemistry at the light microscopic level during fetal and postnatal development of the rat testis. No immunostaining for the 3βHSD enzyme was detected at 15 and 16 days of fetal life, while on day 17 of gestation, only a few weakly stained cells could be observed. On day 18, however, several cells were labeled (Fig. 7.6). Some positive cells appeared without any specific orientation, while others were located around poorly defined

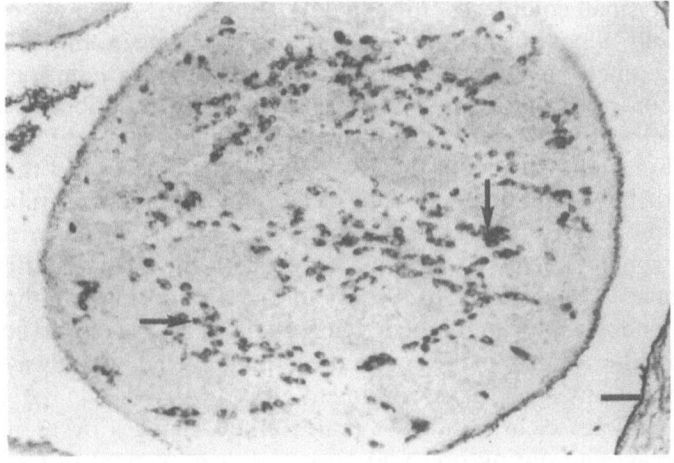

FIGURE 7.6. At 18 days of gestation. Immunostaining can be observed in cells (arrows) that are dispersed throughout the testis. The staining of the capsule does not represent specific labeling and is not related to the presence of immunoreactive material.

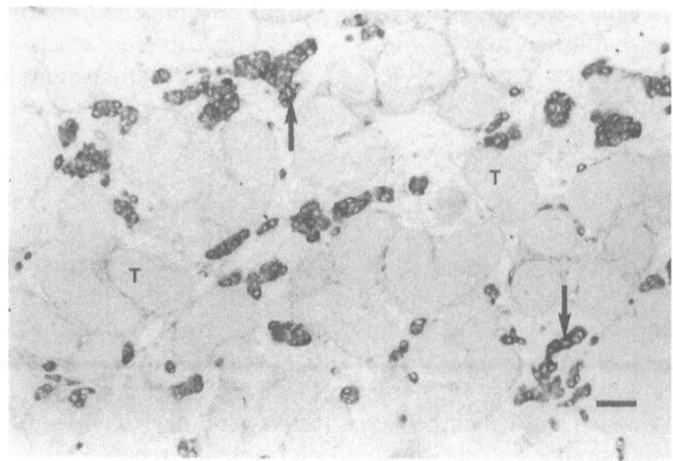

FIGURE 7.7. Six hours after birth. The immunoreaction can be observed in clusters of Leydig cells (arrows). (T = seminiferous tubules; scale bar = 75 μm.)

seminiferous tubules. From days 19 to 21 of fetal life, strongly immuno-positive cells are seen in large clusters occupying the space between the seminiferous tubules.

Shortly after birth, immunopositive Leydig cells are forming clusters in the large intertubular spaces (Fig. 7.7). However, during the days following birth, the number and size of positive cells rapidly decreased. In fact, at 5 and 10 days after birth, the stained cells appeared to be isolated or forming small complexes. Fifteen days after birth, groupings of positive Leydig cells appeared more numerous. At that time, some cells had an elongated shape and were situated alongside the seminiferous tubules. These cells probably correspond to the already described peritubular Leydig cells (57–61). Other positive cells having a round shape were arranged in small clusters in the large peritubular spaces. At the onset of puberty or between 20 and 30 days of age, the seminiferous tubules have grown, and most of the Leydig cells, which are much more abundant, appear strongly labeled and are located in both the peritubular and inter-tubular spaces (Fig. 7.8). At 40 days of age and during adulthood, both strongly and weakly immunolabeled Leydig cells are observed, mostly located in the intertubular spaces. The tubular elements always remain unreactive at any stage of development.

These findings clearly indicate that marked changes occur in 3βHSD immunoreactivity during pre- and postnatal development in Leydig cells. In fact, immunostaining could first be detected on day 17 of gestation, while a plateau was reached between days 19 and 21 of fetal life. At the end of gestation, the Leydig cells appeared well developed and strongly stained. Soon after birth there was a rapid decrease in immunoreactivity

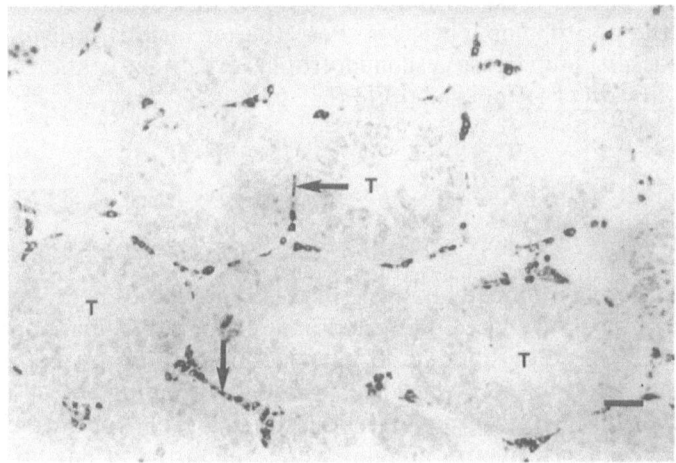

FIGURE 7.8. Twenty-five days after birth. The peritubular (arrow) and intertubular (arrow heads) Leydig cells are more numerous and more strongly stained than at 15 days. (T = seminiferous tubules; scale bar = 75 μm.)

that reached the lowest levels during the first 10 days of postnatal life. During the days preceding the onset of puberty and at the onset of puberty, there was a moderate increase in the number of 3βHSD-containing cells that were located in both peritubular and intertubular spaces. These positive cells continued to increase in number and size during and after puberty to become mostly concentrated in the intertubular spaces during adulthood.

Using histochemical techniques, Ziegler et al. (62) have shown that during fetal development maximal 3βHSD enzyme activity was seen on day 19 of fetal life. In the present study we have observed that 3βHSD immunoreactivity remained maximal from day 19 until birth. There was no detectable decrease in the number and staining of Leydig cells on days 20 and 21 of gestation. The present data support the previous reports on changes in plasma testosterone levels during fetal life (63) that were low on day 17 and increased during the following days of gestation to reach maximal values before birth. Thus, there seems to exist a correlation between plasma testosterone levels and the amount of 3βHSD present in Leydig cells during fetal life.

The progressive atrophy of Leydig cells that takes place during the days following birth is possibly related to the very low levels of circulating gonadotropins observed during the neonatal period (64). Such changes in plasma gonadotropins and the present data on the level of immunoreactive 3βHSD during the neonatal period can provide an explanation for the low levels of plasma testosterone observed on days 3 and 15 after birth (63). The increase in number and size of Leydig cells, especially those of the peritubular type, from day 10 after birth is probably related to the progressive increase in gonadotropin levels that is known to occur after the first week of postnatal life (64–66).

Regulation of 3βHSD Expression and Activity in Rat Testis

Since 3βHSD plays an essential role in the formation of androgens by the mammalian testis, we were interested in studying its regulation in the rat testis by a physiologically important modulator, *luteinizing hormone* (LH). The effect of LH was investigated by treating intact rats with *human chorionic gonadotropin* ([hCG] 10 IU, b.i.d.) for 15 days, while *hypophysectomized* (hypox) animals were treated for 9 days starting 15 days after surgery. Testicular 3βHSD mRNA levels were measured by

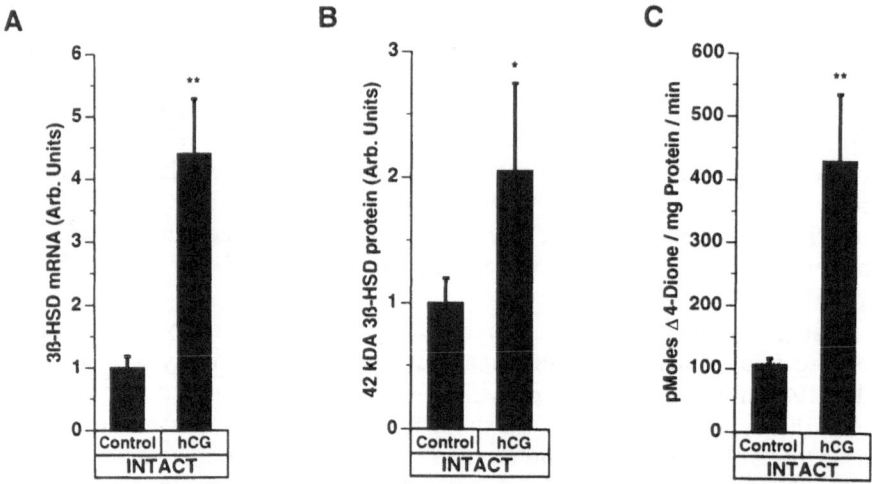

FIGURE 7.9. Effect of treatment with hCG (10 IU) on rat testicular 3βHSD mRNA levels (*A*), immunoreactive 3βHSD protein content (*B*), and enzymatic activity (*C*) in intact rats. Intact adult male rats received twice-daily injections of vehicle (intact control) or hCG for 15 days. Testicular 3βHSD mRNA levels were measured by dot blot hybridization using full-length rat type I [^{32}P]-labeled cDNA probe and were calculated relative to the levels observed in vehicle-treated control rat testis. Testicular 3βHSD activity was assayed by measuring the rate of formation of [4-^{14}C] Δ^4-dione from [4-^{14}C] DHEA. Diluted testicular homogenates were incubated with 10 μM [4-^{14}C] DHEA and 0.8 mM NAD$^+$. The 42-kd testicular 3βHSD protein was revealed by Western blot analysis using antibodies directed against human placental 3βHSD and labeled goat antirabbit IgG. Data are expressed as means ± SEM. (n = 6–7; *P < 0.05; **P < 0.01.)

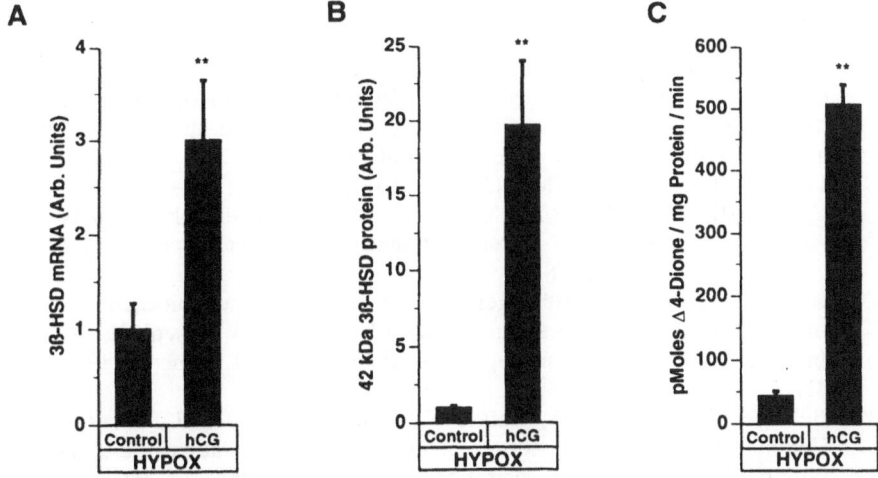

FIGURE 7.10. Effect of treatment with hCG (10 IU) on rat testicular 3βHSD mRNA levels (A), immunoreactive 3βHSD protein content (B), and enzymatic activity (C) in hypophysectomized rats. Hypophysectomized (15 days before) adult male animals received twice-daily injections of vehicle (HYPOX control) or hCG for 9 days. Testicular 3βHSD mRNA levels were measured by dot blot hybridization using full-length rat type I [^{32}P]-labeled cDNA probe and were calculated relative to those levels measured in vehicle-treated control rat testis. Testicular 3βHSD activity was assayed by measuring the rate of formation of [4-^{14}C] Δ^4-dione from [4-^{14}C] DHEA. Diluted testicular homogenates were incubated with 10 μM [4-^{14}C] DHEA and 0.8 mM NAD$^+$. The 42-kd testicular 3βHSD protein was revealed by Western blot analysis using antibodies directed against human placental 3βHSD and labeled goat antirabbit IgG. Data are expressed as means ± SEM. (n = 6–7; **$P < 0.01$.)

dot blot hybridization using a full-length rat 3βHSD I cDNA (10), while 3βHSD activity was determined by the conversion of ^{14}C-DHEA into ^{14}C-Δ^4-dione. The antibody against human placental 3βHSD was used to measure rat testicular 3βHSD protein by immunoblotting (25, 26).

As illustrated in Figure 7.9A, hCG treatment caused a 4.4-fold increase in 3βHSD mRNA levels in intact rat testis. In parallel, hCG increased immunoreactive testicular 3βHSD by 2.0-fold (Fig. 7.9B) and caused a 4.3-fold increase in enzymatic activity (Fig. 7.9C). In hypox animals, hCG caused a 3.0-, 19.7-, and 11.5-fold stimulation in the steady state levels of testicular 3βHSD mRNA (Fig. 7.10A), immunoreactive 3βHSD protein (Fig. 7.10B), and enzymatic activity (Fig. 7.10C), respectively. The close parallelism observed between changes in 3βHSD mRNA levels, protein content, and enzymatic activity indicates that LH/hCG exerts its action in the rat testis, at least to a large extent, at the level of 3βHSD gene expression and/or 3βHSD mRNA stability.

146   F. Labrie et al.

References

1. Labrie F, Simard J, Luu-The V, Bélanger A, Pelletier G. Structure, function and tissue-specific gene expression of 3β-hydroxysteroid dehydrogenase/-5-ene-4-ene isomerase enzymes in classical and peripheral intracrine steroidogenic tissues. J Steroid Biochem Mol Biol 1992;43:805–26.
2. Labrie F. Intracrinology. Mol Cell Endocrinol 1991;78:C113–8.
3. Cutler GB, Glenn M, Bush M, Hodgen GD, Graham CE, Loriaux DL. Adrenarche: a survey of rodents, domestic animals and primates. Endocrinology 1978;103:2112–8.
4. Labrie F, Dupont A, Bélanger A. Complete androgen blockade for the treatment of prostate cancer. In: De Vita V, Hellman S, Rosenberg SA, eds. Important advances in oncology. Philadelphia: J.B. Lippincott, 1985:193–217.
5. Bélanger B, Bélanger A, Labrie F, Dupont A, Cusan L, Monfette G. Comparison of residual C-19 steroids in plasma and prostatic tissue of human, rat and guinea pig after castration: unique importance of extratesticular androgens in men. J Steroid Biochem 1989;32:695–8.
6. Luu-The V, Lachance Y, Labrie C, et al. Full length cDNA structure and deduced amino acid sequence of human 3β-hydroxy-5-ene steroid dehydrogenase. Mol Endocrinol 1989;3:1310–2.
7. Lorence MC, Murry BA, Trant JM, Mason JL. Human 3β-hydroxysteroid dehydrogenase/Δ^5-Δ^4 isomerase from placenta: expression in nonsteroidogenic cells of a protein that catalyses the dehydrogenation/isomerization of C21 and C19 steroids. Endocrinology 1990;126:2493–8.
8. Rhéaume E, Lachance Y, Zhao HF, et al. Structure and expression of a new cDNA encoding the almost exclusive 3β-hydroxysteroid dehydrogenase/Δ^5-Δ^4 isomerase in human adrenals and gonads. Mol Endocrinol 1991;5:1147–57.
9. Zhao HF, Rhéaume E, Trudel C, Couet J, Labrie F, Simard J. Structure and sexual dimorphic expression of a liver-specific rat 3β-hydroxysteroid dehydrogenase/isomerase. Endocrinology 1990;127:3237–9.
10. Zhao HF, Labrie C, Simard J, et al. Characterization of 3β-hydroxysteroid dehydrogenase/Δ^5-Δ^4 isomerase cDNA and differential tissue-specific expression of the corresponding mRNAs in steroidogenic and peripheral tissues. J Biol Chem 1991;266:583–93.
11. Lorence MC, Naville D, Graham-Lorence SE, et al. 3β-hydroxysteroid dehydrogenase/Δ^5-Δ^4 isomerase expression in rat and characterization of the testis isoform. Mol Cell Endocrinol 1991;80:21–31.
12. Naville D, Keeney DS, Jenkin G, Murry BA, Head JR, Mason JI. Regulation of expression of male-specific rat liver microsomal 3β-hydroxysteroid dehydrogenase. Mol Endocrinol 1991;5:1090–100.
13. Simard J, Zhao HF, Labrie C, et al. Molecular cloning of rat 3β-HSD: structure of two types of cDNAs and differential expression of corresponding mRNAs in the ovary. In: Gibori G, ed. Signaling mechanisms and gene expression in the ovary. New York: Springer-Verlag, 1991:274–9.
14. Simard J, Melner MH, Breton N, et al. Characterization of macaque 3β-hydroxy-5-ene steroid dehydrogenase/Δ^5-Δ^4 isomerase: structure and expression in steroidogenic and peripheral tissues in primates. Mol Cell Endocrinol 1991;75:101–10.

15. Zhao HF, Simard J, Labrie C, et al. Molecular cloning, cDNA structure and predicted amino acid sequence of bovine 3β-hydroxy-5-ene steroid dehydrogenase/Δ^5-Δ^4 isomerase. FEBS Lett 1989;259:153–7.

16. Bain PA, Yoo M, Clarke T, Hammond SH, Payne AH. Multiple forms of mouse 3β-hydroxysteroid dehydrogenase/$\Delta^5\Delta^4$ isomerase and differential expression in gonads, adrenal glands, liver, and kidneys of both sexes. Proc Natl Acad Sci USA 1991;88:8870–4.

17. Lachance Y, Luu-The V, Labrie C, et al. Characterization of human 3β-hydroxysteroid dehydrogenase/Δ^5-Δ^4 isomerase gene and its expression in mammalian cells. J Biol Chem 1990;265:20469–75.

18. Lachance Y, Luu-The V, Verreault H, et al. Structure of the human type II 3β-hydroxysteroid dehydrogenase/Δ^5-Δ^4 isomerase (3β-HSD) gene: adrenal and gonadal specificity. DNA Cell Biol 1991;10:701–11.

19. Lorence MC, Jo Corbin C, Kamimura N, Mahendroo MS, Mason JI. Structural analysis of the gene encoding human 3β-hydroxysteroid dehydrogenase/Δ^5-Δ^4-isomerase. Mol Endocrinol 1990;4:1850–5.

20. Dupont E, Luu-The V, Labrie F, Pelletier G. Ontogeny of 3β-hydroxysteroid dehydrogenase/Δ^5-Δ^4 isomerase (3β-HSD) in human adrenal gland performed by immunocytochemistry. Mol Cell Endocrinol 1990;74:R7–10.

21. Dupont E, Zhao HF, Rhéaume E, et al. Localization of 3β-hydroxy-5-ene steroid dehydrogenase/Δ^5-Δ^4 isomerase in the rat gonads and adrenal glands by immunocytochemistry and in situ hybridization. Endocrinology 1990;127:1394–403.

22. Dupont E, Luu-The V, Labrie F, Pelletier G. Ontogeny of 3β-hydroxysteroid dehydrogenase/Δ^5-Δ^4 isomerase (3β-HSD) in human testis as studied by immunochemistry. J Androl 1991;12:161–4.

23. Dupont E, Labrie F, Luu-The V, Pelletier G. Immunochemical localization of 3β-hydroxysteroid dehydrogenase/Δ^5-Δ^4 isomerase (3β-HSD) in human ovary. J Clin Endocrinol Metab 1992;74:994–8.

24. Riley SC, Dupont E, Walton JC, et al. Immunohistochemical localization of 3β-hydroxy-5-ene-steroid dehydrogenase/Δ^5-Δ^4 isomerase in human placenta and fetal membranes. J Clin Endocrinol Metab 1992;96:127–34.

25. Luu-The V, Côté J, Labrie F. Purification and characterization of human placental 3β-hydroxysteroid dehydrogenase/Δ^5-Δ^4 isomerase. Clin Invest Med 1988;11:C200.

26. Luu-The V, Takahashi M, Labrie F. Purification of mitochondrial 3β-hydroxysteroid dehydrogenase/Δ^5-Δ^4 isomerase from human placenta. Ann NY Acad Sci 1990;595:386–8.

27. Bongiovanni AM, Kellenbenz G. The adrenogenital syndrome with deficiency of 3β-hydroxysteroid dehydrogenase. J Clin Invest 1962;41:2086–92.

28. New M, White P, Pang S, Dupont B, Speiser PW. The adrenal hyperplasias. In: Scriver CR, Beaudet A, Sly WS, Valle D, eds. The metabolic basis of inherited diseases. 6th ed. New York: McGraw-Hill, 1989:1881–917.

29. Bérubé D, Luu-The V, Lachance Y, Gagné R, Labrie F. Assignment of the human 3β-hydroxysteroid dehydrogenase gene to the p13 band of chromosome 1. Cytogen Cell Genet 1989;52:199–200.

30. Luu-The V, Lachance Y, Leblanc G, Labrie F. Human 3β-hydroxysteroid dehydrogenase/Δ^5-Δ^4 isomerase: characterization of three additional related genes [#1499]. Proc 74th meet Endocr Soc, 1992:426.

31. de Launoit Y, Zhao HF, Bélanger A, Labrie F, Simard J. Expression of liver-specific member of the 3β-hydroxysteroid dehydrogenase family, an isoform possessing an almost exclusive 3-keto steroid reductase activity. J Biol Chem 1992;267:4513–7.
32. Luu-The V, Takahashi M, de Launoit Y, Dumont M, Lachance Y, Labrie F. Evidence for distinct dehydrogenase and isomerase sites within a single 3β-hydroxysteroid dehydrogenase/5-ene-4-ene isomerase protein. Biochemistry 1991;30:8861–5.
33. Griffin JE, Wilson JD. The testis. In: Browdy PK, Rosenberg LE, eds. Metabolic control and disease. 8th ed. Philadelphia: W.B. Saunders, 1980: 1535–78.
34. Forest MG, Cathiard AM. Pattern of plasma testosterone and Δ^4-androstenedione in normal newborns: evidence for testicular activity at birth. J Clin Endocrinol Metab 1975;41:977–84.
35. Bidlingmainer F, Dorr HG, Eisenmenger W, Kuhnle U, Knorr D. Testosterone and androstenedione concentrations in human testis and epididymis during first two years of life. J Clin Endocrinol Metab 1983;57:311–5.
36. Winter JSD, Hughes IA, Reyes FI. Pituitary-gonadal relations in infancy, 2. Patterns of serum gonadal steroid concentrations in man from birth to two years of age. J Clin Endocrinol Metab 1976;42:679–86.
37. Frasier SD, Gafford F, Horton R. Plasma androgens in childhood and adolescence. J Clin Endocrinol Metab 1969;29:1404–8.
38. August GP, Grumbach MM, Kaplan SL. Hormonal changes in puberty, III. Correlation of plasma testosterone, LH, FSH, testicular size, and bone age with male pubertal development. J Clin Endocrinol Metab 1972;34: 319–26.
39. Baillie AH, Niemi M, Ikanen M. 3β-hydroxysteroid dehydrogenase activity in the human fetal testis. Acta Endocrinol (Copenh) 1965;48:429–38.
40. Morel Y, Miller WL. Clinical and molecular genetics of congenital hyperplasia due to 21-hydroxylase deficiency. Adv Hum Genet 1991;20:1–68.
41. Bongiovanni AM. Acquired adrenal hyperplasia with special reference to 3β-HSD. Fertil Steril 1981;35:599–608.
42. de Peretti E, Forest MG. Pattern of plasma dehydroepiandrosterone sulfate levels in human from birth to adulthood: evidence for testicular production. J Clin Endocrinol Metab 1978;47:572–7.
43. Zachmann M, Völlmin JA, Mürset G, Curtius HCH, Prade A. Unusual type of congenital adrenal hyperplasia probably due to deficiency 3β-hydroxysteroid dehydrogenase: case report of a surviving girl and steroid studies. J Clin Endocrinol 1970;30:719–26.
44. Zachmann M, Forest MG, de Peretti E. 3β-hydroxysteroid dehydrogenase deficiency follow-up study in a girl with pubertal bone age. Horm Res 1979;11:292–302.
45. Zachmann M, Kempken B, Anner I, Pezzoli V. Age-related differences in the excretion of delta 5 steroids (D5S) in two related girls with 3β-hydroxysteroid dehydrogenase deficiency. Pediatr Res 1988;24:543.
46. Parks GA, Bermudez JA, Anast CS, Bongiovanni AM, New MI. Pubertal boy with the 3β-hydroxysteroid dehydrogenase defect. J Clin Endocrinol 1971;33:269–78.

47. Cara JF, Moshang T, Bongiovanni AM, Marx BS. Elevated 17-hydroxypro-
gesterone and testosterone in a newborn with 3β-hydroxysteroid dehydroepiandrosterone sulfate levels in human from birth to adulthood: evidence for
testicular production. J Clin Endocrinol Metab 1985;47:572–7.
48. Rhéaume E, Simard J, Morel Y, et al. Congenital adrenal hyperplasia due to
point mutations in the type II 3β-hydroxysteroid dehydrogenase gene. Nature
Genet 1992;1:239–45.
49. Simard J, Rhéaume E, Sanchez R, et al. Molecular basis of congenital
adrenal hyperplasia due to 3β-hydroxysteroid dehydrogenase deficiency. Mol
Endocrinol 1993.
50. Baker ME, Blasco R. Expansion of the mammalian 3β-hydroxysteroid dehy-
drogenase/plant dihydroflavonol reductase superfamily to include a bacterial
cholesterol dehydrogenase, a bacterial UDP-galactose-4-epimerase and open
reading frames in vaccinia virus and fish lymphocystis disease virus. FEBS
Lett 1992;301:89–93.
51. Goldman AS, Yakovac WC, Bongiovanni AM. Development of activity of
3β-hydroxysteroid dehydrogenase in human fetal tissues and in two anen-
cephalic newborns. J Clin Endocrinol Metab 1966;26:14–22.
52. Milewich L, Shaw CE, Doody KM, Rainey WE, Mason JI, Carr BR. 3β-
hydroxysteroid dehydrogenase activity in glandular and extraglandular human
fetal tissues. J Clin Endocrinol Metab 1991;73:1134–40.
53. Simard J, de Launoit Y, Labrie F. Characterization of the structure-activity
of rat type I and type II 3β-hydroxysteroid dehydrogenase/Δ^5-Δ^4 isomerase by
site-directed mutagenesis and expression in HeLa cells. J Biol Chem 1991;
266:14842–5.
54. de Launoit Y, Simard J, Durocher F, Labrie F. Androgenic 17β-hydroxy-
steroid dehydrogenase activity of expressed rat type I 3β-hydroxysteroid
dehydrogenase/Δ^5-Δ^4 isomerase. Endocrinology 1992;130:553–5.
55. Bogovich K, Payne AH. Purification of rat testicular 17-ketosteroid reductase, evidence that 17-ketosteroid reductase and 17β-hydroxysteroid dehydrogenase are distinct enzymes. J Biol Chem 1980;255:5552–9.
56. Inano H, Tamaoki BI. Purification and properties of NADP$^+$-dependent 17β-
hydroxysteroid dehydrogenase solubilized from porcine testicular microsomal
fraction. Eur J Biochem 1974;44:13–23.
57. Goslar HG, Passia D, Hilscher W, Hilscher B. Enzymhistochemische unter-
suchungen zur kinetik des inter-und peritubulären Leydig-zellsystem der ratte.
Acta Histochem Suppl (Jena) 1980;Band XXI:243–6.
58. Haider SG, Passia D, Overmeyer G. Studies on the fetal and postnatal
development of rat Leydig cells employing 3β-hydroxysteroid dehydrogenase
activity. Acta Histochem Suppl (Jena) 1986;Band XXXII:197–202.
59. Kerr JB, Knell CM. The fate of fetal Leydig cells during the development of
the fetal and postnatal rat testis. Development 1988;103:535–44.
60. Weisz J, Ward I. Plasma testosterone and progesterone titers of pregnant
rats, their male and female fetuses and neonatal offspring. Endocrinology
1980;106:306–16.
61. Passia D, Hahner J, Hilscher B, Hilscher W. Enzymhistochemische unter-
suchungen an dem inter-und peritubularen Leydig-zellsystem der ratte. Verh
Dtsch Ges Anat 1978;73:699–700.

62. Ziegler HG, Haider SG, Passia D, Hilscher W. Enzymohistochemical and morphometrical studies on Δ^5-3β-hydroxysteroid dehydrogenase during the fetal and neonatal development of rat Leydig cells. Andrologia 1983;15: 392–7.
63. Wilson JD. Sexual differentiation. Annu Rev Physiol 1978;40:279–306.
64. Chowdhury M, Steinberger E. Pituitary and plasma levels of gonadotrophins in foetal and newborn male and female rats. J Endocrinol 1976;69:381–4.
65. Huhtaniemi I, Warren DW, Catt KJ. Regulation of infant and developing rat testicular gonadotropin and prolactin receptors and steroidogenesis by treatments with human chorionic gonadotropin, gonadotropin-releasing hormone analogs, bromocriptine, prolactin and estrogens. Biol Reprod 1985;32:721– 32.
66. Huhtaniemi IT, Warren DW, Catt KJ. Functional maturation of rat testis Leydig cells. Ann New York Acad Sci 1984;438:283–303.

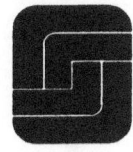

8

Regulation of DNA Synthesis in Leydig Cells

Shafiq A. Khan, Katja Teerds, and Jennifer Dorrington

In the testis, the gonadotropins LH and FSH are required in vivo for the normal growth and differentiation of the Leydig cells and the Sertoli cells, respectively (1, 2). As a result of the actions of the gonadotropins on the somatic cells of the testis, the appropriate microenvironment is established in the seminiferous tubule that is conducive to the initiation and progression of spermatogenesis (3). LH acts on Leydig cells to stimulate the synthesis and secretion of testosterone, which then acts as a paracrine factor that is essential for spermatogenesis to proceed (4, 5). In intact rats that lack detectable levels of LH, administration of testosterone alone can maintain normal spermatogenesis (6, 7).

Leydig cells in the rat testis undergo two clearly defined periods of proliferation and differentiation, the first occurring during fetal life and the second proceeding during the prepubertal period (8, 9). In the fetal testis, a population of Leydig cells is generated whose morphology and profile of secreted steroids are unique to this phase of testicular development (10). After birth the number of fetal Leydig cells decreases; however, some fetal-type Leydig cells persist even in the adult testis (11).

The population of adult-type Leydig cells is derived from the differentiation of mesenchymal-like precursor cells and the proliferation of these newly formed Leydig cells (9). This wave of Leydig cell proliferation is initiated around 14 days of postnatal life, and as this progresses there is a rapid increase in the total number of Leydig cells per testis (9). The increase in cell number is particularly dramatic between days 21 and 28 of age (Fig. 8.1).

Puberty in the male is dependent on the elevated production of androgen by the testis, and this is achieved by increasing the total number of Leydig cells in the testis and by stimulating androgen synthesis in these cells (Fig. 8.1) (12). The process of proliferation and differentiation of Leydig cells is LH dependent (13). In hypophysectomized immature rats, Leydig cells can be stimulated to proliferate by the administration of LH;

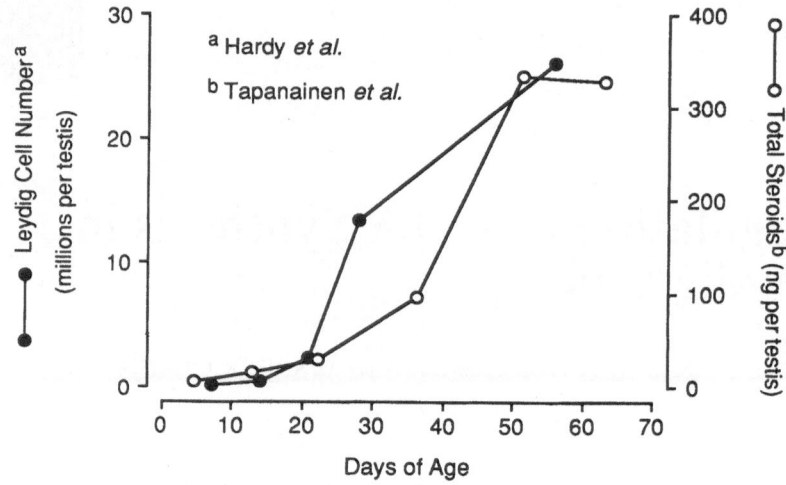

FIGURE 8.1. Leydig cell numbers and steroid content during development of the testis. Leydig cell numbers are from reference 9. Steroid content of testis is from reference 12.

however, the requirement for LH cannot be replaced by testosterone (13, 14). This suggests that LH stimulates Leydig cell proliferation directly or indirectly by regulating the production of growth-promoting factors that act in an autocrine manner.

Growth Factor Requirements for DNA Synthesis by Immature Leydig Cells

Since Leydig cell proliferation is a prerequisite for the onset of puberty, we have looked for this factor, or a combination of factors, that is required for Leydig cells to enter the S-phase of the cell cycle and initiate DNA synthesis in vitro. Since it is well established that growth factors acting at multiple sites are essential for the progression through G1 of the cell cycle leading to DNA synthesis and cell division (15) (Fig. 8.2), we have focused on the identification of these growth factors and their interactions with LH (16).

Transforming Growth Factor α

Leydig cells isolated from the testis of 21-day-old rats were maintained in culture in a serum-free medium for 48 h with no treatment to allow the cells to become quiescent. Subsequent treatment of the cells with *transforming growth factor α* (TGFα) for 18 h significantly stimulated the

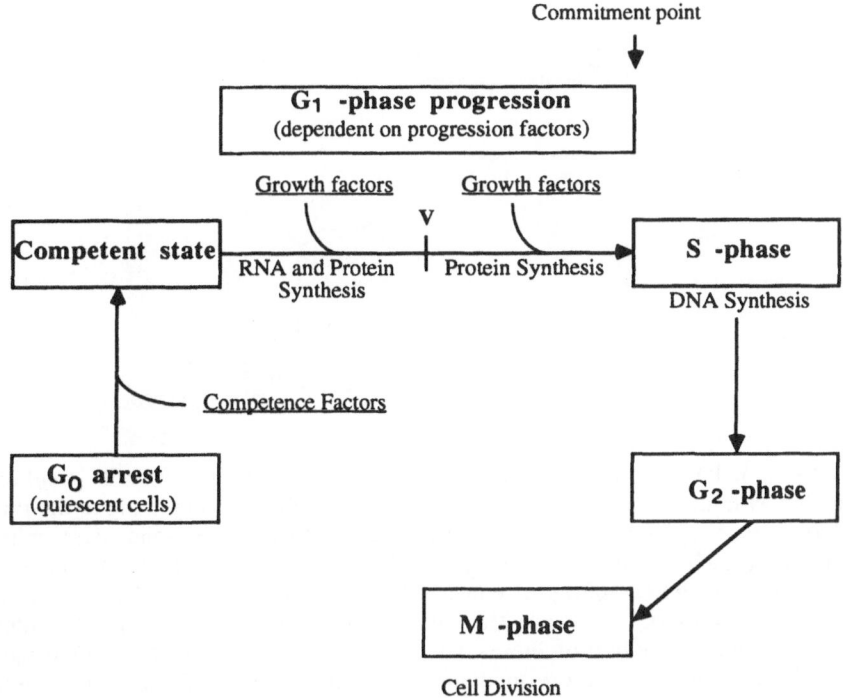

FIGURE 8.2. A schematic diagram of the mammalian cell cycle.

incorporation of [³H]thymidine into DNA (Fig. 8.3). The proliferative responses of the Leydig cells to TGFα were dependent on the cell densities and on the concentration of TGFα (Fig. 8.4) (16).

Interaction of TGFα with IGF-I

Since IGF-I has been detected by immunohistochemical techniques in Leydig cells of the immature rat testis (17), we examined the possibility that this growth factor may be involved in the proliferation of Leydig cells. IGF-I or insulin alone stimulated [³H]thymidine incorporation into DNA of cultured Leydig cells from 21-day-old rats. The effect of insulin (or IGF-I) was not as great as that of TGFα, but when the cells were treated with both growth factors, there was an additive or greater than additive effect on DNA synthesis and on the number of labeled cells after incubation with [³H]thymidine (Fig. 8.3) (16).

Interaction of LH with the Growth Factors TGFα and IGF-I

Since proliferation of Leydig cells in vivo is an LH-dependent process, quiescent cells were treated with LH (100 ng/mL) for 18 h. This treat-

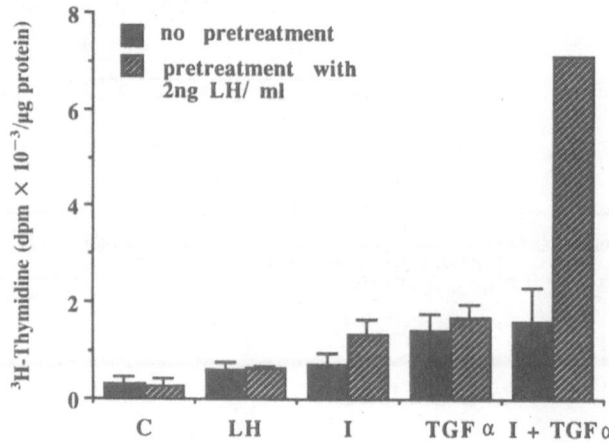

FIGURE 8.3. [³H]Thymidine incorporation into DNA by rat Leydig cells in culture treated with insulin and TGFα. During the 48 h of culture from plating to the time of treatment, the Leydig cells were cultured in the absence (solid bars) or the presence (hatched bars) of LH (2 ng/mL). The cells were then treated for the subsequent 18 h with LH (100 ng/mL), insulin ([I], 1 μg/mL), and/or TGFα (10 ng/mL). Control cultures (C) were not treated during this 18-h period. [³H]Thymidine incorporation per μg of protein was determined after a 4-h incubation with 0.5 μCi [³H]thymidine/0.5 mL culture medium. Data are means ± SD of 3 determinations. Reprinted with permission from Khan, Teerds, and Dorrington (16).

ment caused a small but reproducible increase in the incorporation of [³H]thymidine into DNA (Fig. 8.3). Leydig cells, however, are known to lose their steroidogenic responses to LH in culture, but the responsiveness can be maintained in cells cultured in the presence of LH (18). When immature Leydig cells were cultured from the time of plating in the presence of LH (2 ng/mL), no effect on the subsequent responses to the higher concentrations of LH (100 ng/mL) were found. Pretreatment of the cells with LH, however, did cause a dramatic increase in the response to TGFα plus insulin (Fig. 8.3) (16).

Interleukin-1

In recent years there has been a considerable amount of interest in the physiological significance of the high levels of the interleukin-1-like factor found in human and rat testis (19, 20). *Interleukin-1* (IL-1) increased the basal level of synthesis of testosterone in Leydig cells from adult rats; however, when the cells were stimulated with hCG, IL-1 inhibited cAMP production, cholesterol side-chain cleavage P450 mRNA gene expression, and testosterone synthesis. *Tumor necrosis factor α* (TNFα) enhanced the inhibitory effects of interleukin-1β on Leydig cell steroidogenesis (21). It

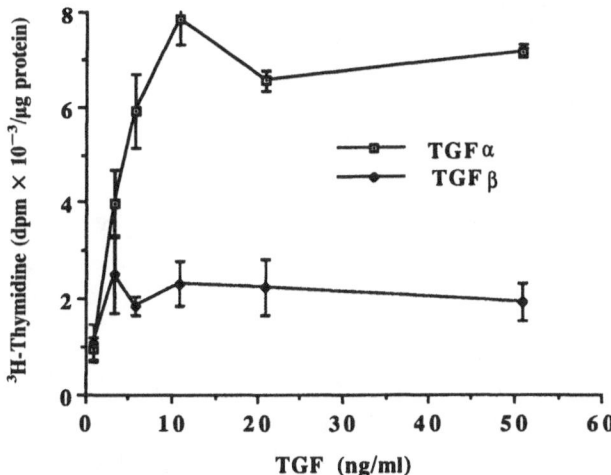

FIGURE 8.4. [³H]Thymidine incorporation into DNA by rat Leydig cells treated with a range of concentrations of TGFα and TGFβ. After 48 h in culture with no treatment, the cells were cultured for a further 18 h with a range of concentrations of TGFα and TGFβ. [³H]Thymidine incorporation per μg of protein was determined after a 4-h incubation with 0.5 μCi [³H]thymidine/0.5 mL culture medium. Data are means ± SD of 3 determinations. Reprinted with permission from Khan, Teerds, and Dorrington (16).

has been suggested that IL-1α may stimulate DNA synthesis in intermediate and type B spermatogonia and during meiosis (22, 23).

Since IL-1 stimulates the proliferation of a variety of cells and IL-1 receptors have been localized predominantly in the interstitial compartment of the adult testis (24), we explored the possible effects of IL-1 on DNA synthesis in rat Leydig cells in culture (25). Recombinant IL-1β stimulated the incorporation of [³H]thymidine into DNA of Leydig cells from 10- and 20-day-old rats, but had no effect on DNA synthesis of Leydig cells from adult rats (Fig. 8.5). Treatment with IL-1β was shown by autoradiography to increase the number of labeled cells, indicating that IL-1β recruited more cells to enter the G1 to S-phase of the cell cycle and initiate DNA synthesis (25).

To examine the interactions of IL-1β with the other known regulators of DNA synthesis, Leydig cells from 10-day-old rats were stimulated maximally with hCG or TGFα plus insulin and treated concomitantly with IL-1β. As shown in Figure 8.6, IL-1β significantly enhanced the effects of hCG and TGFα plus insulin on DNA synthesis.

It has been suggested that a number of different cell types in the testis may contribute to the high levels of IL-1. Sertoli cells in culture secreted an IL-1-like factor (26). Macrophages known to produce interleukins and

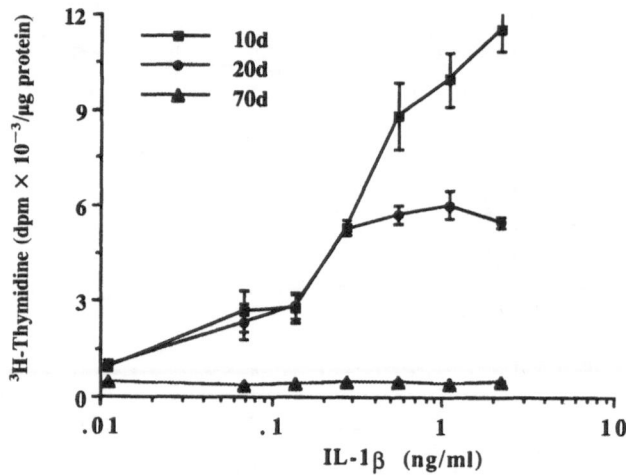

FIGURE 8.5. Dose-dependent effects of IL-1β on in vitro DNA synthesis in Leydig cells obtained from the testes of immature (10- and 20-day-old) and adult (70-day-old) rats. Cells were cultured for 48 h in serum-free medium, washed, and treated with different concentrations of IL-1β for 18 h. The medium was removed, fresh medium containing 0.5 μCi [³H]thymidine was added, and cells were incubated for a period of 4 h, after which the incorporation of [³H]thymidine was determined. Each point represents mean ± SD of triplicate determination. Reprinted with permission from Khan, Khan, and Dorrington (25), © The Endocrine Society, 1992.

other cytokines are present in the interstitial compartment of the testis in close association with Leydig cells through specialized junctions (27), and their secretion products affect testosterone production by Leydig cells in vitro (28). Of particular interest was the demonstration that Leydig cells contained IL-1α mRNA and that its level was increased after treatment of the cells with IL-1β (29). The latter study indicated that Leydig cells may be a cellular source of interleukin-1 in the adult testis. Whether IL-1 is synthesized by Leydig cells in the prepubertal rat to act as an autocrine regulator of DNA synthesis remains to be investigated.

Steroidogenesis-Inducing Protein

Steroidogenesis-inducing protein (SIP) has been isolated from follicular fluid aspirated from women at the time of egg retrieval for in vitro fertilization. SIP stimulates steroid production in early luteal cells, adrenal cells, and Leydig cells from a number of species at the level of the cholesterol side-chain cleavage enzyme, which is the rate-limiting step in steroidogenesis (30–32). Even though SIP stimulates steroidogenesis at the same site of regulation as LH/hCG, the steroidogenic response of

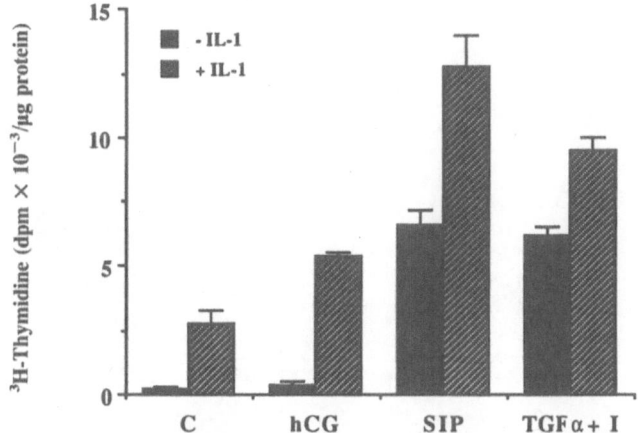

FIGURE 8.6. Effects of hCG (100 ng/mL), SIP (50 µg/mL), and TGFα (10 ng/mL) plus insulin ([I], 5 µg/mL) on in vitro DNA synthesis in Leydig cells from 10-day-old rats in the absence (solid bars) or presence (hatched bars) of IL-1β (1 ng/mL). Reprinted with permission from Khan, Khan, and Dorrington (25), © The Endocrine Society, 1992.

Leydig cells to SIP is greater than that achieved in the presence of maximum concentrations of LH/hCG. SIP was found to be distinct from LH/hCG in physicochemical, immunological, and functional properties. SIP has a molecular weight of ~60,000 and an isoelectric point of 4.8. It copurifies with human albumin in several purification steps, but upon removal of the albumin with antibodies, the SIP-like activity is retained. Based on the above findings, it appeared that SIP was a unique protein that may play a role in the regulation of steroidogenesis in the gonads.

Based on the observations that a number of intragonadal peptides, such as growth factors and cytokines, have a wide variety of activities on gonadal cells, we examined the possibility that SIP may also be multifunctional and influence proliferation of cells in addition to modulating their differentiated functions. As shown in Figure 8.7, SIP had a pronounced effect on DNA synthesis by Leydig cells from 21-day-old rats. Pretreatment with 2-ng LH/mL augmented the response to subsequent treatment with SIP. By radioautography we showed that 22.4% ± 5.5% of the cells were labeled with [³H]thymidine after stimulation with SIP compared to 0.93% ± 0.6% of the cells in control wells (30). The stimulation of DNA synthesis by SIP was greater than that obtained with maximum concentrations of insulin and/or TGFα. The effects of SIP plus TGFα were not significantly different from those observed with SIP or TGFα alone. In contrast, insulin enhanced the effects of SIP in a synergistic manner (Fig. 8.8).

158   S.A. Khan et al.

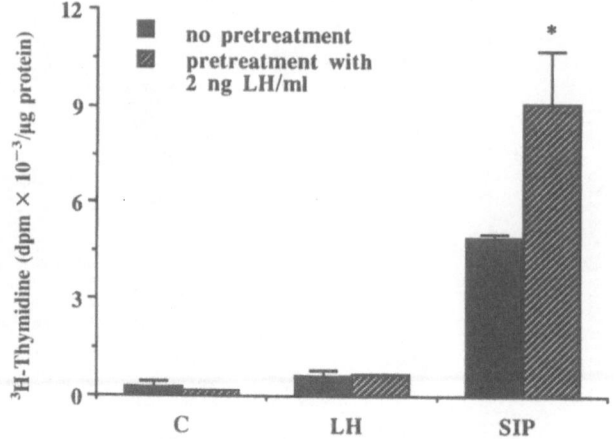

FIGURE 8.7. Potentiating effect of low levels of oLH (2 ng/mL) on SIP-stimulated DNA synthesis in 10-day-old Leydig cells. During the 48 h of culture from plating to the time of treatment, the cells were cultured in the absence (solid bars) or the presence (hatched bars) of LH (2 ng/mL). The cells were then treated with either a higher dose of LH (100 ng/mL) or SIP (50 µg/mL) for the subsequent 18-h culture period. [³H]Thymidine incorporation into DNA was determined as described in Fig. 8.3. (* = significantly different [P < 0.01] from cells precultured in the absence of oLH [C].) Reprinted with permission from Khan, Teerds, and Dorrington (33), © The Endocrine Society, 1992.

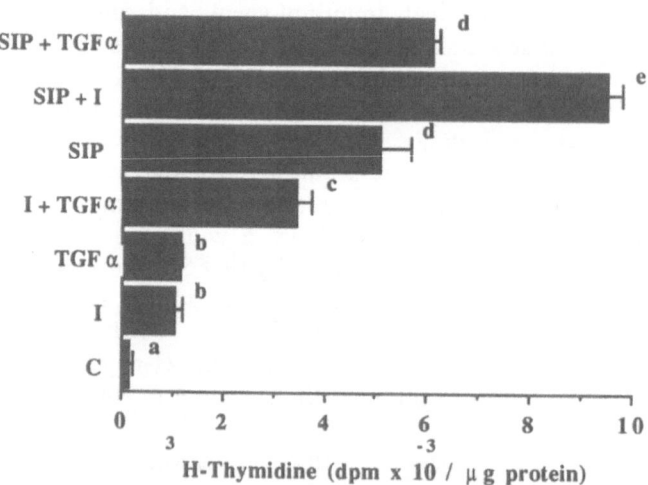

FIGURE 8.8. Influence of SIP on Leydig cell DNA synthesis in the presence or absence of insulin ([I], 0.5 µg/mL) and TGFα (10 ng/mL). Different superscripts indicate significant differences (P < 0.05) among different treatment groups. (C = control.) Reprinted with permission from Khan, Teerds, and Dorrington (33), © The Endocrine Society, 1992.

Striking changes in the morphology of Leydig cells in culture were observed after treatment with SIP; the cells became flattened and developed extended projections connecting adjacent cells. We concluded, therefore, that SIP not only stimulated steroidogenesis in Leydig cells, but also stimulated DNA synthesis and induced morphological changes. The latter properties of SIP, combined with the magnitude of the responses, identify SIP as a protein with multiple biological activities that is functionally distinct from other gonadal regulators.

Transforming Growth Factor β

Transforming growth factors β (TGFβs) are multifunctional regulators exerting a wide range of responses in normal and transformed cells. The actions of TGFβ can be divided into those that are related to growth—either stimulatory or inhibitory—and those that are unrelated to growth—for example, differentiation and chemotaxis. In the mouse (34) and rat (35) testis, TGFβ1 and TGFβ2 mRNAs are expressed in Sertoli cells and peritubular myoid cells throughout testicular development. TGFβ2 mRNA was expressed predominantly in the prepubertal testis (35). In the adult mouse testis, TGFβ1 mRNA was also found in the germ cells (27). Sertoli cells and peritubular/myoid cells secrete a TGFβ-like factor as assessed by binding to TGFβ receptors and from its bioactivity (36). In porcine Sertoli cells the amount of TGFβ-like activity secreted is increased after treatment with estradiol and decreased after treatment with

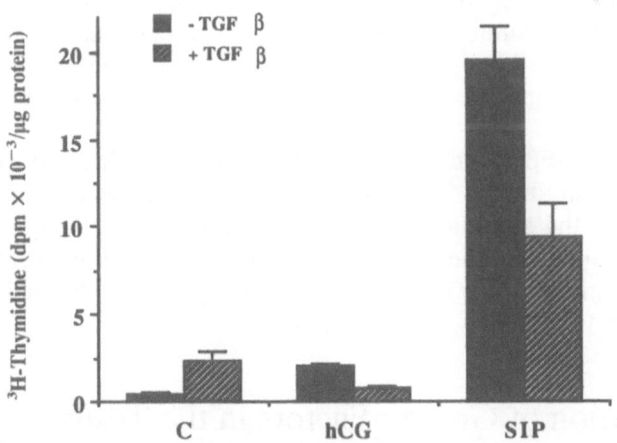

FIGURE 8.9. Incorporation of [³H]thymidine into DNA by Leydig cells isolated from 10-day-old rats and cultured with hCG (30 mU/mL) or SIP (50 μg/mL) in the presence or absence of TGFβ (10 ng/mL). Khan and Dorrington, unpublished data.

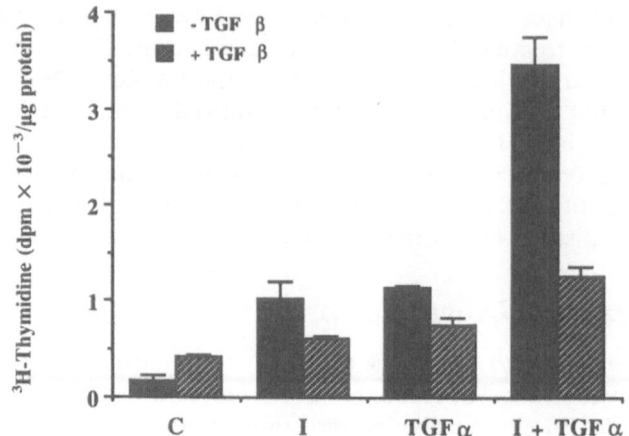

FIGURE 8.10. Effects of insulin ([I], 0.5 μg/mL), TGFα (10 ng/mL), and a combination of insulin and TGFα on Leydig cell DNA synthesis in the presence or absence of TGFβ (10 ng/mL). Khan and Dorrington, unpublished data.

FSH. TGFβ had no effect on the basal production of steroids by Leydig cells, but inhibited the steroidogenic response of the cells to LH (37).

In a number of systems, TGFα and TGFβ have opposing actions. For example, in the rat ovary TGFα attenuates the actions of TGFβ plus FSH on granulosa cell DNA synthesis and aromatase activity (38). In the bovine granulosa cells, TGFβ inhibits the growth-promoting effects of TGFα (39). To determine if TGFβ could influence Leydig cell growth, its effects alone and together with other growth-promoting factors on DNA synthesis were examined. TGFβ alone caused a significant stimulation of DNA synthesis in Leydig cells from 10- and 20-day-old rats (Figs. 8.4 and 8.9). The response to TGFβ was small compared to the response to TGFα.

Whereas TGFβ alone had a limited stimulatory capacity, it exerted a pronounced inhibitory effect on DNA synthesis when the cells were treated with the stimulatory agents identified earlier. We initially found that TGFβ abolished the small but significant effect of hCG (Fig. 8.9). On further examination TGFβ was shown to attenuate the actions of SIP, TGFα, and IGF-I (Figs. 8.9 and 8.10).

Localization of Growth Factors in the Testis

TGFα

The studies mentioned above showed that TGFα was able to stimulate DNA synthesis in Leydig cells isolated from immature rats, raising the possibility that this growth factor may be involved in increasing the size of the Leydig cell population during the prepubertal period. To determine if TGFα is present in the testis at various stages of development, the peptide was localized by immunoperoxidase staining using a monoclonal antibody that does not cross-react with *epidermal growth factor* (EGF) (40). In sections of testes from 21-day-old rats, approximately 50% of Leydig cells stained positively for TGFα; however, in some Leydig cells TGFα immunoreactivity could not be detected. At this age the adult-type Leydig cell population is expanding as a result of the increase in the number of mesenchymal cells differentiating into Leydig cells (9). In the adult rat testis, all Leydig cells stained intensely for TGFα. To confirm that TGFα was present in the Leydig cells of the adult and not in other interstitial cell types, rats were treated with EDS to destroy Leydig cells. After destruction of the Leydig cells, no staining for TGFα could be detected in the interstitial cell compartment (40).

TGFβ1 and TGFβ2 in Leydig Cells

As discussed above, TGFβ has the ability to inhibit DNA synthesis in Leydig cells in culture. Since the response of cells depends on the local concentration of TGFβ, we have localized TGFβ in the interstitial cell compartment to determine if there is a correlation between the presence of the peptide and the degree of proliferation of Leydig cells (41).

In sections of the testes from 7-day-old neonatal rats, appreciable staining for TGFβ1 was seen, whereas TGFβ2 was low or undetectable. Immunoreactivity for TGFβ1 and TGFβ2 was detected in Leydig cells of testes from 21-day-old rats; however, the proportion of positive cells was variable. Leydig cells that were in clusters located in close proximity to the seminiferous tubule stained positively, whereas isolated Leydig cells were often negative (41). During subsequent testicular development characterized by a rapid increase in the number of Leydig cells, the proportion of TGFβ1- and TGFβ2-positive Leydig cells decreased until they could no longer be seen by 35 days of age and throughout adult life.

Discussion

Our recent studies have given us new insights into the biology of Leydig cells. The main focus was to identify the factors that were required for the proliferation of the Leydig cells in the rat testis prior to puberty. During this prepubertal period the wave of mitotic activity in the Leydig cells results in an increase in the total population per testis from 2 million on day 21 of age, to 13 million on day 28 of age, and to approximately 26 million in the adult animal. This large increase in the population of

Leydig cells is essential for the completion of the first wave of spermato-genesis and its maintenance during adult life.

The generation of the full complement of adult Leydig cells that are steroidogenically active is dependent on LH in vivo. Treatment of imma-ture Leydig cells with LH in vitro, however, was ineffective in stimulating DNA synthesis. An important finding was that the ability of LH to stimulate DNA synthesis was manifested when the cells were incubated in the presence of both TGFα and IGF-I. Either TGFα or IGF-I alone did not allow LH to exhibit this stimulatory effect. Based on the latter observations that pointed to the requirement for interactions between TGFα and IGF-I for LH to be effective in promoting DNA synthesis in immature rats, we examined Leydig cells in vivo to determine if, indeed, TGFα was present during the prepubertal period. Using immunohisto-chemical techniques we showed that approximately 50% of the Leydig cells present in the testis at 21 days of age—that is, about 1 million cells—contained TGFα peptide. As the wave of proliferation proceeded, all the Leydig cells acquired intense staining for TGFα. During the period of growth and differentiation of the adult-type of Leydig cells, therefore, the population of TGFα-containing cells increases from 1 million to 26 million (Fig. 8.11). The increase in the appearance of TGFα-containing cells coincides with the time at which LH actions are essential in the growth-promoting process. IGF-I is also present in Leydig cells during the prepubertal period (17). Based on the latter observations, we would like to propose that LH is able to increase the numbers of Leydig cells in the testis prior to puberty due to the continued presence of IGF-I and due to the expression of TGFα peptide as this new population of Leydig cells develops.

In addition to their growth-promoting actions, TGFα and IGF-I have positive effects on androgen production (18). It is possible, therefore, that the generation of elevated levels of androgens that are required at puberty for the maintenance of spermatogenesis may involve interactions between LH, TGFα, and IGF-I. These interactions may be essential to achieve an increase in the total population of cells and increased total steroidogenic capacity (Fig. 8.11).

Two other very effective stimulators of DNA synthesis in Leydig cells were identified in the study, namely, SIP and interleukin-1β. SIP was isolated from human follicular fluid; however, its pronounced effects on both steroidogenesis and DNA synthesis in immature Leydig cells suggest a physiological role in the testis. SIP stimulated tyrosine phosphorylation of proteins in immature Leydig cells, suggesting that SIP may be similar to the growth factors that exert their effects on target cells by stimulating tyrosine kinase activity associated with their receptors (42). We have previously detected SIP-like activity in conditioned medium from imma-ture rat Sertoli cells; however, this activity has not been further charac-terized. Similarly, the pronounced effects of IL-1β on Leydig cells from

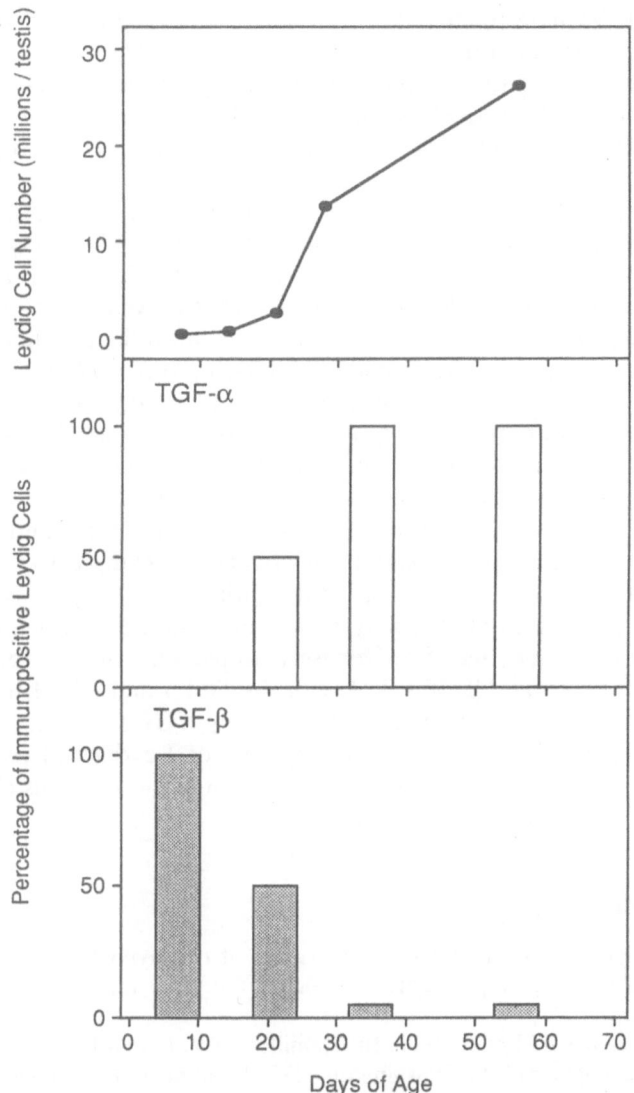

FIGURE 8.11. Percentage of Leydig cells that stain positively for TGFα and TGFβ during the development of the testis, as described in reference 41.

the immature testis suggest a physiological role; however, a developmental study of the expression of IL-1β mRNA and peptide in the testis is necessary to identify potential stages at which Leydig cells may be influenced by this cytokine.

Whereas LH and the intragonadal factors—TGFα, IGF-I, IL-1β, and SIP—stimulated DNA synthesis in Leydig cells, TGFβ was the only factor that we investigated that attenuated the response of the Leydig cells to these stimulatory agents. Localization of TGFβ in the rat testis showed that all the Leydig cells contained TGFβ1 at 7 days of age, but the percentage of positively staining cells declined so that by 21 days of age, approximately 50% of the Leydig cells contained TGFβ. As the wave of Leydig cell proliferation progressed to generate the adult-type Leydig cell, all the cells ceased to express TGFβ so that by 35 days of age, no TGFβ could be detected in the interstitial space (Fig. 8.11).

Since TGFβ inhibits the actions of LH, TGFα, and IGF-I on DNA synthesis in immature cells, we propose that it plays a physiological role to hold the growth of Leydig cells in abeyance until the prepubertal stage, even when exposed to intragonadal stimulators (e.g., TGFα and IGF-I). TGFβ also inhibits androgen production, and its presence in the interstitium prior to puberty may suppress steroidogenesis as well as proliferation, thereby avoiding precocious puberty. The decline in TGFβ levels during the prepubertal period and its subsequent loss from the interstitial space (Fig. 8.11) would allow LH, together with TGFα and IGF-I, to produce a population of steroidogenically active Leydig cells.

In conclusion, the growth and differentiation of Leydig cells in the interstitial compartment is strictly regulated. Whereas LH is essential for proliferation in vivo, its effectiveness is dependent on the relative proportions of TGFα and TGFβ. LH requires TGFα and IGF-I to stimulate growth; however its actions are abolished by TGFβ. The loss of TGFβ from Leydig cells at the onset of the wave of Leydig cell proliferation during prepuberty and the generation of a population of cells enriched in TGFα are consistent with this hypothesis.

References

1. Greep RO, Fevold HL, Hisaw FL. Effect of two hypophyseal gonadotropic hormones on the reproductive system of the male rat. Anat Rec 1936;65: 261–71.
2. Christensen AK. Leydig cells. In: Hamilton DW, Greep RO, eds. Handbook of physiology; vol 5. Washington, DC: American Physiological Society, 1975:57–94.
3. Dorrington JH, Armstrong DT. Effects of FSH on gonadal functions. Recent Prog Horm Res 1979;35:301–42.
4. Ahmed N, Haltmeyer GC, Eik-Nes KB. Maintenance of spermatogenesis in rats with intratesticular implants containing testosterone or dihydrotestosterone (DHT). Biol Reprod 1973;8:411–9.
5. Sharpe RM. Testosterone and spermatogenesis. J Endocrinol 1987;113:1–3.
6. Santulli R, Sprando RL, Awoniyi CA, Ewing EL, Zirkin BR. To what extent can spermatogenesis be maintained in the adult hypophysectomized rat testis with exogenously administered testosterone? Endocrinology 1990;126: 95–101.
7. Awoniyi CA, Sprando RL, Santulli R, Chandrashekar V, Ewing LL, Zinlin BR. Restoration of spermatogenesis by exogenously administered testosterone

in rats made azoospermic by hypophysectomy of withdrawal of luteinizing hormone alone. Endocrinology 1990;127:177–84.

8. Mendis-Handagama SMLC, Risbridger GP, de Kretser DM. Morphometric analysis of the components of the neonatal and the adult rat testis interstitium. Int J Androl 1987;10:525–34.

9. Hardy MP, Zirkin BR, Ewing LL. Kinetic studies on the development of the adult population of Leydig cells in testis of the pubertal rat. Endocrinology 1989;124:762–70.

10. de Kretser DM, Kerr JB. The cytology of the testis. In: Knobil E, Neill J, eds. The physiology of reproduction. New York: Raven Press, 1988:837–932.

11. Kerr JB, Knell CM. The fate of fetal Leydig cells during the development of the fetal and postnatal rat testis. Development 1988;103:535–44.

12. Tapanaien J, Kuopio T, Pelliniemi LJ, Huhtaniemi I. Rat testicular endogenous steroids and number of Leydig cells between the fetal period and sexual maturity. Biol Reprod 1984;31:1027–35.

13. Teerds KJ, Closset J, Rommerts FFG, et al. Effects of pure FSH and LH preparations on the number and function of Leydig cells in immature hypophysectomized rats. J Endocrinol 1989;120:97–106.

14. Sharpe RM, Maddocks S, Kerr JB. Cell-cell interaction in the control of spermatogenesis as studied using Leydig cell destruction and testosterone replacement. Am J Anat 1990;188:3–20.

15. Pledger WJ. Regulation of cell proliferation: serum growth factors control an ordered series of G_1 events. In: Veneziale CM, ed. Control of cell growth and proliferation. New York: Van Nostrand-Reinhold, 1985:108–31.

16. Khan S, Teerds K, Dorrington J. Growth factor requirements for DNA synthesis by Leydig cells from the immature rat. Biol Reprod 1992;46:335–41.

17. Handelsman DJ, Spaliviera JA, Scott CD, Baxter RC. Identification of insulin-like growth factor-I and its receptors in the rat testis. Acta Endocrinol 1985;109:543–9.

18. Verhoeven G, Gailleau J. Stimulatory effects of epidermal growth factor on steroidogenesis in Leydig cells. Mol Cell Endocrinol 1986;47:99–106.

19. Khan SA, Schmidt K, Hallin P, DiPauli R, De Geyter Ch, Nieschlag E. Human testis cytosol and ovarian follicular fluid contain high amount of interleukin-1-like factor(s). Mol Cell Endocrinol 1988;58:221–30.

20. Khan SA, Soder O, Syed V, Gustafsson K, Lindh M, Ritzen EM. The rat testis produces large amount of an interleukin-1-like factor. Int J Androl 1987;10:495–503.

21. Calkins JH, Guo H, Sigel MM, Lin T. Tumor necrosis factor-alpha enhances inhibitory effects on interleukin-1 beta on Leydig cell steroidogenesis. Biochem Biophys Res Commun 1990;166:1313–8.

22. Pollänen P, Söder O, Parvinen M. Interleukin-1α stimulation of spermatogonial proliferation in vivo. Reprod Fertil Dev 1989;1:85–7.

23. Parvinen M, Söolev O, Mali P, Fröysa B, Ritzen EM. In vitro stimulation of stage-specific deoxyribonucleic acid synthesis in rat seminiferous tubule segments by interleukin-1α. Endocrinology 1991;129:1614–20.

24. Takao T, Mitchell WM, Tracey DE, DeSouza EB. Identification of interleukin-1 receptors in the mouse testis. Endocrinology 1990;127:251–8.

25. Khan SA, Khan SJ, Dorrington JH. Interleukin-1 stimulates deoxyribonucleic acid synthesis in immature rat Leydig cell in vitro. Endocrinology 1992;131:1853–7.

26. Gerard N, Syed V, Bardin W, Genelet N, Jegou B. Sertoli cells are the site of interleukin-1α synthesis in rat testis. Mol Cell Endocrinol 1991;82:R13–6.
27. Christensen AK, Gillim SW. The correlation of fine structure and function in steroid-secreting cells with emphasis on those of the gonads. In: Mckerns KW, ed. The gonads. New York: Appleton-Century-Crofts, 1969:415–88.
28. Yee JB, Hutson JC. Effects of testicular macrophage-conditioned medium on Leydig cells in culture. Endocrinology 1985;116:2682–4.
29. Wang D, Nagpal M, Calkins JH, Chang W, Sigel MM, Lin T. Interleukin-1β induces interleukin-1α messenger ribonucleic acid expression in primary cultures of Leydig cells. Endocrinology 1991;129:2862–6.
30. Khan SA, Hallin P, Barlett J, DeGeyter Ch, Nieschlag E. Characterization of a factor from ovarian follicular fluid which stimulates Leydig cell testosterone production. Acta Endocrinol (Copenh) 1988;188:283.
31. Khan SA, Keck C, Gudermann T, Nieschlag E. Isolation of a protein from human ovarian follicular fluid which exerts major stimulatory effects on in vitro steroid production of testicular ovarian and adrenal cells. Endocrinology 1990;126:3034–52.
32. Khan SA, Keck C, Nieschlag E. Stimulation of Leydig cells steroidogenesis by a factor from human ovarian follicular fluid. In: Findlay J, Haseltine H, eds. Growth factors in fertility regulation. UK: Cambridge University Press, 1991:185–93.
33. Khan SA, Teerds KJ, Dorrington JH. Steroidogenesis-inducing protein promotes deoxyribonucleic acid synthesis in Leydig cells from immature rats. Endocrinology 1992;130:599–606.
34. Watrin T, Scotto L, Assoian RK, Wolgemuth DJ. Cell lineage specificity of expression of the murine transforming growth factor-β3, and transforming growth factor-β1 genes. Cell Growth Differ 1991;2:77–83.
35. Mullaney BP, Glenn B, Skinner MK. Cell-cell interactions in the testis: the role of transforming growth factors [Abstract 520]. Biol Reprod 1991; 44(suppl 1).
36. Skinner MK, Moses HL. Transforming growth factor-β gene expression and action in the seminiferous tubule: peritubular cell-Sertoli cell interactions. Mol Endocrinol 1989;3:625–34.
37. Benahmed M, Sondoillet C, Chauvin MA, DePeretti E, Morena AM. On the mechanisms involved in the inhibitory and stimulating actions of transforming growth factor-β on porcine testicular steroidogenesis: an in vitro study. Mol Cell Endocrinol 1989;67:155–64.
38. Bendell JJ, Dorrington JH. Epidermal growth factor influences growth and differentiation of rat granulosa cells. Endocrinology 1990;127:533–40.
39. Lobb DK, Dorrington JH. Intraovarian regulation of follicular development. Anim Reprod Sci 1992;28:343–54.
40. Teerds KJ, Rommerts FFG, Dorrington JH. Immunohistochemical detection of transforming growth factor-α in Leydig cell during the development of the rat testis. Mol Cell Endocrinol 1990;69:R1–6.
41. Teerds KJ, Dorrington JH. Localization of transforming growth factor β_1 and β_2 during testicular development in the rat. Biol Reprod 1993;48:40–5.
42. Khan SA, Dorrington JH, Moran MF. Steroidogenesis-inducing protein stimulates protein-tyrosine kinase activity in rat Leydig cells. Endocrinology 1993;132:109–14.

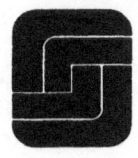

9

Use of Two-Compartment Cultures of Sertoli Cells for Investigating Blood-Testis Barrier Physiology

ANNA STEINBERGER, ANDRZEJ JANECKI, AND ANDRZEJ JAKUBOWIAK

The epithelial layer formed by *Sertoli cells* (Sc) within the seminiferous tubules of the mammalian testis is one example of a functionally polarized epithelium. The ability of absorptive, secretory, and transporting epithelia to perform their specialized functions depends on polarization of their plasma membranes into distinct apical and basolateral domains and on the integrity of intercellular tight (occluding) junctions (1–4).

Significant progress in clarifying the nature and regulation of polarized cell functions can be attributed, in large part, to the development of cell cultures on permeable membranes (1, 4, 5) that allowed long-term maintenance of cell polarity in vitro (6, 7). Permeable membranes have been used to culture a variety of epithelial cells, including monkey kidney cells (3), mouse mammary epithelial cells (7), and uterine epithelial cells (8). Their use as cell supports in two-compartment chambers was adapted for culture of mammalian Sc that under appropriate culture conditions become morphologically and functionally polarized and resemble, in many respects, their in vivo counterparts (9–12). Sertoli cell cultures have been utilized to explore vectorial secretion, iron transport, and the formation and regulation of Sc tight junctions (13, 14), as well as the mode of action of reproductive toxicants suspected to cause testicular damage by altering the blood-testis barrier (15).

During testicular maturation at the time of puberty, the Sc develop highly specialized tight junctions that divide the seminiferous tubular space into the basal and adluminal compartments. They also provide the structural basis for the so-called blood-testis barrier that restricts the passage of ions and macromolecules from interstitial fluid into the adluminal compartment. Consequently, germ cells located in the adluminal compartment (e.g., spermatocytes and spermatids) are exposed to a unique microenvironment that is maintained by the blood-testis barrier and apical Sc secretions and that is believed to be essential for the differentiation of germ cells into mature spermatozoa. Thus, the functional integrity of the blood-testis barrier is very critical for normal testicular function, particularly the spermatogenic process.

The formation of Sc tight junctions during puberty coincides with the development of seminiferous tubular lumen and the first appearance of spermatocytes in the adluminal compartment; however, despite their physiological importance, little is known about the regulatory mechanisms involved in the initial formation, maturation, and subsequent maintenance of the Sc tight junctions. Reports in the literature suggest that in humans (16–17) and rodents (18–20), *follicle stimulating hormone* (FSH) may be necessary for their initial appearance during puberty, whereas their maintenance after the blood-testis barrier has been established may be independent of hormones (16, 18, 19). Other publications, however, suggest that the formation of Sc tight junctions can occur, although with a considerable delay, in the absence of measurable FSH (21–24). In addition to FSH, testosterone has been considered to be a hormonal regulator of Sc tight junctions because treatment of men with idiopathic hypogonadotropic hypogonadism with *human chorionic gonadotropin* (hCG) stimulated the formation of Sc tight junctions and germ cell differentiation (25). Thus, clarification of the factors and regulatory mechanisms involved in the formation and maintenance of the blood-testis barrier has considerable clinical relevance to male fertility (16, 17, 25).

Use of Two-Compartment Cultures for In Vitro Studies of Sc Tight Junctions

The development of two-compartment Sc cultures provided a very useful model that allowed for the first time exploration of the formation of Sc tight junctions and their regulation by hormones and other factors in vitro (13–15). Although the results from in vitro experiments should not be directly extrapolated to normal physiological situations (12), they have provided important new information and clues as to what hormones/factors may affect the blood-testis barrier in vivo.

For our studies Sc from 18-day-old rats were cultured on Matrigel-coated Nucleopore membranes (Nucleopore Co., Pleasanton, CA) in two-compartment chambers, as previously described (10, 11). The chemically defined culture medium, with or without added hormones/factors, was replenished at 24- to 48-h intervals, and spent media, collected separately from the apical and basal compartments, were stored frozen for subsequent radioimmunoassay of Sc secretory products (e.g., transferrin and inhibin).

The formation of Sc tight junctions was monitored by two methods: (i) diffusion of ^{3}H-inulin across the epithelial layer (9–11) and (ii) daily

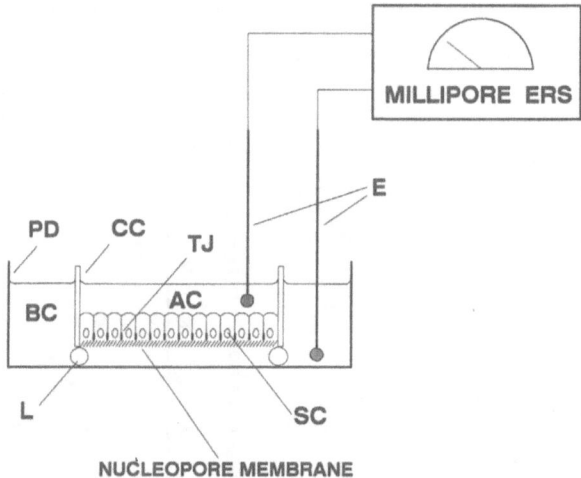

FIGURE 9.1. Schematic diagram of TER measurements in two-compartment culture of Sertoli cell (SC) monolayer using the Millipore ERS device. Shown are electrodes (E), apical compartment (AC), basal compartment (BC), tight junctions (TJ), culture chamber (CC), Petri dish (PD), and legs of culture chamber (L). The measurements were carried out at 27°C under standard conditions. The readout was in ohms.

measurements of *transepithelial electrical resistance* (TER) using the Millipore ERS device (Millipore Corp., Bedford, MA) (13–15). Although both of these methods have been used as indicators of tight junction formation, the TER values are considered to be the most sensitive parameter of tight junction status (26, 27). Figure 9.1 shows diagrammatically the use of TER measurements in two-compartment cultures of Sc, and Figure 9.2 illustrates the inverse relationship between changes in ^{3}H-inulin diffusion and TER of Sc monolayers maintained in the two-compartment culture system. Although the absolute values of these measurements will vary depending on cell density per surface area of the filter support and the tight junction status, initially, there is a good reciprocal correlation between the decline in monolayer permeability to ^{3}H-inulin and increase of TER with time in culture. Subsequently, however, the TER values continue to increase without any further decline in ^{3}H-inulin diffusion.

These observations generally agree with the results reported by Onoda et al. (28). The increasing TER values are believed to reflect greater complexity of the developing tight junctions (26, 27), beyond that needed to maximally (by 80%) reduce the ^{3}H-inulin diffusion. An additional advantage of the TER method is that repeated measurements can be performed on the same sets of cultures without adversely affecting cell viability or culture sterility (30). TER measurements were, therefore,

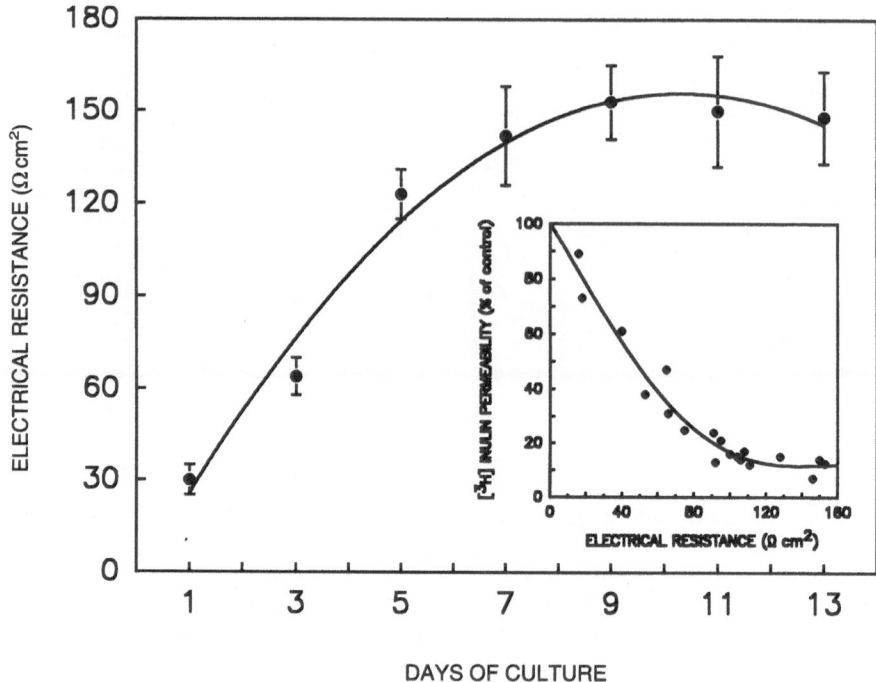

FIGURE 9.2. Changes in TER with time of culture and the inverse relationship between the changes in ³H-inulin permeability and TER (insert). Note the increase of electrical resistance after ³H-inulin permeability has plateaued.

utilized in our laboratory as the main parameter for monitoring the formation of Sc tight junctions. These measurements were carried out daily at 27 ± 1°C under standardized conditions.

We explored the effects of FSH and steroid hormones (testosterone, dihydrotestosterone, and 17β-estradiol) on the development of tight junctions by Sc cultured in the two-compartment system for up to 11 days. The hormones were added at different concentrations, either alone or in combination, on day 1 or 5 of culture. Moreover, since the Sc tight junctions are formed in male rats at the time of testicular descent from an abdominal to scrotal position (temperature difference of 2°C–5°C), we also compared the effects of two incubation temperatures: 36.5°C (abdominal) and 33°C (scrotal). The changes in TER were, in some experiments, correlated with the vectorial secretion of inhibin and transferrin measured by radioimmunoassay methods, as previously described (29, 30). The results are shown in Figures 9.3 through 9.9.

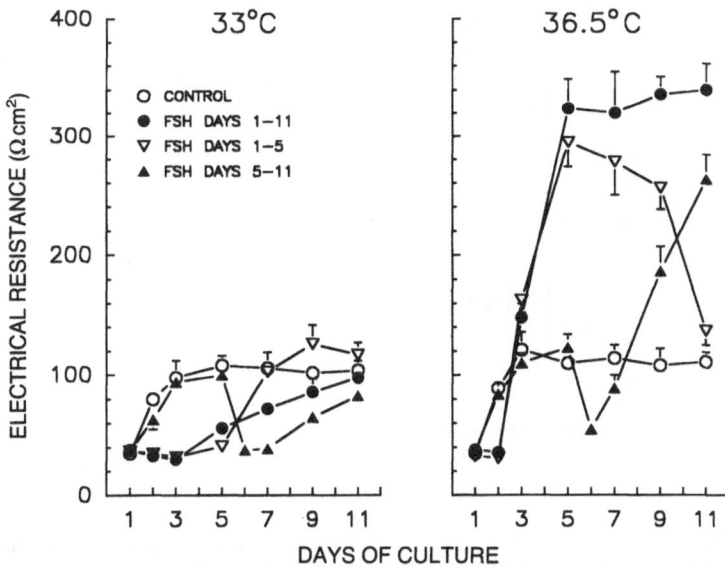

FIGURE 9.3. Changes in TER at 33°C and 36.5°C in the absence (CONTROL) or presence of FSH (200 ng/mL) at various times of culture. The hormone was present either during the entire culture period (FSH DAYS 1–11), only on days 1–5 of culture (FSH DAYS 1–5), or from day 5 onward (FSH DAYS 5–11). Values are mean ± SD from triplicate cultures in 1 representative experiment. Reprinted with permission from Janecki, Jakubowiak, and Steinberger (13), © The Endocrine Society, 1991.

Effects of FSH on TER of Sc Monolayers Incubated at 33°C and 36.5°C

In the absence of added hormones, the TER values increased at both temperatures from $30–40\,\Omega\,cm^2$ to approximately $100\,\Omega\,cm^2$ after 3–4 days of culture, then remained stable for up to 11 days (Fig. 9.3). At 33°C the continuous presence of FSH (NIH-oFSH-S17: 200 ng/mL) from day 1 of culture initially slowed the increase of TER that then approached the untreated control levels by days 9–11. A 24-h decline in TER with gradual recovery to control level by day 11 was also observed when FSH was added on day 5 of culture. Removal of FSH on day 5, following its presence from day 1, caused an increase of TER to control values within the subsequent 48 h.

The responses to FSH at 36.5°C were much more pronounced (Fig. 9.3). In the continuous presence of FSH, the TER values increased dramatically (2- to 4-fold of control) by days 3–4 of culture after an initial 24-h delay, then remained stable for up to 11 days. As observed at the

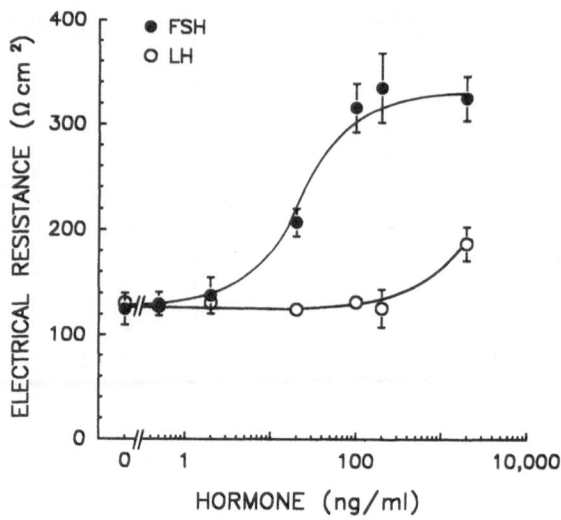

FIGURE 9.4. TER values for Sc monolayers exposed for 7 days to increasing doses of FSH or LH. The Sc cultures were incubated at 36.5°C, and the media + hormones were replenished every 48 h. Values are the mean ± SD from triplicate cultures in 1 representative experiment. Reprinted with permission from Janecki, Jakubowiak, and Steinberger (13), © The Endocrine Society, 1991.

lower incubation temperature, the addition of FSH on day 5 resulted in an initial decline and then a rise of TER, which at 36.5°C exceeded the control level by 2.5-fold. Removal of FSH on day 5 caused a gradual decline of TER to control value by day 11.

The TER response to FSH was dose dependent, with ED_{50} = 23 ng/mL and 80 ng/mL eliciting maximal TER increase. It was also hormone specific, as it could not be reproduced by using similar doses of *luteinizing hormone* (LH) (Fig. 9.4). The differences in TER response to FSH at different temperatures in vitro suggest that the elevated temperature to which the testes are exposed prior to their descent into the scrotum may be important for more efficient formation of Sc tight junctions.

Failure of Antiproteases to Prevent FSH-Induced Transient Drop of TER

Since FSH is known to stimulate the secretion of Sc proteases (e.g., plasminogen activator) that may have caused the initial decline in TER observed after the addition of FSH to the cultures (31), we explored the effects of several protease inhibitors on the TER. Preincubation of cultures for 1 h with either aprotinin (serine protease inhibitor: 0.5 mg/mL), e-aminocaproic acid (plasminogen activator inhibitor: 10 µM, Sigma

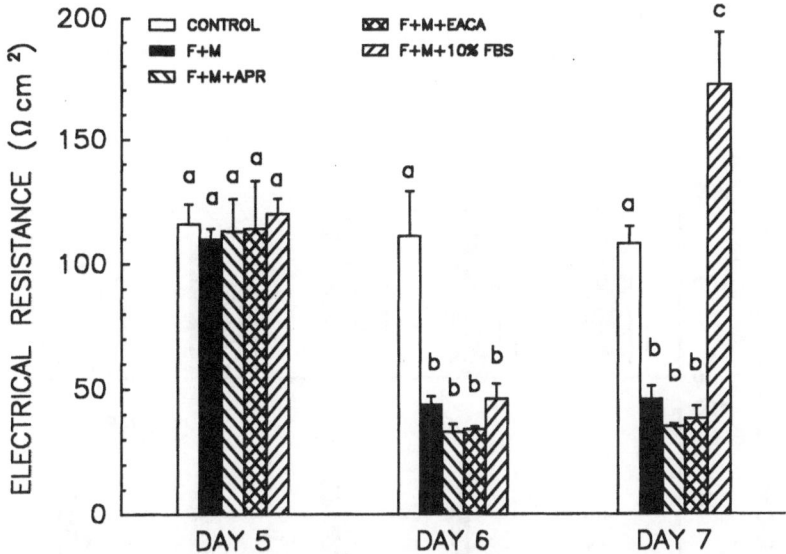

FIGURE 9.5. Effect of antiproteases on the FSH-induced decrease in TER in Sc monolayers cultured for 5 days at 36.5°C in the absence of FSH. On day 5, FSH (200 ng/mL) + MIX (0.2 mM) were added to triplicate cultures either without (F + M) or with 0.5-mg/mL aprotinin (F + M + APR); 10 μM e-aminocaproic acid (F + M + EACA); or 10% FBS (F + M + 10% FBS). TER was measured after 24 and 48 h. Values are mean ± SD from triplicate cultures in 2 separate experiments. Different superscripts indicate a significant ($P < 0.01$) difference. Reprinted with permission from Janecki, Jakubowiak, and Steinberger (13), © The Endocrine Society, 1991.

Chemical Co.), or 10% fetal bovine serum (nonspecific protease inhibitor)—all of which prevented a rapid drop in TER following brief exposure of the Sc cultures to trypsin—failed to eliminate the early decline of TER in response to FSH (Fig. 9.5). Thus, the reason for this initial decline of TER remains unclear, although Ailenberg and Fritz (31) found an increase of plasminogen activity and rate of ^{3}H-inulin diffusion in Sc cultures treated with FSH.

Effects of Gonadal Steroids on TER of Sc Monolayers

In the presence of testosterone (10 μM) alone, a rapid increase of TER was observed at either temperature, but it was more pronounced and longer lasting at 36.5°C (up to 380 Ω cm^2). The effect of testosterone was dose dependent within a concentration range of 0.001 to 10 μM and ED_{50} of ~0.1 μM (Fig. 9.6). The concomitant presence of testosterone (10 μM) and FSH at 36.5°C resulted in several-fold-higher TER values than those

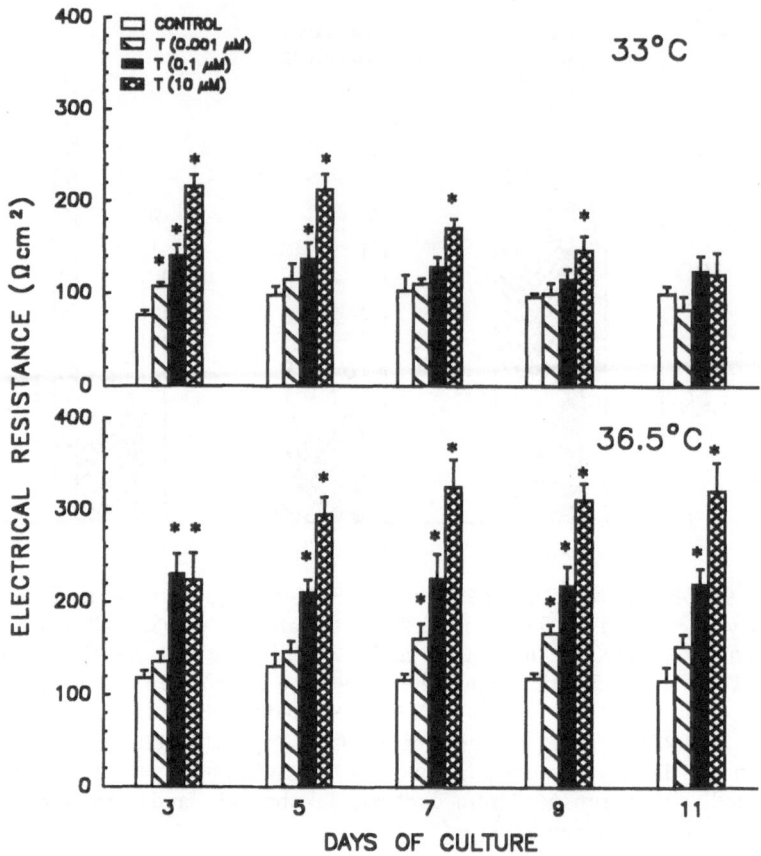

FIGURE 9.6. Effect of testosterone (T) on TER development in Sc monolayer cultures. The cultures were incubated for 11 days at 33°C or 36.5°C in the absence (CONTROL) or presence of increasing concentrations of T (0.001, 0.1, or 10 μM). Fresh hormone was added at each medium change. Values are mean ± SD from triplicate cultures in 1 representative experiment. (* = significant difference [$P <$ 0.01] from control on the same day.) Reprinted with permission from Janecki, Jakubowiak, and Steinberger (13), © The Endocrine Society, 1991.

achieved with either hormone alone, suggesting synergistic action between these two hormones. FSH at a dose of 2 ng/mL significantly enhanced the effect of testosterone, but was ineffective when used alone, and 200 ng/mL showed the greatest effect (Fig. 9.7).

When other gonadal steroids were tested, dehydrotestosterone was 30%–45% more effective than testosterone or androstenediol, whereas 17β-estradiol used alone or in combination with FSH had no effect on the TER (Fig. 9.8). It should be pointed out that the increase of TER in Sc cultures incubated at 33°C with testosterone plus FSH (not shown) was

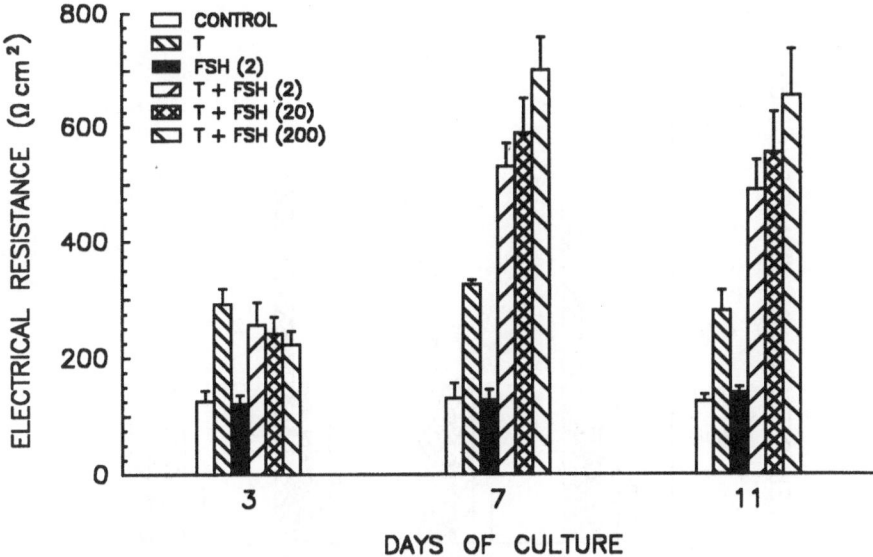

FIGURE 9.7. Effects of increasing FSH doses on the testosterone (T)-stimulated development of TER. Sc monolayers were cultured for 7 days at 36.5°C in control medium (CONTROL) or medium supplemented with T (10 µM), 2-ng/mL FSH (FSH [2]), T (10 µM) + 2-ng/mL FSH (T + FSH [2]), T (10 µM) + 20-ng/mL FSH (T + FSH [20]), or T (10 µM) + 200-ng/mL FSH (T + FSH [200]). Media with fresh hormones were replaced every other day. Values are mean ± SD from triplicate cultures in 1 representative experiment. Reprinted with permission from Janecki, Jakubowiak, and Steinberger (13), © The Endocrine Society, 1991.

comparable to that observed with testosterone alone, again emphasizing the importance of elevated temperature in the hormonal regulation of the developing Sc tight junctions.

The effects of incubation temperature and hormonal treatments on the changes of TER did not appear to be due to changes in Sc number, as the average DNA contents of cultures treated at either 33°C or 36.5°C were not significantly different from each other or from untreated control values. This is consistent with the previously reported lack of detectable Sc proliferation in 18-day-old rat testes in vivo or in vitro (32). Also, cell viability was not significantly affected by the different treatments.

Since the effects of hormonal treatments and incubation temperature on TER could be due to alterations of metabolic activity, we compared the changes of TER with the vectorial secretion of inhibin and transferrin. Although the total inhibin secretion during the first 5 days of culture was considerably higher (35%–60%) at 36.5°C compared to 33°C, the rates of vectorial secretion at 33°C and 36.5°C for days 5–7 were comparable (Fig. 9.9) and similar to those previously reported (30). Moreover, the

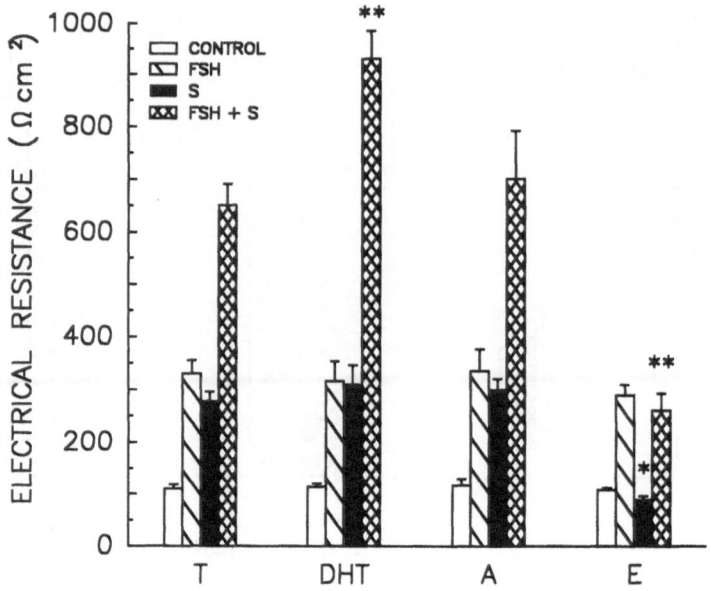

FIGURE 9.8. Development of TER by Sc monolayers incubated for 7 days at 36.5°C in either control medium (CONTROL) or medium supplemented with FSH (200 ng/mL), one of the 4 tested steroids (S: 10 μM), or with FSH + steroid (FSH + S), 200 ng/mL and 10 μM, respectively. The steroids tested were testosterone (T), dihydrotestosterone (DHT), androstenedione (A), and 17β-estradiol (E). Values are the mean ± SD from triplicate cultures in 1 representative experiment. (* = significant difference [$P < 0.01$] from other groups treated with single steroid; ** = significant difference [$P < 0.01$] from FSH + T group.) Reprinted with permission from Janecki, Jakubowiak, and Steinberger (13), © The Endocrine Society, 1991.

basal:apical ratios remained essentially unaltered during the entire culture period despite the changing TER values. A similar pattern was observed with transferrin secretion.

These results suggest that vectorial secretion of Sc products may be only partly dependent on the presence of tight junctions, as was reported for other polarized epithelia (33). These results are in contrast with the observations published by Onoda et al. (28), who found that vectorial secretion of transferrin in two-compartment Sc cultures increased with an increase of TER. This discrepancy is most likely due to differences in the experimental conditions employed. The TER values achieved in our laboratory have been much higher (up to $1200 \, \Omega \, cm^2$) than those reported by Onoda and colleagues ($85 \, \Omega \, cm^2$). The high TER values are a clear indication of monolayer patency and high complexity of the tight junctions (26, 27). Indeed, confluent monolayers of cells that do not form

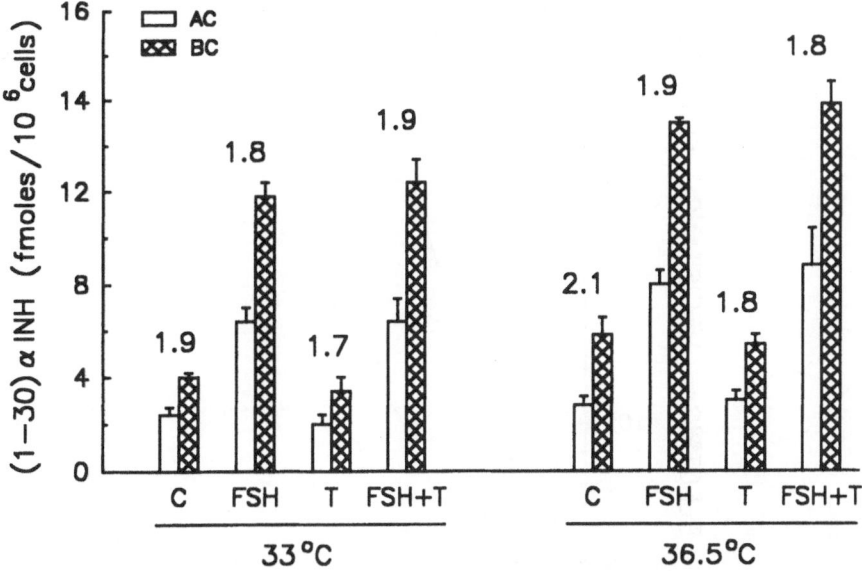

FIGURE 9.9. Vectorial secretion of inhibin (INH) by Sc monolayers incubated at 33°C or 36.5°C in control medium (C) or in presence of FSH (200 ng/mL), T (10 μM), or FSH (200 ng/mL) + T (10 μM). Fresh hormones were added at each medium change, and inhibin that accumulated in the basal (BC) and apical (AC) compartment between days 5 and 7 was measured by RIA. The numbers above the bars indicate the BC:AC ratio. Values are mean ± SD from 3 cultures in 1 representative experiment. Reprinted with permission from Janecki, Jakubowiak, and Steinberger (13), © The Endocrine Society, 1991.

tight junctions—that is, peritubular myoid cells or 3T3 cells—seldom showed TER values greater than $25 \, \Omega \, cm^2$. This is also the case when Sc isolated from very young rats (≤10 days old) prior to the appearance of tight junctions are cultured in vitro.

Mechanisms of FSH and Testosterone Action on the Formation of Sc Tight Junctions In Vitro

In an attempt to clarify the molecular mechanism(s) by which FSH and androgens regulate the development of Sc tight junctions, we explored the effects of *dibutyryl cyclic AMP* ([dbcAMP], 0.8–500 μM), phorbol ester (100 nM), *1-methyl-3-isobutylxanthine* ([MIX], 0.2 mM), cholera toxin (10 ng/mL), and forskolin (50 μM) in the absence or presence of FSH (200 ng/mL) and/or testosterone (10 μM) using two-compartment cultures of immature (18-day-old) rat Sc (14). An increase of TER was utilized as an indicator of tight junction formation and maturation. The results are shown in Figures 9.10 to 9.12.

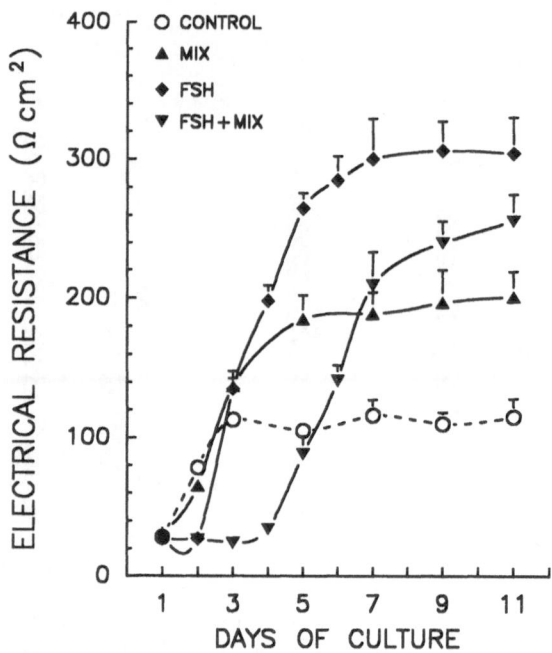

FIGURE 9.10. Patterns of TER development in monolayer cultures of immature rat Sc incubated at 36.5°C in the continuous presence of either MIX, FSH, or FSH + MIX. The control cultures were incubated in medium without these additives (CONTROL). The data are mean ± SD from 3 monolayers in a representative experiment. Reprinted with permission from Janecki, Jakubowiak, and Steinberger (14).

As mentioned earlier, the continuous presence of FSH from day 1 of culture at 36.5°C significantly increased the TER after an initial 24-h decline. This initial delay was prolonged to 3–4 days by the concomitant presence of MIX (0.2 mM), and the subsequent FSH-stimulated increase of TER remained below that achieved with FSH alone. When MIX only was added to the cultures, the TER values were about halfway between the FSH-stimulated and untreated (control) levels. The addition of dbcAMP alone or in combination with FSH and/or testosterone had a concentration-dependent biphasic effect. Low concentrations (4–20 µM) increased the TER, whereas high concentrations (100–500 µM) had an opposite effect. As shown in Figure 9.11, cholera toxin and forskolin at the indicated amounts closely mimicked the effects of FSH. Phorbol ester alone did not alter the TER, and its combined use with FSH did not affect the FSH-stimulated increase of TER. However, the addition of phorbol ester together with testosterone increased the TER and produced

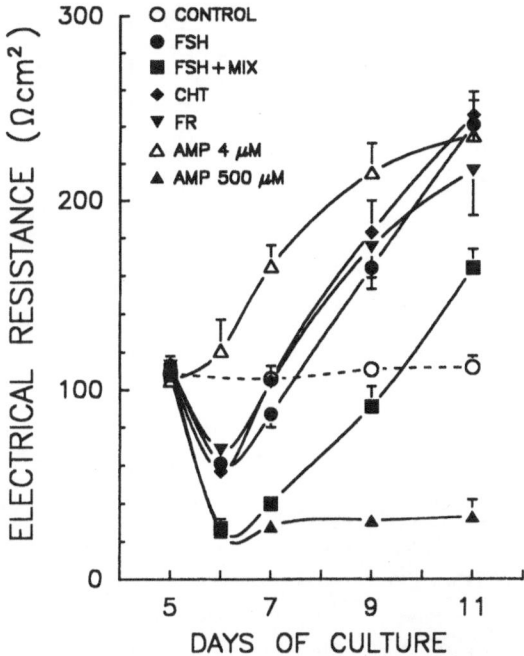

FIGURE 9.11. Changes in TER of Sc monolayers exposed from day 5 of culture to either FSH, FSH + MIX, cholera toxin ([CHT] 10 ng/mL), forskolin (FR), 4 μM dbcAMP (AMP 4 μM), or 500 μM dbcAMP (AMP 500 μM). The monolayers were cultured in control medium for 5 days. The indicated supplements were added on day 5 and were replaced every 48 h. The values are mean ± SD from a representative experiment. Reprinted with permission from Janecki, Jakubowiak, and Steinberger (14).

TER values similar to those observed with FSH plus testosterone (Fig. 9.12).

These results indicate that the FSH-induced rise of TER in Sc monolayers cultured on permeable membranes in two-compartment chambers is, at least partly, mediated by cAMP since agents that increase endogenous cAMP, such as cholera toxin and forskolin, closely mimicked the effects of FSH. Surprisingly, only low concentrations of dbcAMP significantly increased the TER and, like FSH, enhanced the testosterone-stimulated TER increase. On the other hand, phorbol ester, known to stimulate protein kinase C activity, enhanced the testosterone-induced TER increase, but had no effect alone or on TER increase in response to FSH.

The mechanism by which phorbol ester affects the formation of tight junctions is not clear. Results of a study on frog red blood cells published by Yoshimasa et al. (34) suggested a direct link between increased poly-

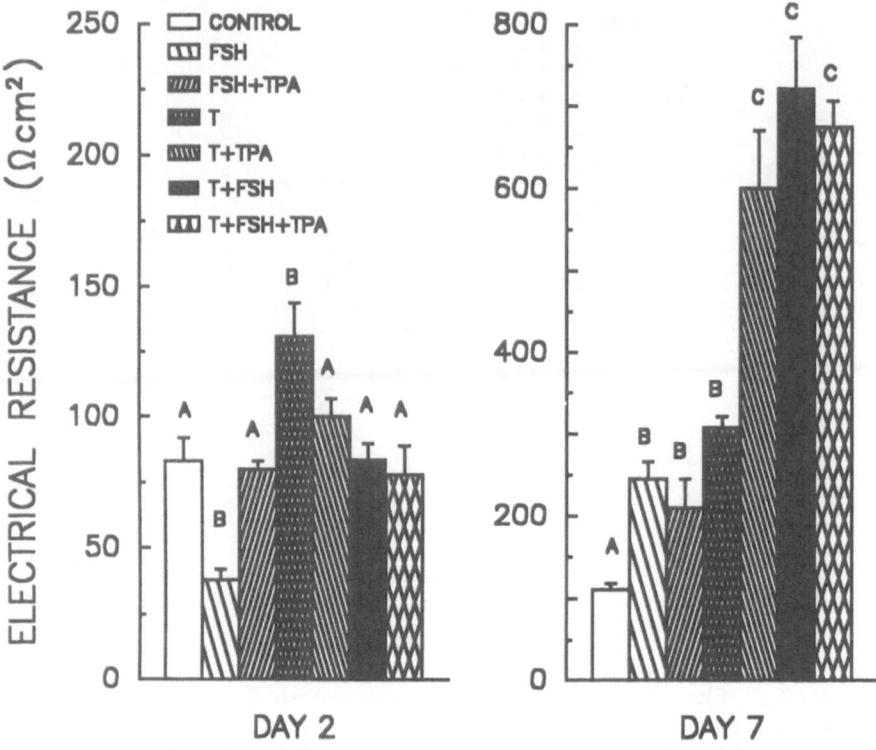

FIGURE 9.12. TER values observed in 2- and 7-day Sc cultures incubated in the continuous presence of either FSH; FSH + phorbol ester (FSH + TPA); testosterone (T); or testosterone + either phorbol ester (T + TPA), FSH (T + FSH), or FSH + phorbol ester (T + FSH + TPA). The values are mean ± SD from a representative experiment. Different letters above the bars indicate significant differences between the experimental groups. Reprinted with permission from Janecki, Jakubowiak, and Steinberger (14).

merase kinase C activity and the level of cAMP, probably due to phosphorylation of the catalytic unit of adenylate cyclase. Assuming that a similar mechanism operates in the Sc, a moderate increase in endogenous cAMP level could explain the potentiating effect of phorbol ester on the testosterone-induced increase of TER in Sc cultures. Taken together, our results suggest that FSH may regulate the formation of Sc tight junctions via a cAMP-mediated mechanism, whereas protein kinase C may be involved in modulation of the testosterone effect, implying a possible role of the phosphoinositol pathway in the formation of the blood-testis barrier.

Use of Two-Compartment Sc Cultures for Toxicological Studies

A number of testicular toxicants, including *cadmium chloride* ($CdCl_2$), have been reported to cause spermatogenic damage (35), apparently by increasing the permeability of the blood-testis barrier (35). Due to the complexities of performing toxicologic studies in vivo, the mechanism of $CdCl_2$ toxicity is still not known. Using two-compartment cultures of immature rat Sc, we investigated the effect of different concentrations ($0.75-24\,\mu M$) of $CdCl_2$ on several Sc functions: development of tight junctions (TER), vectorial secretion of inhibin (RIA), cell number (DNA content), and cell viability (MTT test).

As previously reported (15), the effects of $CdCl_2$ depended on the concentration of the toxicant, as well as on the onset and duration of exposure. At the highest cumulative doses employed (concentration × duration of exposure), the TER values decreased significantly and irreversibly during 13 days of culture, and this decrease was accompanied by a significant and irreversible drop in inhibin secretion, cell viability, and cell number. Within a narrow range of lower doses, the irreversible, or partially reversible, decrease of TER was accompanied by a transient decrease, or no change, of secretory activity and no significant changes in Sc cell number and/or viability. With still lower cumulative doses, the TER values rapidly decreased and then returned to control level within 3–4 days. In this group no changes in either inhibin secretion or cell viability were observed. These results suggest that $CdCl_2$ may selectively compromise, at least in vitro, the development and maintenance of the inter-Sc tight junctions without affecting the secretory activity or the cell number and viability.

Future Directions

It is evident that more work is needed to clarify the regulation of Sc tight junctions by hormones and other factors. The paracrine interactions between Sc and other cells in the testis must not be ignored, as it was shown both in vivo (37) and in vitro that germ cells (38–40) and peritubular myoid cells (29, 41) can significantly alter Sc functions and modulate the effects of hormones. Undoubtedly, the use of two-compartment Sc cultures will continue to provide a valuable approach for exploring the role of these interactions in the physiology of the blood-testis barrier. This in vitro model also offers a unique opportunity for exploring the function of such specific substances as Z0-1 (42, 43), testins (44), and other proteins believed to participate in the formation of Sc tight junctions. The ability to monitor the tight junction status by TER measurements under controlled conditions in vitro should yield valuable new information in this area. In addition, one of the most intriguing functional aspects of Sc junctions—that is, the process by which germ cells become translocated from the basal to adluminal compartment—may be clarified using cocultures of Sc and germ cells.

Finally, Sc cultures in two-compartment chambers are well suited for mechanistic studies of testicular toxicants that have been demonstrated, or implied, to affect Sc function(s) in vivo. The ability to test direct effects of suspected toxicants on specific Sc functions, including vectorial secretion and formation of tight junctions, provides a useful additional approach to clarifying the site and molecular mechanism of action of toxic substances that may be difficult to explore using living animals. The in vitro study of $CdCl_2$ effects (15), briefly described in this chapter, is one example of how the results from culture experiments can complement information derived from observations in vivo.

References

1. Cereijido M, Robbins ES, Dolan WJ, Rotonno A, Sabatini DD. Polarized monolayer formed by epithelial cells on a permeable and translucent support. J Cell Biol 1978;77:853–80.
2. Sabatini DD, Griepp EB, Boulan-Rodriguez E, et al. Biogenesis of epithelial cell polarity. Mol Cell Biol 1983;2:419–50.
3. Roth MG, Gundersen D, Patil N, Rodriguez-Boulan E. The large external domain is sufficient for the correct sorting of secreted or chimeric influenza virus hemagglutinins in polarized monkey kidney cell. J Cell Biol 1987;104: 769–82.
4. Simons K, Fuller SD. Cell surface polarity in epithelia. Annu Rev Cell Biol 1985;1:243–88.
5. Matlin KS, Simons K. Sorting of an apical plasma membrane glycoprotein occurs before it reaches the cell surface in cultured cells. J Cell Biol 1984; 99:2131–9.
6. Emmerman JT, Burwen SJ, Pitelka DR. Substrate properties influencing ultrastructural differentiation of mammary epithelial cells in culture. Tissue Cell 1979;11:109–19.
7. Li ML, Aggeler J, Farson DA, Hatter C, Hassell J, Bissell MJ. Influence of a reconstituted basement membrane and its components on casein gene expression and secretion in mouse mammary epithelial cells. Proc Natl Acad Sci USA 1987;84:136–40.
8. Glasser SR, Julian J, Decker GL, Tang J, Carson DD. Development of morphological and functional polarity in primary cultures of immature rat uterine epithelial cells. J Cell Biol 1988;107:2409–23.
9. Janecki A, Steinberger A. Polarized Sertoli cell functions in a new two-compartment culture system. J Androl 1986;7:69–71.
10. Janecki A, Steinberger A. Bipolar secretion of androgen-binding protein and transferrin by Sertoli cells cultured in a two-compartment culture chamber. Endocrinology 1987;120:291–8.

11. Byers SW, Hadley MA, Djakiew D, Dym M. Growth and characterization of polarized monolayers of epididymal epithelial cells and Sertoli cells in dual environment culture chambers. J Androl 1986;7:59–68.
12. Kelly CW, Janecki A, Steinberger A, Russell LD. Structural characteristics of immature rat Sertoli cells in vivo and in vitro. Am J Anat 1991;192: 183–93.
13. Janecki A, Jakubowiak A, Steinberger A. Regulation of transepithelial electrical resistance in two-compartment Sertoli cell cultures: in vitro model of the "blood-testis" barrier. Endocrinology 1991;129:1489–96.
14. Janecki A, Jakubowiak A, Steinberger A. Effects of cyclic AMP and phorbol ester on transepithelial electrical resistance of Sertoli cell monolayers in two-compartment culture. Mol Cell Endocrinol 1991;82:61–9.
15. Janecki A, Jakubowiak A, Steinberger A. Effect of cadmium chloride on transepithelial resistance of Sertoli cell monolayers in two-compartment cultures, a new model for toxicological investigations of the "blood-testis barrier" in vitro. Toxicol Appl Pharmacol 1992;112:51–7.
16. DeKretser D, Burger HG. Ultrastructural studies of the human Sertoli cell in normal men and males with hypogonadotropic hypogonadism before and after gonadotropin treatment. In: Saxena BRIJB, Beling CG, Gandy HM, eds. Gonadotropins. New York: Wiley-Interscience, 1972:640–56.
17. Setchell BP, Brooks DE. Anatomy, vasculature, innervation, and fluids of the male reproductive tract. In: Knobil E, Neill J, eds. Physiology of reproduction. New York: Raven Press, 1988:753–836.
18. Bressler R. Dependence of Sertoli cell maturation on the pituitary gland in the mouse. Am J Anat 1976;147:447–56.
19. Hagenas L, Ploen L, Ekwall H. Blood-testis barrier: evidence for intact inter-Sertoli cell junctions after hypophysectomy in the adult rat. J Endocrinol 1978;76:87–91.
20. Russell LD, Bartke A, Goh JC. Postnatal development of the Sertoli cell barrier, tubular lumen, and cytoskeleton of Sertoli and myoid cells in the rat, and their relationship to tubular fluid secretion and flow. Am J Anat 1989; 184:179–89.
21. Vitale R, Fawcett DW, Dym M. The normal development of the blood-testis barrier and the effects of clomiphene and estrogen treatment. Anat Rec 1973;176:333–44.
22. Steinberger A, Heindel JJ, Lindsey JN, Elkington JSH, Sanborn BM, Steinberger E. Isolation and culture of FSH responsive Sertoli cells. Endocr Res Commun 1975;2:261–72.
23. Janecki A, Lukaszyk A, Jakubowiak A. Culture of rat Sertoli cells isolated with a modified procedure: morphological identification of cell population and cell reactivity. Folia Histochem Cytobiol 1981;19:135–42.
24. Meyer R, Posalaky Z, McGinley D. Intercellular junction development in maturing rat seminiferous tubules. J Ultrastruct Mol Struct Res 1977;61: 217–83.
25. Faruya S, Kumamoto Y, Ikegaki S. Blood-testis barrier in men with idiopathic hypogonadotropic eunuchoidism and postpuberal pituitary failure. Arch Androl 1980;5:361–7.
26. Gumbiner B. Structure, biochemistry, and assembly of epithelial tight junctions. Am J Physiol 1987;253:C747–58.

27. Madara J, Hecht G. Tight (occluding) junctions in culture (and native) epithelia. In: Matlin KS, Valentich JD, eds. Functional epithelial cells in culture. New York: Liss, 1989:131–63.
28. Onoda M, Suarez-Quian CA, Djakiew D, Dym M. Characterization of Sertoli cells cultured in bicameral chamber system: relationship between formation of permeability barriers and polarized secretion of transferrin. Biol Reprod 1990;43:672–83.
29. Janecki A, Steinberger A. Vectorial secretion of transferrin and androgen binding protein in Sertoli cell cultures: effect of extracellular matrix, peritubular myoid cells and medium composition. Mol Cell Endocrinol 1987;52: 125–36.
30. Janecki A, Jakubowiak A, Steinberger A. Vectorial secretion of inhibin by immature rat Sertoli cells in vitro: reexamination of the previous results. Endocrinology 1990;127:1896–1903.
31. Ailenberg M, Fritz IB. Influences of follicle-stimulating hormone, proteases, and antiproteases on permeability barrier generated by Sertoli cells in a two-chambered assembly. Endocrinology 1989;124:1399–1407.
32. Steinberger A, Steinberger E. Replication pattern of Sertoli cells in maturing rat testis in vivo and in organ culture. Biol Reprod 1971;4:84–7.
33. Cereijido M, Ponce A, Gonzalez-Mariscal L. Tight junctions and apical/ basolateral polarity. J Membr Biol 1989;110:1–9.
34. Yoshimasa T, Sibley DR, Bouvier M, Lefkowitz RJ, Caron MG. Cross-talk between cellular signaling pathways suggested by phorbol ester induced adenylate cyclase phosphorylation. Nature 1989;327:67–70.
35. Singhal RL, Vijayvargiya R, Shukla GS. Toxic effects of cadmium and lead on reproductive functions. In: Thomas JA, ed. Endocrine toxicology. New York: Raven Press, 1985:149–79.
36. Setchell BP, Waites GMH. Changes in the permeability of the testicular capillaries and of the "blood-testis barrier" after injection of cadmium chloride in the rat. J Endocrinol 1970;47:81–6.
37. Parvinen M. Regulation of the seminiferous epithelium. Endocr Rev 1982; 3:404–17.
38. Galdieri M, Monaco L, Stefanini M. Secretion of androgen binding protein by Sertoli cells is influenced by contact with germ cells. J Androl 1984;5: 409–15.
39. Janecki A, Jakubowiak A, Steinberger A. Effect of germ cells on vectorial secretion of androgen binding protein and transferrin by immature rat Sertoli cells in vitro. J Androl 1988;9:126–32.
40. Castellon E, Janecki A, Steinberger A. Influence of germ cells on Sertoli cell secretory activity in direct and indirect co-cultures with Sertoli cells from rats of different ages. Mol Cell Endocrinol 1989;64:169–78.
41. Skinner MK, Fetterolf PM, Anthony CT. Purification of a paracrine factor, P-Mod-S, produced by testicular peritubular cells that modulates Sertoli cell function. J Biol Chem 1988;263:2884–90.
42. Stevenson BR, Siliciano JD, Mooseker MS, Goodenough DA. Identification of Z0-1: a high molecular weight polypeptide associated with the tight junction (zonula occludens) in a variety of epithelia. J Cell Biol 1986;103: 755–66.

43. Byers S, Graham R, Dai HN, Hoxter B. Development of Sertoli cell junctional specializations and the distribution of the tight-junction-associated protein Z0-1 in the mouse testis. Am J Anat 1991;191:35–47.
44. Zong SD, Bardin CW, Phillips D, Cheng CY. Testins are localized to the junctional complexes of rat Sertoli and epididymal cells. Biol Reprod 1992; 47:568–72.

10

Mediation of the Hormonal Stimulation of Steroidogenesis by the Polypeptide Diazepam Binding Inhibitor

NOUREDDINE BOUJRAD, JAMES R. HUDSON, JR., AND VASSILIOS PAPADOPOULOS

Eukaryotic steroid hormones, derived from cholesterol, are responsible for the maintenance of the organism's homeostasis, adaptability to the environment, and developmental and reproductive functions. The mechanisms by which pituitary trophic hormones, such as adrenocorticotropin and *luteinizing hormone* (LH), act on their respective target organs, adrenal and gonads, to stimulate steroidogenesis have been under extensive investigation during the last 40 years (1–5). Both hormones bind to cell surface receptors that activate adenylyl cyclase that then initiates a complex series of events in which cAMP-dependent protein kinase and cholesterol esterase are involved, ultimately increasing delivery of cholesterol to the cytochrome *P450 side-chain cleavage enzyme* ($P450_{scc}$). Cholesterol is liberated from extramitochondrial stores, transported to mitochondria, incorporated into the outer mitochondrial membrane, and finally delivered to the inner mitochondrial membrane where it is converted to pregnenolone by $P450_{scc}$ and auxiliary electron-transferring proteins (1–5).

The primary point of control in the acute stimulation of steroidogenesis by hormones involves the first step in this biosynthetic pathway, where cholesterol is converted to pregnenolone (1–5). More detailed studies have shown that the reaction catalyzed by $P450_{scc}$ is not rate determining in the synthesis of steroid hormones, but rather it is the transport of the precursor, cholesterol, from intracellular stores to the inner mitochondrial membrane where steroid production begins (1–5). Thus, the rate-determining step in the regulation of steroidogenesis and the primary site of acute hormone action is the process of cholesterol delivery to $P450_{scc}$

across the mitochondrial membranes. This hormone-dependent transport mechanism was shown to be mediated by cAMP, to be sensitive to protein synthesis inhibitors, and to be localized in the mitochondrion where it regulates the intramitochondrial transport of cholesterol from the outer to the inner membrane (1–5). Thus arose the concept of the existence of a labile (of very short half-life), intracellular, and presumably cytoplasmic protein mediator of hormone action on steroidogenic tissues (1–5).

In search of such a steroidogenesis-stimulating factor, a protein was purified (6, 7) that was shown to be identical to the polypeptide *diazepam binding inhibitor* (DBI), except for the loss of two amino acids (Gly-Ile) from the carboxyterminus (8). DBI was originally purified from brain by monitoring its ability to displace diazepam from its recognition sites in synaptosomes (9, 10). Purified rat brain and bovine adrenal DBI was shown to stimulate intramitochondrial cholesterol transport and increase pregnenolone formation by isolated mitochondria (11–13). Similar results were also obtained using purified rat testis DBI (Table 10.1). In order to exclude the possibility that the stimulatory effect of DBI was due to the α-helical structure of the protein, we used as control the protein β-endorphin, which also possesses α-helical structures (12). The β-endorphin did not affect the mitochondrial steroid synthesis (Table 10.1).

In conjunction with these findings, a drug binding site, the mitochondrial *peripheral-type benzodiazepine receptor* (PBR), was shown to regulate cholesterol delivery to the inner membrane (11, 14–18). Later on, it was demonstrated that the action of DBI on mitochondrial steroidogenesis was mediated by the mitochondrial PBR in all tissues examined thus far, which include testicular Leydig, adrenocortical, ovarian granulosa, and brain glial cells (8, 11–13, 18).

However, it has also been reported that DBI can directly stimulate $P450_{scc}$ activity in an in vitro reconstituted system (19). These data suggest that in vivo DBI would enter the mitochondria to activate the $P450_{scc}$

TABLE 10.1. Effect of rat testis DBI on MA-10 mitochondrial pregnenolone formation.

Treatment	Pregnenolone (ng/mg protein)
Control	59.6 ± 2.5
DBI (0.3 μM)	136.6 ± 15.5
β-endorphin (0.3 μM)	64.2 ± 4.6

Note: Mitochondria were prepared from MA-10 Leydig cells cultured as described in reference 12. Rat testis DBI, as described in reference 32, and β-endorphin were added to the mitochondria, and the reaction proceeded as described in reference 12. Pregnenolone formed after 15-min incubation time was determined by radioimmunoassay. Results shown are mean ± SD of triplicate determinations.

TABLE 10.2. Effect of DBI on side-chain cleavage of cholesterol.

P450$_{scc}$	Additions	Pregnenolone (nmol/nmol P450$_{scc}$/min)
Control	—	Nondetectable
P450$_{scc}$ (50 nM)	—	512 ± 120
P450$_{scc}$ (50 nM)	DBI (0.3 μM)	825 ± 263

Note: Reconstitution of the P450$_{scc}$ system was performed as described in reference 19. Pregnenolone was measured by radioimmunoassay. Results are mean ± SD of triplicate determinations.

located on the inner membrane. However, we were unable to show any DBI immunoreactivity present in the mitochondria, in cells treated with or without hormones. Furthermore, immunolocalization of DBI in Leydig cells by electron microscopy indicated that DBI was localized around, but not inside, the mitochondria (20). In order to examine the direct effects of DBI on P450$_{scc}$, we purified bovine adrenal P450$_{scc}$, adrenodoxin, and adrenodoxin reductase using published methods (21, 22). In the presence of 0.3 μM DBI, pregnenolone formation by the reconstituted P450$_{scc}$ was increased by 50% (Table 10.2). This effect is minor when compared to the 250% stimulatory effect of DBI on mitochondrial pregnenolone formation (Table 10.1) (12, 13). Nevertheless, we cannot exclude the possibility that PBR may function as a channel, thus allowing DBI and/or its processing products (some of them being bioactive) (12) to enter the mitochondria and affect P450$_{scc}$ activity.

Although the role of DBI in the intracellular regulation of steroidogenesis, as determined using cell-free systems (6–8, 11–13), is now well established, there is no information about its in situ role in hormone-stimulated steroidogenesis. In order to determine the in situ role of DBI, we suppressed DBI levels in steroidogenic cells using antisense oligodeoxynucleotides. The hormone-responsive MA-10 mouse tumor Leydig cell line (23) that was previously shown to be an excellent model to study the role of PBR and DBI in steroidogenesis (12, 14–16, 24) was used. To overcome the usually encountered oligodeoxynucleotide uptake problems (25) and knowing that steroidogenic cells, including the MA-10 cells, can utilize exogenous cholesterol via the lipoprotein endocytotic pathway (26, 27), we constructed *cholesterol-linked phosphorothioate oligodeoxynucleotides* (CHOL-ODNs) (28) where cholesterol could act as a carrier to facilitate their uptake. CHOL-ODNs were produced using β-cyanoethyl phosphoramidate chemistry with a sulfurizing reagent in place of the normal oxidizer to produce the phosphorothioate backbone (29). A cholesterol phosphoramidite was used to add a single cholesterol functionality to the 5′ end (28). The oligodeoxynucleotides were purified by

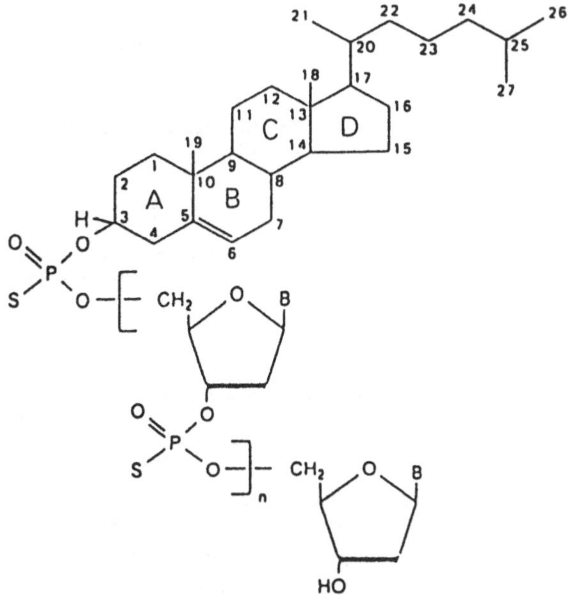

FIGURE 10.1. Structure of CHOL-ODNs.

reverse phase chromatography. The structure of CHOL-ODNs is shown in Figure 10.1.

CHOL-ODNs complementary to either the sense or the antisense strand of the 24 nucleotides encoding mouse DBI (Fig. 10.2) (30) were constructed (29). The nucleotide sequences of both the sense and antisense CHOL-ODNs that were constructed, as well as that of a nonrelated CHOL-ODN, are shown in the legends of Figures 10.3 and 10.4. Figure 10.3 shows that DBI levels in MA-10 cells treated with CHOL-ODN antisense to DBI were dramatically decreased. In contrast, treatment of MA-10 cells with CHOL-ODN sense to DBI did not affect DBI expression. The addition of saturating amounts of hCG (1 nM) to MA-10 cells resulted in a 150-fold increase in progesterone production. The addition of increasing concentrations of CHOL-ODNs sense to DBI or of a non-related sequence did not reduce the MA-10 cell response to hCG (Fig. 10.4).

In contrast, a 2-fold increase in the amount of steroids produced was observed due to the cholesterol linked to the oligodeoxynucleotide, liberated in the cells, and used as substrate for steroid synthesis (29). A similar increase in the amount of steroids produced by MA-10 cells treated with 30–100 nM CHOL-ODN antisense to DBI was also observed. However, the addition of higher concentrations of CHOL-ODN antisense to DBI caused a reduction in DBI levels in the cells, and the MA-10 cells

                                                         20
                                                          •
                              TTGCTCCCGCGCATTCGGCATCCGTATCAC<u>CTCACCAGT</u>

40   60   80
 •   •   •
<u>ATG TCT CAG GCT GAA</u> TTT GAC AAA GCC GCT GAG GAG GTG AAG CGC CTC AAG ACT CAG CCA
Met Ser Gln Ala Glu Phe Asp Lys Ala Ala Glu Glu Val Lys Arg Leu Lys Thr Gln Pro   20

100   120   140
 •   •   •
ACT GAT GAA GAG ATG CTG TTC ATC TAC AGT CAC TTC AAA CAA GCT ACC GTG GGC GAT GTA
Thr Asp Glu Glu Met Leu Phe Ile Tyr Ser His Phe Lys Gln Ala Thr Val Gly Asp Val   40

160   180   200
 •   •   •
AAT ACA GAT CGG CCG GGG CTC TTG GAC CTC AAG GGC AAA GCA AAG TGG GAC TCG TGG AAC
Asn Thr Asp Arg Pro Gly Leu Leu Asp Leu Lys Gly Lys Ala Lys Trp Asp Ser Trp Asn   60

220   240   260
 •   •   •
AAG CTG AAA GGG ACT TCC AAG GAA AGT GCC ATG AAG ACC TAT GTG GAA AAG GTA GAC GAG
Lys Leu Lys Gly Thr Ser Lys Glu Asn Ala Met Lys Thr Tyr Val Glu Lys Val Asp Glu   80

280   300   320   340
 •   •   •   •
CTA AAG AAG AAA TAC GGA ATA TAA ATCACCAGATTTGGTGGCCAGCCACACGTGTGACCTGTGAGGACATAA
Leu Lys Lys Lys Tyr Gly Ile ***

FIGURE 10.2. Nucleotide sequence of the sense strand of cloned mouse DBI cDNA as described in reference 30. CHOL-ODNs were constructed based on the underlined sequence of the mouse DBI cDNA, 9 bases immediately 5' to the initiator codon ATG and 12 downstream of the ATG codon.

lost their ability to respond to hCG; 50% inhibition was obtained using 1 µM antisense CHOL-ODN (Fig. 10.4).

It should be noted that the potency of cholesterol-linked oligodeoxynucleotides ($ED_{50} = 1$ µM) is 10–30 times higher than that of other types of oligodeoxynucleotides ($ED_{50} = 10-30$ µM) (25) used to inhibit protein expression. It is evident that cholesterol greatly facilitated the oligodeoxynucleotide uptake via the active lipoprotein-mediated endocytotic pathway. Further, in a recently published study, Stein and coworkers (31) constructed CHOL-ODNs and demonstrated that CHOL-ODNs are capable of binding to LDL, are sequence specific, and have increased potency as compared to unmodified oligodeoxynucleotides. These data further support our results and suggest that this novel approach may be useful in other cell systems to study the physiological role of specific proteins in cells.

Two controls were performed in order to validate the specificity of the results obtained with CHOL-ODNs: the hCG-dependent intracellular cAMP accumulation and the 22R-hydroxycholesterol-supported steroidogenesis. Both functions were not affected by the CHOL-ODN treatment,

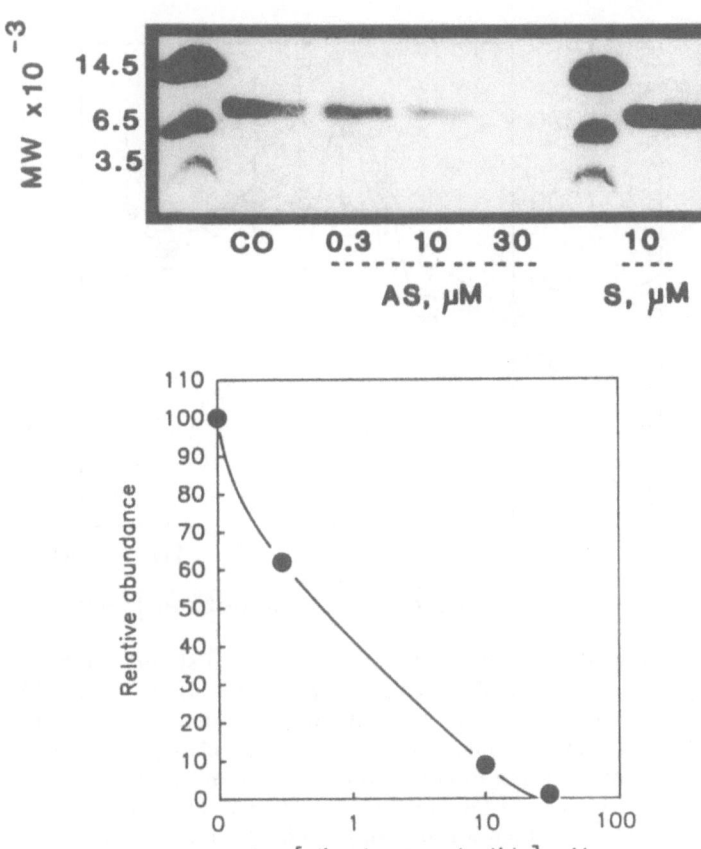

FIGURE 10.3. Inhibition of DBI expression in MA-10 Leydig cells by CHOL-ODNs antisense but not sense to DBI. MA-10 cells were plated for 24 h in modified Waymouth's MB 752/1 medium, as described in references 14 and 23. After 24 h the media were changed, and fresh media containing the indicated amounts of oligodeoxynucleotides antisense (5′-cholesterol-TTCAGCCTGAGA CATACTGGTGAG-3′) or sense (5′-cholesterol-AAGTCGGACTCTGTATGA CCACTC-3′) to DBI were added for 48 h. DBI levels were determined by immunoblot analysis of MA-10 cell sodium acetate extracts enriched in DBI by Sep-Pak C_{18} chromatography as described in reference 24. DBI levels were then quantified by densitometry. (CO = control; AS = antisense; S = sense.) Adapted with permission from Boujrad, Hudson, and Papadopoulos (29).

which shows that the cells respond to hCG by synthesizing cAMP and have the capacity to synthesize steroids if the substrate is available (29).

These data unequivocally demonstrate that DBI plays a vital role and probably mediates the hormone-stimulated steroid synthesis. However, since we previously demonstrated that DBI has a long life (24) and is not

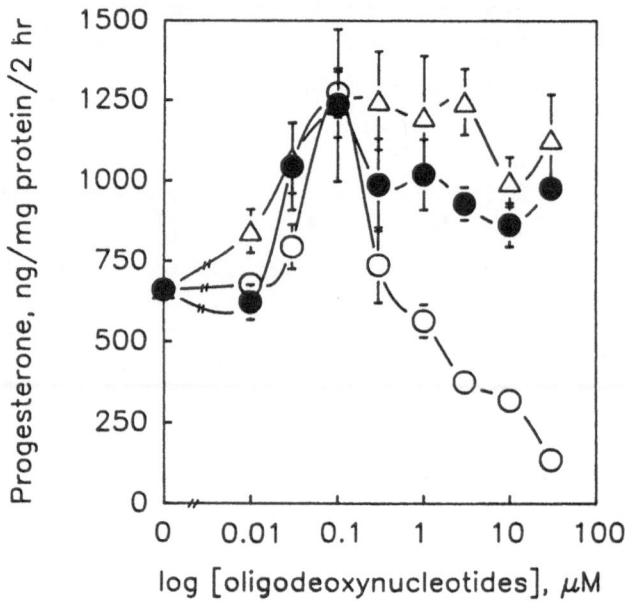

FIGURE 10.4. Inhibition of trophic hormone-stimulated steroid synthesis by MA-10 Leydig cells by CHOL-ODNs antisense but not sense to DBI. MA-10 cells were plated as described in Fig. 10.3, and after 24 h the media were changed and fresh media containing the indicated amounts of oligodeoxynucleotides antisense to DBI (open circles), sense to DBI (closed circles), or of a nonrelated sequence (5'-cholesterol-CGCGCCCCGCCGCCGCCATG-3') (open triangles) were added for 48 h. The cells were then washed with serum-free medium and stimulated for 2 h with saturating amounts of hCG (1 nM). At the end of the incubation period, the cell media were saved, and progesterone was measured by radioimmunoassay as described in reference 14. Results shown are mean ± SD of 2 experiments (n = 6) performed on cells from the same passage. Similar results were obtained in 2 other independent experiments performed on cells of different passages. Reprinted with permission from Boujrad, Hudson, and Papadopoulos (29).

induced by hormones nor cAMP (24), the possibility that DBI is the long-sought cycloheximide-sensitive factor can be excluded. These findings, together with the previous observations that DBI stimulates mitochondrial steroidogenesis acting via PBR (8, 11–14, 18), suggest that PBR may be the site of hormone action. In that case, hormones will alter PBR to enhance the affinity for its ligands—that is, DBI—and thereby activate cholesterol delivery to the inner mitochondrial membrane.

Acknowledgment. This work was supported by NIH Grant DK-43358.

References

1. Simpson ER, Waterman MR. Regulation by ACTH of steroid hormone biosynthesis in the adrenal cortex. Can J Biochem Cell Biol 1983;61:692–707.
2. Hall PF. Cellular organization for steroidogenesis. Int Rev Cytol 1984; 86:53–95.
3. Kimura T. Transduction of ACTH signal from plasma membrane to mitochondria in adrenocortical steroidogenesis. Effects of peptide, phospholipid, and calcium. J Steroid Biochem 1986;25:711–6.
4. Orme-Johnson NB. Distinctive properties of adrenal cortex mitochondria. Biochim Biophys Acta 1990;1020:213–31.
5. Jefcoate CR, McNamara BC, Artemenko I, Yamazaki T. Regulation of cholesterol movement to mitochondrial cytochrome P450scc in steroid hormone synthesis. J Steroid Biochem Mol Biol 1992;43:751–67.
6. Hall PF, Papadopoulos V, Yanagibashi K. On the mechanism of action of ACTH. In: Imura H, Shizume K, Yoshida S, eds. Progress in endocrinology. Amsterdam: Elsevier, 1988:253–8.
7. Yanagibashi K, Ohno Y, Kawamura M, Hall PF. The regulation of intracellular transport of cholesterol in bovine adrenal cells: purification of a novel protein. Endocrinology 1988;123:2075–82.
8. Besman MJ, Yanagibashi K, Lee TD, Kawamura M, Hall PF, Shively JE. Identification of des-(Gly-Ile)-endozepine as an effector of corticotropin-dependent adrenal steroidogenesis: stimulation of cholesterol delivery is mediated by the peripheral benzodiazepine receptor. Proc Natl Acad Sci USA 1989;86:4897–901.
9. Guidotti A, Forchetti CM, Corda MG, Konkel D, Bennet CD, Costa E. Isolation, characterization, and purification to homogeneity of an endogenous polypeptide with agonistic action on benzodiazepine receptors. Proc Natl Acad Sci USA 1983;80:3531–3.
10. Shoyab M, Gentry LE, Marquardt H, Todaro G. Isolation and characterization of a putative endogenous benzodiazepinoid (Endozepine) from bovine and human brain. J Biol Chem 1986;261:11968–73.
11. Papadopoulos V, Guarneri P, Krueger KE, Guidotti A, Costa E. Pregnenolone biosynthesis in C6 glioma cell mitochondria: regulation by a diazepam binding inhibitor mitochondrial receptor. Proc Natl Acad Sci USA 1992; 89:5118–22.
12. Papadopoulos V, Berkovich A, Krueger KE, Costa E, Guidotti A. Diazepam binding inhibitor (DBI) and its processing products stimulate mitochondrial steroid biosynthesis via an interaction with mitochondrial benzodiazepine receptors. Endocrinology 1991;129:1481–8.
13. Yanagibashi K, Ohno Y, Nakamichi N, et al. Peripheral-type benzodiazepine receptors are involved in the regulation of cholesterol side chain cleavage in adrenocortical mitochondria. J Biochem 1989;106:1026–9.
14. Papadopoulos V, Mukhin AG, Costa E, Krueger KE. The peripheral-type benzodiazepine receptor is functionally linked to Leydig cell steroidogenesis. J Biol Chem 1990;265:3772–9.
15. Papadopoulos V, Nowzari FB, Krueger KE. Hormone-stimulated steroidogenesis is coupled to mitochondrial benzodiazepine receptors. J Biol Chem 1991;266:3682–7.

16. Krueger KE, Papadopoulos V. Peripheral-type benzodiazepine receptors mediate translocation of cholesterol from outer to inner mitochondrial membranes in adrenocortical cells. J Biol Chem 1990;265:15015–22.
17. Amsterdam A, Suh BS. An inducible functional peripheral benzodiazepine receptor in mitochondria of steroidogenic granulosa cells. Endocrinology 1991;128:503–10.
18. Papadopoulos V. Peripheral-type benzodiazepine/diazepam binding inhibitor receptor: biological role in steroidogenic cell function. Endocr Rev 1993; 14:222–40.
19. Brown AS, Hall PF. Stimulation by endozepine of the side-chain cleavage of cholesterol in a reconstituted enzyme system. Biochem Biophys Res Commun 1991;180:609–14.
20. Schultz R, Pelto-Huikko M, Alho H. Expression of diazepam binding inhibitor-like immunoreactivity in rat testis is dependent on pituitary hormones. Endocrinology 1992;130:3200–6.
21. Seybert DW, Lancaster JR, Lambeth JD, Kamin H. Participation of the membrane in the side chain cleavage of cholesterol. Reconstitution of cytochrome P-450scc into phospholipid vesicles. J Biol Chem 1979;254:12088–98.
22. Lambeth JD, Kamin H. Adrenodoxin reductase-adrenodoxin complex. J Biol Chem 1979;254:2766–74.
23. Ascoli M. Characterization of several clonal lines of cultured Leydig tumor cells: gonadotropin receptors and steroidogenic responses. Endocrinology 1981;108:88–95.
24. Brown AS, Hall PF, Shoyab M, Papadopoulos V. Endozepine/diazepam binding inhibitor in adrenocortical and Leydig cell lines: absence of hormonal regulation. Mol Cell Endocrinol 1992;83:1–9.
25. Stein CA, Cohen JS. Oligodeoxynucleotides as inhibitors of gene expression: a review. Cancer Res 1988;48:2659–68.
26. Brown MS, Kovanen PT, Goldstein JL. Receptor-mediated uptake of lipoprotein-cholesterol and its utilization for steroid synthesis in the adrenal cortex. Recent Prog Horm Res 1979;35:215–57.
27. Freeman DA, Ascoli M. The low-density lipoprotein pathway of cultured Leydig tumor cells. Biochim Biophys Acta 1983;754:72–81.
28. MacKellar C, Graham D, Will DW, Burgess S, Brown T. Synthesis and physical properties of anti-HIV antisense oligonucleotides bearing terminal lipophilic groups. Nucleic Acids Res 1992;20:3411–7.
29. Boujrad N, Hudson JR Jr, Papadopoulos V. Inhibition of hormone-stimulated steroidogenesis in cultured Leydig tumor cells by a cholesterol-linked phosphorothioate oligodeoxynucleotide antisense to diazepam binding inhibitor. Proc Natl Acad Sci USA 1993.
30. Owens GP, Sinha AK, Sikela JM, Hahn WE. Sequence and expression of the murine diazepam binding inhibitor. Mol Brain Res 1989;6:101–8.
31. Krieg AM, Tonkinson J, Matson S, et al. Modification of antisense phosphodiester oligodeoxynucleotides by a 5′ cholesterol moiety increases cellular association and improves efficacy. Proc Natl Acad Sci USA 1993;90:1048–52.
32. Garnier M, Boujrad N, Oke BO, et al. Diazepam binding inhibitor is a paracrine/autocrine regulator of Leydig cell proliferation and steroidogenesis: action via peripheral-type benzodiazepine receptor and independent mechanisms. Endocrinology 1993;132:444–58.

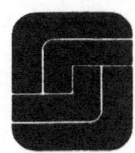

11

Inhibition of Spermatogenesis by Tetanus Toxin Expression in Sertoli Cells of Transgenic Mice

ULRICH EISEL, KAY REYNOLDS, MICHELLE RIDDICK, ANNE ZIMMER, AND ANDREAS ZIMMER

The event of homologous recombination provides a very powerful tool to introduce mutations into possibly every gene (1, 2). Once a gene is cloned, a mutation can be introduced, first, by standard recombinant DNA methodology and, subsequently, with the help of embryonic stem cells into the mouse germline. The challenge then remains to find out how the animals can cope with the genetic defect and to deduce the role of the gene in the intact organism from the results obtained. This technique now makes it feasible to deploy a genetic approach for the molecular analysis of mammalian development and physiology. An obvious disadvantage of homologous recombination, however, is the difficulty in targeting a specific tissue. For example, if one were to introduce a mutation into the β-actin gene, it seems likely that this mutation would affect the viability of the whole organism. Thus, such a mouse model might not be useful for studying the role of the actin cytoskeleton in Sertoli cells.

An alternative approach takes advantage of the fact that several bacterial strains produce toxins that interfere specifically with the function of certain proteins. These toxins can be expressed under the control of tissue-specific promoter in transgenic mice and inhibit specifically the function of their target molecules in those tissues where the promoter is active. As some toxins are cytotoxic, expressing cells are ablated. For example, diphtheria toxin has been used to ablate acinar pancreatic cells (3), nuclear fiber cells of the lens (4), and growth hormone-producing cells in the pituitary (5). Other toxins interfere with proteins that are not essential for cell survival. Cholera toxin, for example, interferes only with the function of certain G-proteins. When expressed in the pituitary gland, it induces pituitary hyperplasia and gigantism (6).

In this chapter we show that the expression of the tetanus toxin light chain gene in Sertoli cells impairs spermatogenesis. Tetanus toxin is produced by the anaerobic spore-forming bacterium *Clostridium tetani* (7). Upon infection it causes convulsions by blocking the release of inhibitory neurotransmitters in the central nervous system. Tetanus toxin consists of of two polypeptides that are generated by proteolytic cleavage of a single precursor molecule (8). The heavy chain mediates the uptake of the light chain into neurons. The light chain is a Zn^{++}-dependent protease that has been shown to cleave specifically synaptobrevin-2, but not the closely related synaptobrevin-1 (9, 10). Synaptobrevin-2 was initially identified as an integral membrane protein of small synaptic vesicles (11), but it is also found in other cell types, particularly endocrine cells.

Results and Discussion

The codon usage of Clostridium is significantly different from eukaryotic cells or even gram-negative bacteria, resulting in a very high [A+T] content. This feature has prevented the expression of Clostridial genes in eukaryotes (12). To express the *tetanus toxin light chain* in transgenic animals, we have therefore synthesized a new gene (tetLC) and adapted the codon usage by raising the [G+C] content to fit the requirements of eukaryotic cells. This synthetic gene could be expressed efficiently in cultured cells.

We used three different promoters to express the tetLC gene in transgenic mouse lines: (i) the proopiomelanocortin promoter (13), (ii) the L7 gene promoter (14), or (iii) the SV40 promoter. All three constructs expressed tetLC at high levels in the seminiferous epithelium, but not in other tissues. This was unexpected, as the L7 promoter was supposed to be specifically active in Purkinje cells (14) and not in testis. In Northern blot analysis, L7 transcripts were not detected in the testis (15), while the endogenous proopiomelanocortin gene was expressed in the male gonad (16). The proopiomelanocortin promoter was also supposed to direct expression in the pituitary gland (13). It is possible that the coding sequence of the tetLC gene itself contains a testis-specific enhancer.

Most of the male founder animals and all expressing F2 males were sterile. Females were fertile and were used to establish several transgenic lines. All male transgenic offspring from these lines were sterile as well. This result was very surprising because synaptobrevin-2 is not expressed in the testis. Also, tetanus toxin so far has failed to reveal any effects on nonneuronal cells.

Semithin sections of transgenic mice testes showed defects at late stages of germ cell differentiation (Fig. 11.1). Mature sperm cells were completely absent, and the number of late spermatocyes was significantly

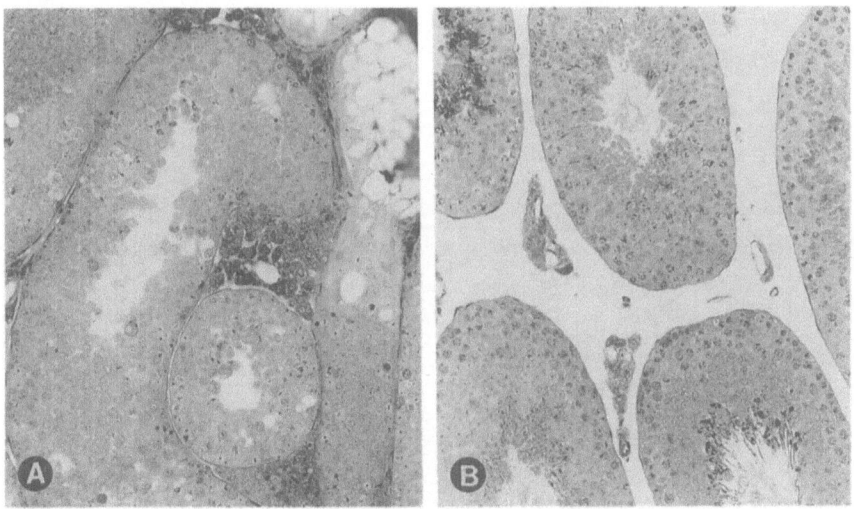

FIGURE 11.1. Histological analyses of testes from transgenic and nontransgenic animals. *A*: Shown is a semithin section through a seminiferous tubule, interstitium, and neighboring tubules of a transgenic animal. Note that the tubule is atrophic and that the interstitium is almost completely filled with Leydig cells. Spermatogonia and round spermatids, however, appeared to be normal, while mature sperm cells were completely absent, and the number of late spermatocyes was significantly reduced. The heads of the few late spermatids had an irregular, pleomorphic morphology; contained invaginations, intranuclear vacuoles, and inclusions; and had an irregularly shaped acrosome. *B*: Shown is an overview of a stage VIII seminiferous tubule of a nontransgenic animal. Only a few Leydig cells can be seen in the interstitium. Round spermatids, beginning to form acrosomes, and step 16 spermatids prior to release can be seen in the adluminal part of the seminiferous epithelium.

reduced. Spermatogonia and round spermatids, however, appeared to be normal. The number of residual bodies was increased in all atrophic tubules. In some of these residual bodies, we found structures resembling remnants of late spermatids. The heads of these spermatids had an irregular, pleomorphic shape. They contained invaginations, intranuclear vacuoles, and inclusions. The acrosome was irregularly shaped and often hyperplastic. Nuclear condensation in clusters of spermatids was not synchronized. We also found a large number of vacuoles interspersed throughout the seminiferous epithelium. Transgenic seminiferous tissue exhibited distorted sperm heads, acrosomal defects and intranuclear vacuoles in late spermatids.

Similar observations have been made in a number of human andrological disorders characterized by oligo-, terato- or azoospermia (17), as well as in infertile patients with inflammatory bowel disease (18). Con-

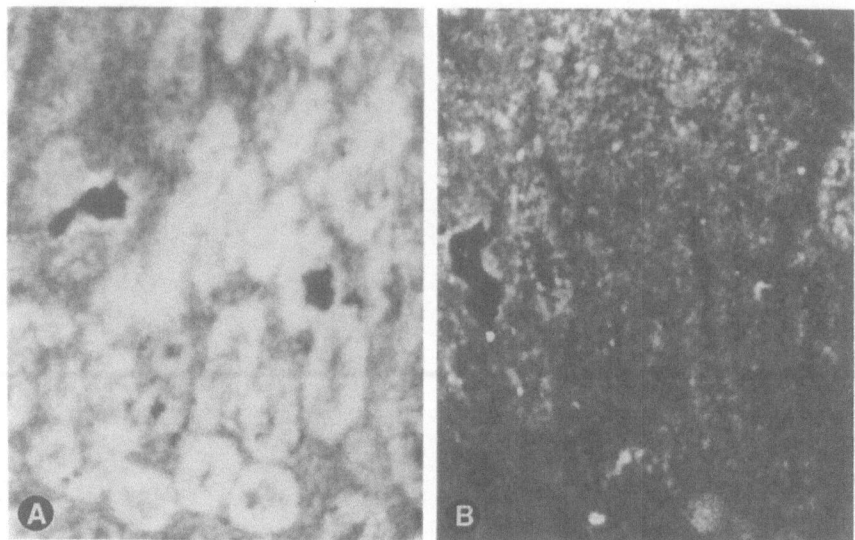

FIGURE 11.2. TetLC expression in seminiferous tubules. *A*: Shown is a darkfield photograph of testis cryosections from a transgenic founder hybridized with an antisense riboprobe against the synthetic tetLC gene. Strong expression can be observed in all tubules, but not in the interstitium. *B*: Darkfield photograph of a testis from control mouse shows nonhybridization with the same probe.

comitant with these defects were dramatic changes in the pattern of the actin cytoskeleton. Actin filaments in normal Sertoli cells are most prominent in the basal region, as well as in ectoplasmic specializations and tubulobulbar complexes that are closely associated with spermatids (19, 20). During spermatogenesis the actin pattern surrounding the developing spermatids undergoes characteristic changes. This process is regulated by *follicle stimulating hormone* (FSH) and is modulated by other factors, such as testosterone (21). In transgenic testes, the amount of F-actin was markedly increased at all spermatogenic stages. The actin cytoskeleton was mostly affected in the adluminal region where actin filaments were not associated with spermatids. These data suggest that ectoplasmic specializations and tubulobulbar complexes are altered in transgenic animals.

Interestingly, the number of Leydig cells in the interstitium of transgenic testis was increased. As tetLC expression was not detected in the interstitium, this might be caused by paracrine disturbances due to the loss of late spermatids or impaired Sertoli cell functions. Paracrine factors, such as angiotensin II, epidermal growth factor, macrophage-derived growth factor, interleukin-1, insulin-like growth factor, and a *luteinizing hormone-*

releasing hormone (LHRH)-like factor, have been reported to be involved in Leydig cell regulation (22, 23). An increased number of Leydig cells can also be found in infertile patients suffering from Klinefelter syndrome or cryptorchidism (24, 25).

Expression of the tetLC gene was monitored by in situ hybridization analysis (Fig. 11.2). Strong expression was detected in the seminiferous tubule, but not in the interstitium or in any other tissue. As germ cells are imbedded in Sertoli cells, it was not possible to unequivocally correlate the in situ hybridization signal with expression in either one of the two cell types.

Examination of a mosaic transgenic founder animal, however, supported the idea that expression in Sertoli cells rather than germ cells was the cause for the defects observed. This particular founder was fertile and did transmit the transgene to his offspring. We found a patchy expression pattern in the seminiferous tubules, with 30% of all tubules in any given section expressing the transgene at very similar levels. Expressing regions showed characteristic defects, while nonexpression regions were phenotypically normal. This expression pattern makes it unlikely that both cell types express tetLC, as in that case, one would expect a more diffuse pattern with different levels of expression. These data also indicate that transgenic spermatozoa are viable. Hence, we must assume that only Sertoli cells express tetLC and that the patchy expression pattern in this mosaic animal reflects the clonal origin of Sertoli cells. All transgenic F2 males derived from this founder were sterile and expressed in the entire seminiferous epithelium.

Further investigations are required to elucidate the molecular and cellular basis for the observed defects more precisely. As synaptobrevin-2 is not expressed in Sertoli cells, we must postulate a second target for tetanus toxin. All intracellular-acting bacterial toxins are highly substrate specific. Such a substrate specificity has also been reported for tetanus toxin (9, 10). Therefore, it seems likely that a yet unknown protein closely related to synaptobrevin-2, and perhaps another member of the same gene family, is the target in testes.

Synaptobrevin-2 is an essential component for the fusion of synaptic vesicles and may function as a receptor for *soluble N-ethylmaleimide-sensitive fusion protein* (NSF) *attachment proteins* (SNAPs). It has been suggested that the same molecular machinery may be involved in regulated exocytosis and constitutive fusion processes (26). Therefore, it seems possible that tetanus toxin expression interferes with constitutive and/or regulated vesicular transport mechanisms in Sertoli cells. Sertoli cells produce and secrete a large number of nutritional and regulatory products under the control of pituitary gonadotropins (27, 28). One possible explanation for the defects found in tetLC transgenic mice is that these secretory Sertoli cell functions are primarily affected and that the observed changes in the actin cytoskeleton are a secondary attribute caused by altered paracrine and/or mechanical influences of Leydig and/ or germ cells.

An alternative hypothesis takes into account that synaptobrevins may be involved in intracellular vesicle targeting (26). Thus, tetLC expression may interfere with the intracellular membrane traffic and perhaps with the formation of such specialized structures as ectoplasmic specializations. This could also explain the observed changes in the actin cytoskeleton. It is interesting to note that the early stages of spermatogenesis are largely unaffected and that the appearance of morphological defects correlates with the development of ectoplasmic specialization. The transgenic mouse lines should be very valuable tools for studying the role of synaptobrevin-related proteins and vesicular transport during spermatogenesis.

Acknowledgments. Anne Z. was supported by the National Alliance for Research on Schizophrenia and Depression. U.E. was supported by the Boehringer Ingelheim Fonds, Stuttgart, Germany.

References

1. Capecchi MR. The new mouse genetics: altering the genome by gene targeting. TIG 1989;5:70–6.
2. Zimmer A. Manipulating the genome by homologous recombination in embryonic stem cells. Annu Rev Neurosci 1992;15:115–37.
3. Palmiter RD, Behringer RR, Quaife CJ, Maxwell F, Maxwell IH, Brinster R. Cell lineage ablation in transgenic mice by cell-specific expression of a toxin gene. Cell 1987;50:435–43.
4. Breitman ML, Clapoff S, Rossant J, et al. Genetic ablation: targeted expression of a toxin gene causes microphthalmia in transgenic mice. Science 1987;238:1563–5.
5. Behringer RR, Mathews LS, Palmiter R, Brinster R. Dwarf mice produced by genetic ablation of growth hormone-expressing cells. Genes Dev 1988; 2:453–61.
6. Burton FH, Hasel KW, Bloom FE, Sutcliffe JG. Pituitary hyperplasia and gigantism in mice caused by a cholera toxin transgene. Nature 1991;350:74–7.
7. Habermann E, Dreyer F. Clostridial neurotoxins: handling and action at the molecular level. Curr Top Microbiol Immunol 1986;129:93–179.
8. Helting TB, Parschat S, Engelhardt H. Structure of tetanus toxin. J Biol Chem 1979;254:10728.
9. Schiavo G, Benfenati F, Poulain B, et al. Tetanus and botulinum-B neurotoxins block neurotransmitter release by proteolytic cleavage of synaptobrevin. Nature 1992;359:832–5.
10. Schiavo G, Poulain B, Rossetto O, Benfenati F, Tauc L, Montecucco C. Tetanus toxin is a zinc protein and its inhibition of neurotransmitter release and protease activity depends on zinc. EMBO J 1992;11:3577–83.
11. Baumert M, Maycox PR, Navone F, DeCamili P, Jahn R. Synaptobrevin: an integral membrane protein of 18,000 daltons present in small synaptic vesicles of rat brain. EMBO J 1989;8:379–84.

12. Eisel U, Jarausch W, Goretzki K, et al. Tetanus toxin: primary structure, expression in *E. coli*, and homology with botulinum toxins. EMBO J 1986; 5:2495–502.
13. Hammer GD, Fairchild-Huntress V, Low MJ. Pituitary-specific and hormonally regulated gene expression directed by the rat proopiomelanocortin promoter in transgenic mice. Mol Endocrinol 1990;4:1689–97.
14. Oberdick J, Smeyne RJ, Mann JR, Zackson S, Morgan JI. A promoter that drives transgene expression in cerebellar Purkinje cells and retinal bipolar neurons. Science 1990;248:223–6.
15. Oberdick J, Levinthal F, Levinthal C. A Purkinje cell differentiation marker shows a partial DNA sequence homology to the cellular sis/PDGF2 gene. Neuron 1988;1:367–76.
16. Gizang-Ginsberg E, Wolgemuth DJ. Expression of the proopiomelanocortin gene is developmentally regulated and affected by germ cells in the male mouse reproductive system. Proc Natl Acad Sci USA 1987;84:1600–4.
17. Holstein AF. Morphologische studien an abnormen spermatiden und spermatozoen des menschen. Virchows Arch [A] 1975;367:93–112.
18. Hrudka F, Singh A. Sperm nucleomalacia in men with inflammatory bowel disease. Arch Androl 1984;13:37–57.
19. Maekawa M, Nagano T, Kamimura K, Murakami T, Ishikawa H, Dezawa M. Distribution of actin-filament bundles in myoid cells, Sertoli cells, and tunica albuginea of rat and mouse testis. Cell Tissue Res 1991;266:295–300.
20. Vogl AW. Distribution and function of organized concentrations of actin filaments in mammalian spermatogenic cells and Sertoli cells. Int Rev Cytol 1989;119:1–56.
21. Kerr JB, Maddocks S, Sharpe RM. Testosterone and FSH have independent, synergistic and stage-dependent effects upon spermatogenesis in the rat testis. Cell Tissue Res 1992;268:179–89.
22. Chubb C, Ewing LL. Leydig cell. In: Lipschultz LI, Howard SS, eds. Infertility in the male. 2nd ed. St. Louis: Mosby-Year Book, 1991:37–53.
23. Lejeune H, Skalli M, Chatelain PG, Avallet O, Saez JM. The paracrine role of Sertoli cells on Leydig cell function. Cell Biol Toxicol 1992;8:73–83.
24. Wong TW, Strauss FH, Warner NE. Testicular biopsy in the study of male infertility, I. Testicular causes of infertility. Arch Pathol 1973;95:151–9.
25. Lipschultz LI. Cryptorchidism in the subfertile male. Fertil Steril 1976; 27:609–20.
26. Söllner T, Whiteheart SW, Brunner M, et al. SNAP receptors implicated in vesicle targeting and fusion. Nature 1993;362:318–24.
27. Jegou B. The Sertoli cell. Baillieres Clin Endocrinol Metab 1992;6:273–311.
28. Skinner MK. Cell-cell interactions in the testis. Endocr Rev 1991;12:45–77.

Part III

Signaling Molecules and Their Receptors

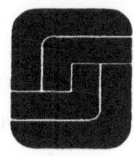

12

Pituitary Control of Fetal and Adult Leydig Cells and Physiological Regulation of Gonadotropin Gene Expression and Secretion in the Male

I. Huhtaniemi, P. Pakarinen, M. Bergendahl, A. Perheentupa, T. Matikainen, and A.-M. Haavisto

The cloning of genes for the gonadotropins has shifted the emphasis of physiological research on these hormones from their synthesis and secretion to the expression of the gonadotropin subunit mRNAs. These methodological advancements have undoubtedly broadened our view. In addition, important developments have taken place in the methods for gonadotropin measurement. These include the new supersensitive and specific immunoassays based on use of monoclonal antibodies and the immunometric assay principles (1, 2). The methods for measurement of gonadotropin bioactivity have also developed recently, especially as regards FSH (3–5). It is obvious that the molecular biological approaches, combined with the novel methods of gonadotropin measurement, will provide an improved arsenal for further elucidation of the physiology of gonadotropin synthesis, secretion, and actions.

The first part of this chapter summarizes some of our findings on the development of gonadotropin secretion with regard to its onset, action, and feedback regulation in the perinatal male rat. In the second part we address some aspects of the physiological regulation of gonadotropin gene expression and secretion in the adult male rat, including the direct effects of gonadal steroids at the pituitary level and the effects of food restriction. In the third section we present some of our findings on the hormonal regulation of serum bioactive and immunoreactive LH and FSH in human males.

Ontogeny of Gonadotropin Secretion, Action, and Feedback Regulation in the Male Rat

Supersensitive Immunoassay for Rat LH

For the last 25 years, the measurement of LH in rats has been carried out almost exclusively using the NIH radioimmunoassay reagent kit (6). Although it has provided reliable information on a variety of samples of peripheral serum, pituitary homogenates, and cell culture media, its sensitivity is insufficient for some applications. For example, the normal circulating levels of LH in the rat are at the lower limit of detectability of the assay, and suppressed levels are often nondetectable. A large volume of serum is needed (usually $2 \times 200 \mu L$), making it impossible to monitor individual samples from fetal and neonatal animals. Realizing these methodological limitations of the current rat LH RIA methods, we recently developed a more sensitive and specific *immunofluorometric assay* (IFMA) for rat LH (7).

The novel IFMA assay for rat LH is based on two monoclonal antibodies, one to bovine and one to human LH. Signal detection occurs by time-resolved fluorescence evoked by a europium label (Delfia®, Wallac, Turku, Finland) (1). The method is fast in comparison to the standard RIA (4 h vs. 3 days). The sensitivity of the IFMA assay is superior in comparison to the RIA, 0.5–0.75 vs. 20–30 pg per assay tube (NIDDK RP-2 standard). Using 25-μL samples we can detect a concentration of 0.02 μg/L, and with 100 μL, the limit of detection is 0.005 μg/L. There is good correlation ($r = 0.93$) between the IFMA and in vitro bioassay measurements of rat LH.

A linear correlation between the RIA and IFMA levels is seen above concentrations of 0.4 μg/L, whereas below this level only IFMA is able to detect concentration differences between samples (Fig. 12.1). As Table 12.1 demonstrates, the levels of LH measured by IFMA in intact adult males are about 30% lower than those measured by RIA. The high postcastration concentrations are similar in both assays. In contrast, when low LH levels were measured—for example, after acute starvation or testosterone treatment—the RIA results were 2- to 5-fold higher than those obtained by IFMA. Likewise, the LH immunoreactivity measured in hypophysectomized animals was negligible in the IFMA measurements, while the posthypophysectomy concentrations of LH determined by RIA were 10-fold higher. Very recently (Pakarinen et al., unpublished observations), we have shown that IFMA is sensitive enough for measurement of LH levels in samples of fetal serum as small as 25 μL. We expect that the new assay will prove useful in a number of experimental conditions assessing LH secretion of the rat and mouse. The small sample volume needed and the high sensitivity of the assay are real assets in measure-

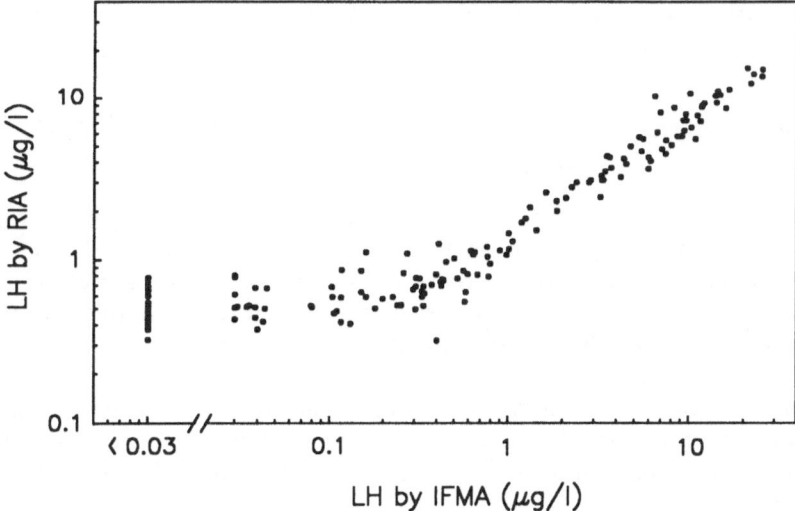

FIGURE 12.1. Correlation of the LH levels measured in a total of 165 serum samples from intact and castrated male and female rats and hypophysectomized male rats using both IFMA (x-axis) and RIA (y-axis). The correlation coefficient (r) at LH levels >0.4 μg/L by IFMA ($n = 89$) was 0.91 ($P < 0.001$; slope $y = 0.64x + 1.0$) Reprinted with permission from Haavisto, Pettersson, Bergendahl, Perheentupa, Roser, and Huhtaniemi (7), © The Endocrine Society, 1993.

TABLE 12.1. Concentrations of serum LH in different rat peripheral serum samples using the IFMA assay; for comparison, some of the levels were also measured using conventional RIA.

Sample	(n)	IFMA	RIA
Adult males	(12)	0.670 ± 0.0920	0.94 ± 0.076
Randomly cycling females	(10)	0.570 ± 0.1000	
Castrated males	(6)	5.570 ± 0.5000	5.77 ± 0.550
Testosterone-treated males[a]	(6)	0.086 ± 0.0570*	0.44 ± 0.048*
Sham controls	(6)	0.560 ± 0.0570	0.73 ± 0.057
Starved males[b]	(6)	0.300 ± 0.0500*	0.66 ± 0.130
Fed controls	(6)	0.570 ± 0.1000	0.80 ± 0.070
Hypophysectomized males[c]	(8)	0.035 ± 0.0033	0.36 ± 0.012

Note: Concentrations are in μg/L of the NIDDK RP-2 standard (mean ± SEM).
[a] Implanted s.c. for 1 week with a 2-cm Silastic implant containing testosterone.
[b] Five days with only water.
[c] A concentration of 0.03 μg/L (i.e., the limit of detection) assigned to nondetectable levels in IFMA.
* $P < 0.01$ vs. respective controls (unpaired Student's t-test).
Source: Reprinted with permission from Haavisto, Pettersson, Bergendahl, Perheentupa, Roser, and Huhtaniemi (7), © The Endocrine Society, 1993.

ments of serial blood samples, in conditions with suppressed gonadotropin secretion, and in studies of fetal and neonatal animals.

Developmental Onset of Gonadal Negative Feedback Regulation of Gonadotropin Secretion

The onset of gonadal negative feedback regulation of gonadotropin
secretion starts early in life in the male rat. At the age of 7 days, we
have demonstrated a clear increase in both LH and FSH secretion after
castration (8), which is in contrast to the inability to demonstrate such
an effect consistently in the female before puberty (8, 9). The early
castration-associated increase of gonadotropins in male rats could be
totally suppressed with *testosterone* (T) replacement. The cognate gonadotropin subunit mRNAs (LHβ, FSHβ, and common α) showed similar
patterns of increase and decrease after orchidectomy and T-replacement,
respectively.

Very recently, we have extended these observations to the fetal period
and observed that fetal orchidectomy at 20.5 days of gestation results in
a significant elevation of LH 24 h later (Pakarinen et al., unpublished
observation). This finding is in keeping with the contention that the
endocrine activity of the testis starts during the fetal period and is apparently under pituitary control. It was surprising that the ovary attained
the capability to consistently regulate gonadotropin secretion only after
puberty (9), although it is clearly responsive to LH and FSH by the end
of the first week of postnatal life (10).

Postnatal Changes in the Response of Gonadotropin Gene Expression to Castration and T-Replacement

Pituitary-gonadal interaction has been shown to undergo conspicuous
changes during postnatal maturation (11, 12). To examine to what extent
these changes are due to changes at the level of transcription or translation, we castrated male rats at the ages of 10, 30, 60, and 90 days and
treated some animals at each age with T-replacement (13). After 7 days
the secretion of both LH and FSH had increased 2- to 6-fold in all age
groups. In general, no quantitative differences were found between the
increased levels of serum gonadotropins at any of the ages studied. In the
case of serum FSH in the 30-day group, the fold increase was lowest
because of the characteristically high basal serum FSH concentration at
that age. Testosterone was able to suppress the elevated serum gonadotropin levels to normal or below at all ages studied.

Although the serum and pituitary gonadotropin responses to castration
and T-treatment were similar at the various ages, the measurement of
simultaneous mRNA responses revealed clear differences (Fig. 12.2). The

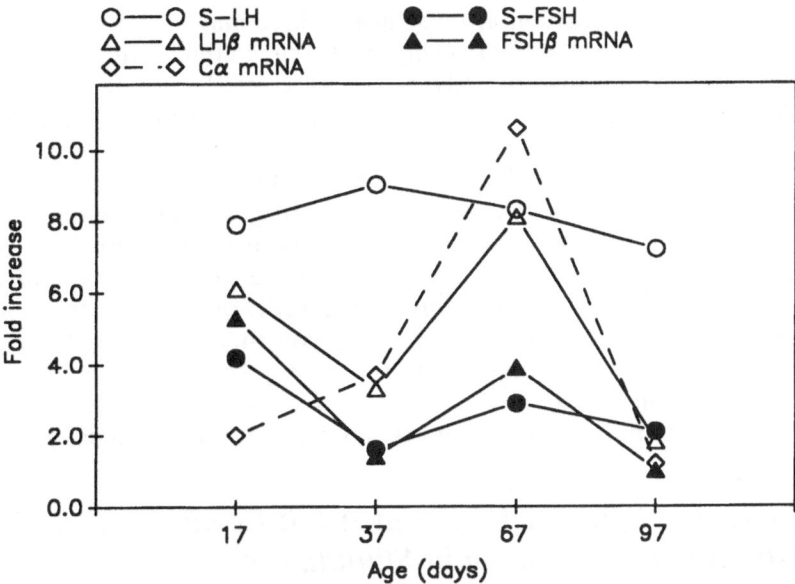

FIGURE 12.2. Fold increases in serum (S) LH and FSH and the pituitary mRNAs of the common α (Cα) subunit and the LH and FSH β-subunits 7 days after castration in male rats of different ages. The means of measurements of 6–12 individual samples were used to calculate each point presented. Data from reference 13.

fold increases in gonadotropin β-subunit mRNAs were largest in the 10-day and 60-day groups (5- to 6-fold and 4- to 8-fold, respectively). The increases in common α-subunit mRNAs were low in the 10-, 30-, and 90-day groups (2.0-, 3.7-, and 1.2-fold, respectively), but high in the 60-day group (10.6-fold) (Fig. 12.2). Hence, the profile of the age-dependent responses to castration was parallel in serum FSH and the FSH β-subunit mRNA.

In contrast, the consistently high increase (about 8-fold) in serum LH after castration at all the ages studied was not always correlated with a similar increase in the LH β-subunit mRNA. Response of the α-subunit mRNA correlated even less with those of the gonadotropin proteins and their β-subunit mRNAs. The ages of 10 and 60 days were more responsive to gonadal feedback regulation than ages of 30 and 90 days at the level of specific β-subunit mRNAs. The often reported similarity of changes in serum gonadotropins and the cognate mRNA levels in adult rats was not observed for LH in the different age groups.

Since the level of LHβ mRNA did not respond with the same intensity as predicted from serum LH concentrations at the less responsive ages (30 and 90 days), some additional posttranscriptional events may be involved in gonadotropin regulation at those ages. These may include increases in translation efficiency and/or stability of the circulating hormone. In fact, it has been shown that the LH isoforms secreted after castration are biologically more active and may have a longer half-time in circulation (14, 15).

In conclusion, the qualitative aspects of the pituitary responses to gonadal feedback regulation are altered during postnatal maturation of male rats (13), and these changes involve both transcription and post-transcriptional events. When these responses are compared to those of the female rat (9), the most conspicuous difference is that the female pituitary responds to gonadectomy with a consistent increase of gonado-tropin output only after puberty. In the male gonadal feedback response seems to be functional from the fetal period onward, and no quiescent period akin to that of the human prepubertal age can be detected.

Differences Between Fetal and Adult Leydig Cells in Response to Gonadotropin Stimulation

Two morphologically distinct Leydig cell growth phases (populations) occur during development. The first, fetal population, appears during the latter half of gestation and is gradually replaced around the age of 2 weeks of postnatal life with the adult growth phase (16, 17). Both Leydig cell populations produce androgens, but their responses to gonadotropin stimulation are vastly different. High doses of gonadotropin (LH or hCG) induce down-regulation of LH receptors and blockade of androgen production (steroidogenic lesion) in adult Leydig cells, but in the fetal/neonatal testis, the corresponding effect is stimulatory (16–19).

We have studied these effects further by measuring the responses of mRNAs of the LH receptor and some key steroidogenic enzymes 1, 2, and 3 days after I.M. injection of a high dose (600 IU/kg b.w.) of hCG to adult and 2-day-old rats (20). The blockade of the 17-hydroxylase/C17-20 lyase step of steroidogenesis, as a sign of the steroidogenic lesion (16–18), was seen in adult rat testis as a transient accumulation of progesterone in serum and as a corresponding decrease of the 17-hydroxylase/17,20-lyase cytochrome P450 mRNA level (Fig. 12.3). In neonatal rats serum pro-gesterone did not rise after hCG injection, and the mRNA level of the above enzyme increased (Fig. 12.3). The responses of the other two steroidogenic enzyme mRNAs analyzed—*cholesterol side-chain cleavage cytochrome P450* ($P450_{scc}$) and *aromatase cytochrome P450* ($P450_{arom}$)—were similar in neonatal and adult rat testis: There was an increase in the mRNA of $P450_{scc}$ and no change in that of $P450_{arom}$ (results not shown). It has been suggested that aromatization of T to estradiol plays a role in the steroidogenic lesion, but our results showed that at least increased transcription of the aromatase gene is not involved.

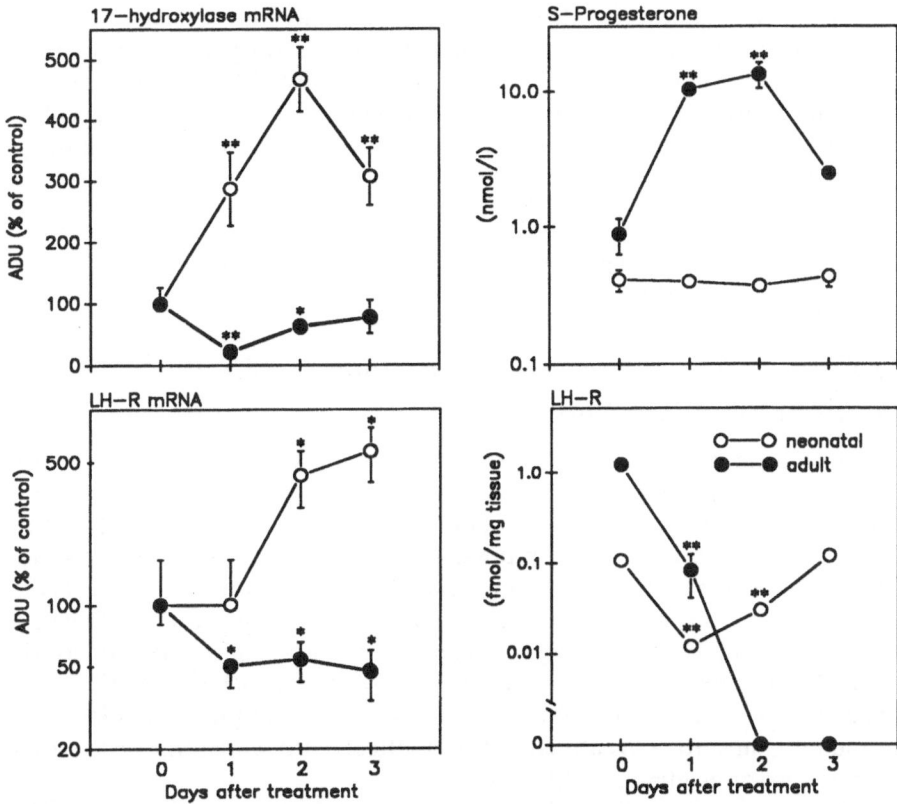

FIGURE 12.3. Effects of a single I.M. injection of a high dose of human chorionic gonadotropin (hCG: 600 IU/kg b.w.) on testicular 17-hydroxylase/17,20-lyase cytochrome P450 (17-hydroxylase) and LH receptor (LH-R) mRNAs and testicular LH receptor (R) binding and serum (S) progesterone in adult and neonatal (5-day-old) rat testis. Measurements were carried out before (0) and 1, 2, and 3 days after the injection. Each point represents the mean ± SEM of measurements from 5–12 animals. (ADU = arbitrary densitometric units; * = $P < 0.05$; ** = $P < 0.01$ as compared with the control levels on day 0.) Data from reference 20.

High-dose stimulation with hCG in adult rats results in the disappearance of testicular LH receptor binding for 7–10 days (18, 19). In contrast, in the neonatal testis a similar dose of hCG brings about only transient occupancy of receptors, and the binding fully recovers in 3 days (Fig. 12.3) (20). By following the LH receptor mRNA levels, we could show that the down-regulation observed in the adult testis involves not only increased receptor occupancy, sequestration, and intracellular degradation (18), but is also partly due to suppression of the receptor mRNA levels. In contrast, in neonatal testes the receptor mRNA level was significantly increased in response to hCG (Fig. 12.3). The molecular mechanisms behind these developmental differences in testicular function still remain obscure. A tempting hypothesis for further studies is that different mRNA species could encode functionally differing LH receptor proteins at the different stages of development. A variety of LH receptor splice variants have been identified, but their functions are still largely unknown (21–23).

Steroidal and Nutritional Control of Gonadotropin Gene Expression

Effects of Testosterone on Gonadotropin Gene Expression at the Pituitary Level

Testosterone plays a major role as the feedback regulator of pituitary gonadotropin gene expression, synthesis, and secretion in males. It acts primarily by decreasing the secretion of hypothalamic GnRH (24), but it also has direct effects at the pituitary level (25). These direct effects are normally overshadowed by the quantitatively more pronounced effects on GnRH secretion, but if the secretion and/or action of GnRH is blocked (e.g., by GnRH antagonist treatment), the direct pituitary effects of gonadal steroids are apparent. We have carried out such studies (25–27), and they revealed vastly differential responses of the three gonadotropin subunit genes to T-treatment. These effects are apparent only when GnRH action is eliminated.

In one of our studies (26), a 10-day treatment regimen with GnRH antagonist (Ac-D-pClPhe-D-pClPhe-D-Trp-Ser-Tyr-D-Arg-Leu-Arg-Pro-D-Ala-NH$_2$CH$_3$COOH, Organon 30276, 1 mg/kg b.w. per day [ANT]) decreased pituitary FSHβ mRNA contents in intact and castrated adult male rats by nearly 90% (Fig. 12.4). These levels were increased to 80% of controls in both groups by combined treatment with ANT and T. When a similar experiment was carried out in intact immature (20-day-old) male rats (26), ANT treatment decreased the FSHβ mRNA contents to a similar extent (80%), and a partial recovery to 40% of controls was seen with combined ANT + T-treatment. In castrated immature male rats, ANT treatment was less effective, resulting in a 50% decrease of the pituitary FSHβ mRNA level, but combined treatment with ANT + T resulted in full recovery of this mRNA. These changes in gonadotropin gene expression were similarly reflected by pituitary and serum gonadotropin levels.

In contrast to the FSH β-subunit mRNA, no direct pituitary effects of T were found in the mRNA of the common α or LH β-subunits. A decrease in serum immunoreactive inhibin was observed from 8.3 ± 2.0 U/mL to 2.1 ± 0.4 U/mL ($P < 0.01$) by the ANT treatment, and T reversed it to 5.8 ± 0.6 U/mL. These results indicate that T is able to

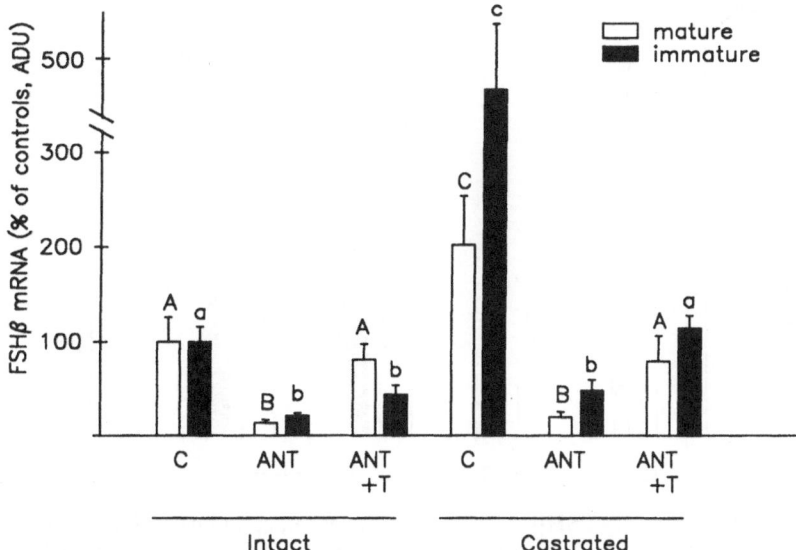

FIGURE 12.4. Pituitary FSHβ mRNA contents of intact and castrated mature and immature male rats (n = 7–9 per group) treated with GnRH antagonist (ANT) or GnRH antagonist and testosterone (ANT + T). A value of 100% was assigned to the means of the controls (C). (ADU = arbitrary densitometric units.) The different letters (capital for mature and small for immature) above the bars indicate a statistically significant difference between these groups (P at least <0.05). Data from reference 26.

stimulate FSH gene expression directly in immature and adult rats. The T-response is clearly enhanced in immature rats by orchidectomy, which corroborates earlier findings on age differences in the role of testicular inhibin in the regulation of FSH gene expression (26).

The direct pituitary effect of T on gonadotropin gene expression was also brought about by using high doses of T alone (27). We treated intact and castrated male rats with increasing doses of T (Silastic implants), and the highest doses resulted in serum T-levels 3-fold higher than in intact control animals. As expected, the low doses of T suppressed pituitary FSHβ mRNA contents by about 50% (Fig. 12.5). However, the highest dose (16-cm implant of T) returned this mRNA level back to control level. Castration alone increased the FSHβ mRNA level 2-fold, and after T-treatments the FSHβ message returned to control levels, but this time the effect was monophasic. Pituitary LHβ mRNA displayed a dose-dependent suppression in response to T, to 30%–35% of controls with the 8- and 16-cm implants. The common α mRNA contents were suppressed by all T-doses to 30%–50% of controls. These effects of T on gonadotropin gene expression were reflected in the pituitary contents of

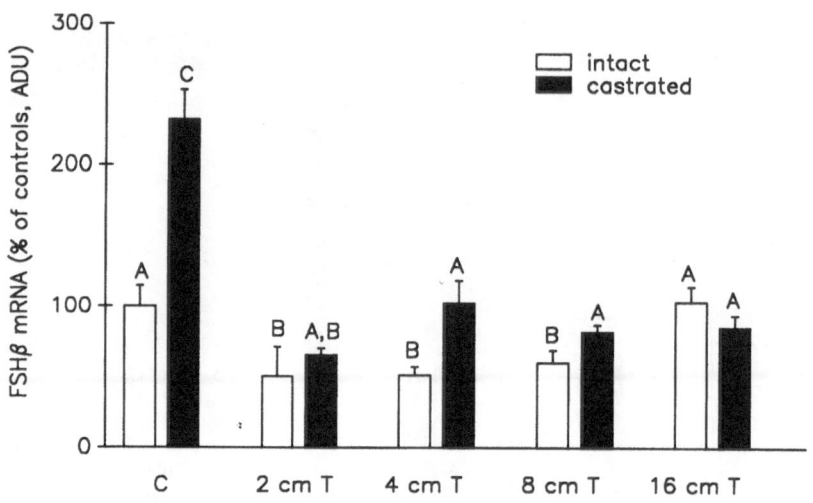

FIGURE 12.5. Pituitary FSHβ mRNA contents of intact and castrated mature male rats ($n = 8$ per group) treated with increasing lengths (cm) of Silastic inplants containing testosterone (T). A value of 100% was assigned to the mean of intact controls (C). (ADU = arbitrary densitometric units.) Different letters above the bars indicate a statistically significant difference between these groups (P at least <0.05). Data from reference 27.

gonadotropins in a similar manner. In contrast, serum FSH showed no increase after the treatments, a finding that is in contrast to that made in the ANT + T-treated animals.

In conclusion, the findings show that T-treatment has a biphasic effect on the gene expression of FSH. However, the secretion of FSH was not affected by the high doses of T, which may be due to the fact that T suppresses GnRH secretion so profoundly that there was no stimulus left for FSH secretion. ANT treatment may not fully block GnRH action, and the residual activity may be sufficient for FSH release. Although the biphasic effect of T on FSH β-subunit expression is clear, the physiological significance of this finding still awaits further investigation.

Effects of Food Restriction on Gonadotropin Gene Expression and Secretion in Male Rats

Inadequate nutrient intake impairs the function of the hypothalamic-pituitary-gonadal axis in mammals (28–30). Although there are some conflicting results, most investigators agree that the synthesis and secretion of gonadotropins and other anterior pituitary hormones are reduced during fasting (29, 31, 32). Reduced hypothalamic release of GnRH has

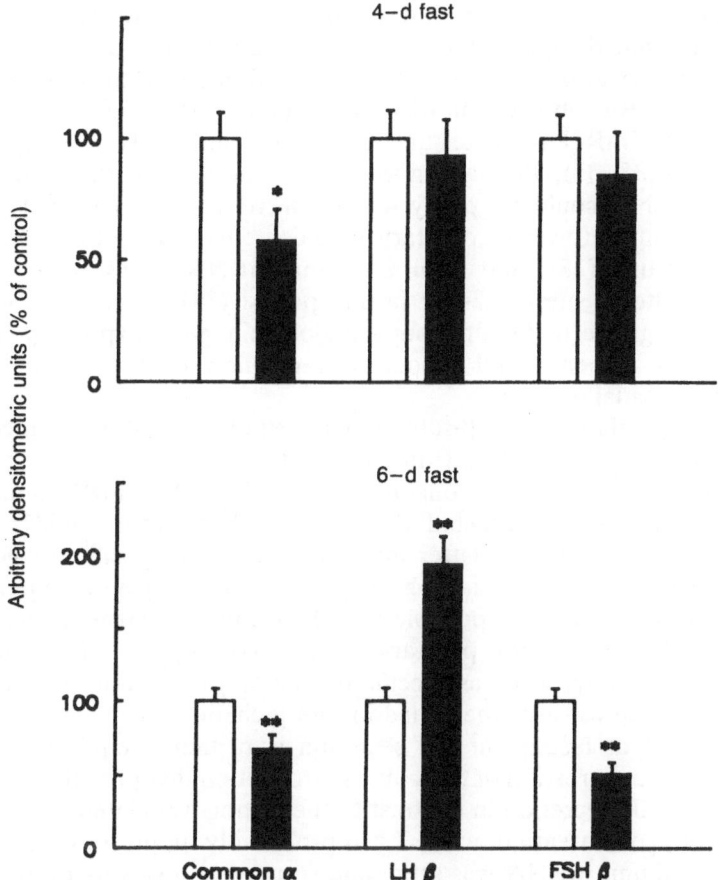

FIGURE 12.6. Effect of a 4- (upper panel) or 6-day (lower panel) fasting period on pituitary mRNA levels of the common α-subunit and the LH and FSH β-subunits in adult male rats ($n = 6$ per group). The control (open bars) and fasting groups (solid bars) consist of 6 animals (mean ± SEM). The levels of the mRNAs are plotted as arbitrary densitometric units, related to the intensity of hybridization with the α-actin cDNA, and the mean of each control group is taken as 100%. Data from reference 34.

been suggested as the major factor responsible for the gonadotropin decrease, but only on the basis of indirect evidence (30, 33).

In order to elucidate the effects of food restriction on gonadotropin gene expression, we studied pituitary gonadotropin subunit mRNA levels in adult male rats after a 4-day fast (only water allowed) (34) (Fig. 12.6). A significant 42% decrease was only seen in the pituitary mRNA levels of the common α-subunit after fasting, but not in the LH and FSH β-subunits. At the same time serum LH levels decreased by 30%–50%, and

FSH by 25%–45%. After 6 days of fasting, the common α-subunit mRNA remained suppressed (Fig. 12.6), and a further decrease to 50% of controls was also seen in the FSH β-subunit mRNA. Surprisingly, a 2-fold increase was observed in the LH β-subunit mRNA levels.

Although GnRH regulates the common α and FSH β-subunit gene expression (35–37), the fasting-associated decrease in the common α-subunit mRNA could be partly due to decreased mRNA levels of the TSH β-subunit. However, pituitary TSH contents were not changed after a 3-day fasting (38), and TSH β-subunit mRNA levels were not significantly altered after a 2-day fasting period (39). To our knowledge, these findings were the first on gonadotropin gene expression in this widely used experimental model of regulation of the hypothalamic-pituitary-gonadal axis.

The finding that the LH β-subunit gene expression did not change after the fasting period of 4 days is in contrast to some previous observations where the suppression of gonadotropin synthesis by GnRH antagonist or steroid treatments affected the levels of LH β-chain mRNA more profoundly (40, 41). The latter manipulations were specific for gonadotropin secretion, whereas fasting resulted in a more general suppression of the pituitary (31). It is possible that the fasting-associated general suppression of the anterior pituitary affects transcription of the α-chain gene preferentially, whereas specific regulation of gonadotropins is more closely directed towards the gonadotropin β-chains.

The effect of fasting on LH secretion is further complicated by the concomitant suppression of testicular T-production that provides a positive stimulus to LH secretion in contrast to the fasting-associated suppression. After 6 days of starvation, when T was profoundly suppressed, an increase of LH β-subunit mRNA was seen, and serum LH levels had returned to the control range (34). Moreover, short-term fasting is unable to suppress gonadotropin levels in castrated rats (28, 29). It therefore seems that starvation is able to suppress LH secretion only in the presence of normal T-levels.

The second study was carried out to test the hypothesis that reduced hypothalamic GnRH release is responsible for the suppression of reproductive functions during fasting. Adult male rats were kept for 4 days under total fasting (only water allowed) and injected during this time at 2-h intervals with 100 or 500 ng/kg b.w. of GnRH or vehicle (42). The lower dose of GnRH increased the levels of common α-subunit mRNA in both control and starved animals, but did not change the difference between the two groups (Fig. 12.7). The treatment with the higher GnRH dose totally reversed the decreased levels of the common α-subunit mRNA in the starved animals. Fasting alone did not affect the mRNA levels of LH β-subunit (Fig. 12.7), but the slight up-regulation effect of the lower GnRH dose in this parameter in the control animals could not be demonstrated in the starved rats.

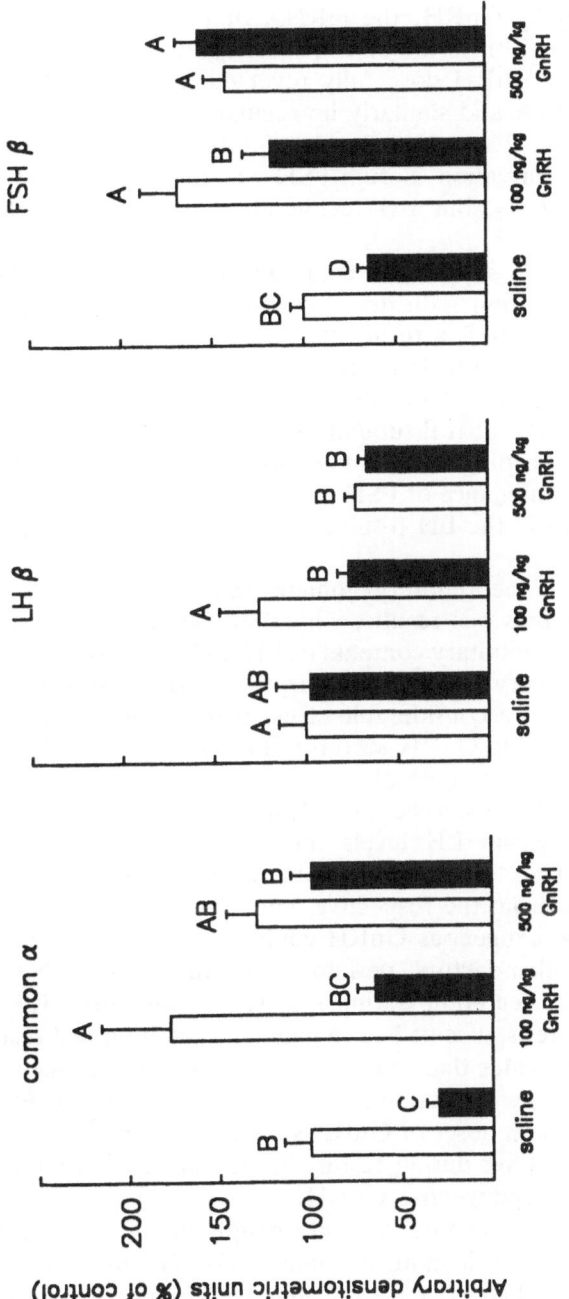

FIGURE 12.7. Effect of GnRH substitution during a 4-day fasting period on pituitary mRNAs for common α-subunit and LH and FSH β-subunits. The animals were treated with either vehicle (100 μL of saline) or 100 ng/kg or 500 ng/kg b.w. of GnRH. The injections were given at 2-h intervals during the whole fasting period. If different letters are above the bars, the difference between these groups is statistically significant (P at least <0.05). Data from reference 42. For further details, see the legend of Fig. 12.6.

Fasting decreased the pituitary mRNA contents of the FSH β-subunit compared to saline-treated controls (Fig. 12.7). In the starved groups treated with 100 ng/kg GnRH, the mRNA of FSH β-subunit remained suppressed in comparison to the up-regulated level of the controls fed ad libitum. The higher GnRH dose fully reversed the effect of fasting on FSH β-subunit mRNA and similarly up-regulated these levels in starved and control animals. The serum levels of FSH and LH decreased by about 30% during starvation. Serum FSH was fully reversed in GnRH-treated starved animals, but LH recovered only partially (results not shown).

The recovery of the suppressed common α-subunit mRNA by exogenous GnRH substitution indicates that the loss of gonadotropin gene expression during fasting is a result of the decrease in the hypothalamic release of GnRH (see above). Moreover, the suppressed α-subunit expression occurs in gonadotrope and not thyrotrope cells. The fasting-induced decrease in the FSH β-subunit mRNA and its recovery by GnRH substitution are in keeping with findings that endogenous GnRH is also required for the maintenance of FSH β-subunit mRNA levels (36). Again no change was seen in the LH β-subunit mRNA after the fasting period of 4 days.

The responses of the pituitary gonadotropin levels were somewhat different from our previous results, where starvation resulted in significant decreases in the pituitary contents of LH and FSH (34). It is obvious that the injections every second hour during the whole starvation period of 4 days represented a considerable stress to the animals. Stress alone can inhibit hypothalamic GnRH secretion (43, 44); then, the magnitude of differences between the control and starved animals is not as great as between undisturbed and starved animals in our previous experiment.

The suppressed serum LH levels remained on the starvation level after both doses of GnRH substitution in starved animals, although the absolute difference from the respective control was reduced (results not shown). Obviously, exogenous GnRH could not affect LH secretion, or there were some other factors, perhaps neurotransmitters, that directly inhibit pituitary LH secretion during fasting. As with the LHβ mRNA levels, the relative resistance of LH changes to starvation and restoration of GnRH pulses indicates that other factors are involved in regulation of this gonadotropin. The starvation-suppressed serum levels of FSH were totally reversed by both doses of GnRH substitution. Thus, the decline in circulating levels of FSH during fasting seems mainly to be due to the decreased release of endogenous GnRH.

Reduced food intake also delays and disrupts the development of the reproductive processes of immature animals (45–47). On the other hand, in conditions with enhanced gonadotropin secretion (e.g., after castration), the antigonadotropic effect of fasting is not observed in male rats (28, 29). The effects of short-term fasting on pituitary-testicular function in other physiological conditions with increased gonadotropin secretion (e.g., during puberty) have not been elucidated. We therefore decided to study the influence of fasting on the activation of the reproductive functions in pubertal male rats. It was hypothesized that the regulation of gonadotropin secretion during fasting at puberty could differ from the situation before and after this developmental stage.

In the third study the effects of short-term fasting (3 or 4 days) on pituitary-testicular functions during sexual maturation were studied in male rats at 25, 35, 45, 55, and 65 days of age (48). After 3 days of fasting, significant 28%–55% decreases were seen in the pituitary mRNA levels of the common α-subunit at 25, 55, and 65 days of age (65-day-old animals were starved for 4 days) (Fig. 12.8), but not in the 35- and 45-day-old animals. The pituitary mRNA levels of LH β-subunit were again unchanged in all starved groups (Fig. 12.8). The mRNA levels of the FSH β-subunit decreased in starved animals of 25, 55, and 65 days of age (Fig. 12.8), but not in the 35- and 45-day-old animals. Serum LH decreased after starvation at 25, 55, and 65 days of age, whereas pituitary LH levels increased slightly at 35 days. Serum FSH decreased only in the 65-day-old rats, whereas pituitary FSH also declined at 25 days.

In conclusion, gonadotropin gene expression and secretion decreased consistently during short-term fasting only in the prepubertal (25 days) and adult (65 days) rat, but not in peripubertal animals (35 and 45 days). Hence, the pubertal rise in gonadotropins is due to such a strong positive induction that it overrides the antigonadotropic effect of fasting. Before and after this period, fasting effectively suppresses gonadotropin gene expression.

Regulation of Bioactive LH and FSH Secretion in the Human Male

Reevaluation of Bioactive/Immunoreactive Ratios of Serum LH Using Immunoassays with Improved Sensitivity and Specificity

In vitro bioassays of LH have been available for the measurement of plasma LH levels for more than 10 years. A wide array of findings have been made in various clinical and experimental conditions, demonstrating that the *bioactive* (B) and *immunoreactive* (I) concentrations of LH do not always vary in parallel (4, 49, 50). One such finding has been the increased B:I ratio of LH during the endogenous and exogenously GnRH-induced LH secretion pulses (51, 52). This phenomenon has been attributed to differences in secreted LH in the distribution of the differentially glycosylated isoforms with varying intrinsic biological activities

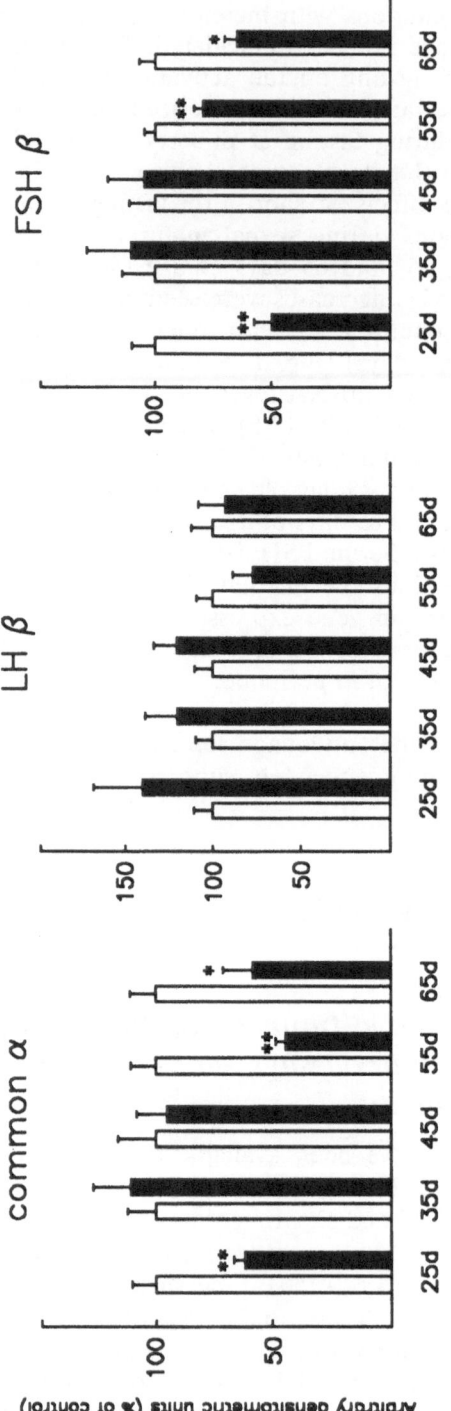

FIGURE 12.8. Effect of short-term fasting on pituitary steady state levels of mRNAs for the common α-subunit and LH and FSH β-subunits in male rats of 25–65 days (d) of age. The control (open bars) and starvation (solid bars) groups consisted of 5–6 animals (mean ± SEM). * = $P < 0.05$; ** = $P < 0.01$ compared with control animals of the same age by Student's t-test.) Data from reference 48. For further details, see legend of Fig. 12.6.

(4, 49, 50). Another example of the altered B:I ratio of LH is the decrease in this parameter when LH secretion is suppressed by GnRH agonist treatment; for example, in prostatic cancer patients (53, 54). In general, the B:I ratio is increased when the LH levels increase, and vice versa.

The techniques for the measurement of I-LH have improved recently with the advent of monoclonal antibodies and the immunometric assay principle (see above for the improved rat LH assay). The sensitivity of the new assays is up to 100-fold greater than that of the conventional RIA measurements. Additionally, their specificity has improved, and their correlation with bioassay results is also good at low levels of LH, which is not the case with the RIA measurements (50, 55).

We have studied the regulation of serum LH and its B:I ratio in men in a variety of physiological and pathophysiological conditions using a sensitive and specific immunofluorometric assay (Delfia®, Wallac, Turku, Finland) for I-LH measurements. The findings agree rather well with earlier measurements using RIA at normal to high hormone concentrations, but great discrepancies occur when the B:I ratios and their changes are compared at normal to low levels of LH. We were able to reproduce the B:I ratio increase that was previously demonstrated during the endogenous LH secretion pulses with RIA, but not with the IFMA assay (Delfia®) (Fig. 12.9) (56). A similar observation was made after exogenous GnRH injection in normal men (50). Additionally, no decrease was observed in the B:I ratios of LH during GnRH agonist (Fig. 12.10) and estrogen treatments of prostatic cancer patients (57 and Matikainen et al., unpublished observation). Also, in contrast to previous findings (58), we were unable to show differences in the B:I ratio of LH between prepubertal and pubertal boys (55).

Besides the above negative findings, we have also identified cases where altered B:I ratios can be documented irrespective of the method used for I-LH measurement. Such conditions are, for example, the postgonadectomy increase of LH in men and premenopausal women, where the B:I ratio increases (15, 50), and the higher B:I ratio of LH in women with polycystic ovarian disease (59). What is common to these conditions is that LH levels are very high.

What can be concluded from our B- and I-LH measurements is that many of the conditions with reported alterations in B:I ratios of LH should be reevaluated. The earlier findings may be erroneous due to the nonlinearity of bioactive and immunoreactive LH levels, especially at low concentrations. This fact was not realized before the novel sensitive and specific immunoassays became available. It seems that the quality of circulating LH (i.e., the B:I ratio) is more stable than assumed before. Thus, although there are genuine changes in the isoform distribution of LH in various physiological and clinical conditions, further studies are needed for their verification.

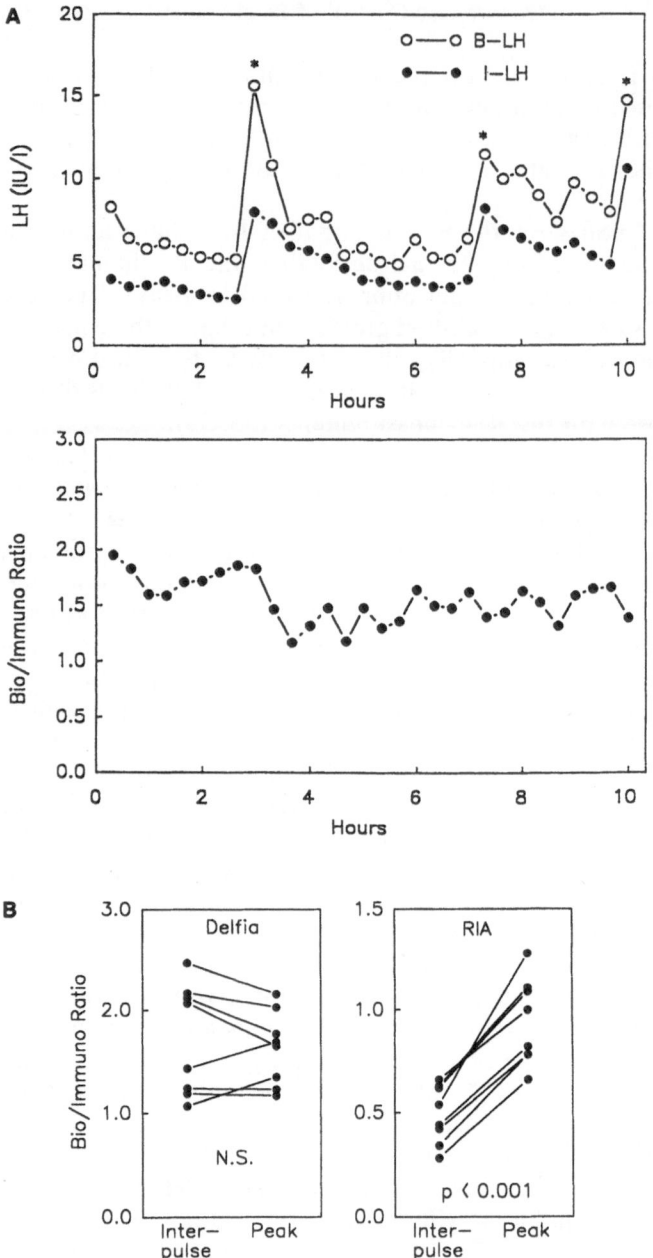

FIGURE 12.9. Stability of the B:I ratio of serum LH during the endogenous secretion pulses. An example is shown of the pulsatile secretion of bioactive (B) and immunoreactive (I) LH (*A*, upper panel) and variation of the B:I ratio of LH (*A*, lower panel) detected in a representative male subject. (Open circles = B-LH; solid circles = I-LH, measured by IFMA.) The B:I ratios, measured at the LH secretion peaks and during the interpulse nadirs in 8 men using IFMA (Delfia) and RIA are presented in *B*. Reprinted with permission from Huhtaniemi, Ding, Tähtelä, and Välimäki (56), © The Endocrine Society, 1992.

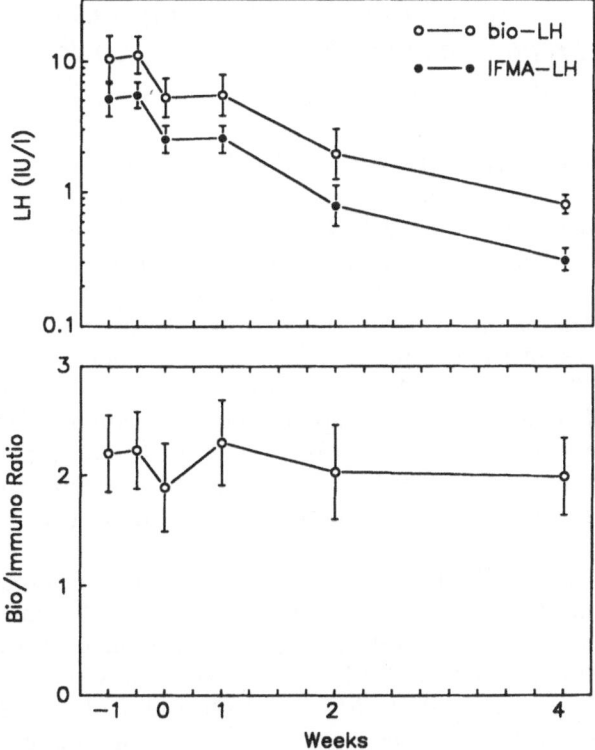

FIGURE 12.10. Serum concentrations of bioactive (bio-LH) and immunoreactive (IFMA-LH) LH (upper panel) and the B:I ratio of LH (lower panel) in patients treated with the depot form of the GnRH agonist Buserelin (mean ± SEM; $n = 5$). The first 2 samples were taken 7 and 3 or 4 days before insertion of the first Buserelin implant; the last time point is 4 weeks after the first implant. One of the 4-week levels of bio-LH was undetectable (<0.5 IU/L); hence, $n = 4$ in this group. After 4 weeks of treatment, all bio-LH concentrations were undetectable. Reprinted with permission from Huhtaniemi, Venho, Jacobi, and Rannikko (57).

Experimental research on the regulation of LH isoform distribution has been hampered by the insufficient sensitivity of the rat and mouse LH assays (both in vitro bioassay and immunoassay) when applied to serum samples. In this respect, our new IFMA assay for rat LH (see above) should be useful. Moreover, Debertin and Pomeranz (60) recently reported an improved in vitro bioassay method for LH with 30-fold increased sensitivity. This assay is suitable for measurements of bioactive LH in rat peripheral circulation. It is evident that the currently confusing view about the real meaning and diagnostic value of variations of the LH isoform distribution can be better resolved with thorough experimental observations than with the fragmentary clinical data currently available.

Response of Serum Bioactive and Immunoreactive FSH Levels to Treatments with Testosterone, GnRH Agonist, and Orchidectomy in Men

The assay of bioactive FSH in serum has been technically more difficult, and satisfactory in vitro bioassays for this purpose have been available for only a few years (3, 5). As with LH, the novel, improved immunoassay methods have changed the picture of I-FSH measurements, providing more accurate assessments of this hormone at low concentrations. An example of this improved accuracy is provided in Figure 12.11 (upper panel), which shows the response of I-FSH to treatment with testosterone enanthate (200 mg/week I.M.) for contraceptive purposes in normal male volunteers (Huhtaniemi et al., unpublished observation). The difference in the suppressed I-FSH levels is 10-fold with the two assays and indicates the overestimation of low levels by RIA. A similar picture emerges from LH measurements with the two immunoassay methods (Fig. 12.11, lower panel).

We have assessed the FSH responses in two groups of prostatic cancer patients during GnRH agonist (Buserelin, Hoechst) treatment. One of the groups received the agonist as the primary treatment; the other group was orchidectomized first (50, 61). These two groups allowed us then to compare the responses of B-FSH and I-FSH to GnRH agonist in the presence and absence of testicular tissue. During treatment of both groups, the level of I-FSH decreased after the initial stimulatory phase. One of the most conspicuous findings was that the suppression of FSH was consistently much less than that of LH. When we used a more potent depot preparation of GnRH agonist in another study (57), the difference was even clearer; the greater the suppression of LH, the less FSH was affected (Fig. 12.12).

It still remains a paradox why LH and FSH respond so differently to GnRH agonist treatment, although with both hormones the main driving force of their release, GnRH, should be blocked. One explanation could be that the inhibin:activin ratio of the testis changes in favor of the latter, which then could maintain the release of FSH. In support of this we have found that both I-FSH and B-FSH levels decrease immediately after orchidectomy in GnRH-treated patients, whereas a concomitant increase of FSH occurs in nontreated orchidectomized men (61).

The pattern of response of B-FSH to GnRH agonist treatment in prostatic cancer patients was similar to that of I-FSH, but when the B:I ratios were calculated, there was a transient increase in this ratio about 1 month after the initiation of treatment that subsequently returned to the pretreatment levels (Fig. 12.13). As speculated above for the recovery of the I-FSH levels during the agonist treatment, the initial increase of the B:I ratio may also be due to suppressed inhibin secretion (61). However,

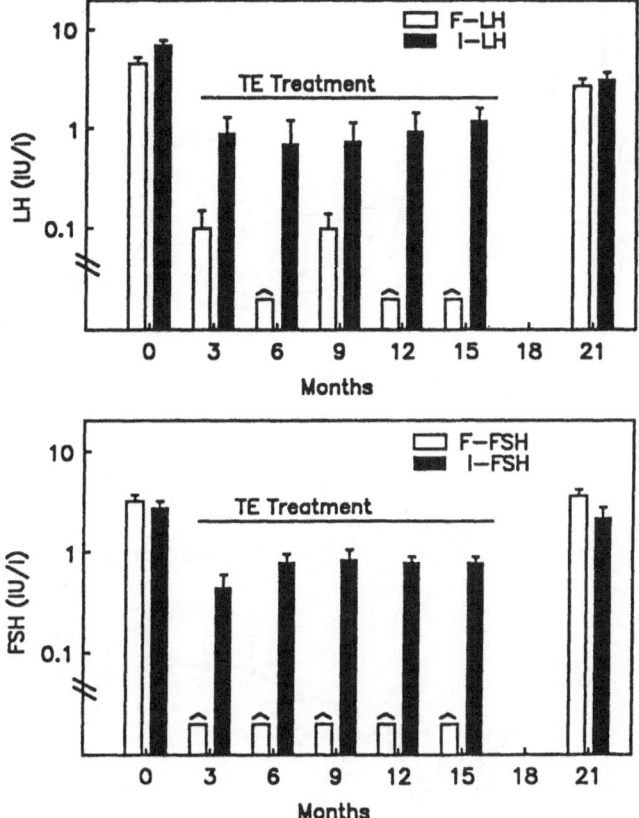

FIGURE 12.11. Serum immunoreactive LH and FSH measured by a standard RIA (I-LH and I-FSH, WHO reagents) or IFMA (F-LH and F-FSH, Delfia, Wallac) in 15 healthy male volunteers treated with 200-mg I.M. injections of testosterone enenthate (TE) once weekly for contraceptive purposes (mean ± SEM). The treatment lasted for 12 months, and the first pair of data (0 months) represent pretreatment levels, and the last pair (21 months) represent levels 6 months after discontinuation of the treatment. The caret symbol above the bars indicates that at this time point, all levels were below the limit of detection (0.03 IU/L).

this could not be documented when inhibin levels were monitored in the intact GnRH-treated men (Huhtaniemi, de Kretser, unpublished observation). An increase, possibly transient, in the B:I ratio of FSH is also seen after orchidectomy (61) and during estrogen treatment of prostatic cancer patients (Matikainen et al., unpublished observation).

The mechanisms of these increases and their physiological significance still remain unclear. When this effect is compared to that evoked by

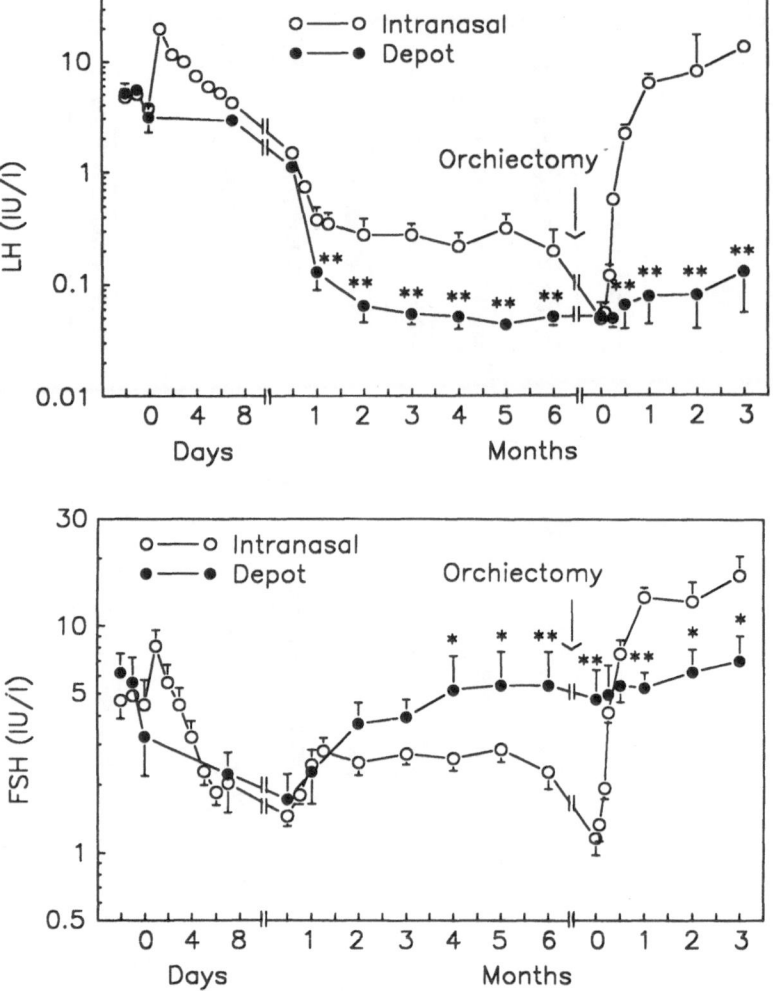

FIGURE 12.12. The IFMA-LH (upper panel) and IFMA-FSH (lower panel) responses to intranasal (open circles, $n = 8$) and depot (solid circles, $n = 5$) regimens of Buserelin treatment in prostatic cancer patients. After 6 months of Buserelin treatment, all patients were orchidectomized. LH levels of the intranasal group were significantly ($P < 0.01$) suppressed between week 2 of treatment and week 1 after orchidectomy. All LH levels of the depot group were suppressed ($P < 0.01$) from week 2 of the treatment onward. The FSH levels of the intranasal group were significantly suppressed ($P < 0.01-0.05$) from pretreatment levels between day 5 and month 1 of treatment and from month 6 to day 4 after orchidectomy. The FSH levels in the depot group were suppressed ($P < 0.05$) between weeks 1 and 4 of the treatment. Asterisks in the figures indicate times when the gonadotropin concentrations in the Buserelin depot group differ significantly from those of the intranasal treatment group (* = $P < 0.05$; ** = $P < 0.01$.) Reprinted with permission from Huhtaniemi, Venho, Jacobi, and Rannikko (57).

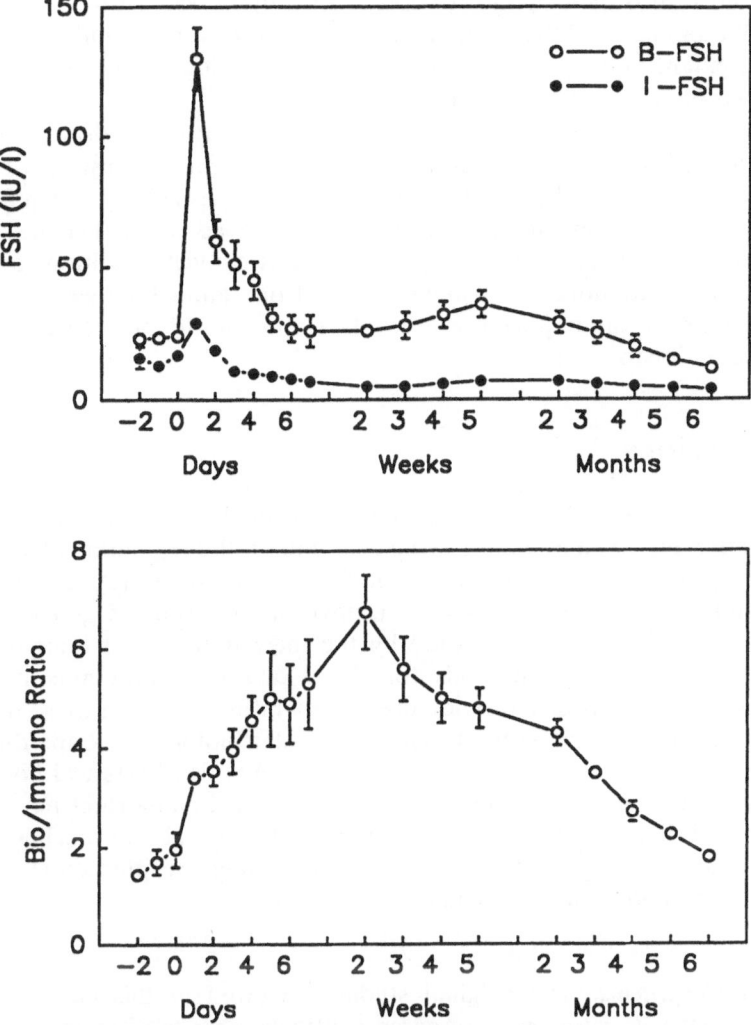

FIGURE 12.13. Mean (± SEM) serum immunoreactive (I, solid circles) and bio-active (B, open circles) FSH levels (upper panel) and the respective B:I ratios (lower panel) in 5 men treated for 6 months with intranasal application of the GnRH agonist Buserelin. Data from reference 61.

GnRH antagonist, a consistent difference prevails: Antagonist treatments suppress the B:I ratio of FSH both in men and women in a variety of clinical conditions (62–65). In conclusion, there are clear differences in the GnRH agonist and antagonist effects on LH and FSH. The quantity, but not the quality, of LH released from the pituitary is critically dependent on GnRH action. Concerning FSH, the quality of the hormone

228   I. Huhtaniemi et al.

released is also dependent on GnRH, and suppressed GnRH action has variable effects on FSH. Our findings also indicate profound differences in the extent of suppressive effects of GnRH analogs on the secretion of LH and FSH and in the mechanisms involved.

Even more than with LH, it is currently difficult to draw clear-cut conclusions about the mechanisms that regulate the isoform distribution of FSH (as reflected by the B:I ratio). Also, the physiological significance of the phenomenon still remains unknown. It is apparent that the observations on FSH B:I ratio changes in various clinical conditions are not sufficient, and more fundamental data from animal experiments are needed before the importance of the B:I ratios in the physiology of FSH can be understood.

Conclusions

These studies summarize some of our recent findings on the pituitary regulation of fetal and adult Leydig cells and on the regulation of gonadotropin gene expression, as well as qualitative and quantitative aspects of LH and FSH secretion. Besides the novel information on gonadotropin gene expression that is provided by the measurements of their mRNA levels, the improved methods for the assays of immunoreactive and bioactive gonadotropins provide useful new information. In these days of emphasis on research methods using molecular biology, we hope that the present chapter provides evidence that the methods of classical hormone measurements have also improved and can provide important new information. When combined with molecular biological approaches, they may help us obtain new perspectives concerning the physiology of the hypothalamic-pituitary-testicular axis.

Acknowledgments. The original studies reviewed in this chapter were carried out with support of research contracts from the Finnish Life and Pension Insurance Companies and The Academy of Finland and a grant from the Sigrid Jusélius Foundation.

References

1. Lövgren T, Hemmilä I, Pettersson K, Eskola JU, Bertoft E. Determination of hormones by time-resolved fluoroimmunoassay. Talanta 1984;31:909–16.
2. Petersson K, Söderholm JR-M. Ultrasensitive two-site immunometric assay of human lutropin by time-resolved fluorometry. Clin Chem 1990;36:1928–33.
3. Wang C. Bioassays of follicle-stimulating hormone. Endocr Rev 1988;9:374–7.
4. Tsatsoulis A, Shalet SM, Robertson WR. Bioactive gonadotropin secretion in man. Clin Endocrinol (Oxf) 1991;35:193–206.

5. Beitins IZ, Padmanabhan V. Bioactivity of gonadotropins. Endocrinol Metab Clin North Am 1991;20:85–120.
6. Monroe SE, Parlow AF, Midgley AR Jr. Radioimmunoassay of rat luteinizing hormone. Endocrinology 1968;83:1004–12.
7. Haavisto A-M, Pettersson K, Bergendahl M, Perheentupa A, Roser JF, Huhtaniemi I. A supersensitive immunofluorometric assay for rat luteinizing hormone. Endocrinology 1993.
8. Pakarinen P, Huhtaniemi I. Gonadal and sex steroid feedback regulation of gonadotrophin mRNA levels and secretion in neonatal male and female rats. J Mol Endocrinol 1989;3:139–44.
9. Pakarinen PA, Huhtaniemi IT. Development of gonadal feedback regulation of gonadotropin gene expression and secretion in female rats. Acta Endocrinol (Copenh) 1992;127:545–8.
10. Sokka T, Huhtaniemi I. Ontogeny of gonadotrophin receptors and gonadotrophin stimulated cAMP production in the neonatal rat ovary. J Endocrinol 1990;127:297–303.
11. Negro-Vilar A, Ojeda SR, McCann SM. Evidence for changes in sensitivity to testosterone negative feedback on gonadotropin release during sexual development in the male rat. Endocrinology 1973;93:729–35.
12. Ojeda SR, Andrews WW, Advis JP, Smith White S. Recent advances in the endocrinology of puberty. Endocr Rev 1980;1:228–57.
13. Pakarinen P, Huhtaniemi I. Age-related discrepancies between serum and pituitary gonadotrophin, and pituitary gonadotrophin subunit mRNA responses to castration and testosterone replacement in male rats. J Endocrinol 1992;135:507–15.
14. Keel BA, Grotjan HE Jr. Characterization of rat pituitary luteinizing hormone charge microheterogeneity in male and female rats using chromatofocusing: effects of castration. Endocrinology 1985;117:354–60.
15. Haavisto A-M, Simberg N, Huhtaniemi I. The bio/immuno ratio of serum luteinizing hormone increases after orchiectomy in prostatic cancer patients and is associated with decreased molecular weight and appearance of isohormones with alkaline pI values. Biol Reprod 1990;42:597–602.
16. Huhtaniemi IT, Warren DW, Catt KJ. Functional maturation of rat testis Leydig cells. Ann NY Acad Sci 1984;438:283–303.
17. Huhtaniemi I, Pelliniemi LJ. Fetal Leydig cells: cellular origin, morphology, life span, and special functional features. Proc Soc Exp Biol Med 1992; 201:125–40.
18. Catt KJ, Harwood JP, Clayton RN, et al. Regulation of peptide hormone receptors and gonadal steroidogenesis. Recent Prog Horm Res 1980;36: 557–622.
19. Huhtaniemi IT, Katikineni M, Catt KJ. Regulation of luteinizing hormone receptors and steroidogenesis in the neonatal rat testis. Endocrinology 1981; 109:588–95.
20. Pakarinen P, Vihko K, Voutilainen R, Huhtaniemi I. Differential response of luteinizing hormone receptor and steroidogenic enzyme gene expression to human chorionic gonadotropin stimulation in the neonatal and adult rat testis. Endocrinology 1990;127:2469–74.
21. Wang H, Ascoli M, Segaloff DL. Multiple luteinizing hormone/chorionic gonadotropin receptor messenger ribonucleic acid transcripts. Endocrinology 1991;129:133–98.

22. Aatsinki JT, Pietilä EM, Lakkakorpi JT, Rajaniemi HJ. Expression of the LH/CG receptor gene in rat ovarian tissue is regulated by extensive alternative splicing of the primary transcript. Mol Cell Endocrinol 1992;84:127–35.
23. Sokka T, Hämäläinen T, Huhtaniemi I. Functional LH receptor appears in the neonatal rat ovary after changes in the alternative splicing pattern of the LH receptor mRNA. Endocrinology 1992;130:1738–40.
24. Steiner RA, Bremner WJ, Clifton DK. Regulation of luteinizing hormone pulse frequency and amplitude by testosterone in the adult male rat. Endocrinology 1982;111:2055–60.
25. Perheentupa A, Huhtaniemi I. Gonadotropin gene expression and secretion in gonadotropin-releasing hormone antagonist-treated male rats: effect of sex steroid replacement. Endocrinology 1990;126:3204–9.
26. Perheentupa A, Bergendahl M, de Jong FH, Huhtaniemi I. Diffential regulation of FSH and inhibin gene expression and synthesis by testosterone in immature and mature male rats. J Endocrinol 1993.
27. Perheentupa A, de Jong FH, Huhtaniemi I. Biphasic effect of exogenous testosterone on FSH gene expression and synthesis in the male rat. Mol Cell Endocrinol 1993.
28. Root AW, Russ RD. Short-term effects of castration and starvation upon pituitary and serum levels of luteinizing hormone and follicle stimulating hormone in male rats. Acta Endocrinol (Copenh) 1972;70:665–75.
29. Pirke KM, Spyra B. Influence of starvation on testosterone-luteinizing hormone feedback in the rat. Acta Endocrinol (Copenh) 1981;96:413–21.
30. Badger TM, Lynch EA, Fox PH. Effects of fasting on luteinizing hormone dynamics in the male rat. J Nutr 1985;15:788–97.
31. Campbell GA, Kurcz M, Marshall S, Meites J. Effects of starvation in rats on serum levels of follicle-stimulating hormone, luteinizing hormone, thyrotropin, growth hormone and prolactin; response to LH-releasing hormone and thyrotropin-releasing hormone. Endocrinology 1977;100:580–7.
32. Xie Q-W. Experimental studies on changes of neuroendocrine functions during starvation and refeeding. Neuroendocrinology 1991;53(suppl 1):52–9.
33. Bronson FH. Food-restricted, prepubertal, female rats: rapid recovery of luteinizing hormone pulsing with excess food, and full recovery of pubertal development with gonadotropin-releasing hormone. Endocrinology 1986; 118:2483–7.
34. Bergendahl M, Perheentupa A, Huhtaniemi IT. Effect of short-term starvation on reproductive hormone gene expression, secretion and receptor levels in male rats. J Endocrinol 1989;121:409–17.
35. Haisenleder DJ, Khoury S, Zmeili SM, et al. The frequency of gonadotropin-releasing hormone secretion of α and luteinizing hormone β-subunit messenger ribonucleic acids in male rats. Mol Endocrinol 1987;1:834–8.
36. Rodin DA, Lalloz MRA, Clayton RN. Gonadotropin-releasing hormone regulates follicle-stimulating hormone β-subunit gene expression in the male rat. Endocrinology 1989;125:1282–9.
37. Wierman ME, Rivier JE, Wang C. Gonadotropin-releasing hormone-dependent regulation of gonadotropin subunit messenger ribonucleic acid levels in the rat. Endocrinology 1989;124:272–8.
38. Hugues J-N, Epelbaum J, Voirol M-J, Modigliani E, Sebaoun J, Enjalbert A. Influence of starvation on hormonal control of hypophyseal secretion in rats. Acta Endocrinol (Copenh) 1988;199:195–202.

39. Ramsden DB, Ahlqvist JAO, Fitch NJ, Sheppard MC. The influence of starvation on gene expression in the pituitary gland [Abstract]. Abstracts of the 178th meet Br Soc Endocrinol, 1988:62.
40. Abbot SD, Docherty K, Clayton RN. Gonadal regulation of pituitary hormone mRNA levels in male rats. J Mol Endocrinol 1988;1:61–8.
41. Lalloz MRA, Detta A, Clayton RN. Gonadotropin-releasing hormone is required for enhanced luteinizing hormone subunit gene expression in vivo. Endocrinology 1988;122:1681–8.
42. Bergendahl M, Perheentupa A, Huhtaniemi I. Starvation-induced suppression of pituitary-testicular function in rats is reversed by pulsatile gonadotropin-releasing hormone substitution. Biol Reprod 1991;44:413–9.
43. Gray GE, Smith ER, Ehrenkrantz JRL, Davidson JM. Neuroendocrine mechanisms mediating the suppression of circulating testosterone in male rats following chronic stress. Neuroendocrinology 1978;25:247–56.
44. Tache Y, DuRuisseau P, Ducharme JR, Collu R. Pattern of adenohypophyseal hormone changes in male rats following chronic stress. Neuroendocrinology 1978;26:208–19.
45. Glass AR, Swerdloff RS. Nutritional influences on sexual maturation in the rat. Fed Proc 1980;39:2360–4.
46. Glass AR, Herbert DC, Anderson J. Fertility onset, spermatogenesis, and pubertal development in male rats: effect of graded underfeeding. Pediatr Res 1986;20:1161–7.
47. Sisk CL, Bronson FH. Effects of food restriction and restoration on gonadotropin and growth hormone secretion in immature male rats. Biol Reprod 1986;35:554–61.
48. Bergendahl M, Huhtaniemi I. Acute fasting is ineffective in suppressing pituitary-gonadal function of pubertal male rats. Am J Physiol 1993.
49. Dufau ML, Veldhuis J. Pathophysiological relationships between the biological and immunological activities of luteinizing hormone. Baillieres Clin Endocrinol Metab 1987;1:153–76.
50. Jaakkola T, Ding Y-Q, Kellokumpu-Lehtinen P, et al. The ratios of serum bioactive/immunoreactive LH and FSH in various conditions with increased and decreased gonadotropin secretion: reevaluation by an ultrasensitive immunometric assay. J Clin Endocrinol Metab 1990;70:1496–505.
51. Veldhuis J, Johnson ML, Dufau ML. Preferential release of bioactive luteinizing hormone in response to endogenous, and low dose exogenous gonadotropin-releasing hormone pulses in man. J Clin Endocrinol Metab 1987;64:1275–83.
52. Talbot JA, Stuart R, Rodger C, Shalet SM, Littley M, Robertson WR. The pulsatile secretion of bioactive luteinizing hormone in normal adult men. Acta Endocrinol (Copenh) 1990;122:643–50.
53. Evans RM, Doelle GC, Linder J, Bradley V, Rabin DA. A luteinizing hormone-releasing hormone agonist decreases activity and modifies chromatographic behavior of luteinizing hormone in man. J Clin Invest 1984;73:262–6.
54. St Arnaud RM, Lachance R, Dupont A, Labrie F. Serum luteinizing hormone (LH) biological activity in castrated patients with cancer of the prostate receiving a pure antiandrogen and in estrogen-pretreated patients with an LH-releasing hormone agonist and antiandrogen. J Clin Endocrinol Metab 1986;63:297–302.

55. Haavisto A-M, Dunkel L, Pettersson K, Huhtaniemi I. LH measurements by in vitro bioassay and a highly sensitive immunofluorometric assay improve the distinction between boys with constitutional delay of puberty and hypogonadotropic hypogonadism. Pediatr Res 1990;27:211–4.

56. Huhtaniemi I, Ding Y-Q, Tähtelä R, Välimäki M. The bio/immuno ratio of plasma luteinizing hormone does not change during the endogenous secretion pulse: reanalysis of the concept using improved immunometric techniques. J Clin Endocrinol Metab 1992;75:1442–5.

57. Huhtaniemi I, Venho P, Jacobi G, Rannikko S. Response of circulating gonadotropin levels to GnRH agonist treatment in prostatic cancer. J Androl 1991;12:46–53.

58. Lucky AW, Rich BH, Rosenfeld RL, Fang VS, Roche-Bender N. LH bioactivity increases more than immunoreactivity during puberty. J Pediatr 1980;97:205–13.

59. Ding Y-Q, Anttila L, Ruutiainen K, Erkkola R, Irjala K, Huhtaniemi I. Exogenous and endogenous GnRH stimulation do not alter the bio/immuno ratio of LH in healthy women and in polycystic ovarian disease. Acta Obstet Gynecol Scand 1991;70:211–7.

60. Debertin WJ, Pomerantz DK. Improved sensitivity of the mouse interstitial cell testosterone assay with the addition of forskolin. Can J Physiol Pharmacol 1992;70:866–71.

61. Huhtaniemi IT, Dahl KD, Rannikko S, Hsueh AJ. Serum bioactive and immunoreactive follicle-stimulating hormone in prostatic cancer patients during gonadotropin-releasing hormone agonist treatment and after orchidectomy. J Clin Endocrinol Metab 1988;66:308–13.

62. Dahl KD, Bicsak TA, Hsueh AJW. Naturally occurring antihormones: secretion of FSH antagonists by women treated with a GnRH analog. Science 1988;239:72–4.

63. Kessel B, Dahl KD, Kazer RR, et al. The dependency of bioactive follicle-stimulating hormone secretion on gonadotropin releasing hormone in hypogonadal and cycling women. J Clin Endocrinol Metab 1988;66:361–6.

64. Pavlou SN, Wakefield G, Schlechter NL, et al. Mode of suppression of pituitary and gonadal function after acute or prolonged administration of a luteinizing-hormone releasing hormone antagonist in normal men. J Clin Endocrinol Metab 1989;68:446–54.

65. Matikainen T, Ding Y-Q, Vergara M, Huhtaniemi I, Couzinet B, Schaison G. Differing responses of plasma bioactive and immunoreactive follicle-stimulating hormone and luteinizing hormone to gonadotropin-releasing hormone antagonist and agonist treatments in postmenopausal women. J Clin Endocrinol Metab 1992;75:820–5.

13

Expression of the FSH Receptor in the Sertoli Cells

MICHAEL D. GRISWOLD, LESLIE HECKERT, TAMARA GOETZ, AND CAROL LINDER

The regulation of cell-specific genes could result from the interaction of unique transcription factors with specific promoter sequences. In the testes of mammals, *follicle stimulating hormone receptors* (FSH-Rs) are apparently located only on the Sertoli cells. Thus, the gene for the FSH-R and the gene for the *mullerian inhibiting substance* (MIS) are the only gene products known that are unique to the Sertoli cells. Our interest in gene expression in the Sertoli cells has stimulated our studies of the expression of the FSH-R gene in an effort to dissect important cell-specific regulatory elements.

Physiological Response of Testis to FSH

The role of the Sertoli cells in the process of spermatogenesis includes the physical and biochemical support for germ cell development into spermatozoa. Sertoli cells create an environment where germ cells are provided with the appropriate stimulants, metabolites, nutrients, and physical support. Presumably, FSH influences spermatogenesis solely as a result of effects on the Sertoli cells and their functions (1–5). For a recent review on the role of FSH in spermatogenesis, see (1). The initial response of the Sertoli cells to FSH is an increased level of intracellular cAMP, and although there is some evidence that FSH can alter intracellular calcium levels, it is generally accepted that the control of most Sertoli cell functions by FSH involves the cAMP/protein kinase A pathway (1–5).

In the rat the response of the testis to FSH is contingent on the age of the animal. FSH produced by the fetal and early postnatal rat pituitary seems to be a mitogenic stimulant for Sertoli cells. The Sertoli cell population is expanding in fetuses, and the proliferation continues, but steadily declines, until the second week after birth (6). In advance of the onset of meiosis in germ cells, the number of Sertoli cells in each testis is set at a final fixed number. If the Sertoli cells fail to receive sufficient FSH in the early postnatal period, the numbers of Sertoli cells and the number of mature spermatozoa produced by the testis are reduced (7).

Most considerations of the role of FSH in regulating Sertoli cell functions have been done in cells isolated from 10- to 30-day-old rats and maintained in culture. Sertoli cells from rats of this age will respond to FSH with increased levels of cAMP that in turn increase the synthesis of a number of proteins (1, 8). There are few mitotic cells observed in cultures of Sertoli cells obtained from rats of 20 days of age or older, indicating that Sertoli cells lose their ability to divide even in the presence of FSH. Thus, the role of FSH may change to a stimulation of the maturation of the Sertoli cells. It has been shown that FSH can stimulate the synthesis of a large number of proteins in Sertoli cells from 20-day-old rats (1–3). In addition, the action of FSH is required for the development of morphological features of mature Sertoli cells, including the pattern of chromatin condensation, the development of large nucleoli and nuclear infoldings, and the accumulation of smooth endoplasmic reticulum (9).

In the adult rat (greater than 40 days of age), the role of FSH changes again since there is a large increase in the phosphodiesterase activity in the Sertoli cells, and the accumulation of cAMP and subsequent stimulation of specific protein synthesis is diminished or absent (10–12). Long-term passive immunization of adult rats with antiserum to FSH did not inhibit spermatogenesis (13). In the rat, but not in the primate, it appears that FSH is not required for qualitative adult spermatogenic function. In primates it is more easily demonstrated that spermatogenesis in the adult is dependent on the action of FSH (14).

FSH-R

FSH, LH/CG, and *thyroid stimulating hormone* (TSH) have similar structures and have been categorized as glycoprotein hormones. The receptors of the glycoprotein hormones—along with other members of this receptor family that includes the adrenergic, muscarinic cholinergic, dopamine, and the substance K receptors—and the visual pigment rhodopsin have similar structures and act through interactions with G-proteins (15–22). In this receptor family the primary amino acid sequence in the regions that traverse the membrane is highly conserved. This region is characterized by 7 membrane-spanning α-helices, an extracellular aminoterminus and an intracellular carboxyterminus (17, 20, 23). The receptors for the glycoprotein hormones differ from the other members of this family in that they have large external aminoterminal domains that are presumably involved in the interaction of the receptors with the hormones. The cDNA for the FSH-R was first cloned by Sprengel et al. in 1990 (22), and the derived amino acid sequence encodes a 675-amino acid, 75-kd protein that has a 348-amino acid extracellular domain (Fig. 13.1) (22). While the transmembrane domains of the FSH, LH/CG, and TSH receptors share about 70% sequence homology, the extracellular domains are only 40%–45% similar and consist of a series of repeated 25-residue leucine-rich motifs that dictate the protein-protein interactions leading to hormone receptor binding (reviewed in 24).

We utilized cloned genomic DNA as a probe on Northern blots to quantify the mRNA for FSH-R in cultured Sertoli cells and total testis. Sertoli cells and testis were found to contain a major transcript of 2.6 kb and a minor transcript of 4.5 kb that hybridized to the FSH-R DNA sequence (25). The 4.5-kb transcript appears to have an extended 3′ terminus, suggesting a possible alternative polyadenylation site. When a variety of tissues were screened for FSH-R mRNA, we detected hybridizing bands only in tissue from the testis or from the ovary. We found that both the 2.6- and the 4.5-kb transcripts were present in testes of rats from 10 to 60 days of age. In the adult rats the relative amount of FSH-R mRNA was decreased due to the increased numbers of germ cells, but the overall amount of FSH-R mRNA per testis remained high in the adult.

Utilizing mRNA isolated from rat testes that were synchronized by a scheme of vitamin A depletion and repletion, we were able to measure the steady state levels of FSH-R mRNA during the different stages of the cycle of the seminiferous epithelium by Northern blots (25). We found that the relative levels of FSH-R mRNA varied in a cyclic manner, with low levels in stages V–IX and 5-fold-higher levels in stages XIII, XIV, and I. The binding of labeled FSH to Sertoli cells and the ability of FSH to stimulate cAMP production in these cells were determined in dissected seminiferous tubules of defined stages, and the results correlate very well with our analysis of mRNA levels (26, 27). These studies suggest that the primary action of FSH in the adult rat may be regulated by the cyclic changes in receptor numbers and may be most effective in stages XII–IV. These are the stages that encompass a number of the spermatogonial divisions, and an effect of FSH on these mitotic divisions would explain the decrease in the efficiency of spermatogenesis in animals deprived of FSH (28).

Another factor that influences the action of FSH in the adult testis is phosphodiesterase. Since the known actions of FSH are modulated by cAMP, then the presence of a cAMP phosphodiesterase could be inhibitory to this action. The phosphodiesterase activity of Sertoli cells in the adult rat appears to be maximal during stages VII–VIII of the cycle (29). This result would be consistent with a lack of FSH action during stages VII–VIII.

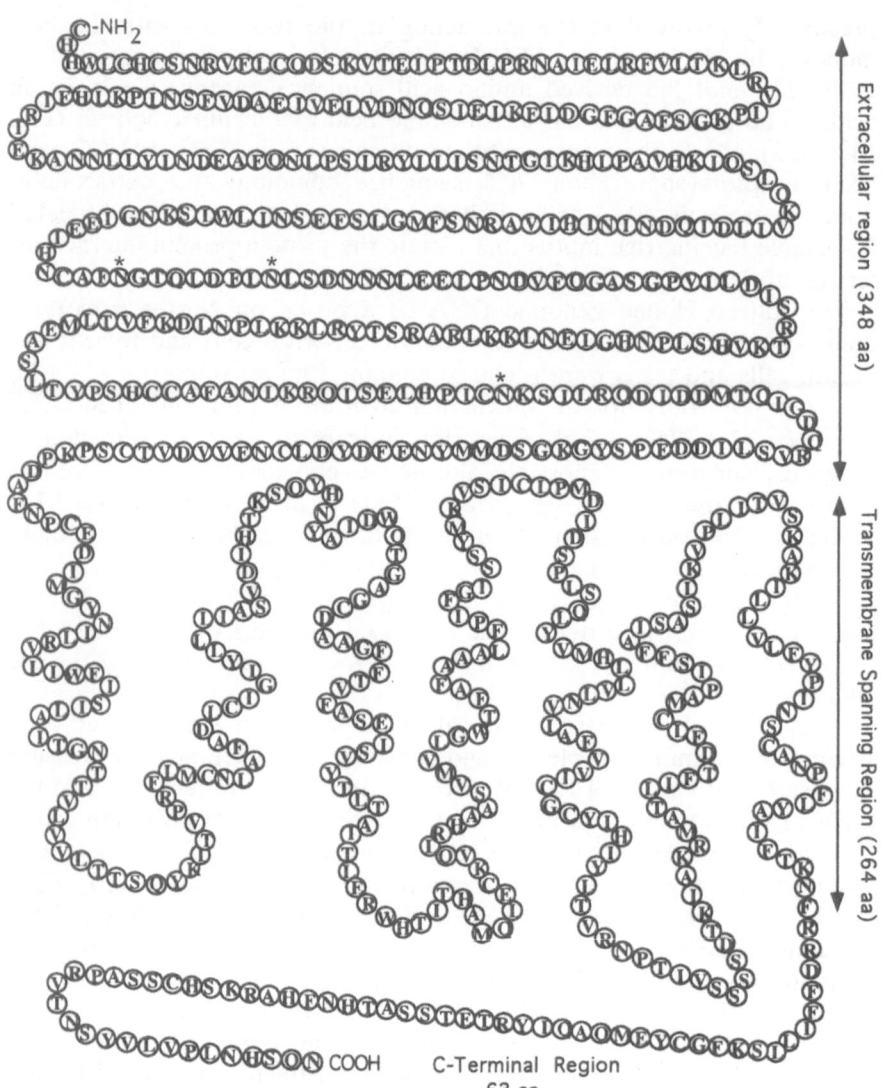

FIGURE 13.1. Primary amino acid sequence and derived functional domains of the FSH-R. The sequence was obtained from reference 22. The external domain, transmembrane region, and internal domain were determined from comparison to other members of this receptor family and from analysis of the primary sequence. The asterisks denote potential N-glycosylation sites.

FIGURE 13.2. Comparison of the basic structures of the genes for the rat FSH-R (described in reference 31), rat LH-R (described in references 32 and 33), and human TSH-R (described in reference 30). The FSH-R and TSH-R genes have 10 exons, while the LH-R gene has 1 extra exon denoted by X. In each case exons 2–9, shown in black (10 for LH-R), code for a series of leucine-rich repeats while the most 3' exon codes for the entire membrane-spanning region. The regions of the genes that show the greatest sequence differences between the 3 genes, which include the first exon and parts of the 3' exon (regions A and B), are shown in gray. Region B of the 3' exon is the transmembrane region that is highly conserved among members of the glycoprotein hormone family.

Gene for FSH-R

The details of the structures of many of the members of the G-protein-coupled family of receptors have been reported. The genes for several members of this family contain no introns or lack introns within the coding region for the protein. In contrast, the genes for the *LH receptor* (LH-R), *TSH receptor* (TSH-R), and FSH-R are very large and consist of many exons and introns (30–33).

The gene for FSH-R consists of at least 85 kb of DNA that is divided into 10 exons and contains some very large (20–30 kb) introns (31). The first 9 exons are relatively small (68–185 bp) and code for the aminoterminal extracellular domain that contains the leucine repeat regions. The region of the molecule that encodes the 7 transmembrane-spanning regions is contained within a single large exon (exon 10). The genes coding for FSH-R, TSH-R, and LH-R are all very large genes (60–85 kb) with 10 (TSH-R and FSH-R) or 11 (LH) exons (Fig. 13.2). The last exon in each gene codes for the entire membrane-spanning region, and the preceding exons code for the aminoterminal repeats. It has been proposed that the glycoprotein hormone receptors constitute a closely related subfamily of the G-protein-coupled receptors that have a relatively recent evolutionary origin (24).

Promoter Regions of the FSH-R

A few basic types of upstream control regions have been described for mammalian genes. One type is characterized by the presence of consensus CAAT and/or TATAA sequences that bind known transcription factors (reviewed in 34). This type of promoter has been associated with genes that code for inducible genes, and they are generally associated

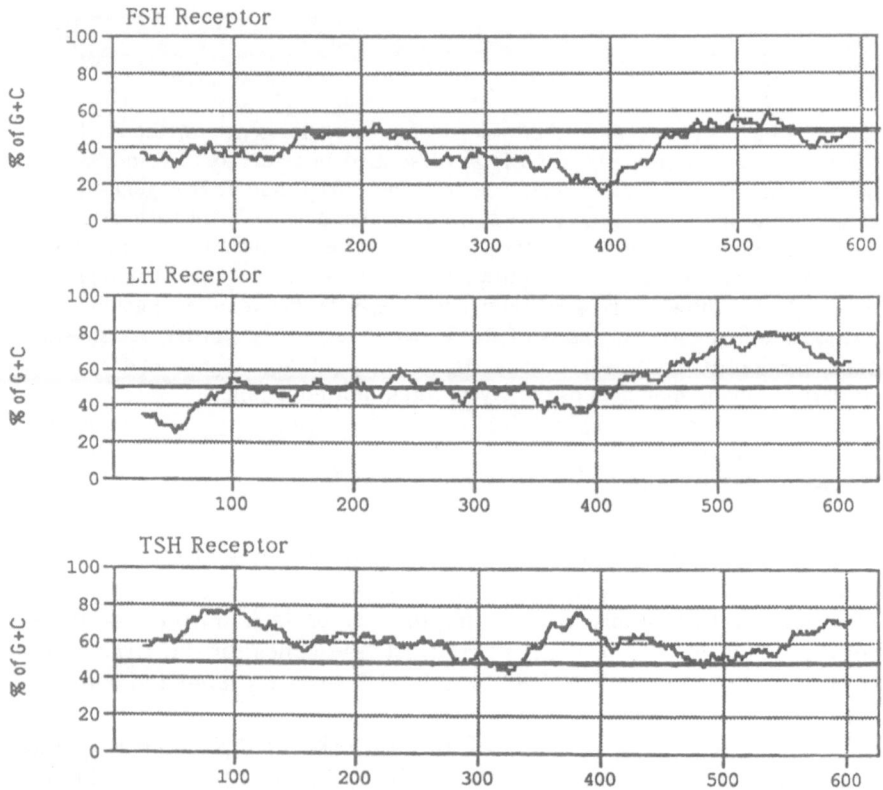

FIGURE 13.3. Relative GC content of proximal promoter regions of the rat FSH-R (described in reference 31), rat LH-R (described in references 32 and 33), and human TSH-R (described in reference 30) genes. The percentage of GC bases along the promoter was determined using the MacVector program and a 50-base window. The approximate transcriptional start sites are on the right, and the GC content of ~600 bp upstream from those sites is shown.

with a unique transcriptional start site. A second type of promoter sequence generally lacks CAAT and/or TATAA sequences and has regions that are relatively rich in CpG sequence (the CpG islands) (35). The CpG sequence is a methylation substrate and is generally under-represented in most parts of the genome. Promoters that lack TATAA sequences and contain CpG islands were first associated with genes that are ubiquitously expressed, are unregulated, and contain several closely spaced transcriptional start sites. These genes became known as *house-keeping* genes.

When the genes for the glycoprotein receptors were sequenced, the putative promoter regions did not contain TATAA or CAAT elements (30–33). In addition, primer extension and S1 nuclease experiments

-242 ATAAAGCAGAAGATTATTGACACACA‾T‾TAGTCACATATTAATATATATAA

-192 TCACTATTGACACATATTAATTTTACTTGCCTGGAAGCGACTAAAAAAAA

∇
-142 AGGCATCTCTTGGTGGGTCACGTGACTTTGCCCGTTCTCCAAGCAGATCT

            ∇
-92 CTCTTAT‾C‾C‾GGACAGTGTGTGGAGGAGCCTGGGGAATCTGTGGAAGTTTT

                                        +1
-42 CGCGCTGATGCAGAAAGAAAGTCGGTGAATGGATAAATAAGGATG

FIGURE 13.4. Features of the sequence of the immediate proximal promoter region of the rat FSH-R gene. The 2 major transcriptional start sites are designated with arrowheads. The putative AP-1 site is shown with an overline, and the short region of sequence similarity to the TSH-R gene promoter is underlined. The GGCC sequence that is methylated in some thyroid cell lines and that is also present in a similar position in the LH-R and FSH-R gene promoters is shown between an overline and an underline.

revealed the presence of multiple transcriptional start sites. The promoter regions for the LH-R and TSH-R genes contained large regions of high GC content that constituted CpG islands (Fig. 13.3). The promoter region of the FSH-R has only a very small region that could be construed to be a CpG island (31). There are no sequence similarities in the FSH-R gene promoter to the LH-R gene promoter and only a short region of sequence similarity to the TSH-R gene promoter (Fig. 13.4).

The FSH-R promoter has two major transcriptional start sites at positions -80 and -98 relative to the translational start site and contains a consensus AP-1 binding site at position -214. AP-1 is a transcriptional factor that interacts with the promoters of phorbol ester-inducible genes. There is evidence that treatment of Sertoli cells with phorbol esters in culture results in a decreased response of the cells to FSH (36).

The promoter for the TSH-R gene has been characterized using various regions of the promoter to drive *chloramphenicol transferase* (CAT) genes in transient transfection assays. The general conclusions from these studies were that the proximal 200 bp of the TSH-R gene promoter contained the elements for transcriptional activation and for cell specificity (37). In addition, in cell lines derived from thyroid cells, it was shown that some lines expressed the receptor and that some lines failed to express the receptor. In the lines that failed to express the receptor, a CCGG sequence near the transcriptional start sites was found to be methylated. It has been shown that a protein factor can bind to these methylated sequences and prevent transcription. This CCGG sequence is

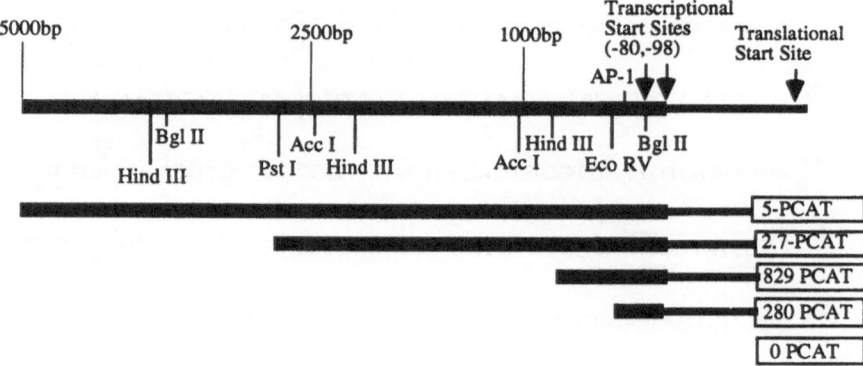

FIGURE 13.5. Restriction map of 5 kb of the proximal promoter region of the FSH-R gene and representation of the reporter gene constructs used in transient transfection assays. The CAT reporter gene was linked approximately 30 bp upstream from the translational start site to 5 kb (5-PCAT), 2.7 kb (2.7-PCAT), 829 bp (829-PCAT), and 280 bp (280-PCAT) of the proximal promoter region. The CAT reporter gene with no promoter region was used for control (0-PCAT).

present in the promoters of all three glycoprotein receptor genes and opens the possibility of regulation by methylation (30–33).

We have initiated similar studies with the FSH-R promoter using chimeric constructs in which 5 kb of 5' flanking region of the FSH-R promoter, or deletions thereof, was ligated to the CAT reporter gene (Fig. 13.5). Each DNA construct was transiently transfected into either primary cultures of Sertoli cells, MSC-1 cells—a mouse Sertoli cell line (38)—or COS-7 cells using calcium phosphate precipitation. Each construct was transfected in triplicate within an experiment, and each experiment was repeated a minimum of 3 times with at least 2 different DNA preparations to control for differences in transfection efficiencies. In each case the level of expression of each construct was compared to the level of expression of control vector (0-PCAT) that contained no promoter or enhancer elements (Table 13.1).

Each of the four promoter constructs was active in both the primary cultures of Sertoli cells and MSC-1 cells, but none of the FSH-R promoter constructs was functional in the COS-7 cells. These results suggested that similar to the TSH-R gene promoter, the elements controlling transcriptional initiation and some aspects of the cell specificity were contained within the proximal 250 bp of the promoter region. Deletion of parts of the proximal 3' region of the promoter eliminated the activity of the promoter. Recent studies where we transfected the reporter constructs into other cell types have shown that the FSH-R promoter was active in MA-10 cells (a Leydig tumor cell line) (39) and was partially active in NIH 3T3 cells. These results suggest that in transiently trans-

TABLE 13.1. Relative levels of promoter activity of the reporter gene constructs in transient transfection assays in several types of cells.

Construct	Cell type				
	Sertoli cells	MSC-1	COS-7	3T3	MA-10
280-PCAT	++	++	−	±	++
829-PCAT	++	++	−	±	++
2.7-PCAT	++	++	−	±	++
5-PCAT	±	±	−	−	−

Note: The cells used in this analysis were cultured Sertoli cells from 20-day-old rats, a mouse transformed Sertoli cell line (MSC-1), COS-7, NIH 3T3, and MA-10, a Leydig cell line. If the CAT activity of the transfected construct was usually 3- to 10-fold or more greater than the control (0-PCAT), the activity is shown as (++). If the CAT activity was equal to 0-PCAT, the activity is shown as (−). Intermediate levels of CAT activity are shown as (±).

fected cells the FSH-R gene promoter is not completely cell specific and is somewhat promiscuous.

Nuclei were isolated from the primary Sertoli cells, MSC-1 cells, and COS-7 cells. The DNA binding proteins were extracted from these nuclei according to standard protocols. The ability of proteins in these extracts to bind specifically to the proximal 200 bp of the FSH-R gene promoter was examined using gel retardation assays. Extracts from nuclei of all three cell types were able to bind and retard the DNA to some extent. However, analysis of the extracts from MSC-1 and Sertoli cells revealed unique bands not present in the binding protein extracts from the COS-7 cells. These bands were shown to be specific by competition with unlabeled ligand and may represent the interaction of Sertoli cell-specific factors with the promoter region of the FSH-R gene.

During the course of the transfection studies, we observed that when the Sertoli cells were incubated with calf serum prior to and following transfection, much higher levels of CAT activity were observed. When total mRNA was isolated from Sertoli cells cultured in the presence and absence of serum and the FSH-R mRNA levels were determined by Northern blots, increased levels of FSH-R mRNA were found in the serum-treated cells. The promoter region of the FSH-R gene does not contain a known consensus serum response element in the proximal 1000 bp that we have sequenced.

It is clear that Sertoli cells cultured in minimal medium in the absence of serum have a decreased ability to utilize the transfected FSH-R promoter. It was possible that this decreased ability resulted from a diminished availability of the appropriate transcription factors or mRNA precursors in the serum-starved cells. Thus, the gel retardation assays described above were done with extracts of Sertoli cells cultured in the presence and absence of serum. There was a dramatic increase in one specific band in the extracts from the cells cultured in the presence of serum. In total, these results suggest that serum-deprived Sertoli cells may have decreased levels of important transcription factors. Many published results have examined the response of Sertoli cells cultured in minimal medium to hormones or other stimulatory factors. Our results suggest that some of these studies may have detected only the minimum responses of the serum-starved cells.

Summary

The action of FSH is very important in the development of the testis, the onset of spermatogenesis, and in some species the maintenance of spermatogenesis. The FSH-R is confined to the Sertoli cells in the testis, and the FSH-R gene has been sequenced and characterized. The promoter region of the FSH-R gene may contain elements that are recognized by transcription factors unique to Sertoli cells. Analysis of the promoter region reveals very little sequence homology with promoters for the TSH-R or LH-R genes. Transient transfection assays using the CAT reporter gene have shown that the transcriptional activity is present in the first 280 bp. The transfection assays showed that the promoter had partial, but not complete, cell specificity since it was active in MA-10 cells, a Leydig cell line. Gel retardation assays revealed potential Sertoli cell-specific interactions of nuclear protein extracts with the promoter. Both the transfection assays and the gel retardation assays revealed that Sertoli cells cultured in the absence of serum had a decreased ability to transcriptionally activate, and to bind to, the FSH-R promoter.

References

1. Griswold MD. Action of FSH on mammalian Sertoli cells. In: Griswold MD, Russell L, eds. The Sertoli cell. Clearwater, FL: Cache River Press, 1993: 493–508.
2. Fritz I. Sites of actions of androgens and follicle stimulating hormone on cells of the seminiferous tubule. In: Litwack G, ed. Biochemical actions of hormones. New York: Academic Press, 1978:249–78.
3. Griswold MD, Morales C, Sylvester SR. Molecular biology of the Sertoli cell. Oxf Rev Reprod Biol 1988;10(124):124–61.
4. Means AR, Fakunding JL, Huckins C, Tindall DJ. Follicle-stimulating hormone, the Sertoli cell, and spermatogenesis. Recent Prog Horm Res 1976; 32:477–527.
5. Means AR, Fakunding JL, Tindall DJ. Follicle stimulating hormone regulation of protein kinase activity and protein synthesis in testis. Biol Reprod 1976;14:54–63.

6. Orth JM. The role of follicle-stimulating hormone in controlling Sertoli cell proliferation in testes of fetal rats. Endocrinology 1984;115:1248–55.

7. Orth JM, Gunsalus GL, Lamperti AA. Evidence from Sertoli cell-depleted rats indicates that spermatid number in adults depends on numbers of Sertoli cells produced during perinatal development. Endocrinology 1988;122:787–94.

8. Fritz KB, Rommerts FG, Louis BG, Dorrington JH. Regulation by FSH and dibutyryl cyclic AMP of the formation of androgen-binding protein in Sertoli cell-enriched cultures. J Reprod Fertil 1976;46:17–24.

9. Solari AJ, Fritz IB. The ultrastructure of immature Sertoli cells. Maturation-like changes during culture and the maintenance of mitotic potentiality. Biol Reprod 1978;18:329–45.

10. Hugly S, Roberts K, Griswold MD. Transferrin and sulfated glycoprotein-2 messenger ribonucleic acid levels in the testis and isolated Sertoli cells of hypophysectomized rats. Endocrinology 1988;122:1390–6.

11. Means AR, Dedman JR, Tash JS, Tindall DJ, van Sickle M, Welsh MJ. Regulation of the testis Sertoli cell by follicle stimulating hormone. Annu Rev Physiol 1980;42:59–70.

12. Means AR, Fakunding JL, Huckins C, Tindall DJ, Vitale R. Follicle-stimulating hormone, the Sertoli cell, and spermatogenesis. Recent Prog Horm Res 1976;32:477–527.

13. Dym M, Raj HGM, Lin YC, et al. Is FSH required for maintenance of spermatogenesis in adult rats? J Reprod Fertil 1979;26(suppl):175–81.

14. Wickings EJ, Usadel KH, Dathe G, Nieschlag E. The role of follicle stimulating hormone in testicular function of the mature rhesus monkey. Acta Endocrinol (Copenh) 1980;95:117–28.

15. Dal Toso R, Sommer B, Ewert M, et al. The dopamine D2 receptor: two molecular forms generated by alternative splicing. EMBO J 1989;8:4025–34.

16. Kobilka BK, Frielle T, Collins S, et al. An intronless gene encoding a potential member of the family of receptors coupled to guanine nucleotide regulatory protein. Nature 1987;329:75–9.

17. O'Dowd BF, Lefkowitz RJ, Caron MG. Structure of the adrenergic and related receptors. Annu Rev Neurosci 1989;12:67–83.

18. Sunahara RK. Human dopamine D1 receptor encoded by an intronless gene on chromosome 5. Nature 1990;347:80–3.

19. Loosfelt H, Misrahi M, Atger M, et al. Cloning and sequencing of porcine LH-hCG receptor cDNA: variants lacking transmembrane domain. Science 1989;245:525–8.

20. McFarland KC, Sprengel R, Phillips HS, et al. Lutropin-choriogonadotropin receptor: an unusual member of the G protein-coupled receptor family. Science 1989;245:494–9.

21. Parmentier M. Molecular cloning of the thyrotropin receptor. Science 1989;246:1620–2.

22. Sprengel R, Braun T, Nikolics K, Segaloff DL, Seeburg PH. The testicular receptor for follicle stimulating hormone: structure and functional expression of cloned cDNA. Mol Endocrinol 1990;4:525–30.

23. Johnson GL, Dhanasedaran N. The G protein family and their interaction with receptors. Endocr Rev 1989;10:317–31.

24. Vassart G, Parmentier M, Libert F, Dumont J. Molecular genetics of the thyrotropin receptor. Trends Endocrinol Metab 1991;2:151–6.
25. Heckert LL, Griswold MD. Expression of follicle-stimulating hormone receptor mRNA in rat testes and Sertoli cells. Mol Endocrinol 1991;5:670–7.
26. Kangasniemi M, Kaipia A, Toppari J, Mali P, Huhtaniemi I, Parvinen M. Cellular regulation of basal and FSH-stimulated cyclic AMP production in irradiated rat testes. Anat Rec 1990;227:32–6.
27. Kangasniemi M, Cheng CY, Toppari J, et al. Cyclic secretion and FSH stimulation of clusterin, alpha2-macroglobulin, and testibumin in rat seminiferous tubules at defined stages of the epithelial cycle. J Androl 1992;13: 208–13.
28. Sharpe RM. Follicle-stimulating hormone and spermatogenesis in the adult male. J Endocrinol 1989;121:405–7.
29. Parvinen M. Regulation of the seminiferous epithelium. Endocr Rev 1982; 3:404–17.
30. Gross B, Misrahi M, Sar S, Milgrom E. Composite structure of the human thyrotropin receptor gene. Biochem Biophys Res Commun 1991;177:679–87.
31. Heckert LL, Daley I, Griswold MD. Structural organization of the follicle hormone receptor gene. Mol Endocrinol 1992;6:70–80.
32. Koo YB, Ji I, Slaughter RG, Ji TH. Structure of the leuteinizing hormone receptor gene and multiple exons of the coding sequence. Endocrinology 1991;128:2297–650.
33. Tsai-Morris CH, Buczko E, Wei W, Xie XZ, Dufau ML. Structural organization of the rat leuteinizing hormone (LH) receptor gene. J Biol Chem 1991;266:11355–8.
34. Mitchell P, Tijan R. Transcriptional regulation in mammalian cells by sequence-specific DNA binding proteins. Science 1989;245:371–8.
35. Larsen F, Gunderson G, Lopez R, Prydz H. CpG islands as markers in the human genome. Genomics 1992;13:1095–107.
36. Monaco L, Conti M. Inhibition by phorbol esters and other tumor promoters of the response of the Sertoli cell to FSH: evidence for dual site of action. Mol Cell Endocrinol 1987;49:227–36.
37. Ikuyama S, Niller H, Shimura H, Akamizu T, Kohn LD. Characterization of the 5'-flanking region of the rat thyrotropin receptor gene. Mol Endocrinol 1992;6:793–804.
38. Peschon JJ, Behringer RR, Cate RL, et al. Directed expression of an oncogene to Sertoli cells in transgenic mice. Mol Endocrinol 1992.
39. Ascoli M. Characterization of several clonal lines of cultured Leydig tumor cells: gonadotropin receptors and steroidogenic responses. Endocrinology 1981;108:88–95.

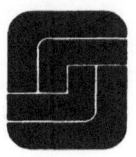

14

Regulation of Germ Cell Proliferation In Vitro by Activins, Follistatin, and Other Growth Factors

JENNIE P. MATHER, PENELOPE E. ROBERTS,
AND LYNNE A. KRUMMEN

Inhibin A and B are heterodimers made up of an α-subunit combined with a β_A- or β_B-subunit. Inhibin is secreted by testicular Sertoli cells in the male and causes a decrease in FSH secretion from the pituitary (reviewed in 1–4). Activin A and B are homodimers of the β_A- or β_B-subunits, respectively. Activin has been shown to increase FSH secretion from pituitary cultures in vitro (5) and to increase circulating FSH levels in vivo in adult macaques (6) and rats (7). More recently, it has been demonstrated that testicular Leydig cells produce activin (8). Messenger RNA and protein subunits of the α- and β-chains have been localized in multiple cell types in the testis of rats of various ages (9–12); however, these data cannot differentiate between inhibin and activin production if both α- and β-subunits are present. These data led us, and others, to suggest that inhibin and activin act as paracrine and/or autocrine regulators of gonadal function and may play a role in regulating Sertoli-Leydig cell interaction or Sertoli-germ cell interaction (8–18).

Franchimont et al. (19) reported a decrease in ^{3}H-thymidine incorporation in the testes of animals treated with follicular fluid. It was suggested that the decrease in spermatogenesis was secondary to a decrease in circulating FSH in the treated animals. More recently, van Dissel-Emiliani et al. have shown a decrease in spermatogonial numbers in hamster testis injected locally with both impure and purified inhibin (18). The latter data supported the hypothesis that inhibin may act locally in the testis, but did not distinguish which cell type(s) may be the primary targets of the hormone.

It has recently been recognized that at least two binding proteins exist for inhibin and activin: *α2 macroglobulin* (α2M) and follistatin (20, 21).

Therefore, it is important to elucidate the possible role of follistatin as well as the inhibin family in the local endocrine or paracrine regulation of the testis. We used cocultures of Sertoli and germ cells derived from immature rat testes to determine the in vitro effects of purified recombinant activin and inhibin, their binding proteins, and other vitamins, hormones, and growth factors on spermatogonial proliferation.

Effects of Activin and Inhibin on Sertoli-Germ Cell Cocultures

There is a marked proliferation of germinal elements in cocultures of immature Sertoli and germ cells treated with activin in vitro. The increase in clusters of spermatogonia seen in phase contrast microscopy is reflected by an increase in ^{3}H-thymidine incorporation in the total culture after treatment with activin. Moreover, autoradiography of the treated cultures shows that label was predominantly incorporated into the clusters of germ cells attached to the monolayer (22). These data are consistent with the hypothesis that activin is stimulating division of spermatogonia in vitro.

There is also a marked morphologic change in the cultures treated with activin. Previous reports of Sertoli, Sertoli-peritubular, or Sertoli-peritubular-germ cell cocultures have shown that reassociation of cells in the cultures was dependent on the presence of peritubular cells (23) or basement membrane components (24). We have shown that a single hormone, activin, can rapidly cause the reassociation of the Sertoli and germ cells in the cultures into tubular elements (22, 25). This reassociation begins within 24 h of the addition of the hormone. After 3 days of hormone treatment, electron microscopic observation could detect no peritubular cells and no formation of basement membrane in the cultures. Thus, the reassociation seems to be occurring via a different mechanism than that reported previously; or activin, presumably produced in vitro, is involved in the reassociation seen in the Sertoli-peritubular coculture system. These effects seem to be specific for activins A and B.

Inhibin had no effect on thymidine incorporation or the morphology in the cocultures. Inhibin at concentrations equal to or up to 4-fold greater than activin was unable to antagonize the effects of activin.

Effects of Binding Proteins on Activin Activity in Cocultures

As mentioned above, at least two binding proteins exist for inhibin and activin: α2M and follistatin (20, 21). The α2M is a high-capacity, low-affinity binding protein present in human serum that is known to bind a number of growth factors, including activin, TGFβ, and NGF (20). The follistatins are heterogeneous glycoproteins, originally isolated from follicular fluids as inhibitors of FSH release, that bind activin A with high affinity (26). Several lines of evidence indicate that binding to follistatin may block activin activity (20, 21). These observations suggest that the local secretion of follistatin in vivo may act as a regulator of activin activity. Since follistatin, as well as activin, is produced in the bone marrow, the pituitary, and the testis, it is of importance to elucidate the role that the interaction of activin and follistatin might have on the local endocrine or paracrine activity in these tissues.

We have recently shown (25) that follistatin acts to inhibit in vitro FSH release by pituitary cultures and hemoglobin synthesis in K562 cultures, while α2M, another activin binding protein, has no effect on these activities. In contrast, follistatin has no effect on thymidine incorporation, but does block cell reaggregation, in mixed Sertoli-germ cell cultures. This ^{3}H-thymidine incorporation has previously been shown to be predominantly into spermatogonia in these cultures (22). In contrast, activin is able to direct a clonal Sertoli-derived cell line, TM4, to reaggregate in the absence of any germ cells, suggesting that the morphological effect is directly on Sertoli cells. The ability of an inactivating antiserum directed against activin B to block both activin B-stimulated thymidine incorporation and germ cell aggregation further suggests that both of these effects are direct responses to activin.

The binding of activin to follistatin is a high-affinity interaction with a reported kd of 800–900 pM (26). The binding of activin to α2M is of lower affinity than its affinity for follistatin (20). The activin II and IIB receptors also have different affinities for activin (27, 28). It is possible that follistatin forms a reservoir of activin in the testis and other tissues that differentially regulates the availability of activin to the various subclasses of receptor. The differential effect in these cultures would suggest that activin in the presence of a 2- to 4-fold (w:w) excess of follistatin is biologically available to germ cells, but not Sertoli cells, in the mixed cultures. The regulation of follistatin levels may then selectively inhibit activin-stimulated Sertoli cell reaggregation while allowing stimulation of spermatogonial proliferation through a separate high-affinity receptor. Thus, follistatin concentration, along with differential expression of activin receptors with varying affinities for activin, is capable of conveying both tissue and cell type specificity on activin action in vitro. This is schematically represented in Figure 14.1.

Effects of Other Vitamins and Growth Factors on Cocultures

The germ-Sertoli cell coculture system from 21-day-old rats (25) was used to screen a number of vitamins, growth factors, and hormones for their effect on thymidine incorporation into these cultures. The results could be separated into four categories: (i) factors that always increased incor-

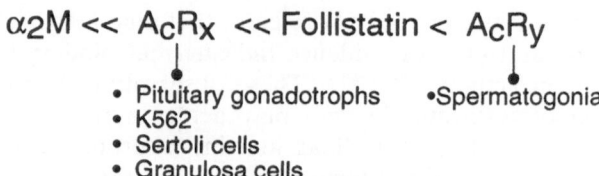

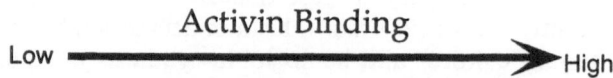

FIGURE 14.1. Schematic representation of how the relative affinity of activin for follistatin and various activin receptors might allow the modulation of activin effects in tissues by follistatin.

TABLE 14.1. Factors tested for activity in the 21-day germ-Sertoli cell coculture system.

Factors that significantly stimulate thymidine incorporation
    Activin A
    Activin B
    TGFβ1
    Retinoic acid
    Forskolin
Factors that significantly inhibit thymidine incorporation
    IL-1β
Factors having no effect on thymidine incorporation
    NT3
    BDNF
    NGF
    IL-1α
    IL-2
    IL-3
    IL-4
    IL-5
    IL-6
    Ceruloplasmin
    aFGF
    bFGF
    SGF

poration; (ii) factors that significantly increased thymidine incorporation in replicate cultures in some preparations, but not others; (iii) factors that had no effect over the range of concentrations tested; and (iv) factors that inhibited thymidine incorporation. These factors are listed in Table 14.1. The only factors that consistently increased thymidine incorporation in the cocultures were activin A and activin B. As discussed above, the data

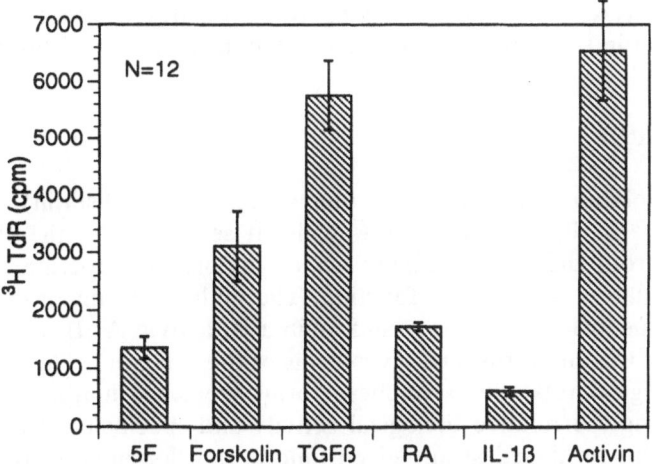

FIGURE 14.2. Effect of activin A (50 ng/mL), TGFβ (50 ng/mL), RA (0.5 μM), forskolin (10 μM), and IL-1β (3 ng/mL) on thymidine incorporation in 21-day-old rat germ-Sertoli cell cocultures. Values shown are the mean ± SE of 12 replicates. Cultures were prepared and assayed as previously described in reference 25. The activin A-, forskolin-, and TGFβ-treated conditions are significantly ($P < 0.01$) different from the control condition.

are consistent with activins stimulating the division of spermatogonia and/or spermatocytes in the cultures. In addition, three other factors— *retinoic acid* (RA), forskolin (a stimulator of cAMP production), and TGFβ—would frequently, but not always, give a significant stimulation of thymidine incorporation.

Such stimulation is seen in the results shown in Figure 14.2. Activin consistently gives a significant stimulation of thymidine incorporation as previously reported. In this case, the stimulation by RA was of small magnitude; however, in the majority of experiments, RA stimulated thymidine incorporation up to 2-fold. TGFβ and forskolin produced significant stimulation (2- to 4-fold) of thymidine incorporation in the experiment shown, although this was not seen in every experiment. The IL-1β was seen to inhibit thymidine incorporation into the cocultures. While the effect of IL-1β shown in Figure 14.2 is not statistically significant (at $P < 0.01$), other experiments showed a significant 2- to 4-fold inhibition of incorporation of thymidine with IL-1β. This inhibition could also be seen in the presence of activin, RA, or forskolin. The IL-1β caused a significant drop in pH in the cultures, presumably by stimulating lactic acid production by Sertoli cells in these cultures. None of the other growth factors tested showed a significant stimulation or inhibition of thymidine incorporation. Work is currently under way to determine whether forskolin, RA, or TGFβ are acting by stimulating endogenous activin production in the cultures, which then stimulates germ cell proliferation.

Discussion

Activin, inhibin, their binding proteins (α2M and follistatin), and their receptors are all produced in the testis. It seems likely, therefore, that proteins from the inhibin family act as important autocrine and paracrine regulators of testicular function. The multiple forms of the inhibins and activins include inhibin A and inhibin B; activin A, B, and AB; and various processing forms and glycoforms of these factors. The genetic and immunological relatedness of these forms creates a major challenge in determining the relative biological activity and specificity of the activins and inhibins produced at any given time in development. However, it is clear that the expression of the α-, β_A-, and β_B-genes is regulated during embryonic and prepubertal development and in the adult seminiferous cycle (12, 14). There are multiple receptors for the activins (27, 28), and current binding data suggest that a distinct receptor should exist for inhibin (29). The expression of these receptors is also regulated during the seminiferous cycle. In addition, follistatin can alter the biological activity of activin on some tissues in the testis and may restrict activity to specific cell types or areas of the testis.

Activins appear to play a major role in regulating spermatogonial proliferation in vitro, with vitamin A, forskolin, and TGFβ also capable of stimulating thymidine incorporation in these cultures. Forskolin stimulates cAMP formation and may be acting as a surrogate for FSH in this system. Retinoic acid has long been implicated as an important factor in the regulation of germ cell proliferation (30). Work is in progress to determine the relationship between these factors in the paracrine control of germ cell development.

References

1. Mather JP, Woodruff TK, Krummen LA. Paracrine regulation of reproductive function by inhibin and activin. Proc Soc Exp Biol Med 1992;201:1–15.
2. De Jong FH. Inhibin. Physiol Rev 1988;68:555–607.
3. Risbridger GP, Robertson DM, de Kretser DM. Current perspective of inhibin biology. Acta Endocrinol 1990;122:673–82.
4. Rivier C, Meunier H, Roberts V, Vale W. Inhibin: role and secretion in the rat. Recent Prog Horm Res 1990;46:231–59.
5. Ling N, Ying S-Y, Ueno N, Shimasaki S, Hotta M, Guillemin SR. Pituitary FSH is released by a heterodimer of the β-subunits from the two forms of inhibin. Nature 1986;321:779–82.
6. McLachlan RI, Dahl KD, Bremner WJ, et al. Recombinant human activin-A stimulates basal FSH and GnRH-stimulated FSH and LH release in the male macaque, *Macaca fascicularis*. Endocrinology 1989;125:2787–9.

7. Schwall R, Schmelzer CH, Matsuyama E, Mason AJ. Multiple actions of recombinant activin-A in vivo. Endocrinology 1989;125:1420–3.
8. Lee W, Mason AJ, Schwall R, Szonyi E, Mather JP. Secretion of activin by interstitial cells in the testis. Science 1989;243:396–8.
9. Lee W, Schwall R, Mason AJ, Mather JP. Interstitial cells secrete an activity with characteristics of the inhibin B-B homodimer, activin. In: Cooke BA, Sharpe RM, eds. The molecular and cellular endocrinology of the testis; vol 50. New York: Raven Press, 1988:21–7.
10. Roberts V, Meunier H, Sawchenko PE, Vale W. Differential production and regulation of inhibin subunits in rat testicular cell types. Endocrinology 1989; 125:2350–9.
11. Shaha C, Morris PL, Chen CL, Vale W, Bardin CW. Immunostainable inhibin subunits are in multiple types of testicular cells. Endocrinology 1989; 125:1941–50.
12. Bhasin S, Krummen LA, Swerdloff RS, et al. Stage dependent expression of inhibin alpha and beta-B subunits during the cycle of the rat seminiferous epithelium. Endocrinology 1989;124:987–91.
13. de Kretser DM, Robertson DM. The isolation and physiology of inhibin and related proteins. Biol Reprod 1989;40:33–47.
14. Hsueh AJW, Dahl KD, Vaughan J, et al. Heterodimers and homodimers of inhibin subunits have different paracrine action in the modulation of luteinizing hormone-stimulated androgen biosynthesis. Proc Natl Acad Sci USA 1987;84:5082–6.
15. Lin T, Calkins H, Morris PL, Vale WW, Bardin CW. Regulation of Leydig cell function in primary culture by inhibin and activin. Endocrinology 1989; 125:2134–40.
16. Drummond AE, Risbridger GP, de Kretser DM. The involvement of Leydig cells in the regulation of inhibin secretion by the testis. Endocrinology 1989; 125:510–5.
17. Sharpe RM, Kerr JB, Maddocks S. Evidence for a role of the Leydig cell in control of the intratesticular secretion of inhibin. Mol Cell Endocrinol 1988; 60:243–7.
18. van Dissel-Emiliani FM, Grootenhuis AJ, de Jong FH, de Rooij DG. Inhibin reduces spermatogonial numbers in testes of adult mice and Chinese hamsters. Endocrinology 1989;125:1899–903.
19. Franchimont P, Hazee-Hagelstein MT, Jaspar JM, Charlet-Renard C, Demoulin A. Inhibin and related peptides: mechanisms of action and regulation of secretion. J Steroid Biochem 1989;32:193–7.
20. Krummen LA, Woodruff TK, DeGuzman G, et al. Identification and characterization of binding proteins for inhibin and activin in human serum and follicular fluids. Endocrinology 1993;132:431–43.
21. DePaolo LV, Bicsak TA, Erikson GF, Shimasaki S, Ling N. Follistatin and activin: a potential intrinsic regulatory system within diverse tissues. Proc Soc Exp Med Biol 1991;198:500–12.
22. Mather JP, Attie KA, Woodruff TK, Rice GC, Phillips DM. Activin stimulates germ cell proliferation in germ-Sertoli cell co-cultures from immature rat testis. Endocrinology 1990;127:3206–14.
23. Mather JP, Phillips DM. Establishment of a peritubular myoid-like cell line and interactions between established testicular cell lines in culture. J Ultrastruct Res 1984;87:263–74.

24. Tung PS, Fritz IB. Extracellular matrix components and testicular peritubular cells influence rate and pattern of Sertoli cell migration in vitro. Dev Biol 1986;113:119.
25. Mather JP, Roberts PE, Krummen LA. Follistatin modulates activin activity in a cell- and tissue-specific manner. Endocrinology 1993;132:2732-4.
26. Kogawa K, Nakamura T, Sugino K, Taiko K, Titani K, Sugino H. Activin-binding protein is present in pituitary. Endocrinology 1991;128:1434-40.
27. Mathews LS, Vale W. Expression cloning of an activin receptor, a predicted transmembrane serine kinase. Cell 1991;65:1-20.
28. Attisano L, Wrana JL, Cheiftez S, Massague J. Novel activin receptors: distinct genes and alternative mRNA splicing generate a repertoire of serine/threonine kinase receptors. Cell 1992;68:97-108.
29. Woodruff TK, Borree J, Attie KA, Cox ET, Rice GC, Mather JP. Stage-specific binding of inhibin and activin to subpopulations of rat germ cells. Endocrinology 1992;130:871-81.
30. Griswold MD, Bishop PD, Kim KH, Ping R, Siteri JE, Morales C. Function of vitamin A in normal and synchronized seminiferous tubules. Ann NY Acad Sci 1989;564:154-72.

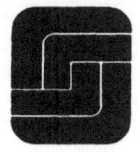

15

Androgen Action in Sertoli and Peritubular Cells

BARBARA M. SANBORN, CHUNG YING KU, AND QINGXIAN LU

The importance of androgens for the maintenance and restoration of spermatogenesis has been recognized for some time (1–5). In the absence of the pituitary, spermatogenesis is arrested at the level of primary spermatocytes. *Testosterone* (T) treatment alone is sufficient to qualitatively maintain spermatogenesis. Although in the absence of LH or GnRH testosterone can quantitatively maintain spermatogenesis in the adult rat, FSH and other hormones can also influence quantitative aspects of the process (1–6).

Considering these facts, a number of questions arise regarding the role of androgens in the control of spermatogenesis:

Which cell types are directly affected by androgens?
What are the primary and secondary effects of androgen stimulation on the affected cell types?
What factors control cellular responsiveness to androgens?
How are the effects of androgens related to the control of spermatogenesis?

Significant progress has been made in this laboratory and others toward answering a number of these questions.

Androgen Target Cells in the Testis

The presence of androgen receptors as measured by direct cytosol binding, nuclear exchange, and whole-cell binding assays has identified Sertoli and peritubular cells as potential androgen targets (7–13). On the other hand, significant concentrations of androgen receptors have not been found in testicular germ cells (10, 14). This is illustrated in Figure 15.1, where our data obtained by nuclear exchange assay in cells from testes in different stages of maturation are summarized. It follows from these data that the effects of androgens on germ cell maturation are either direct and

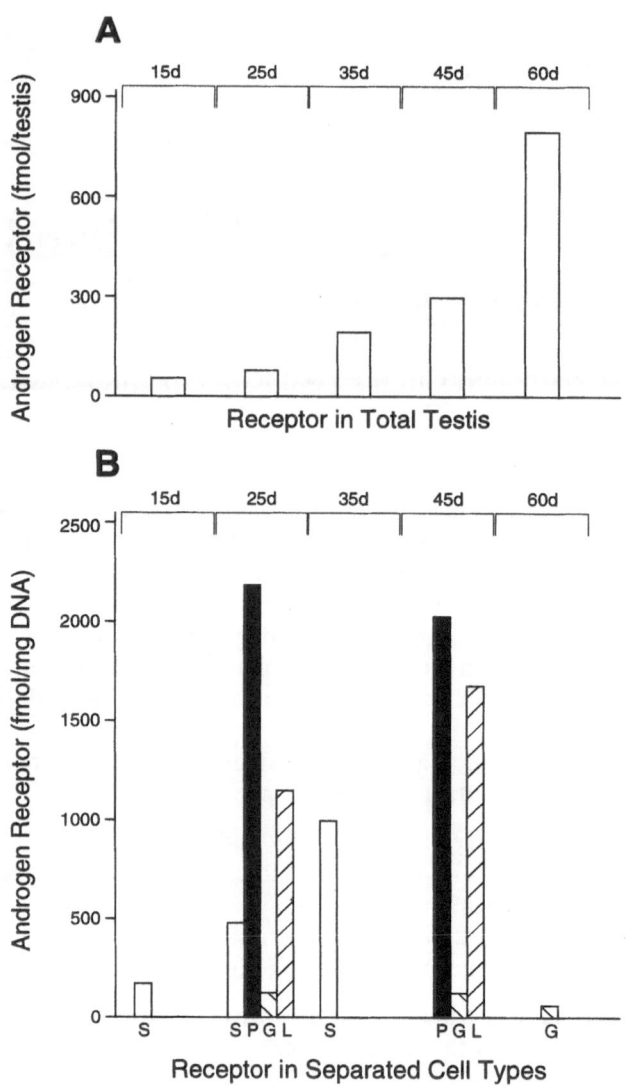

FIGURE 15.1. Changes in nuclear androgen/testis (*A*) and in nuclear androgen receptor per mg DNA (*B*) in Sertoli (S), peritubular (P), total germ cell fractions (G), and Leydig cells (L) as a function of animal age. Data are from reference 10.

independent of receptor or indirect via influence on Sertoli and/or peritubular cell function.

Since Sertoli and peritubular cells were implicated as possible androgen targets, it was important to define the nature of the responses of these cell types to androgens. The manner in which these two cell types dealt with androgens with respect to time course and concentration dependence of uptake, receptor affinity for T (0.7–0.8 nM), and the ability of androgens and antiandrogens to alter receptor distribution was virtually indistinguishable (15). Furthermore, androgen receptor mRNA was elevated in both cell types isolated from hypophysectomized rats and was partially restored to normal by T-replacement in vivo (16). Figure 15.2 illustrates these points.

Both Sertoli and peritubular cells respond to androgens with changes in protein and RNA synthesis, each in a specific manner. Androgens stimulate peritubular cells to secrete PModS (17). This protein in turn stimulates a number of parameters in Sertoli cells, including ABP and transferrin secretion and c-*fos* expression (17, 18). Factors exhibiting similar activities are secreted by prostate stromal cells (19), suggesting that this form of intercellular communication may be a common pathway between mesenchymal and epithelial cells. Androgens also increase alkaline phosphatase activity in peritubular cells (20).

The effects of androgens on Sertoli cell function have been identified by ^{35}S-Met mapping of secreted and intracellular proteins (21–23), as well as in studies of specific proteins. In contrast to a number of other androgen-regulated systems, the response of the Sertoli cell to androgens appears to involve relatively nonabundant proteins (reviewed in 24). Synthesis of testins I and II is stimulated by androgens in vitro, but depressed by germ cells (22, 25). Testosterone has negative effects on plasminogen activator and ornithine decarboxylase activity (26, 27) and a positive effect on RNA polymerase II activity (28).

As mentioned above, the absence of androgens in vivo elevates androgen receptor mRNA in both Sertoli and peritubular cells, and this is reversed by T-treatment (16). To date, however, this effect has not been convincingly demonstrated in vitro. In contrast, in both Sertoli and peritubular cells, androgens increase androgen receptor protein in vitro by mechanisms that may involve an increase in translation and/or protein stability (12, 13, 29–31). These data suggest that these cells may have developed mechanisms to create a *set point* between receptor mRNA and protein in the face of high androgen concentrations.

Androgens have also been shown to decrease the mRNA for NGFβ receptor (32). In an attempt to define other androgen-regulated genes, we have used a subtractive procedure to isolate a number of Sertoli cell cDNA clones hybridizing to mRNAs that are increased greater than 2-fold by androgen treatment of hypophysectomized rats in vivo (Lu, Porter, Sanborn, unpublished observations). We have also observed changes in the respective mRNAs in Sertoli cells treated in vitro with androgens, as illustrated for clones C8 and G7 in Figure 15.3. In peritubular cells, C8 mRNA was increased to a lesser degree, and no change was seen in G7 mRNA. While clone G7 has been identified as hypoxanthine-guanine phosphoribosyl tranferase (Lu, Porter, Sanborn, unpublished observations), which may play a role in promoting the

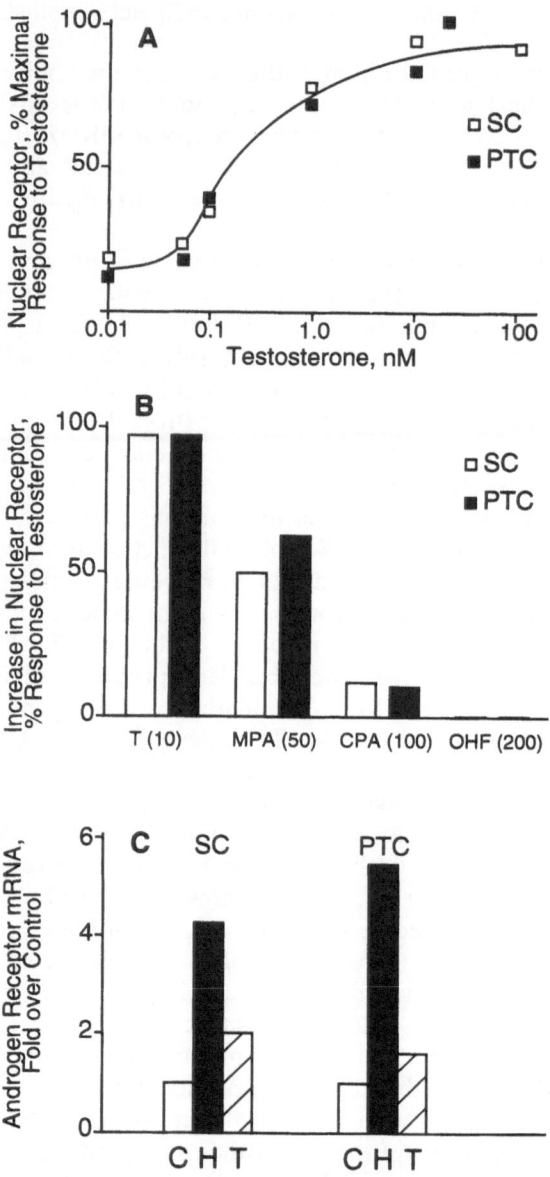

FIGURE 15.2. Similarity of the responses of Sertoli (SC) and peritubular cells (PTC) to T. *A:* Shown is the concentration dependence of nuclear androgen receptor accumulation as measured by exchange assay. Data are from Buzek and Sanborn, reference 15. *B:* Shown is a comparison of the ability of T, medroxyprogesterone acetate (MPA), cyproterone acetate (CPA), and hydroxyflutamide (OHF), in concentrations that inhibit ^{3}H-T binding by 50% (indicated in parentheses), to increase nuclear androgen receptor as measured by exchange assay. Data are from Buzek and Sanborn, reference 15. *C:* Shown is the effect of hypophysectomy (H) and T-treatment (T) in vivo and in vitro on Sertoli and peritubular cell androgen receptor mRNA compared with untreated controls (C), as measured by Northern blot analysis. Data are from reference 16.

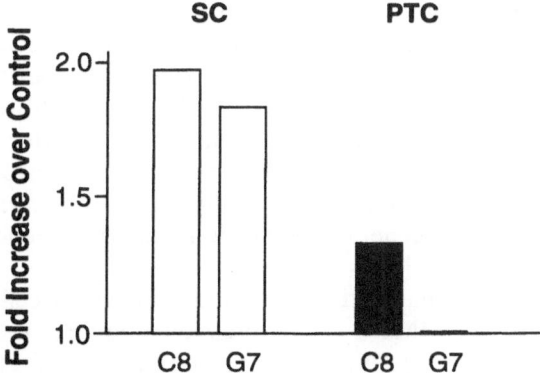

FIGURE 15.3. Increase in mRNA hybridizing to clones C8 and G7 in Sertoli cells (SC) and peritubular cells (PTC) following exposure to 100-ng/mL T for 24 h. Data are expressed as fold increase over control and were obtained by Northern blot analysis.

salvage of nucleotides in Sertoli cells relative to their export to germ cells (33), clone C8 has not been identified to date.

It is clear from these studies that both Sertoli and peritubular cells possess androgen receptors and that both cells are capable of responding to androgen stimulation with changes in specific proteins and mRNAs. Androgens appear to exert complex effects on androgen receptor regulation at the level of both mRNA and protein. However, it has yet to be demonstrated that the effects of androgens on any parameter are the result of direct transcriptional activation involving the androgen receptor. We therefore undertook a study of the transcriptional competence of Sertoli and peritubular cell androgen receptors.

Transcriptional Responses to Androgens in Sertoli and Peritubular Cells

The transcriptional effects of steroid hormones result from the interaction of the steroid-receptor complex via its DNA binding domain with specific response elements in regulated genes (34). In the case of androgen receptors, the interactions are quite complex and may involve interaction with other transcription factors as well (35, 36).

Transcriptional activation of reporter plasmids has been used to study the functional competence of intracellular steroid receptors. The reporter plasmid MMTV-Luc contains the mouse mammary tumor virus long-terminal repeat ligated to the luciferase reporter gene. The MMTV-LTR contains androgen response elements, and MMTV reporters have been shown to respond to endogenous and overexpressed androgen receptors in a number of cell types (37–39). We have used such a reporter to examine the transcriptional response of Sertoli and peritubular cells to androgens. In work described elsewhere (Ku, Loose-Mitchell, Sanborn, submitted), we have found that the expression of reporter plasmids transfected into Sertoli cells decreases markedly with increasing age. This correlates with the decreasing ability of Sertoli cells to divide in vitro and is consistent with the general observation that transfection efficiency is higher in dividing cells.

Figure 15.4 shows that both Sertoli and peritubular cells respond to T with an increase in MMTV-Luc expression and that the antiandrogen hydroxyflutamide can attenuate the response. The MMTV region also contains a glucocorticoid response element. Dexamethasone elicited no response in Sertoli cells, but exhibited a marked response in peritubular cells; however, this response was not inhibited by hydroxyflutamide. These data and other unpublished observations suggest that the androgen receptors in both Sertoli and peritubular cells are competent to regulate transcriptional events.

Factors Affecting Androgen Responsiveness in Sertoli and Peritubular Cells

The data cited above present convincing evidence that both Sertoli and peritubular cells are potential targets capable of mediating the effect of androgens on the seminiferous tubule and spermatogenesis. Responsiveness could be limited at any point in maturation by the concentration of receptor, the concentration of available steroids, and the nature of the steroids present. Although the data regarding these issues are incomplete, they do provide some insights.

The testis undergoes marked maturational changes associated with the first wave of spermatogenesis (1). Over this same period we and others have noted maturational changes in Sertoli cell morphology, secretory patterns, and intracellular proteins (reviewed in 24). Importantly, Sertoli cell androgen receptor concentration markedly increases between 15 and 35 days of age (40). A comparable change is not seen in peritubular cells, at least not between 25 and 45 days of age (10). In accordance with these observations, data described elsewhere (Ku, Loose-Mitchell, Sanborn, submitted) indicate that the androgen-regulated transcriptional response in Sertoli cells also increases with age over this time interval. These data suggest that the Sertoli cell may be increasing in its ability to respond to androgens during testicular maturation, while the peritubular cell response is less affected. This would imply that receptor concentration may be limiting for transcriptional response, at least in Sertoli cells.

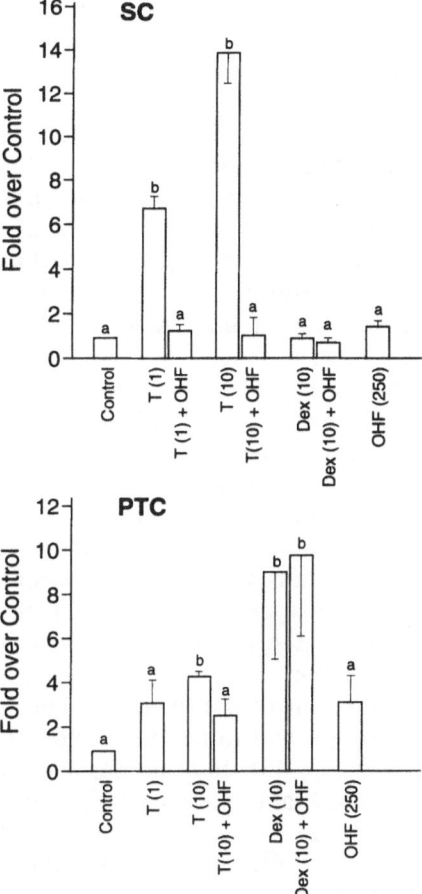

FIGURE 15.4. Ability of T, dexamethasone (DEX), and hydroxyflutamide (OHF) to increase expression of MMTV-Luc in Sertoli (SC) and peritubular cells (PTC) from 23-day-old animals. Cells were exposed to the steroids in the nM concentrations indicated for 24 h after transfection with MMTV-Luc. Data represent the mean ± SE of triplicate determinations. Groups denoted by different letters are significantly different at $P < 0.05$.

Indeed, this has been found to be the case in a number of cell types (41, 42). Under the conditions used in the experiments described in Figure 15.4, the receptor concentration in peritubular cells was significantly lower than that in Sertoli cells (unpublished observations). This may reflect the effects of culture conditions rather than the in vivo condition since androgens stabilize receptor concentrations, and these cells were androgen deprived for a longer time than Sertoli cells prior to challenge. Our preliminary data indicate that cotransfection of a plasmid expressing androgen receptor increases the fold stimulation by androgen of MMTV-Luc in both Sertoli and peritubular cells. Clearly, the role of receptor concentration in limiting response requires additional investigation.

The regulation of genes by androgens appears to involve complex interactions between regulatory factors in addition to receptors (35, 36). Therefore, responsiveness with respect to a given parameter could also be limited by the maturation of other components of the transcriptional regulatory systems. Peritubular cells appear to be functionally androgen responsive during sexual maturation, as indicated by the fact that androgens affect androgen receptor mRNA in cells from animals between 20 and 35 days of age (16), androgen receptor protein at 19 days (11), PModS at 19–20 days (11, 17), and increased expression of MMTV-Luc from 15 to 25 days (Ku, Loose-Mitchell, Sanborn, submitted). Therefore, it is likely that the peritubular cell plays a significant role in androgen-mediated events, even at this early time.

Similarly, androgens affect specific proteins and mRNAs in Sertoli cells from 20- to 35-day-old rats (21–24). Since the changes in androgen receptor concentration and transcriptional responsiveness occur in parallel with the first wave of spermatogenesis, the Sertoli cell could also play a role in early androgen-mediated events, although receptor concentration might well be a limiting factor.

It is possible that androgen responsiveness could also be affected by changes in the concentration and ratios of steroids occuring in the testis during sexual maturation (43–45). In seminiferous tubular fluid from the adult testis, T is the predominant androgen (46). The concentrations of T in seminiferous tubular fluid necessary to quantitatively maintain and reinitiate spermatogenesis ($<50\,nM$) (4, 5) in the absence of LH are considerably lower than those measured in adult animals ($150\,nM$) (46). Turner et al. (46) estimated androgen binding protein concentration to be $<30\%$ of the concentration of T plus dihydrotestosterone in seminiferous tubule fluid of the adult rat. The free T-concentration under these conditions would be more than enough to fulfill the requirement for maintenance or initiation of spermatogenesis, as well as to saturate androgen receptors. Understanding how the Sertoli and peritubular cells deal with this apparent receptor saturation requires additional experimentation.

Consideration of the same question in the immature state is more complex. Although there is considerable disagreement in the literature concerning the maturational pattern and relative amounts of T, dihydrotestosterone, and 3α-androstane-5α,17β-diol in the testis and, more particularly, in the seminiferous tubule, it is generally agreed that 5α-reduced androgens predominate during the period of sexual maturation (43–45). This is probably not a major factor in determining responsiveness in itself since all three androgens bind to androgen receptors—although with different affinities (47, 48)—are capable of maintaining spermatogenesis in the estrogen-treated rat (49), stimulate androgen receptor protein and secretion of PModS-like activity in peritubular cells (11, 19), and stimulate RNA polymerase II activity in Sertoli cells (28). The transcriptional response of an MMTV reporter to androgen receptor complexed with T or dihydrotestosterone is equivalent (48).

Another significant consideration is that at sexual maturation, testicular androgen binding protein concentrations are considerably higher than in the adult (50). Hence, in contrast to the adult, at 25 days of age, free T plus dihydrotestosterone concentrations in seminiferous tubule fluid are probably not in large excess over androgen binding protein. In fact, depending on the assumptions made and values used, it is possible to predict a situation where androgen binding protein might well reduce the concentrations of these steroids below the critical concentration required for spermatogenesis to proceed. Whether the androstanediol concentrations present at this time are high enough to compensate remains to be determined. Consequently, on the basis of available evidence, it seems that responsiveness in the Sertoli and peritubular cells of the maturing testis could be limited both by steroid and receptor concentration, as well as by the maturation of transcriptional regulatory mechanisms.

Conclusions

The available evidence suggests that androgen mechanisms in the seminiferous tubule are mediated via both the Sertoli and the peritubular cells that contribute to the environment necessary for the progression of the spermatogenic process. Exactly how each cell responds and what tnat response contributes to the overall regulatory process remain to be completely delineated. Furthermore, the regulation of androgen responsiveness during sexual maturation and in the presence of high androgen concentrations needs to be addressed. Finally, the possibility that androgens may exert nonreceptor-mediated events directly on germ cells, as suggested by earlier binding studies (51, 52), should not be dismissed.

Acknowledgment. This work was supported in part by HD-17795.

References

1. Steinberger E, Steinberger A. Testis: basic and clinical aspects. In: Balin H, Glasser S, eds. Reproductive biology. Amsterdam: Excerpta Medica, 1972:114–267.
2. Awoniyi CA, Sprando RL, Santulli R, Chandrashekar V, Ewing LL, Zirkin BR. Restoration of spermatogenesis by exogenously administered testosterone in rats made azospermic by hypophysectomy or withdrawal of luteinizing hormone alone. Endocrinology 1990;127:177–84.

3. Awoniyi CA, Santulli R, Chandrashekar V, Schanbacher BD, Zirkin BR. Quantitative restoration of advanced spermatogenic cells in adult male rats made azospermic by active immunization against luteinizing hormone or gonadotropin-releasing hormone. Endocrinology 1989;125:1303–9.

4. Zirkin BR, Santulli R, Awoniyi CA, Ewing LL. Maintenance of advanced spermatogenic cells in the adult rat testis: quantitative relationship to testosterone concentration within the testis. Endocrinology 1989;124:3043–9.

5. Awoniyi CA, Santulli R, Sprando RL, Ewing LL, Zirkin BR. Restoration of advanced spermatogenic cells in the experimentally regressed testis: quantitative relationship to testosterone concentration within the testis. Endocrinology 1989;124:1217–23.

6. Morales C, Griswold MD. Retinol-induced stage synchronization in seminiferous tubules of the rat. Endocrinology 1987;121:432–4.

7. Sanborn BM, Steinberger A, Tcholakian RK, Steinberger E. Direct measurement of androgen receptors in cultured Sertoli cells. Steroids 1977;29:493–502.

8. Schmidt WN, Danzo BJ. Androgen and progesterone binding components in cytosol prepared from cultures enriched in Sertoli cells from immature rat testes. Biol Reprod 1980;23:495–506.

9. Nahkla AM, Mather JB, Janne OA, Bardin CW. Estrogen and androgen receptors in Sertoli, Leydig, myoid and epithelial cells: effects of time in culture and cell density. Endocrinology 1984;115:121–8.

10. Buzek SW, Sanborn BM. Increase in testicular androgen receptor during sexual maturation in the rat. Biol Reprod 1988;39:39–49.

11. Verhoeven G, Cailleau J. Testicular peritubular cells secrete a protein under androgen control that inhibits induction of aromatase activity in Sertoli cells. Endocrinology 1988;123:2100–10.

12. Verhoeven G, Cailleau J. Follicle-stimulating hormone and androgens increase the concentration of the androgen receptor in Sertoli cells. Endocrinology 1988;122:1541–50.

13. Blok LJ, Mackenbach P, Trapman J, Themmen APN, Brinkmann AO, Grootegoed JA. Follicle stimulating hormone regulates androgen receptor mRNA in Sertoli cells. Mol Cell Endocrinol 1989;63:267–71.

14. Grootegoed JA, Peters MJ, Mulder E, Rommerts FFG, van der Molen HJ. Absence of a nuclear androgen receptor in isolated germinal cells of rat testis. Mol Cell Endocrinol 1977;9:159–67.

15. Buzek SW, Sanborn BM. Nuclear androgen receptor dynamics in testicular peritubular and Sertoli cells. J Androl 1990;11:514–20.

16. Sanborn BM, Caston LA, Chang C, et al. Regulation of androgen receptor mRNA in rat Sertoli and peritubular cells. Biol Reprod 1991;45:634–41.

17. Skinner MK, Fritz IB. Testicular peritubular cells secrete a protein under androgen control that modulates Sertoli cell functions. Proc Natl Acad Sci USA 1985;82:114–8.

18. Norton JN, Skinner MK. Regulation of Sertoli cell differentiation by the testicular paracrine factor PModS: potential role of immediate-early genes. Mol Endocrinol 1992;6:2018–26.

19. Swinnen K, Cailleau J, Heyns W, Verhoeven G. Prostatic stromal cells and testicular peritubular cells produce similar paracrine mediators of androgen action. Endocrinology 1990;126:142–50.

20. Anthony CT, Skinner MK. Cytochemical and biochemical characterization of testicular peritubular myoid cells. Biol Reprod 1989;40:811–23.
21. Sanborn BM, Wagle JR, Steinberger A, Greer-Emmert D. Maturational and hormonal influences on Sertoli cell function. Endocrinology 1986;118:1700–9.
22. Cheng CY, Mather JP, Byer AL, Bardin CW. Identification of hormonally responsive proteins in primary Sertoli cells culture medium by anion exchange high performance liquid chromatography. Endocrinology 1986;118:480–8.
23. Roberts K, Griswold MD. Testosterone induction of cell proteins in cultured Sertoli cells from hypophysectomized rats and rats of different ages. Endocrinology 1989;125:1174–9.
24. Sanborn BM, Caston LA, Buzek SW, Ussuf KK. Hormonal regulation of Sertoli cell function. Adv Exp Med Biol 1987;219:561–88.
25. Cheng CY, Grima J, Stahler MS. Lockshin RA, Bardin CW. Testins are structurally related Sertoli cell proteins whose secretion is tightly coupled to the presence of germ cells. J Biol Chem 1989;264:21386–92.
26. Ailenberg M, McCabe D, Fritz IB. Androgens inhibit plasminogen activator secreted by Sertoli cells in culture in a two-chamber assembly. Endocrinology 1990;126:1561–8.
27. Swift TA, Dias JA. Testosterone suppression of ornithine decarboxylase activity in rat Sertoli cells. Endocrinology 1988;23:687–93.
28. Lamb DJ, Wagle JR, Tsai YH, Lee AL, Steinberger A, Sanborn BM. Specificity and nature of the rapid steroid-stimulated increase in Sertoli cell nuclear RNA polymerase activity. J Steroid Biochem 1982;16:653–9.
29. Sanborn BM, Wagle JR, Steinberger A. Control of cytosol androgen receptor concentrations in Sertoli cells: effect of androgens. Endocrinology 1984; 114:2388–93.
30. Kemppainen JA, Lane MV, Sar M, Wilson EM. Androgen receptor phosphorylation, turnover, nuclear transport and transcriptional activation. J Biol Chem 1992;267:968–74.
31. Syms AJ, Norris JS, Panko WB, Smith RG. Mechanism of androgen receptor augmentation: analysis of receptor synthesis and degradation by the density-shift technique. J Biol Chem 1985;260:455–61.
32. Persson H, Ayer-LeLievre C, Soder O, et al. Expression of beta-nerve growth factor receptor mRNA in Sertoli cells downregulated by testosterone. Science 1990;247:704–7.
33. Haugen TB, Fritson P, Hansson V. Effects of cryptorchidism and orchidoplexy on deoxyribonucleoside-activated nucleotidase (DAN) in rat testis. Int J Biochem 1987;19:193–7.
34. Gronemeyer H. Control of transcription activation by steroid hormone receptors. FASEB J 1992;6:2524–9.
35. Adler AJ, Danielsen M, Robins DM. Androgen-specific gene activation via a consensus glucocorticoid response element is determined by interaction with nonreceptor factors. Proc Natl Acad Sci USA 1992;89:11660–3.
36. Rennie PS, Bruchovsky N, Leco KJ, et al. Characterization of two cis-acting DNA elements involved in the androgen regulation of the probasin gene. Mol Endocrinol 1993;7:23–36.
37. Ham J, Thompson A, Needham M, Webb P, Parker M. Characterization of response elements for androgens, glucocorticoids, and progestins in mouse mammary tumor virus. Nucleic Acids Res 1988;16:5263–76.

38. Parker MG, Webb P, Needham M, White R, Ham J. Identification of androgen response elements in mouse mammary tumour virus and the rat prostate C3 gene. J Cell Biochem 1987;35:285–92.
39. Quarmby VE, Kemppainen JA, Sar M, Lubahn DB, French FS, Wilson EM. Expression of recombinant androgen receptor in cultured mammalian cells. Mol Endocrinol 1990;4:1399–407.
40. Buzek SW, Caston LA, Sanborn BM. Evidence for age-dependent changes in Sertoli cell androgen receptor concentration. J Androl 1987;18:83–90.
41. Webb P, Lopez GN, Greene GL, Baxter JD, Kushner PJ. The limits of the cellular capacity to mediate an estrogen response. Mol Endocrinol 1992;6:157–67.
42. Boquel MT, Kumar V, Stricker C, Chambon P, Gronemeyer H. The contribution of the N- and C-terminal regions of steroid receptors to activation of transcription is both receptor and cell-specific. Nucleic Acids Res 1989; 17:2581–91.
43. Podesta EJ, Rivarola MA. Concentration of androgens in whole testis, seminiferous tubules and interstitial tissue of rats at different stages of development. Endocrinology 1974;85:455–61.
44. Corpechot C, Baulieu EE, Robel P. Testosterone, dihydrotestosterone and androstanediols in plasma, testes and prostates of rats during development. Acta Endocrinol (Copenh) 1981;96:127–35.
45. Tapanainen J, Kuopio T, Pelliniemi LJ, Huhtaniemi I. Rat testicular endogenous steroids and number of Leydig cells between the fetal period and sexual maturity. Biol Reprod 1984;31:1027–35.
46. Turner TT, Jones CE, Howards SS, Ewing LL, Zegeye B, Gunsalus GE. On the androgen environment of maturing spermatozoa. Endocrinology 1984;115:1925–32.
47. Shain SA, Boesel RW. Saturation analysis of the binding of androgens, antiandrogens and estrogens by the cytoplasmic high affinity androgen receptor of the rat ventral prostate. J Steroid Biochem 1975;6:43–50.
48. Deslypere JP, Young M, Wilson JD, McPhaul MJ. Testosterone and 5α-dihydrotestosterone interact differently with the androgen receptor to enhance transcription of the MMTV-CAT reporter gene. Mol Cell Endocrinol 1992;88:15–22.
49. Chowdhury AK, Steinberger E. Effect of 5α reduced androgens on sex accessory organs, initiation and maintenance of spermatogenesis in the rat. Biol Reprod 1975;12:609–17.
50. Danzo BJ, Eller BC. The ontogeny of biologically active androgen binding protein in rat plasma, testis, and epididymis. Endocrinology 1985;117:1380–8.
51. Galena HJ, Pillai AK, Terner C. Progesterone and androgen receptors in non-flagellate germ cells of the rat testis. J Endocrinol 1974;63:223–37.
52. Sanborn BM, Steinberger A, Meistrich ML, Steinberger E. Androgen binding sites in testis cell fractions as measured by a nuclear exchange assay. J Steroid Biochem 1975;6:1459–65.

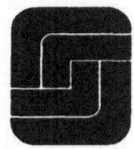

16

Hormonal Control of Apoptotic Cell Death in the Testis

JUHA S. TAPANAINEN AND AARON J.W. HSUEH

The growth and differentiation of somatic cells in the testis, as well as initiation and maintenance of spermatogenesis, are known to be regulated by pituitary gonadotropins and testicular androgens. Under the stimulatory control of *follicle stimulating hormone* (FSH) and androgens, Sertoli cells, the somatic component of the seminiferous epithelium, support spermatogenesis by providing nutrients to the germ cells (1, 2). Sertoli cells possess receptors for FSH (3, 4) and androgens (5) that are secreted by the Leydig cells in response to pituitary *luteinizing hormone* (LH) secretion.

Morphological studies have indicated that spermatogenic cells undergo spontaneous degeneration at specific stages during their development, and up to 75% of spermatogonia has been reported to degenerate before maturation (6–8). The withdrawal of gonadotropins by hypophysectomy enhances the degeneration of both somatic and germ cells in the testis (9–13). In contrast, gonadotropin replacement prevents hypophysectomy-induced testicular cell degeneration (11), but the mechanisms are still poorly understood.

The cellular and molecular mechanisms regulating testicular cell degeneration and survival have not been addressed due to the lack of a sensitive biochemical marker for testicular cell death. In this chapter, the role of gonadotropins and androgens as testicular cell survival factors is reviewed.

Apoptotic Cell Death

There are essentially two mechanism by which the cells die: necrosis and apoptosis (14, 15). Tissue *necrosis* is a consequence of extrinsic cell injury or other traumatization resulting in the damage of cell structures, and *apoptosis* is a process of active cellular destruction induced by specific stimuli, apparently following an intrinsic program. The characteristic features of apoptosis are cell shrinkage, chromatin condensation, formation of pyknotic nuclei, and the formation of apoptotic bodies containing fragmented nucleus and an array of intact organelles, such as mitochondria and lysosomes. This process is induced by the activation of an endogenous calcium/magnesium-dependent endonuclease activity capable of cleaving nuclear DNA at internucleosomal sites. Endonuclease activation appears to occur in virtually all systems in which apoptosis takes place (16, 17). To date, however, it is not clear whether apoptosis involves new synthesis or activation of endonuclease or rearrangement of chromatin proteins to permit DNA cleavage by a preexisting enzyme. The end result of endonuclease activity is the generation of DNA fragments of 185–200 kb in size (18). In contrast to apoptosis, the breakdown of DNA during necrosis is not dependent on endonuclease activation and does not result in cleavage of DNA at regularly spaced sites, but results in a "smear" of DNA following analysis by agarose gel electrophoresis and ethidium bromide staining.

Analysis of Apoptotic DNA Fragmentation in Rat Testis

In the ovary greater than 99.9% of follicles undergo atresia during reproductive life. It has been demonstrated that gonadotropin withdrawal induces atretic degeneration of ovarian follicles (19, 20), and recent studies have indicated that apoptotic cell death is the mechanism underlying follicular atresia (21–23). Using a sensitive autoradiographic method for the detection of apoptotic DNA cleavage (21, 24), we have studied the mechanisms regulating cell death in testis.

To study the hormonal control of apoptotic cell death in rat testis, immature (21-day-old) male hypophysectomized Sprague-Dawley rats were treated for 2 days with s.c. injections of 20-IU FSH-CTP (FSH agonist with a carboxyterminal peptide of hCG appended to the carboxyterminus of the FSHβ subunit), 50-IU hCG, or 5-mg testosterone propionate. To study the effect of a GnRH antagonist (Azaline B) on testicular apoptosis, intact immature rats were given daily injections of the antagonist (10 μg/day) starting at 21 days of age. The operated and hormone-treated animals and the control animals were sacrificed at different time intervals. Testes were removed, decapsulated, and weighed. One testis from each animal was snap frozen and stored at $-70°C$. The other testis was dissociated with 0.2% collagenase in order to separate the interstitial cells and the seminiferous tubules before freezing.

Total DNA was isolated from frozen tissues (24, 25) and analyzed using 3'-end labeling followed by autoradiography. Briefly, 1 μg of DNA from each sample was labeled at 3' ends with $(\alpha^{32}P)$-dideoxy-ATP using 25-U terminal transferase enzyme as described (21, 24). The labeled DNA samples were loaded onto 2% agarose gels (400 ng/lane) and separated by electrophoresis for 3 h at 50 V. The gels were dried for 2 h in a slab-gel drier and exposed to Kodak X-OMAT films at −70°C for 2–6 h for qualitative analysis. After autoradiography the amount of radiolabeled dideoxy-ATP incorporated into low MW (≤15 kb) DNA was measured by excising the corresponding gel portions with a scalpel and counting in a β-counter for quantitative estimation of the degree of internucleosomal DNA cleavage among samples (24).

Hypophysectomy-Induced Apoptotic DNA Fragmentation in the Testis: Effect of Gonadotropin and Androgen Replacement

Hypophysectomy decreased testis weight by 25%, but treatment with FSH-CTP or hCG prevented the effect of hypophysectomy. The testes of intact animals contained predominantly high MW DNA, whereas hypophysectomy increased DNA cleavage into low MW (<15 kb) ladders characteristic of apoptosis. In contrast, treatment with FSH-CTP or hCG inhibited hypophysectomy-induced apoptotic DNA cleavage by 84% and 51%, respectively. Hypophysectomy-induced DNA fragmentation was found in both interstitial cells and seminiferous tubules. Similar to whole testis, treatment with FSH-CTP suppressed hypophysectomy-induced apoptosis by over 90% in seminiferous tubules and interstitial cells. In contrast, hCG treatment was less effective in preventing hypophysectomy-induced DNA cleavage (46% suppression in tubules and 77% suppression in interstitial cells). Furthermore, *testosterone* (T) replacement also suppressed hypophysectomy-induced DNA fragmentation by 75% in the whole-testis tissue, 64% in tubules, and 55% in interstitial cells.

The requirement for pituitary gonadotropins and T-secretion from the Leydig cells during testicular cell differentiation and for optimal spermatogenesis in adult life is well documented (1). However, very little is known about the possible regulation of testicular cell death. Although morphological and histochemical studies have suggested degenerative changes in the testis after gonadotropin deprivation, as well as their prevention by gonadotropins (11, 12, 26, 27), the present results provide conclusive biochemical evidence for apoptotic cell death in testicular cells and evidence that pituitary gonadotropins and androgens play an essential role in preventing this process. Within 2 days after hypophysectomy, the testicular cells showed extensive apoptotic cell death as reflected by the presence of internucleosomal DNA fragmentation, indicating that the immature testis is highly dependent on pituitary gonadotropins. Because gonadotropin replacement effectively prevented hypophysectomy-induced apoptosis in the testis, these findings emphasize the importance of gonadotropins not only for differentiation and growth, but also for the survival of testicular cells.

Although our data indicated that apoptosis occurs in both the interstitial cells and the seminiferous tubules, the exact testicular cell types involved are not clear. Recent morphometric studies on adult rat testis have suggested that apoptosis may be the mechanism of spontaneous spermatogonial depletion (8, 28). Furthermore, hypophysectomy induces the degeneration of germ, Leydig, and Sertoli cells (12, 13) with morphological features resembling apoptosis. Treatment with FSH effectively inhibited hypophysectomy-induced apoptotic cell death in both seminiferous tubule and interstitial cell compartments. FSH is thought to mediate its action on germ cells predominantly via Sertoli cells that are the sole direct target for FSH action (3, 29). In contrast, FSH receptors have not been identified on Leydig cells (30, 31), suggesting that the observed effect of FSH on interstitial cell survival is indirect. Supporting this concept, purified and recombinant FSH has been shown to stimulate Leydig cell functions in vivo in immature hypophysectomized rats by increasing LH/hCG receptor content and androgen biosynthetic potential (32, 33). Furthermore, there is increased evidence suggesting that factors derived from the seminiferous tubules may influence Leydig cell differentiation and T-production through a paracrine mechanism (26, 34, 35).

Treatment with hCG also decreased hypophysectomy-induced DNA fragmentation in both interstitial cells and seminiferous tubules. Since Leydig cells are thought to be the only cells in the testis possessing LH/hCG receptors (36), it is likely that the effect of hCG on cell survival in the seminiferous tubules is mediated by T. This is supported by the finding that treatment with T-propionate also significantly decreased hypophysectomy-induced apoptotic cell death in whole-testis tissue and seminiferous tubules. The direct action of T in the testis may be mediated by androgen receptors found in interstitial cells, Sertoli cells, and peritubular myoid cells (5, 37). It is unclear if germ cells are direct targets of androgen action since observations concerning androgen receptors and androgen binding to germ cells are contradictory (38, 39). Consistent with the present results in the testis, the survival of epithelial cells in the ventral prostate is also dependent on T (17), emphasizing the role of androgens as survival factors in male reproductive organs.

Induction of Apoptotic DNA Fragmentation in the Testis by a GnRH Antagonist

Although the present studies using gonadotropin replacement suggest that the observed increase in testis apoptosis is due to hypophysectomy-induced depletion of circulating gonadotropins, we could not rule out the effect of other pituitary hormones (e.g., growth hormone) in maintaining testicular cell survival. To specifically decrease gonadotropin secretion, immature intact rats were treated with a potent GnRH antagonist, Azaline B. Similar to hypophysectomy, treatment with Azaline B decreased testis weight by 31% in 2 days and induced a 3.5 ± 0.4-fold increase in apoptotic DNA fragmentation in testis tissue within 2 days ($P < 0.01$) and a 4.0 ± 0.5-fold increase within 4 days ($P < 0.001$).

Although one cannot rule out the role of other pituitary factors in maintaining testicular cell survival, these data suggest that the effects of hypophysectomy on apoptosis are due primarily to the removal of FSH and/or LH. Thus, FSH and LH may function as gonadal cell-specific survival factors in the same manner as that described for various peptide factors in several extragonadal cell systems (40, 41).

Summary

Use of the sensitive 3'-end labeling method for detecting apoptotic cell death has made it possible to study the cellular mechanisms of testicular cell demise. Our studies have demonstrated that the removal of gonado-tropin support induces apoptotic cell death in the testis and that the effect of hypophysectomy can be suppressed by treatment with FSH, LH/hCG, or T (42). Therefore, gonadotropins and T are not only important for the regulation of testicular cell proliferation and differentiation; they are also essential for promoting testicular cell survival. Future use of the present immature rat model and the sensitive autoradiographic method to study hormonal regulation of testicular apoptosis should provide new insights on testis physiology and pathophysiology.

References

1. Steinberger E. Hormonal control of mammalian spermatogenesis. Physiol Rev 1991;51:1–22.
2. Parvinen M. Regulation of the seminiferous epithelium. Endocr Rev 1981;3: 404–17.
3. Heckert LL, Griswold MD. Expression of follicle-stimulating hormone receptor mRNA in rat testes and Sertoli cells. Mol Endocrinol 1991;5:670–7.
4. Means AR, Fakunding JL, Huckins C, Tindall DJ, Vitale R. Follicle-stimulating hormone, the Sertoli cell, and spermatogenesis. Recent Prog Horm Res 1976;32:477–527.
5. Sar M, Lubahn DB, French FS, Wilson EM. Immunohistochemical localization of the androgen receptor in rat and human tissues. Endocrinology 1990;127:3180–6.
6. Huckins C. The morphology and kinetics of spermatogonial degeneration in normal adult rats: an analysis using a simplified classification of the germinal epithelium. Anat Rec 1978;180:905–26.

7. De Rooij DG, Lok D. Regulation of the density of spermatogonia in the seminiferous epithelium of the Chinese hamster, II. Differentiating spermatogonia. Anat Rec 1987;217:131–6.

8. Allan DJ, Harmon BV, Roberts SA. Spermatogonial apoptosis has three morphologically recognizable phases and shows no circadian rhythm during normal spermatogenesis in the rat. Cell Prolif 1992;25:241–50.

9. Clermont Y, Morgentaler H. Quantitative study of spermatogenesis in the hypophysectomized rat. Endocrinology 1955;57:369–82.

10. Raj HGM, Dym M. The effects of selective withdrawal of FSH and LH on spermatogenesis in the immature rat. Biol Reprod 1976;14:489–94.

11. Russell LD, Clermont Y. Degeneration of germ cells in normal, hypophysectomized and hormone treated hypophysectomized rats. Anat Rec 1977;187: 347–66.

12. Russell LD, Corbin TJ, Ren HP, Amador A, Bartke A, Ghosh S. Structural changes in rat Leydig cells posthypophysectomy: a morphometric and endocrine study. Endocrinology 1992;131:498–508.

13. Ghosh S, Bartke A, Grasso P, Reihert LE Jr, Russell LD. Structural manifestations of the rat Sertoli cell to hypophysectomy: a correlative morphometric and endocrine study. Endocrinology 1992;131:485–97.

14. Wyllie AH, Kerr JFR, Currie AR. Cell death: the significance of apoptosis. Int Rev Cytol 1980;68:251–306.

15. Wyllie AH. Cell death: a new classification separating apoptosis from necrosis. In: Bowen ID, Lockshin RA, eds. Cell death in biology and pathology. New York: Chapman and Hall, 1981:9–34.

16. Cohen JJ, Duke RC. Glucocorticoid activation of a calcium-dependent endonuclease in thymocyte nuclei leads to cell death. J Immunol 1984;132:38–42.

17. Kyprianou N, Isaacs JT. Activation of programmed cell death in the rat ventral prostate after castration. Endocrinology 1988;122:552–62.

18. Wyllie AH. Glucocorticoid-induced thymocyte apoptosis is associated with endogenous endonuclease activation. Nature 1980;284:555–6.

19. Tsafriri A, Braw RH. Experimental approaches to atresia in mammals. Oxf Rev Reprod Biol 1984;6:226–65.

20. Hirsfield AN. Development of follicles in the mammalian ovary. Int Rev Cytol 1991;124:43–101.

21. Tilly JL, Kowalski KI, Johnson AL, Hsueh AJW. Involvement of apoptosis in ovarian follicular atresia and postovulatory regression. Endocrinology 1991;129:2799–801.

22. Hughes FM Jr, Gorospe WC. Biochemical identification of apoptosis (programmed cell death) in granulosa cells: evidence for a potential mechanism underlying follicular atresia. Endocrinology 1991;129:2415–22.

23. Tilly JL, Billig H, Kowalski KI, Hsueh AJW. Epidermal growth factor and basic fibroblast growth factor suppress the spontaneous onset of apoptosis in cultured rat ovarian granulosa cells and follicles by a tyrosine kinase-dependent mechanism. Mol Endocrinol 1992;6:1942–50.

24. Tilly JL, Hsueh AJW. Microscale autoradiographic method for qualitative and quantitative analysis of apoptotic DNA fragmentation. J Cell Physiol 1993;154:519–26.

25. Gross-Bellard M, Oudet P, Chambon P. Isolation of high-molecular-weight DNA from mammalian cells. Eur J Biochem 1973;36:32–8.

26. Kerr JB, Sharpe RM. Follicle-stimulating hormone induction of Leydig cell maturation. Endocrinology 1985;116:2592–604.
27. Almiron I, Chemes H. Spermatogenic onset, II. FSH modulates mitotic activity of germ and Sertoli cells in immature rats. Int J Androl 1988;11:235–46.
28. Allan DJ, Harmon BV, Kerr JFR. Cell death in spermatogenesis. In: Potten CS, ed. Perspectives on mammalian cell death. London: Oxford University Press, 1987:229–58.
29. Orth J, Christensen AK. Localization of ^{125}I-labeled FSH in the testes of hypophysectomized rats by autoradiography at the light and electron microscope levels. Endocrinology 1977;101:262–78.
30. Means A, Vaitukaitis J. Peptide hormone "receptors": specific binding of ^{3}H-FSH to testis. Endocrinology 1972;90:39–46.
31. de Kretser DM, Catt KJ, Paulsen CA. Studies on the in vitro testicular binding of iodinated luteinizing hormone in rats. Endocrinology 1971;88:332–7.
32. Chen Y-D, Payne AH, Kelch RP. FSH stimulation of Leydig cell function in the hypophysectomized immature rat. Proc Soc Exp Biol Med 1976;153:473–5.
33. Vihko KK, LaPolt PS, Nishimori K, Hsueh AJW. Stimulatory effects of recombinant follicle-stimulating hormone on Leydig cell function and spermatogenesis in immature hypophysectomized rats. Endocrinology 1991;129:1926–32.
34. Verhoeven G, Cailleau A. Leydig cell stimulatory factor produced by human testicular tubules. Mol Cell Endocrinol 1987;49:137–47.
35. Sharpe RM, Maddocks S, Kerr JB. Cell-cell interactions in the control of spermatogenesis as studied using Leydig cell destruction and testosterone replacement. Am J Anat 1990;188:3–20.
36. Catt KJ, Dufau ML. Gonadotropin receptors and regulation of interstitial cell function in the testis. In: Birnbaumer L, O'Malley BW, eds. Receptors and hormone action. New York: Academic Press, 1978:291–339.
37. Sanborn BM, Steinberger A, Tcholakian RK, Steinberger E. Direct measurements of androgen receptors in cultured Sertoli cells. Steroids 1977;29:493–502.
38. Grootegoed JA, Peters MJ, Mulder E, Rommerts FFG, van der Molen HJ. Absence of nuclear androgen receptor in isolated germinal cells of rat testis. Mol Cell Endocrinol 1977;9:159–67.
39. Wright WW, Frankel AI. An androgen receptor in the nuclei of late spermatids in testes of male rats. Endocrinology 1980;107:314–8.
40. Raff MC. Social controls on cell survival and cell death. Nature 1992;356:397–400.
41. Lockshin RA, Zakeri ZF. Programmed cell death: new thoughts and relevance to aging. J Gerontol 1990;45:B135–40.
42. Tapanainen JS, Tilly JL, Vihko KK, Hsueh AJW. Hormonal control of apoptotic cell death in the testis: gonadotropins and androgens as testicular cell survival factors. Mol Endocrinol 1993.

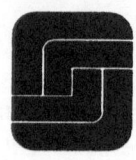

17

Involvement of Liver-Regulating Protein-Like Molecule in Sertoli-Germ Cell Cross Talk

Nadine Gérard, Anne Corlu, Henri Kercret, Bernard Kneip, Maryvonne Rissel, Christiane Guguen-Guillouzo, and Bernard Jégou

Sertoli cells are a key element in the endocrine and paracrine regulation of the spermatogenic process. They are responsible for the constitution of the microenvironment required for germ cell division, differentiation, and metabolism (1, 2). It is also now well established that germ cells control this environment via the regulation of Sertoli cell activity (2, 3). The nature of the molecules and the mechanisms involved in this regulation are beginning to be understood. We have proposed that the germinal control of Sertoli cell function is exerted via germ cell-soluble factors, transfer of germ cell materials to the Sertoli cell (e.g., residual bodies), and plasma membrane molecules (2, 4, 5). Evidence exists that supports this hypothesis: (i) The purification and identification of some germ cell-soluble regulating proteins are in progress (6); (ii) morphologists have described in great detail the various devices involved in the transfer of spermatogonial and spermatid materials to the Sertoli cells (4, 7–9); and (iii) the involvement of residual bodies in the regulation of Sertoli cell IL-1α and IL-6 production has recently been established (10, 11). Furthermore, from recent in vivo studies, it appears that the different molecular families of *cell adhesion molecules* (CAMs)—cadherins, integrins, selectins, and CAMs belonging to the immunoglobulin superfamily—are represented in rodent seminiferous tubules (5, 12, 13). With regard to this latter aspect, it is noteworthy that the direct functional role of cell-cell contact in Sertoli-germ cell cross talk has, thus far, never been investigated. A transmembranous protein named *liver-regulating protein* (LRP) has recently been identified and shown to be involved in the in vitro regulation of the hepatocyte differentiation by cell-cell interaction with primitive *rat liver biliary epithelial cells* (RLEC) (14–16). Our previous studies have shown that both germ cells and RLEC are able to stimulate transferrin and inhibin production by Sertoli cells in vitro (17, 18). Therefore, this study was undertaken to search for the presence of an LRP-like molecule in the testis and to investigate whether LRP could be involved in germ cell-Sertoli cell and RLEC-Sertoli cell interactions.

Materials and Methods

For the immunolocalization studies, 20-day-old and adult Sprague-Dawley rat testes, as well as cultured testicular cells, were fixed in paraformaldehyde solution (4%) buffered in 0.1 M sodium cacodylate. Frozen tissue sections (8 μm) were cut using a Reichert-Jung cryostat, and immunostaining was performed using the indirect immunoperoxidase technique described elsewhere (16). Sertoli cells were isolated from the testes of 20-day-old Sprague-Dawley rats according to a method previously described (19). These cells were exposed to hypotonic shock 2 days after plating (day 2) to eliminate contaminating germ cells (20). *Pachytene spermatocytes* (SPC) were purified by centrifugal elutriation of a crude preparation of germ cells isolated from the testes of adult Sprague-Dawley rats (21). RLEC were established from the liver of 10-day-old Sprague-Dawley rats (22).

The SPC-Sertoli cell and RLEC-Sertoli cell cocultures were run in the presence or absence of the monoclonal antibody directed against the liver form of the LRP (Mab L8) (14) from day 3 to day 5 after Sertoli cell plating. Fresh medium and antibody were added every 24 h. Transferrin and inhibin were assayed in day 4–5 media of both SPC- and RLEC-Sertoli cell cocultures, as well as in control Sertoli cell cultures, using the radioimmunoassays previously described (17, 18). Total RNAs were extracted from cultured cells by the acid guanidium thiocyanate phenol chloroform technique, electrophoresed, and transferred to nylon filters before hybridization with labeled transferrin and inhibin cDNA probes (kindly provided by Drs. A. Khan and R. Sharpe, respectively). Immunoprecipitation studies were performed on cell lysates after peroxydase-catalyzed iodination followed by an SDS-PAGE qualitative analysis.

Results

Using the monoclonal antibody directed against the liver form of the LRP (Mab L8) (14), we located an immunologically related LRP-like molecule in the rat testis. Our study provided evidence for an age- and stage-dependent expression of this protein. The reaction with Mab L8 was detected at the interface of primary spermatocytes and Sertoli cells. The testicular LRP-like protein was first observed when primary spermatocytes

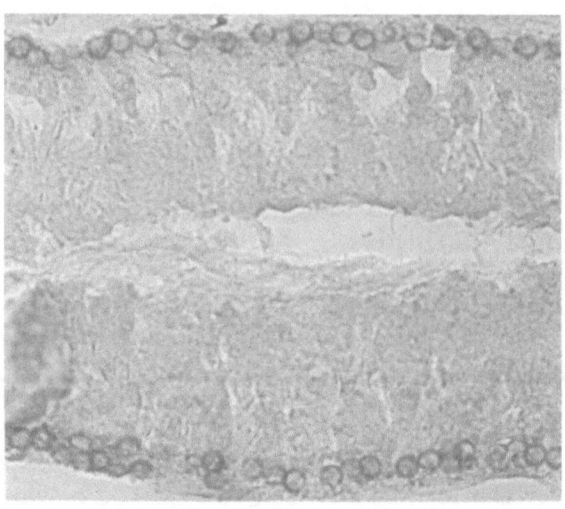

FIGURE 17.1. Location of the LRP-like molecule expressed in the rat testis by immunoreaction with Mab L8 of cryostat sections. A positive reaction is observed at the interface of Sertoli cells and primary spermatocytes: Zygotene is at stage XII of the seminiferous epithelium cycle.

appeared within the seminiferous tubule (not shown). It was weakly expressed on preleptotene spermatocytes and was expressed at a higher rate on leptotene and zygotene spermatocytes. Its expression decreased on pachytene spermatocytes (Fig. 17.1). The LRP-like protein immunoreactivity was also found by electron microscopy analysis on both spermatocyte and Sertoli cell plasma membranes (not shown).

To gain information on the possible role of the testicular LRP-like protein in the Sertoli cell-germ cell communications, the effect of Mab L8 was tested in vitro. The SPC- and RLEC-Sertoli cell cocultures were run in the presence or absence of Mab L8 from day 3 to day 5 of the Sertoli cell cultures. Our results confirm previous observations showing that the addition of both SPC and RLEC to Sertoli cell cultures induced a marked stimulation of transferrin and inhibin radioassayed levels (Fig. 17.2). When Mab L8 was added to the cocultures, an inhibition of the SPC- and RLEC-induced stimulation of these parameters of Sertoli cell function was observed (Fig. 17.2). Whereas an inhibiting effect of Mab L8 was also observed at the level of the transferrin mRNA in cultured Sertoli cells (Fig. 17.3, lane 1 vs. lane 4), this antibody markedly reduced the inhibin α-subunit mRNA levels in SPC- and RLEC-Sertoli cell cocultures (Fig. 17.3, lane 2 vs. lane 5 and lane 3 vs. lane 6, respectively). SDS-PAGE analysis of immunoprecipitates from RLEC and Sertoli cell detergent extracts revealed a double band corresponding to peptides of 85

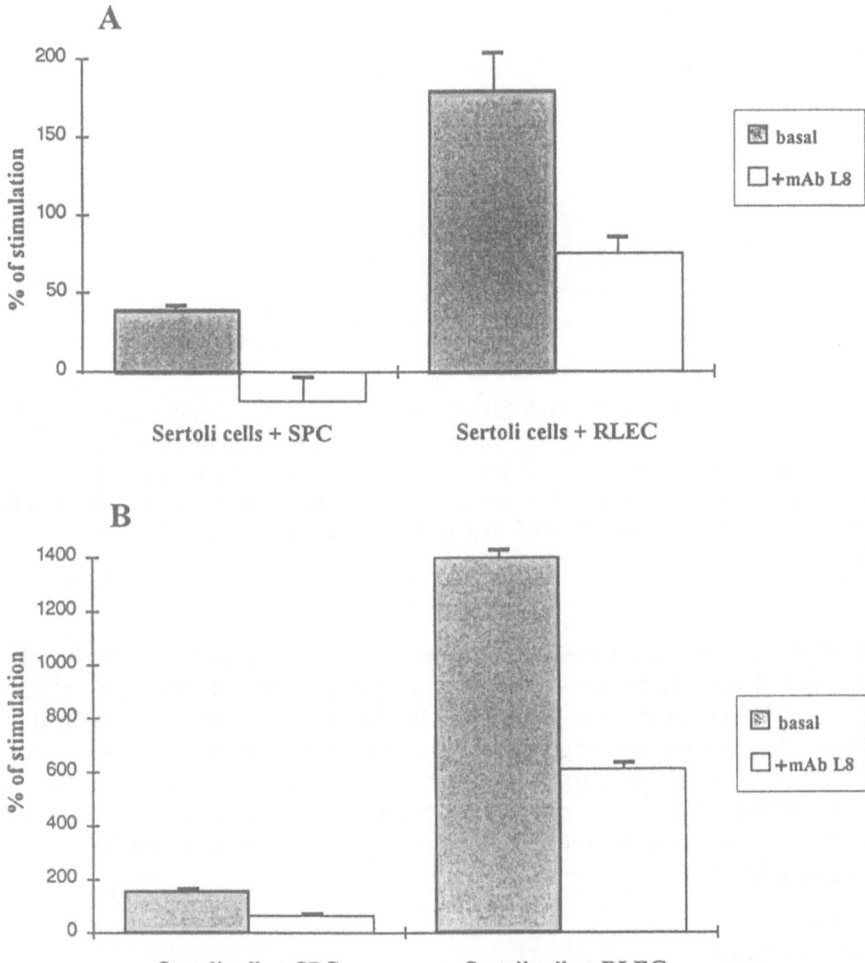

FIGURE 17.2. In vitro RLEC- and SPC-induced production of transferrin (*A*) and inhibin (*B*) secretion by rat Sertoli cells expressed as percent of stimulation of the respective control. Also shown is inhibition by Mab L8.

and 73 kd for both cell types, while a unique specific signal at 56 kd was observed in SPC lysates (not shown).

Conclusion

The results of this study demonstrate the existence of a cell-cell contact-mediated regulation of the Sertoli cell function by primary spermatocytes and provide evidence for the role of a plasma membrane protein antigeni-

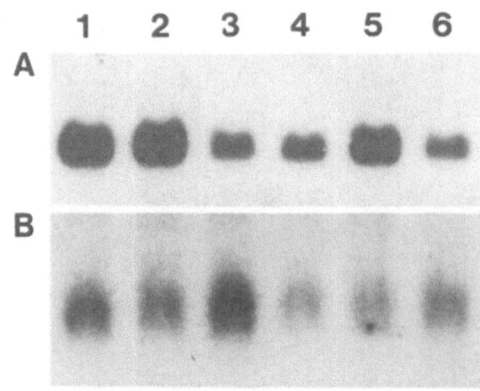

FIGURE 17.3. Inhibition by Mab L8 of transferrin (*A*) and inhibin α-subunit (*B*) mRNA levels in cell cultures by Northern blot analysis. Cultures and cocultures were run either in the absence (lanes 1–3) or in the presence (lanes 4–6) of Mab L8. Lanes 1 and 4 correspond to the Sertoli cells, lanes 2 and 5 to the SPC-Sertoli cell cocultures, and lanes 3 and 6 to RLEC-Sertoli cell cocultures.

cally related to the transmembranous LRP (which plays a role in the functional activity of the mature hepatocytes) in this process. This testicular LRP-like protein is expressed by both spermatocytes and Sertoli cells. Considering its stage-dependent expression, its appearance at the onset of meiosis, and its stimulatory effects on Sertoli cell transferrin and inhibin synthesis in vitro, we postulate that the testicular LRP-like protein plays a crucial role in the cell-contact-dependent control of Sertoli cell activity.

References

1. Fawcett DW. Ultrastructure and function of the Sertoli cell. In: Greep RO, Hamilton DW, eds. Handbook of physiology, male reproduction; vol 5. Bethesda, MD: American Physiology Society, 1975:21–55.
2. Jégou B. The Sertoli cell. In: de Kretser DM, ed. The testes. Baillieres Clin Endocrinol Metab 1992;6:273–311.
3. Skinner MK. Cell-cell interactions in the testis. Endocr Rev 1991;12:45–77.
4. Jégou B, Syed V, Sourdaine P, et al. The dialogue between late spermatids and Sertoli cells in vertebrates: a century of research. In: Nieschlag E, Habenicht UF, eds. Spermatogenesis, fertilization, contraception: molecular, cellular and endocrine events in male reproduction; vol 4. Shering Fondation Workshop, Springer-Verlag, 1992:57–95.
5. Byers S, Jégou B, MacCalman C, Blaschuk OW. Sertoli cell adhesion molecules and the collective organization of the testis. In: Russell LD, Griswold MD, eds. The Sertoli cell. Cache River Press, 1993:461–76.

6. Pineau C, Syed V, Bardin CW, Jégou B, Cheng CY. Identification and partial purification of a germ cell factor that stimulates transferrin secretion by Sertoli cells. Recent Prog Horm Res 1993;48:539–42.
7. Ulvik N. Stage-dependent topographical relationship of spermatogonia and early spermatocytes to Sertoli-Sertoli interspaces in the rat testis. Int J Androl 1983;6:367–74.
8. Kumari M, Duraiswani S. Ultrastructural observations on Sertoli cell-germ cell interaction. Cytologia. 1987;52:111–6.
9. Russell LD. Morphological and functional evidence for Sertoli-germ cell relationships. In: Russell LD, Griswold MD, eds. The Sertoli cell. Cache River Press, 1993:365–91.
10. Gérard N, Syed V, Jégou B. Lipopolysaccharide, latex beads and residual bodies are potent activators of Sertoli cell interleukin-1α production. Biochem Biophys Res Commun 1992:185:154–61.
11. Syed V, Gérard N, Kaipia A, Bardin CW, Parvinen M, Jégou B. Identification, ontogeny and regulation of an interleukin-6-like (IL-6) factor in the rat seminiferous tubule. Endocrinology 1993;132:293–9.
12. Cyr DG, Blaschuk OW, Robaire B. Identification and developmental regulation of cadherin messenger ribonucleic acids in the rat testis. Endocrinology 1992;131:139–45.
13. Newton SC, Millette CF. Sertoli cell plasma membrane involved in spermatogenic cell-Sertoli cell adhesion. J Androl 1992;13:160–71.
14. Guguen-Guillouzo C, Clément B, Baffet G, et al. Maintenance and reversibility of active albumin secretion by adult rat hepatocytes cocultured with another liver epithelial cell type. Exp Cell Res 1983;143:47–54.
15. Fraslin JM, Kneip B, Vaulont S, Glaise D, Munnich A, Guguen-Guillouzo C. Dependence of hepatocyte-specific gene expression on cell-cell interactions in primary culture. EMBO J 1985;110:2487–91.
16. Corlu A, Kneip B, Ladhi C, et al. A plasma membrane protein is involved in cell contact-mediated regulation of tissue-specific genes in adult hepatocyte. J Cell Biol 1991;115:505–15.
17. Le Magueresse B, Pineau C, Guillou F, Jégou B. Influence of germ cells upon transferrin secretion by rat Sertoli cells in vitro. J Endocrinol 1988; 118:R13–6.
18. Pineau C, Sharpe RM, Saunders PTK, Gérard N, Jégou B. Regulation of Sertoli cell inhibin production and of inhibin α-subunit mRNA levels by specific germ cell types. Mol Cell Endocrinol 1990;72:13–22.
19. Mather JP, Phillips DM. Primary culture of testicular somatic cells. In: Barnes D, Sirbasku D, Sato G, eds. Methods in molecular and cell biology; vol 2. Alan R. Liss, 1984:29–45.
20. Galdieri M, Ziparo E, Palombi F, Russo MA, Stefanini M. Pure sertoli cell cultures: a new model for the study of somatic-germ cell interactions. J Androl 1981;5:249–59.
21. Meistrich ML, Longtin J, Brock WA, Grimes SR, Mace ML. Purification of rat spermatogenic cells and preliminary biochemical analysis of these cells. Biol Reprod 1981;25:1065–77.
22. Williams GM, Weisburger EK, Weisburger JS. Isolation and long term cell culture of epithelial like cells from rat liver. Exp Cell Res 1970;69:106–12.

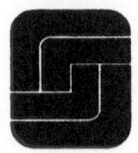

18

5α-Reductase Activity Increased by Oxytocin in the Rat Testis

H.D. Nicholson and L. Jenkin

Oxytocin is produced by the Leydig cells of the rat testis. In vivo chronic administration of the peptide into the testis results in reduced levels of testicular and plasma testosterone, but raised levels of *5α-dihydrotestosterone* (DHT). These experiments were performed to examine whether oxytocin affects the conversion of *testosterone* (T) to DHT in vivo and in vitro.

Adult male rats were treated with oxytocin (0.55 µg/100-g b.w., s.c.) daily for 10 days; control animals received 0.1-mL saline. One hour after the last injection, the animals were killed, and the testes and epididymides were removed and weighed. The 5α-reductase activity was measured in homogenates of these tissues. Oxytocin increased both testicular and epididymal weights. The 5α-reductase activity was significantly increased in the testis ($P < 0.005$) and epididymis ($P < 0.01$) following oxytocin treatment.

In a second experiment the in vitro effects of oxytocin on 5α-reductase activity were measured. Testicular homogenates were incubated with ^{3}H-T and oxytocin (0–50 ng/0.3-mg protein). Oxytocin increased 5α-reductase activity in a dose-related manner at concentrations as low as 10 pg/0.3-mg protein. These experiments provide the first evidence that oxytocin can increase 5α-reductase activity both in vitro and in vivo.

In mammals testicular androgens are necessary for the process of spermatogenesis and the maintenance of epididymal function. Within the testis promotion of spermatogenesis appears to be mediated by T; however, in other tissues, such as the epididymis and prostate, DHT is the active agent (1). Testosterone is converted to DHT by the enzyme 5α-reductase that is localized to the seminiferous tubules of the testis (2) and the principal cells of the rat epididymis (3).

Oxytocin is present in the mammalian testis (4), and in the rat immunoreactive oxytocin has been localized to the Leydig cells (5–7). Recent studies have demonstrated both the presence of oxytocin mRNA in the rat testis (8) and the local production of the peptide by Leydig cells in vitro (9). Two physiological roles for testicular oxytocin have been postulated: (i) as a modulator of seminiferous tubule contractility (10) and (ii) as a paracrine regulator of steroidogenesis.

Oxytocin has been implicated in the regulation of T-production, although the data from in vitro experiments have been controversial. Thus, while Adashi and Hsueh (11) demonstrated that oxytocin decreased T-production from isolated rat Leydig cells and Kwan and Gower (12) found similar effects of oxytocin on rat microsomal fractions, Sharpe et al. (13) found no effect of the peptide on rat Leydig cells. Furthermore, in the mouse oxytocin has been shown to stimulate T-production (14). In vivo, however, chronic administration of oxytocin into the testes of rats resulted in significantly reduced concentrations of T in both the plasma and the testes (15). Surprisingly, however, in these animals spermatogenesis was qualitatively normal, and it was demonstrated that although T-levels were low, concentrations of DHT in the plasma and testes were increased. These data suggest that oxytocin not only affects T-production, but can also modulate the conversion of T to DHT. The following experiments were performed to begin to investigate the effects of oxytocin on 5α-reductase in the testis and epididymis.

Experiment 1

In this experiment the effects of oxytocin were studied in vivo. Groups of 6 adult male Wistar rats (250–300 g) were injected daily for 10 days with oxytocin (0.55 µg/100-g b.w., s.c.); control animals received 0.1-mL saline, s.c. One hour after the last injection, the animals were killed by decapitation, and the trunk blood was collected for hormone measurement. The testes and epididymides were removed and weighed. One testis from each animal was frozen and subsequently extracted for measurement of T and DHT. The 5α-reductase activity was measured in unfractionated homogenates of the remaining testis and the epididymides from each animal according to the method of George et al. (16). Briefly, testicular or epididymal tissue was homogenated in 7 vol of 0.88 M sucrose/1.5 mM $CaCl_2$. Aliquots of the homogenate containing 0.3 mg of protein were incubated in 0.2-mL 0.05 M potassium phosphate buffer at pH 6.5 containing 0.5 mM NADPH with 0.25 µM ^{3}H-T for 1 h at 25°C. The homogenates were extracted with ether, and the tritiated DHT and T were separated using *high-performance liquid chromatography* (HPLC). Chromatography was performed at room temperature on a 250 × 4 mm C18 Techopack column using an isocratic gradient of 40% acetonitrile at a flow rate of 2 mL/min (17).

Plasma concentrations of oxytocin were measured in unextracted plasma using a specific radioimmunoassay (19). Testosterone and DHT were measured in ether extracts of plasma and in testicular tissue extracted

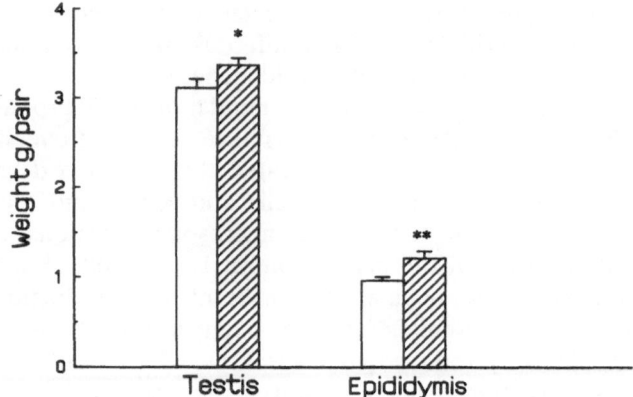

FIGURE 18.1. Effects of oxytocin treatment in vivo on the testicular and epididymal weight of adult rats (values are mean ± SEM; $n = 6$). Open bars represent control animals; hatched bars represent oxytocin-treated animals. (* = $P < 0.05$; ** = $P < 0.001$ [Student's t-test].)

with 70% methanol (18). Testosterone concentrations were measured using a radioimmunoassay (7), and DHT concentrations were measured using a kit supplied by Amersham (TRK 600 Amersham plc, Bucks, UK).

As expected, oxytocin treatment resulted in an elevation of plasma oxytocin levels 1h after the last injection (244 ± 48 pg/mL in oxytocin-treated vs. 7.3 ± 1.9 pg/mL in control animals). Oxytocin increased both testicular and epididymal weights (Fig. 18.1). The 5α-reductase activity was significantly elevated in homogenates of testicular and epididymal tissue (Fig. 18.2), and these results were supported by the presence of raised concentrations of DHT, but not T, in the testes of these animals (Fig. 18.3). Although the plasma levels of both T and DHT appeared raised, these increases were not statistically significant (results not shown).

These data suggest that daily treatment with high doses of oxytocin in vivo can increase 5α-reductase activity in the testis and epididymis of the rat, but give little indication of the site of action of the peptide. Experiments were therefore performed to investigate whether oxytocin has a direct effect on 5α-reductase activity.

Experiment 2

As above, unfractionated homogenates of testicular tissue from adult male rats were prepared (16) and incubated with ^{3}H-T for 1h at 25°C

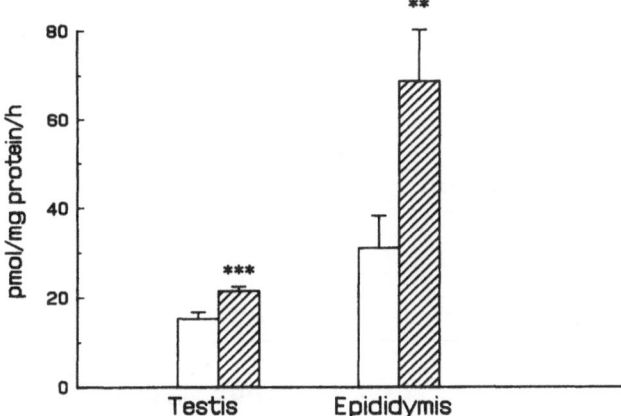

FIGURE 18.2. Effect of oxytocin treatment in vivo on 5α-reductase activity in the rat testis and epididymis. The 5α-reductase activity is expressed as pmol DHT produced/h/mg protein (values are mean ± SEM for 6 animals). Open bars represent control animals; hatched bars represent oxytocin-treated animals. (** = $P < 0.01$; *** $P < 0.005$ [Student's t-test].)

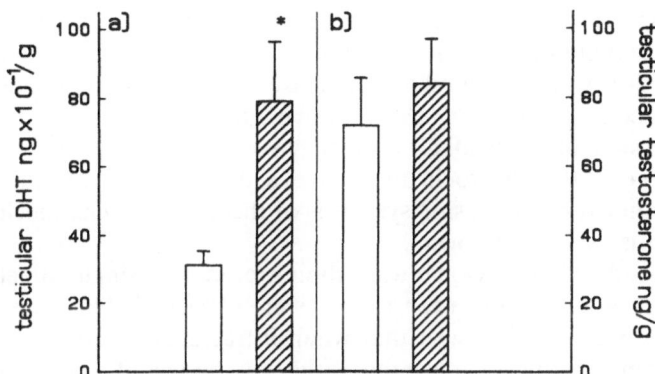

FIGURE 18.3. Effect of oxytocin treatment in vivo on concentrations of testicular DHT (*a*) and testicular T (*b*) (values are mean ± SEM; n = 6). Open bars represent control animals; hatched bars represent oxytocin-treated animals. (* = $P < 0.05$ [Student's t-test].)

with varying concentrations of oxytocin (0–50 ng/0.3-mg protein). The tritiated steroids were separated by HPLC.

Oxytocin increased 5α-reductase activity in a dose-dependent manner (Fig. 18.4). A significant increase in DHT production occurred with concentrations of oxytocin as low as 10 pg/0.3-mg protein ($P < 0.01$).

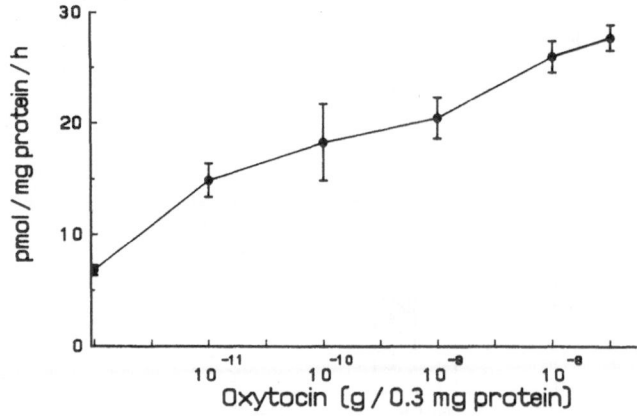

FIGURE 18.4. Dose response of oxytocin on 5α-reductase activity in homogenates of adult rat testis in vitro. The 5α-reductase activity is expressed as pmol DHT produced/h/mg protein.

Discussion

These data provide the first evidence that oxytocin can modulate 5α-reductase activity in the male reproductive tract. DHT has been shown to have a feed-forward effect on 5α-reductase activity, stimulating expression of the 5α-reductase mRNA (16). However, the presence of an increase in the conversion of T to DHT both in vivo and in vitro, at least in the testis, suggests that oxytocin may have a direct action on 5α-reductase activity rather than increasing synthesis of the enzyme. The mechanism of this action is as yet unknown.

In this study when oxytocin was administered by a single injection each day, no alterations in T-concentrations were observed. This contrasts with previous reports that showed that when intratesticular concentrations of oxytocin were chronically elevated, either by the use of a constant-release implant (15) or in the transgenic mouse that overexpresses the bovine oxytocin gene (Ang, Ivell, Nicholson, unpublished observation), plasma and testicular levels of T were significantly reduced. The reasons for this apparent discrepancy are unclear, but may be due to two factors. First, the method of administration may be involved, with prolonged continuously raised levels, rather than intermittent elevations of the peptide, being necessary for inhibition of T-production. Second, the local concentration of oxytocin may be important. In both studies where intratesticular oxytocin was chronically elevated, pharmacological levels of the peptide within the testis were achieved.

While these preliminary findings clearly demonstrate that oxytocin can increase the conversion of T to DHT, the physiological significance of these results is yet to be elucidated. However, the evidence that in vitro 5α-reductase activity is stimulated by low levels of oxytocin similar to those found within the rat testis (10) suggests that oxytocin may play a role in regulating DHT production. Testosterone is thought to be the active androgen involved in promoting spermatogenesis (1), but DHT is necessary in the epididymis for sperm maturation (19). In the epididymis 5α-reductase is most abundant in the ductuli efferentes and the caput region (20). Testosterone alone is unable to maintain levels of this enzyme, but experiments using unilateral orchidectomy have suggested that concentrations of epididymal 5α-reductase may be regulated by a substance entering the epididymis directly from the testis via the lumenal fluid (21). Since the testis contains 5α-reductase activity in the seminiferous tubules (2), testicular oxytocin may act to increase the conversion of T to DHT within the tubules. This DHT may then be secreted into the tubular fluid and transported to the epididymis where it stimulates synthesis of the enzyme.

Another possibility is that oxytocin itself may stimulate epididymal 5α-reductase activity. Oxytocin is present in the rat testis and epididymal fluid of the sheep (22) and in the seminiferous tubular fluid of the rat (Greenfield, Nicholson, unpublished observation). Furthermore, in the sheep oxytocin is actively taken up by the epididymis by a specific receptor-mediated process (23). Endocytosis of oxytocin is highest in the ductuli efferentes and caput epididymis, a regional distribution that is, interestingly, similar to that of maximal 5α-reductase activity. Thus, it is possible that oxytocin might play a role in epididymal function and sperm maturation. Further studies are required to investigate the mechanism of action of oxytocin and the physiological significance of the present results, but these preliminary data may open the way to a greater understanding of the effects of local peptides on testicular and epididymal function.

Acknowledgments. The authors would like to thank Professor B.T. Pickering for his support and encouragement and Mr. T.I. Baker for his technical assistance. This work was supported by the Wellcome Trust.

References

1. Baker HWG, Bailey DJ, Feil PD, Jefferson LS, Santen RJ, Bardin CW. Nuclear accumulation of androgens in perfused rat accessory sex organs and testes. Endocrinology 1977;100:709–21.
2. Folman Y, Ahmad N, Sowell JG, Eik-Nes KB. Formation in vitro of 5 α-dihydrostestosterone by the seminiferous tubules and interstitial tissue from immature and mature rat testes. Endocrinology 1973;92:41–7.
3. Klinefelter GR, Amann RP. Metabolism of testosterone by principal cells and basal cells isolated from the rat epididymal epithelium. Biol Reprod 1980;22:1149–54.

4. Nicholson HD, Swann RW, Burford GD, Wathes DC, Porter DG, Pickering BT. Identification of oxytocin and vasopressin in the testis and in adrenal tissue. Regul Pept 1984;8:141–6.

5. Guldenaar SEF, Pickering BT. Immunocytochemical evidence for the presence of oxytocin in rat testis. Cell Tissue Res 1985;240:485–7.

6. Nicholson HD, Worley RTS, Guldenaar SEF, Pickering BT. Ethan-1,2-dimethanesulphonate reduces testicular oxytocin content and seminiferous tubule movements in the rat. J Endocrinol 1987;112:311–6.

7. Yeung WSB, Guldenaar SEF, Worley RTS, Humphrys J, Pickering BT. Oxytocin in Leydig cells: an immunocytochemical study of Percoll-purified cells from rat testis. Cell Tissue Res 1988;253:463–8.

8. Foo N-C, Carter D, Murphy D, Ivell R. Vasopressin and oxytocin gene expression in rat testis. Endocrinology 1991;128:2118–28.

9. Nicholson HD, Hardy MP. Luteinizing hormone differentially regulates the secretion of testicular oxytocin and testosterone by purified adult rat Leydig cells in vitro. Endocrinology 1992;130:671–7.

10. Pickering BT, Birkett SD, Guldenaar SEF, Nicholson HD, Worley RTS, Yavachev L. Oxytocin in the testis: what, where and why? Ann NY Acad Sci 1989;564:198–209.

11. Adashi EY, Hsueh AJW. Direct inhibition of testicular biosynthesis revealing antigonadal activity of neurohypophysial hormones. Nature 1981;293:650–2.

12. Kwan TK, Gower DB. Inhibition of rat testicular microsomal steroidogenesis by oxytocin and metyrapone. Biochem Int 1988;16:629–37.

13. Sharpe RMS, Cooper I. Comparison of the effects on purified Leydig cells of four hormones (oxytocin, vasopressin, opiates and LHRH) with suggested paracrine roles in the testis. J Endocrinol 1987;113:89–96.

14. Tahri-Joutei A, Pointis G. Developmental changes in arginine vasopressin receptors and testosterone stimulation in Leydig cells. Endocrinology 1989;125:605–11.

15. Nicholson HD, Guldenaar SEF, Boer GJ, Pickering BT. Testicular oxytocin: effects of intratesticular oxytocin in the rat. J Endocrinol 1991;130:231–8.

16. George FW, Russell DW, Wilson JD. Feed-forward control of prostate growth: dihydrostestosterone induces expression of its own biosynthetic enzyme, steroid 5α-reductase. Proc Natl Acad Sci USA 1991;88:8044–7.

17. Cochran RC, Darney KJ, Ewing LL. Measurement of testosterone with high-performance liquid chromatograph equipped with a flow-through ultraviolet spectrophotometer. J Chromatogr 1979;173:349–55.

18. Pierantoni R, Iela L, d'Istria M, Fasano S, Rastogi RK, Delrio G. Seasonal testosterone profile and testicular responsiveness to pituitary factors and gonadotrophin releasing hormone during two different phases of the sexual cycle of the frog (*Rana esculenuta*). J Endocrinol 1984;102:387–92.

19. Orgebin-Crist MC, Danzo BJ, Davies J. Endocrine control of the development and maintenance of sperm fertilizing ability in the epididymis. In: Greep RO, Astwood EB, eds. Handbook of physiology: endocrinology. Washington, DC: American Physiological Society, 1975:319–38.

20. Roselli CE, West NB, Brenner RM. Androgen receptor and 5α-reductase activity in the ductuli efferentes and epididymis of adult rhesus monkeys. Biol Reprod 1991;44:739–45.

21. Robaire B, Zirkin BR. Hypophysectomy and simultaneous testosterone replacement: effects on male rat reproductive tract and epididymal Δ4-5 α-reductase and 3β-hydroxysteroid dehydrogenase. Endocrinology 1981;109: 1225–33.
22. Knickerbocker JJ, Sawyer HR, Amann RP, Tekpetey FR, Niswender GD. Evidence for the presence of oxytocin in the ovine epididymis. Biol Reprod 1988;39:391–7.
23. Veeramachaneni DNR, Amann RP. Oxytocin in the ovine ductuli efferentes and caput epididymis: immunolocalization and endocytosis from the luminal fluid. Endocrinology 1990;126:1156–64.

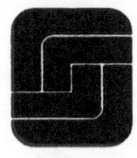

19

Induction by LH/hCG of an Increase in the Rate of Pig Leydig Cell Receptor mRNA Degradation

F. Chuzel, H. Schteingart, O. Avallet, M. Vigier, and J.M. Saez

Luteinizing hormone (LH) is absolutely required for the maintenance of Leydig cell-specific functions, and this hormone is the main factor that under physiological conditions controls Leydig cell *testosterone* (T) secretion. It is clear that the ability of LH or *human chorionic gonadotropin* (hCG), which shares structural, biological, and receptor binding properties with LH (1), to stimulate Leydig cells depends not only on the levels of these hormones, but also on the number and functional activity of the *LH/hCG receptor* (LH/hCG-R). The interaction of LH/hCG with its receptors initiates a sequence of events at the membrane level, followed by the induction of a number of intracellular events, including activation of adenylate cyclase and cAMP-dependent protein kinase, which is believed to be responsible for most of the effects of the hormone. After the initial stimulation of target cell function, many in vivo and in vitro studies have shown that treatment with LH/hCG also decreases the number of its binding sites and causes desensitization (reviewed in 2, 3). However, the molecular bases of this regulation are poorly understood.

The recent cloning of LH/hCG-R cDNAs from pig, rat, and human (4–6) has allowed investigation into the regulation of LH/hCG-R mRNA under conditions in which the receptor number is known to be regulated (7–14). The results from these studies have shown that in most cases there is a correlation between the number of binding sites and the levels of LH/hCG-R mRNA, but the mechanism by which the levels of mRNA are regulated with respect to changes in transcription and/or stability were not investigated. In the present study, we have investigated hCG regulation of LH/hCG-R number and mRNA levels in immature pig Leydig cells cultured in chemically defined media.

Materials and Methods

Leydig cells were isolated, purified, and cultured (15, 16). All the treatments were initiated on the third day of culture. At the end of the experimental period, the medium was removed, the cells were washed with acidic buffer to remove the bound hormone, and the number of binding sites was determined using ^{125}I-hCG (spec. act. $80-100\,\mu Ci/\mu g$) (16). The total RNA was isolated from cells by the method of Chomczynski and Sacchi (17). Then, samples were subjected to Northern blot analysis. Nylon blots were hybridized using ^{32}P-labeled LH/hCG-R cDNA and were analyzed by scanning densitometry. Data were corrected for 28S ribosomal RNA content. Northerns were stripped and reprobed with the ^{32}P-labeled cDNA for *glyceraldehyde-3-phosphatase dehydrogenase* (GAPDH) to verify that effects on the LH/hCG-R mRNA were specific and did not involve all species of mRNA.

Results and Discussion

The treatment of Leydig cells with hCG causes a dose- (16) and time-dependent (Fig. 19.1) down-regulation of its own receptors after a lag period of 6–8 h, even though no lag period exists for the internalization of the hormone-receptor complex (16). Thereafter, the number of binding sites declines to about 20% of control at 36 h and remains at this low level for at least 86 h. Northern blot analysis of pig Leydig RNA reveals 4 major transcripts of 6.7, 4.7, 4.0, and 1.4 kb and 2 minor ones of 7.6 and 2.6 kb (4) (Fig. 19.2), the quantity of which dramatically decreases after a 24-h treatment with 10^{-9} M hCG (Fig. 19.2). The effects of hCG are dose dependent with an ID_{50} of about 10^{-10} M hCG, similar to that required to induce the disappearance of hCG binding sites (data not shown).

Next, we investigated the time course effects of hCG on LH/hCG-R mRNA (Fig. 19.1). Treatment with hCG caused a reduction in the levels of all mRNA transcripts after a lag period of 6–9 h, and by 12 h over 80% of mRNA levels was lost. This decrease similarly affected all the transcripts except that of 1.4 kb, which decreased less than the other transcripts. This rapid, hCG-induced effect appears to be specific for LH/hCG-R mRNA because GAPDH mRNA remains fairly constant (data not shown). This rapid effect of LH/hCG-R mRNA has been previously observed in vivo in both the rat testis and ovary (9–12, 14, 18), but was less marked in the MA-10 Leydig tumor cell line (7).

The decline of LH/hCG-R mRNA levels between 8 and 12 h could be due to either a decreased rate of transcription and/or an increase in the rate of degradation. When pig Leydig cells were incubated in the presence of actinomycin D at concentrations that blocked more than 90% of the RNA synthesis, the half-life of LH/hCG-R mRNA was greater than 18 h

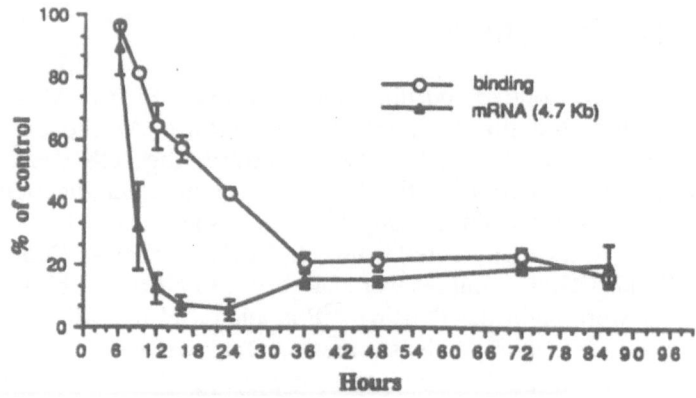

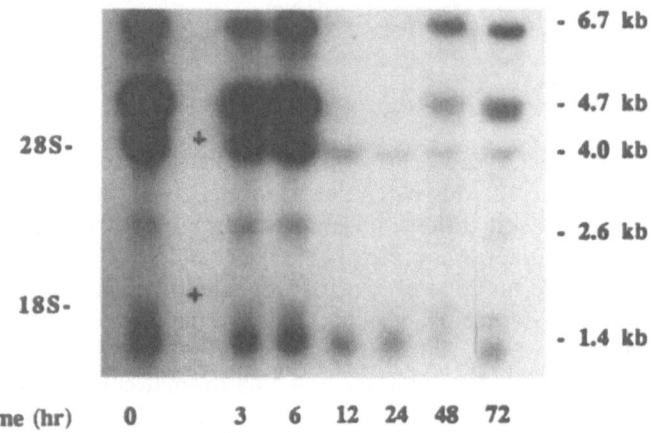

FIGURE 19.1. Time course effects of hCG on both LH/hCG-R number and mRNA levels in pig Leydig cells. Cells were incubated with 10^{-9} M hCG, and at the time indicated the ^{125}I-hCG binding and the LH/hCG mRNA levels were determined as indicated in "Materials and Methods." At the top is shown the mean ± SEM of 3–5 experiments. Only the 4.7-kb transcript is represented. At the bottom is a representative Northern blot.

(data not shown). Therefore, we studied the effects of actinomycin D in combination with hCG on LH/hCG-R number and mRNA. The antibiotic blocked the inhibitory effects of hCG on both binding site number (Table 19.1) and receptor mRNA levels (Fig. 19.3). These results suggest that hCG increased the rate of degradation of its receptor mRNA rather than decreasing the rate of transcription of the LH/hCG-R gene.

These results are similar to those recently reported in the female rat (18) where the hCG-induced down-regulation of LH/hCG-R mRNA in

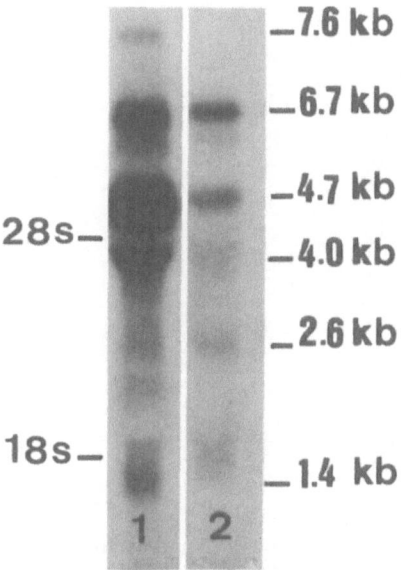

FIGURE 19.2. LH/hCG mRNA transcript in control pig Leydig cells (lane 1) and after 24 h of treatment with 10^{-9} M hCG (lane 2).

TABLE 19.1. Effect of actinomycin D and cycloheximide on hCG-induced receptor down-regulation.

	^{125}I-hCG bound (% of control)		
	Control	Actinomycin D	Cycloheximide
Basal	100	99 ± 9	91 ± 4
hCG (10^{-9} M)	64 ± 2[a]	85 ± 4	91 ± 4

Note: Pig Leydig cells were incubated for 12 h with or without
hCG in the absence or presence of actinomycin D (2 μg/mL)
or cycloheximide (2 μg/mL). After acidic washing the binding
of ^{125}I-hCG was determined.
[a] $P < 0.05$ as compared to basal control.

the ovary is associated with a decrease in the half-life of the message and
not with changes in the rate of transcription. In contrast, in the MA-10
Leydig tumor cell line, the decrease in LH/hCG-R mRNA caused by
treatment with epidermal growth factor, phorbol esters, and cAMP
derivatives is a result of a decrease in transcription rate (19). Because the
decrease of LH/hCG-R binding sites (Table 19.1) and mRNA (Fig. 19.3)
in pig Leydig cells can be blocked by cycloheximide, it is likely that
synthesis of a labile protein is required for this rapid loss of mRNA.

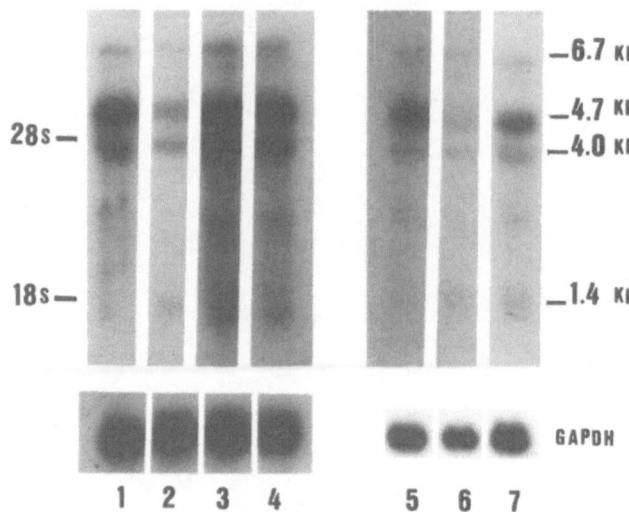

FIGURE 19.3. Effects of actinomycin D and cycloheximide on hCG-induced down-regulation of LH/hCG-R mRNA. Cells were incubated for 12h with or without 10^{-9}M hCG, actinomycin D (2 µg/mL), or cycloheximide (2 µg/mL). (Lane 1 = control; lane 2 = hCG; lane 3 = actinomycin D; lane 4 = hCG + actinomycin; lane 5 = control; lane 6 = hCG; lane 7 = hCG + cycloheximide.)

Summary

Our results strongly suggest that in pig Leydig cells the main mechanism by which hCG reduces its own receptor mRNA is by increasing the rate of degradation of the message. Our results also suggest that hCG-induced down-regulation of LH/hCG mRNA is the main mechanism responsible for the loss of hCG binding sites. Taking into consideration the existence of a lag period and the fact that both actinomycin D and cycloheximide block the effects of hCG, we postulated that hCG induces the transcription of one or several genes encoding the synthesis of protein(s) that specifically decrease LH/hCG-R mRNA.

Acknowledgments. The authors thank Drs. E. Milgrom and P. Fort for the gifts of the cDNA probes for LH/hCG-R and GAPDH, respectively; Dr. A. Clark for reviewing the English manuscript; and J. Bois for secretarial assistance. This work was supported by la Fondation pour la Recherche Médicale Française. Predoctoral fellow F. Chuzel is supported by Ministère de la Recherche et de la Technologie. Postdoctoral fellow H. Schteingart was supported by fellowships from INSERM and the European Economic Community.

References

1. Combarnous Y. Molecular basis of the specificity of binding of glycoproteic hormones to their receptor. Endocr Rev 1992;13:670–91.
2. Saez JM, Benahmed M. Testicular receptor for gonadotropins, prolactin and LHRH. In: Posner BI, ed. Polypeptide hormone receptors. New York: Marcel Dekker, 1985:507–51.
3. Ascoli M. Regulation of luteinizing hormone receptor and action. In: Ascoli M, ed. Luteinizing hormone action and receptor. Boca Raton, FL: CRC Press, 1985:199–217.
4. Loosfelt H, Misrahi M, Atger M, et al. Cloning and sequencing of porcine LH-hCG receptor cDNA; variants lacking transmembrane domain. Science 1989;245:525–8.
5. McFarland KC, Sprengel R, Phillips HS, et al. Lutropin-choriogonadotropin receptor; an unusual member of the G protein coupled receptor family. Science 1989;245:494–9.
6. Minegishi T, Nakamura K, Takakura Y, et al. Cloning and sequencing of human LH/hCG receptor cDNA. Biochem Biophys Res Commun 1990;172: 1049–54.
7. Wang H, Segaloff DL, Ascoli M. Lutropin/choriogonadotropin down regulates its receptor by both receptor mediated endocytosis and a cAMP-dependent reduction in receptor mRNA. J Biol Chem 1991;226:780–5.
8. Wang H, Segaloff DL, Ascoli M. Epidermal growth factor and phorbol esters reduce the levels of the cognate mRNA for the LH/CG receptor. Endocrinology 1991;128:2651–3.
9. Hoffman YM, Peegel H, Sprock MJE, Zhang Q-Y, Menon KMJ. Evidence that human chorionic gonadotropin/luteinizing hormone receptor down-regulation involves decreased levels of receptor messenger ribonucleic acid. Endocrinology 1991;128:388–93.
10. Segaloff DL, Wang H, Richards JS. Hormonal regulation of luteinizing hormone/chorionic gonadotropin receptor mRNA in rat ovarian cells during follicular development and luteinization. Mol Endocrinol 1990;4:1856–65.
11. Piquette GN, LaPolt PS, Oikawa M, Hsueh AJW. Regulation of luteinizing hormone receptor messenger ribonucleic acid levels by gonadotropins, growth factors, and gonadotropin releasing hormone in cultured rat granulosa cells. Endocrinology 1991;128:2449–56.
12. Camp TA, Rahal JO, Mayo KE. Cellular localization and hormonal regulation of follicle-stimulating hormone and luteinizing hormone receptor messenger RNAs in the rat ovary. Mol Endocrinol 1991;5:1405–17.
13. Hu ZZ, Tsai-Morris CH, Buckzo E, Dufau ML. Hormonal regulation of LH receptor mRNA and expression in the rat ovary. FEBS Lett 1990;274:181–4.
14. LaPolt PS, Jia XC, Sincich C, Hsueh AJW. Ligand-induced down-regulation of testicular and ovarian luteinizing hormone (LH) receptor is preceded by tissue-specific inhibition of alternatively processed LH receptor transcripts. Mol Endocrinol 1991;5:397–403.
15. Bernier M, Chatelain P, Mather JP, Saez JM. Regulation of gonadotropin receptor, gonadotropin responsiveness and cell multiplication by somatomedin-C and insulin in cultured pig Leydig cells. J Cell Physiol 1986;129:257–63.

16. Bernier M, Clerget M, Mombrial CF, Saez JM. Processing of human chorio-gonadotropin and its receptors by cultured pig Leydig cells. Eur J Biochem 1986;155:323–30.
17. Chomczynski P, Sacchi U. Single-step method of RNA isolation by acid guanidium thiocyanate-phenol-chloroform extraction. Anal Biochem 1987; 162:156–9.
18. Lu DL, Peegel H, Mosier SM, Menon KMJ. Loss of lutropin (human cho-riogonadotropin) receptor messenger ribonucleic acid during ligand-induced down-regulation occurs post-transcriptionally. Endocrinology 1993; 132:235–40.
19. Nelson S, Ascoli M. Epidermal growth factor, a phorbol ester, and 3′,5′-cyclic adenosine monophosphate decrease the transcription of the luteinizing hormone/chorionic gonadotropin receptor gene in MA-10 Leydig tumor cells. Endocrinology 1992;130:615–20.

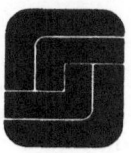

20

Genetics and Luteinizing Hormone Receptors

Armando G. Amador, Artur Mayerhofer, and Andrzej Bartke

Testicular *luteinizing hormone receptors* (LH-Rs) that bind either LH or *human chorionic gonadotropin* (hCG) are required for the differentiation of the Leydig cell (1). This is indicated by the correlation between the onset of Leydig cell differentiation and the appearance of hCG during the development of the human fetus. The number of Leydig cells also correlates with the levels of hCG during pregnancy. Leydig cells, and therefore LH-Rs, are present by week 8 of pregnancy in the human (2–4), and by day 15 in rodents (5).

The gene coding for LH-R is named *LHCGR* and is located on human chromosome 2 at 2p21 (6). For a comparative mapping of the LH-R gene, see Table 20.1. The LH-R belongs to the pituitary glycoprotein hormone receptor family that in turn belongs to the G-protein-coupled receptor superfamily. This latter group includes the receptor families for

TABLE 20.1. Comparative gene mapping of the LH-R gene (*LHCGR*).

Species	Chromosome
Human	2p21
Orangutan	11
Capuchin monkey	4
Owl monkey	2
Mouse lemur	4
House mouse	12A
Norway rat	6
Domestic pig	3
Fox	16

Note: Except for humans and porcine, as described in references 6 and 7, other locations are probable chromosomal locations inferred by conservation of certain chromosomal regions across species.

dopamine, acetylcholine, tachykinin, and adenosine, as well as the can-nabinoid receptor and rhodopsin (8). LH-R is a glycosylated protein having a molecular weight of 85,000–100,000. It has a large 340-amino acid long extracellular domain, 7 hydrophobic α-helices as transmembrane segments, and a shorter intracellular domain (9).

When animals are injected with hCG (or LH), dose-, time-, age-, and genome-dependent changes are observed in the levels of LH-R. The different types of dose-dependent change patterns are given in Figure 20.1. References to these patterns are made in the text below.

The present review is not intended to be an exhaustive analysis of all data in the literature. A review on the ontogeny of LH-R can be found in Chapter 12 of this volume.

Mutant House Mice

Dwarf Mice

The first correlation between a genetically determined condition and altered LH-R levels was reported in *Snell dwarf* (dw/dw) mice (10). It was subsequently demonstrated that whereas the levels of testicular LH-R were reduced in dw/dw mice when compared to their normal siblings, the metabolism of LH-R was normal in this type of dwarf mice (11). Both normal and dw/dw mice had type A LH-R autoregulation patterns. The reduction in LH-R levels was attributed to the absence of circulating PRL and GH in these animals. Deficiency in these hypohyseal hormones is a result of a congenital absence of lactotrophs, somatotrophs, and thyro-trophs due to a mutation in the Pit-1 gene (12–14).

However, when LH-R was studied in *Ames dwarf* (df/df) mice that also lack circulating PRL, GH, and TSH, but in which this is the result of an inherited deficiency that affects the expression of the Pit-1 gene (14) and, consequently, the synthesis of the hormones themselves, different results were observed. The basal testicular levels of LH-R were normal in df/df mice when compared to their normal littermates, and the metabolism of LH-R was regulated differently (15). Whereas in df/df mice low doses of hCG induced positive autoregulation (up-regulation) of LH-R and high doses did not affect receptor levels, in their normal littermates low doses of hCG did not change LH-R levels and high doses caused negative autoregulation (down-regulation). Thus, the result of carrying the df/df genotype was a shift to the right in the LH-R metabolism sensitivity to autoregulation. The df/df mice had a type B LH-R autoregulation pattern, while their normal littermates had a type D pattern. Therefore, it became apparent that a pleiotropic effect of the df locus on the metabolism of testicular LH-R existed and that the altered regulation of LH-R in Ames dwarf mice was capable of counteracting the deleterious effects of the lack of PRL, GH, and TSH stimulation of the Leydig cell.

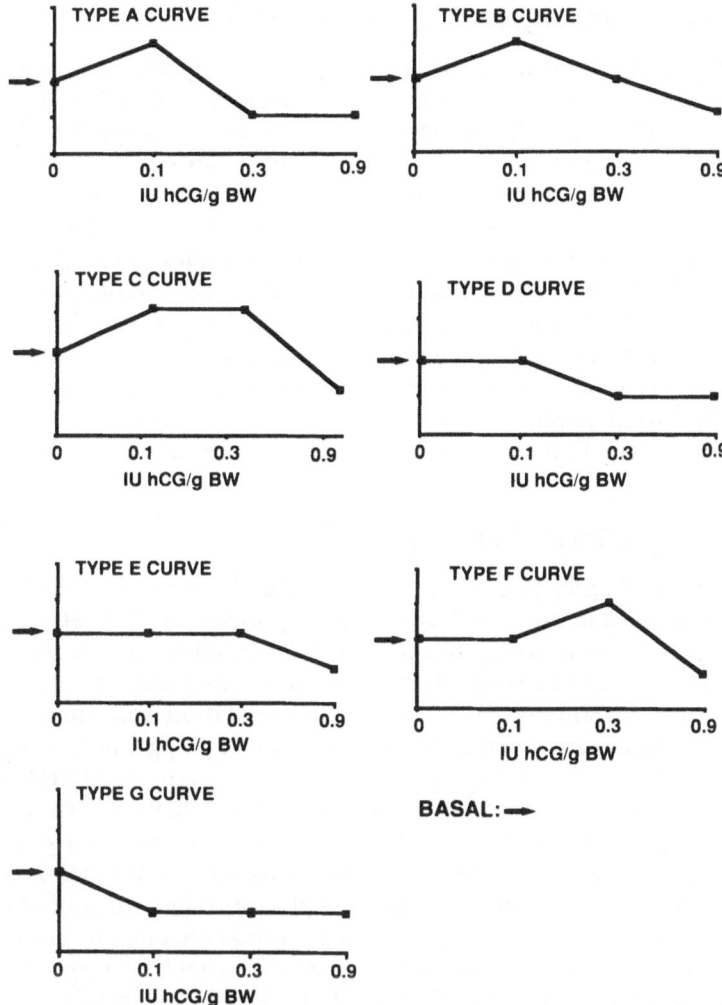

FIGURE 20.1. Patterns of LH-R autoregulation derived from measurements of testicular hCG binding 24 h after administration of various doses of hCG. To facilitate presentation of data in this review chapter, the various patterns of responses were given letter designations (type A curve, etc.). Thus, the patterns of autoregulation of LH-R identified in the text as A, B, C, etc., refer to responses obtained under the conditions of our experiments. Please also note that in animals given the highest dose of hCG (0.9 IU/g body weight), occupancy of LH-R is assumed to be responsible for much of the observed decrease in measured hCG binding, as described in references 18 and 24.

Mice with Congenital Hypothyroidism

Hypothyroid (hyt/hyt) mice suffer from primary hypothyroidism due to a thyroid resistance to TSH, with the consequent deficiency in thyroid hormones and very high circulating TSH levels. Although hyt/hyt mice have normal basal testicular LH-R levels, they did not respond to hCG with the negative autoregulation observed in their normal littermates (16). The data suggest that hyt/hyt mice have a type B or type E LH-R autoregulation pattern, while their normal littermates have either a type A, D, or G LH-R autoregulation pattern. Since the circulating levels of LH, FSH, and PRL were found to be normal in hyt/hyt mice, it was proposed that hypertyrotropinemia might be responsible for a stimulation of the synthesis of testicular LH-R. In *congenital goiter* (cog/cog) mice that suffer from hypothyroidism due to deficient thyroglobulin metabolism, the basal LH-R levels were found to be normal (17).

Mice with Inherited Hypogonadism

Hypogonadal (hpg/hpg) mice are characterized by the presence of tertiary hypogonadism due to a deficiency in hypothalamic GnRH, with the consequent low gonadotropin and gonadal steroid levels. In these mice basal testicular LH-R concentrations were dramatically higher than in their normal littermates (18). When autoregulation was studied, hCG induced negative autoregulation of LH-R in hpg/hpg mice, but not in their normal littermates. The hpg/hpg mice have a type G LH-R autoregulation pattern, whereas their normal littermates have a type E LH-R autoregulation pattern. It was also observed that the circulating PRL levels were significantly higher in hpg/hpg mice than in normal mice (18). Because PRL has been shown to be a trophic hormone for rodent Leydig cells (19, 20), it is postulated that the hyperprolactinemia present in these mice, together with low LH and FSH levels, would be responsible for altering the response of testicular metabolism of LH-R to hCG.

Contrary to what is observed in humans, mice with *testicular feminization* (Tfm/y) suffer from primary hypogonadism, with the consequent low *testosterone* (T) levels and very high circulating LH, FSH, and PRL levels. Basal testicular LH-R concentrations are much higher in Tfm/y mice than normal mice, but the pattern of LH-R autoregulation is similar in both types of mice, both having a type A, D, or G LH-R autoregulation pattern (16). In these mice, contrary to what was observed in hpg/hpg mice, hyperprolactinemia is associated with high LH and FSH levels. Thus, in Tfm/y mice this endocrine environment would increase basal synthesis of LH-R, but not its response to hCG.

Mice with *juvenile spermatogonial depletion* (jsd/jsd) have an abnormal intratubular environment that is unable to support continued germ cell differentiation in adult males (21, 22). In these mice the concentration of testicular LH-R was found to be higher than in normal mice (23).

Perhaps the most interesting findings on the genetic regulation of LH-R have come from studies in mice with mutations of the *stem cell factor* (SCF)-*stem cell factor receptor* (SCF-R) axis. SCF belongs to the platelet-derived growth factor family of cytokines, and SCF-R belongs to the platelet-derived growth factor receptor family of tyrosine kinase receptors. In mice SCF is coded for by the *steel* (*Sl*) locus, and its receptor SCF-R is coded for by the dominant *white* (*W*) locus. Both loci are prone to frequent mutagenesis; thus, a wide variety of mutations have been described for each locus.

When W^x/W^v mice were compared to normal (+/+) mice and heterozygotes ($W^x/+$ and $W^v/+$), basal LH-R levels were found to vary according to the following hierarchy: $W^x/W^v > W^x/+ > W^v/+ = +/+$. Also, whereas hCG induced negative autoregulation in +/+ and $W^v/+$ mice, in W^x/W^v and $W^x/+$ animals, no changes in LH-R levels were observed after treatment with hCG (24). Thus, a type E LH-R autoregulation pattern was observed in W^x/W^v mice versus a type D pattern in +/+ mice. These findings indicated that the W locus regulated LH-R metabolism, at least in mice. When similar studies were conducted in mice with mutations at the *Sl* locus, the data indicated that this locus was also able to regulate LH-R levels. Basal LH-R levels had the following hierarchy: $Sl/Sl^d > Sl^d/+ > Sl/+ = +/+$ (25). Thus, it becomes obvious that the SCF/SCF-R axis plays a major role in the regulation of LH-R metabolism.

Mice with Inherited Diabetes Mellitus

Because *diabetes mellitus* (DM) comprises a heterogeneous group of diseases, many animal models for this disease have been described. Also, one of the least-studied aspects of DM is the multiple reproductive and endocrine alterations that are present. In recent years some data became available regarding Leydig cell function in different types of mice with DM.

Two mutations, db/db and ob/ob, at different loci cause similar presentation of *non-insulin-dependent DM* (NIDDM). If the mutations are present in animals from a C57BL/Ks background, the NIDDM will develop in prepurbertal animals, be very severe, and be accompanied by irreversible obesity. If the mutations are present in mice from a C57BL/6 background, NIDDM will be present in young adult animals and be severe, but the obesity may be reversed with a strict diet. However, most studies have used db/db mice on a C57BL/Ks background and ob/ob mice on a C57BL/6 background, and they are referred to as *diabetic* and *obese* mice, respectively.

No differences in basal testicular LH-R levels were observed between db/db and ob/ob mice and their respective normal littermates (26). However, low doses of hCG induced a significantly smaller positive autoregulation of LH-R in db/db mice than in normal mice, and higher doses of hCG did not produce negative autoregulation of LH-R in db/db animals, while that was the case in their normal siblings. In ob/ob mice low doses of hCG did not produce positive autoregulation of LH-R as in normal mice. However, higher doses of hCG did induce negative autoregulation in both ob/ob mice and their normal littermates.

The previous data, like most of the studies on the regulation of the LH-R metabolism, involved the measurement of receptors 24 h after administration of hCG. When the time course of these changes was studied in db/db and normal mice, the higher doses of hCG produced negative autoregulation of LH-R that was still present 72 h after administration of the hormone in normal mice; whereas in db/db mice no changes were observed at 24 h, while at 72 h negative autoregulation of LH-R was also present in the diabetic animals (26). Thus, the normal littermates for both types of mice have a type A LH-R autoregulation pattern, while db/db mice have a type B LH-R pattern, and ob/ob mice have a type D LH-R pattern.

Some of the known mutations of the *a* locus will also cause NIDDM, albeit of lesser severity. Mutant (A^y/a and A^{vy}/a) *yellow and viable yellow* mice present with minimum or mild NIDDM in adult animals and slowly progressing obesity. As with mutants discussed above, the genetic background plays an important role in the presentation of the diabetic syndrome. The C57BL/6 background supports a very mild presentation, the AY background supports a mild presentation, and the YS background supports a more severe presentation (but less severe than that of ob/ob mice on the C57BL/6 background). Basal testicular LH-R levels and their metabolism were found to be normal in A^y/a mice on a C57BL/6 background (26). Both A^y/a mice and their normal littermates had a type A LH-R autoregulation pattern. In A^{vy}/a mice on a YS background, although basal LH-R levels were normal, their metabolism was different. Lower doses of hCG induced positive autoregulation in A^{vy}/a mice and their normal littermates, but higher doses of hCG caused negative autoregulation only in normal mice and not in A^{vy}/a animals (27). Thus, the A^{vy}/a mice had a type B LH-R autoregulation pattern, and their normal littermates had a type A LH-R pattern.

When the time course of these changes was studied in A^{vy}/a mice on a YS background, the higher doses of hCG produced negative autoregulation of LH-R in normal mice that was still present 72 h after administration of hCG. However, no changes were observed in these A^{vy}/a mice at 24 h, while at 72 h negative autoregulation of LH-R was also present (27). In A^{vy}/a mice on a VY background, both the mutants and the normal littermates had a type A LH-R autoregulation pattern. However, the low hCG dose-induced elevation in LH-R was significantly less pronounced in A^{vy}/a mice than in their normal siblings. After 72 h negative autoregulation of LH-R was still present in A^{vy}/a mice and in their normal littermates, albeit LH-R levels were reduced to a lesser extent than after 24 h (28).

Alterations in LH-R levels are not restricted to mice with NIDDM. Indeed, NOD mice that suffer from *insulin-dependent DM* (IDDM) after 6 months of age have significantly lower levels of LH-R than normal ICR mice. The interesting fact is that low levels of LH-R are detected before the diabetic syndrome becomes manifest. Furthermore, NOD mice have a type E LH-R autoregulation pattern versus the type A pattern observed for their ICR controls (Amador, unpublished). In brief, hCG administration was unable to alter LH-R metabolism in NOD mice. Whereas in normal ICR mice low doses of hCG increased LH-R levels and higher doses decreased them, in NOD mice no changes were observed.

All the data available in diabetic mice tend to indicate that DM is associated with resistance to LH/hCG in addition to the resistance to insulin. It could be that DM causes tissues to be resistant to the action of major trophic hormones. Since insulin is trophic to a large variety of tissues, the resistance to it is observed throughout the body. However, resistance to such hormones as LH/hCG would only be detected in the specific tissues they affect.

Other Mutant Mice

Two types of mutant mice are used to study the effects of deficient *growth hormone* (GH) action. They are lit/lit mice that are deficient in GH and pg/pg mice that have a resistance to somatomedins. Because of the findings in dwarf mice, it was thought that the study of LH-R metabolism and levels in these animals might clarify the role of GH in said functions. However, basal testicular LH-R levels, as well as their pattern of autoregulation, were normal in both pg/pg and lit/lit mice when compared to their normal littermates (15–16). Thus, lit/lit mice and their normal littermates had a type A, D, or G LH-R autoregulation pattern, and pg/pg mice and their littermates had a type B pattern. These results appear to indicate that deficient somatotropic action does not have a major effect on LH-R levels or their metabolism.

Lethargic (lh/lh) mice suffer, among other alterations, from stunted growth and altered reproductive function. When testicular LH-R was studied in these mice, basal receptor levels were found to be normal, and the pattern of the response to autoregulation was similar in lh/lh mice and their normal littermates (29). Both types of mice had a type A LH-R autoregulation pattern. However, low doses of hCG produced a significantly smaller positive autoregulation of LH-R in lh/lh mice than in normal mice. Since circulating LH, FSH, and PRL levels are normal in lh/lh mice, this difference in response to a low dose of hCG is thought to be due to the poor general health of these animals, rather than a specific effect of the lh mutation on LH-R (29).

Dystrophic (dy/dy and dy^{2J}/dy^{2J}) mice have severe alterations of their hypothalamic-pituitary-testicular axis (30, 31). The efficiency of the transduction of the gonadotropic signal was found to be reduced in dy/dy, but not in dy^{2J}/dy^{2J}, mice. However, no significant differences in LH-R levels were observed between dystrophic mice and their normal littermates (31).

Transgenic House Mice

The advent of transgenic mice has opened a new realm of possibilities for studies in biomedicine. Mice carrying different gene constructs, including various growth hormones, have allowed the study of the effects of hyper-somatotropinemia and hyperlactotropinemia on endocrine function.

In mice carrying a *mouse metallothionein I/human growth hormone I fusion gene* (TghGH1), hGH excess is expressed in utero and during postnatal life, and in mice this is equivalent to hypersomatotropinemia and hyperprolactinemia. When testicular LH-R was studied in TghGH1, it was observed that basal levels were similar in TghGH1 mice and their normal littermates (32). However, although the lower doses of hCG produced positive autoregulation of testicular LH-R in both types of mice, a higher dose induced negative autoregulation only in normal mice, with no change in LH-R levels versus basal being observed in TghGH1 mice. Thus, TghGH1 mice had a type B LH-R autoregulation pattern, and normal mice, a type A pattern. Since hyperprolactinemia causes a reduction in testicular LH-R in mice (33), it is proposed that the stimu-latory effects of hypersomatotropinemia counteract the inhibitory effects of hyperlactotropinemia on LH-R metabolism. The somatotropic action is probably mediated by IGF-I, which is known to stimulate Leydig cell function, including the levels of LH-R (34, 35).

In mice carrying a *mouse metallothionein I/bovine growth hormone I fusion gene* (TgbGH1), the bGH excess is also expressed beginning in utero, but the effects of high bGH in mice are equivalent to isolated hypersomatotropinemia. In these mice basal testicular LH-R levels are similar to those of their normal littermates. However, whereas lower doses of hCG caused a positive autoregulation of testicular LH-R in both types of mice, higher doses produced positive autoregulation in TgbGH1 mice and negative autoregulation in their normal littermates (36). Thus, TgbGH1 mice had a type C LH-R autoregulation pattern, and normal mice, a type A pattern. Therefore, it can be concluded that as was suspected from data obtained in TghGH1 mice, the hypersomatotropinemia in TgbGH1 mice stimulates LH-R metabolism. Apparently, GH can stimulate, probably through IGF-I mediation, the synthesis of LH-R.

In mice carrying a *mouse metallothionein I/human growth hormone II* (placental GH variant) *fusion gene* (TghGH2), the hGH has both somatotropic and lactotropic activities, albeit with different combined characteristics than GH-I. This was reflected in the results observed for LH-R in TghGH2 mice. Basal LH-R levels were significantly lower in TghGH2 mice than in their normal littermates. However, when hCG was injected, it produced a positive autoregulation of LH-R in TghGH2 mice and a negative autoregulation in their normal littermates. Thus, TghGH2 mice had a type C or F LH-R autoregulation pattern, and normal mice had a type A, D, or G pattern (37).

Furthermore, as stated above, this same dose of hCG did not change the levels of LH-R in TghGH1 mice. While this tends to confirm that GH stimulates LH-R metabolism, probably through the action of IGF-I, it also underlines the different action of GH-II versus that of GH-I. Since it is thought that GH-II has a relatively smaller lactogenic potency than GH-I (38), the results obtained in TghGH2 mice might be the result of hypersomatotropinemia in the presence of a hyperprolactinemia of lesser severity. Furthermore, since hyperprolactinemia tends to exhibit a bimodal curve of action depending on its severity (39), it is possible that mild hyperprolactinemia actually synergizes with, rather than opposes, the stimulatory action of hypersomatotropinemia on LH-R metabolism.

In mice carrying a *rat phosphoenolpyruvate carboxykinase/bovine growth hormone I fusion gene* (TgPbGH1), bGH excess is expressed postnatally. In these mice basal testicular LH-R levels were higher than those of their normal littermates. Lower doses of hCG caused a positive autoregulation of testicular LH-R in both types of mice, but higher doses produced positive autoregulation in TgPbGH1 mice and negative autoregulation in their normal littermates (40). Thus, TgPbGH1 mice had a type C LH-R autoregulation pattern, and normal mice, a type B pattern. These data confirm the stimulatory influence of hypersomatotropinemia with regard to LH-R metabolism that had been observed in other transgenic mice.

Normal House Mice

The existence of differences in various physiological parameters between mutant animals and their normal littermates is not unexpected. However, in the case of LH-R, differences are often found among stocks and strains of normal mice (41–43). Basal testicular LH-R concentrations were found to vary among several strains and stocks of mice (C57BL/10J > DW/B ≥ C57BL/6J ≥ DBA/2J and DF/B) (42). Similar results were reported by another laboratory (C57BL/10 ≥ C3H/HeJ > C57BL/6J > DBA/2J) without any significant differences in binding affinity (44). Large differences were even found between two different stocks of Swiss albino mice. Not only were LH-R levels 3 times higher in CD-1 than ICR mice, but hCG caused negative autoregulation of LH-R only in ICR mice (43).

However, the clearest demonstration of interstrain differences in LH-R in mice emerged from the study of C57BL/10J, DBA/2J, and their F_1 progeny (B10D2F$_1$ and D2B10F$_1$) (42). Basal testicular LH-R levels were higher in C57BL/10J than in DBA/2J mice. Low doses of hCG 24 h after injection produced positive autoregulation of LH-R in all the mice, but higher doses of hCG induced negative autoregulation in C57BL/10J, but not in DBA/2J, mice. In B10D2F$_1$ mice basal LH-R levels were found to be intermediate between those measured in the parental strains, and in D2B10F$_1$ mice, the levels were similar to DBA/2J. However, hCG caused negative autoregulation in both types of F_1 mice. Thus, C57BL10/J and B10D2F$_1$ mice had a type A LH-R autoregulation pattern, and DBA/2J and D2B10F$_1$ mice had a type B pattern. When autoregulation was studied 72 h after hCG administration, it was found that hCG had produced negative autoregulation in all the mouse types.

These results suggested the existence of at least two genetic factors that regulate testicular LH-R levels and metabolism in mice. Presumably, one of these would be the structural gene coding for the LH-R molecule, and it would be responsible for the differences in basal levels. This locus appears to have at least 2 codominant alleles and could be subject to imprinting, as evidenced by the differences between B10D2F$_1$ and D2B10F$_1$ mice. At least one other factor would regulate the pattern and temporal characteristics of LH-R autoregulation. This locus appears to have at least 2 alleles, with the dominant one being present in C57BL/10J.

Age or developmental stage also appears to have an effect on LH-R metabolism. Whereas adult DW/B mice had a type A LH-R autoregulation pattern, immature mice from the same stock had a type E LH-R autoregulation pattern (11).

Although much less is known about the genetic regulation of LH-R in the ovaries, it appears that similar mechanisms are in place since differences in LH-R levels could be detected among several mouse lines (45, 46). Among the most common strains, the differences were as follows: 129/SV > C57BL/6J > A/J = SJL/J > DBA/J (46). Interestingly enough, in female mice it appears that at least 129/SV animals have significantly greater Ka than the other strains. This would be the only instance known of differences in the affinity of LH-R (46).

Norway Rats

In general, the status of rat genetics lags far behind that of mice or even humans. This is in spite of the rat being a major, if not the main, animal model used in biomedical reseach. The endocrine genetics of the rat is no exception. Thus, in spite of the rat being the animal of choice for the study of LH-R, much less is known about genetics and LH-R in the rat than in the mouse. However, there is some evidence that the situation is similar to that found in mice.

Mutant Rats

Spontaneously hypertensive rats (SHR) are a popular model for the study of hypertension. These animals also have several endocrine alterations, including a progessive increase in circulating PRL that can already be detected by 4 months of age (47, 48). When LH-R levels were determined in SHR, they were found to be reduced versus normotensive Wistar-Kyoto control rats (which have the same genetic origin as SHR) in spite of an increase in testis size (49). This was unexpected because data from rats with pituitary graft-induced hyperprolactinemia showed that PRL increased testicular LH-R levels (50). Possible explanations for the findings in SHR include the following: (i) The slower rise of PRL levels in SHR might regulate LH-R differently from the faster rise observed in grafted animals; (ii) the severity of the hyperprolactinemia, which is lesser in SHR, might determine its effects on LH-R levels; or (iii) the genome of the strain (or in other cases of the species) modulates the heteroregulation of LH-R by PRL. Probably, the latter mechanism, in conjuction with one or both of the other two, is in play (51).

Rats of the *Fisher 344* (F344) strain have a high susceptibility to develop neoplasias. As a matter of fact, most old male F344 rats will develop Leydig cell tumors, most of which become malignant. This is rare in other strains of rats (52–54). The presence of tumors was accompanied by an increase in LH-R (55). The development of these Leydig cell tumors was found to be prevented by DES-induced hyperprolactinemia (56). When it was attempted to differentiate between the effects of DES and those of prolactin in F344 rats, two major observations were made. First, it became evident that such animals as F344 rats and BALB/c mice that have a high susceptibility to tumorigenesis have elevated post-estrogen-treatment LH-R levels. In contrast, in animals with relative low tumor susceptibility, such as Sprague-Dawley rats, C3H mice, and humans, estrogen treatment induces a decrease in LH-R levels (51). Thus, there appears to be a correlation between the tumor susceptibility phenotype and the LH-R response to estrogen treatment. The second observation was that a hierarchy exists for the impact of different factors on LH-R levels. The impact of PRL was the lowest, the impact of estrogens was intermediate, and the impact of the genomic status was the highest (51).

In *restricted hooded* (H^{re}/h) rats that have a mutation of the SCF-R gene (57), the basal LH-R levels were much higher than those measured in *normal hooded* (h/h) rats (58). Their autoregulation was also found to be different. The H^{re}/h rats have a type G LH-R autoregulation pattern, while h/h rats have a type D pattern (58). That alterations in the LH-R metabolism were found in H^{re}/h rats is not surprising since as stated above, mice with mutations of the genes coding for SCF and SCF-R have altered testicular LH-R metabolism.

Although LH-R has not been measured in male *hypogonadism* mutant (hgn/hgn) rats, experimental data indicates resistance to gonadotropin action. This could be due, at least in part, to alterations in LH-R and/or FSH-R (59).

Normal Rats

When different stocks and strains of normal rats were studied, it was observed that significant differences in basal testicular LH-R levels existed, just as had been observed in mice (60). Among the different types of rats, the hierarchy was Long-Evans = Wistar > Sprague-Dawley > Brown-Norway = Lewis. When the autoregulation was analyzed in outbred stocks of rats, differences were also observed. Whereas Wistar and Sprague-Dawley rats had a type D LH-R autoregulation pattern, Long-Evans rats had a type G pattern. In the latter stock even low doses of hCG produced negative autoregulation of testicular LH-R (60).

Other Animal Species

Very little is known about the genetics of LH-R in animal species other than murid rodents. Studies done in lambs revealed that animals from the Romanov stock had about twice as many LH-R per Leydig cell as was found in animals from the Iles-de-France stock (61, 62).

In *cardiomyopathic hamsters* (CMH), which have myotonic dystrophy due to a defect in the voltage-sensitive calcium channels and consequent increase in intracellular calcium (63, 64), alterations in the hypothalamic-pituitary-testicular axis have been described (30, 65). One such alteration is an altered LH-R metabolism. CMH had basal LH-R levels twice as high as those in normal hamsters. Also, whereas normal (F$_1$B) hamsters had a type F pattern of LH-R autoregulation, in CMH the pattern was type B; that is, a low dose of hCG caused an elevation of LH-R in CMH, but not in normal hamsters. Conversely, a higher dose of hCG elevated LH-R in normal hamsters, but not in CMH (65).

Furthermore, when CMH and normal hamsters were exposed to a *short photoperiod* (SPP) for 8 weeks, in normal hamsters the expected photoperiod-related testicular regression was accompanied by a decrease in the total content of LH-R. In contrast, in CMH the decrease in LH-R was less dramatic, and it was accompanied by only a small decrease in testis size without any evidence of gonadal regression (Amador,

Mayerhofer, Bartke, Steger, data presented at the XIIth North American Testis Workshop, 1993). Therefore, it can be concluded that the calcium overload in CMH is responsible for making the LH-R metabolism of CMH more sensitive to gonadotropic regulation and more resistant to SPP-induced changes.

Differences in the metabolism of LH-R can also be observed among stocks of outbred Syrian hamsters. As stated above, F_1B hamsters have a type F pattern of LH-R autoregulation (65). In contrast, LAK hamsters have been shown to exhibit an LH-R autoregulation of the type D pattern (66, 67).

Studies in Humans

In humans LH-R studies have been hampered by the fact that gonadectomy is not performed unless gonadal pathology is known to occur. Moreover, biopsies of healthy gonads are not used frequently in research. Most of the "healthy" gonads available for research are those from patients who underwent gonadectomy as part of the management of their neoplasia of the reproductive tract. Thus, the functional status of these gonads may have been compromised. Therefore, there is a lack of adequate controls in most studies involving LH-R in humans. It should also be noted that the human thyroid has LH-R (68, 69).

A few studies have, however, provided us with some insight on the status of LH-R in humans. In one such study LH-R levels were investigated in the testes of men with prostatic cancer, men with idiopathic infertility, and men with germ cell aplasia (70). In the first two groups, levels were similar (208 vs. 271 fmol/g testis), but in men with germ cell aplasia LH-R levels were significantly higher (436 fmol/g testis). Another study looked at a heterogeneous group of patients (ages 19–54), most of whom had either varicocele or undescended testes (71). In this group the LH-R levels averaged 733 fmol/g testis (75 fmol/mg protein). A study of patients with advanced prostatic carcinoma (ages 53–93) reported LH-R levels of 692 fmol/g testis (72). In patients with primary or idiopathic infertility, LH-R levels were reported to be 7.5 fmol/mg protein, and administration of hCG indicated that these patients might have a type A, D, or G LH-R autoregulation pattern (73, 74), supporting similar data reported previously by another group (70). In men with hypogonadotropic hypogonadism LH-R levels of 8.3 fmol/mg protein have been reported (75).

There are a few other indications that LH-R might be involved in human andrological disorders, just as is the case in rodents. In patients with Leydig cell aplasia/hypoplasia, evidence indicates that the absence of LH-R or the presence of abnormal LH-R might be responsible for at least some of the cases of this disease (1, 76–78).

One would expect patients with *Del Castillo syndrome* (Sertoli cell only syndrome) to have LH-R alterations when the disease is the result of mutation in the SCF or SCF-R genes (79). However, so little is known about the molecular etiology and details of the endocrine state of these patients that the role of LH-R in this disorder remains a mystery. One study showed that this type of patient had LH-R levels twice as high as those observed in other men (70). Another study demonstrated a defect in in vitro 17α-hydroxylase activity in 10 patients with Del Castillo syndrome (80). However, the endocrine profiles of these patients did not indicate the presence of resistance to gonadotropins. Thus, not all patients with Del Castillo syndrome will have abnormal LH-R levels and/or metabolism.

Age and developmental stage appear to have an effect on LH-R levels in human testes. The LH-R levels are 698–1150 fmol/g testis in the fetus, 110 fmol/g testis in young men, 140 fmol/g testis in middle-aged men, and 43 fmol/g testis in old men, respectively (81).

Comparative Studies

Comparisons with Humans

Some studies have compared LH-R levels in humans with those observed in other species. Comparison of human and rat testes appears to indicate very low LH-R levels in humans. Testes from men with advanced prostatic cancer were compared to those of healthy Wistar rats. The levels in humans were 1/14 of those observed in rats (570 vs. 2700 fmol/g testis; 7.8 vs. 120 fmol/mg protein; and 2000 vs. 24,000 LH-R/cell) (82, 83). This indeed would put humans in a category of species having low LH-R levels, such as the Syrian hamster that has only 1400 LH-R/cell (84). When the testicular concentration of LH-R in human testis is compared to that reported for other species, the following hierarchy emerges: Boar > Norway rat = house mouse = bank vole = Siberian hamster > men = Syrian hamster = green monkey = rhesus monkey > bull = ram (81, 85–89).

Mammals

Two studies have compared LH-R levels and metabolism among the most frequently used species of laboratory animals (88, 89). In the first study outbred Sprague-Dawley rats were found to have a numerically, but not significantly, higher concentration of LH-R than random-bred DW/B mice. In turn, both rats and mice had at least 10-fold higher concentrations of LH-R than outbred LAK Syrian hamsters (88). In the second study Sprague-Dawley rats were found to have significantly higher LH-R levels than random-bred DF/B mice (a different stock than DW/B), and both rats and mice had higher levels than another stock of Syrian hamsters (F_1B). Injection of a high dose of hCG caused a decrease in LH-R levels in both mice and rats, but in Syrian hamsters a significant increase in the concentration of LH-R was observed (89). What these two studies proved was that under controlled circumstances the interspecies differences suspected from studies performed on separate occasions by us and others were real.

Measurements in our laboratory indicate that the levels of LH-R in the testes of Siberian hamsters are in the range of those seen in normal mice (88, 90). Another study indicates that in mice the ligand-induced proteolytic cleavage of LH-R appears to be involved in the negative autoregulation of LH-R, whereas in rats this process is a continuous one (91).

When LH-R was compared for differences among testes from bulls, pigs, and rats, no differences in binding affinity or physicochemical characteristics were observed among the species. However, the concentration of LH-R was different among the three species (pig > rat > bull), with approximately 2-fold differences between each species (92). In another study LH-R levels were compared in cultured Leydig cells from rats, mice, and pigs. The levels were much higher in porcine than in murine Leydig cells, with cells from rats and mice having similar levels (93). Testes from stallions and male donkeys appear to have lower concentrations of LH-R than those seen in rat testes (94).

Other Species

In quails, the concentration of LH-R was reported to be 280 fmol/g testis (95). In fish, which have developmental phase-related gonadotropins GTH I and II instead of LH and FSH, LH-R-like receptors capable of binding hCG have been described. Most of the studies have been done in salmonids, specifically in the common trout (*Salmo gairdneri*). In this species the GTH-R levels have been reported to be in the range of 45 to 880 fmol/g testis (96, 97). In coho salmon (*Oncorhynchus kisutch*), GTH-R levels were measured at 113 fmol/mg protein in immature animals (98).

In bullfrogs the concentration of LH-R is reported as being 320 fmol/g testis (99). Interestingly, amphibians are characterized by the presence of specific LH-R in the liver, and in bullfrogs their concentration is 40%–50% greater than that measured in the testes of the same animals (100). These same investigators report the presence of hepatic LH-R in a total of 7 amphibian species, but were unable to detect hepatic LH-R in rats, chickens, or gobies. The only nonamphibian species in which hepatic LH-R has been detected is the Siberian hamster (*Phodopus sungorus sungorus*) in which the hepatic concentration of specific LH-R is 2.5 fmol/mg protein versus a testicular LH-R concentration of 4.6 fmol/mg protein in the same animals (Amador, Bartke, unpublished).

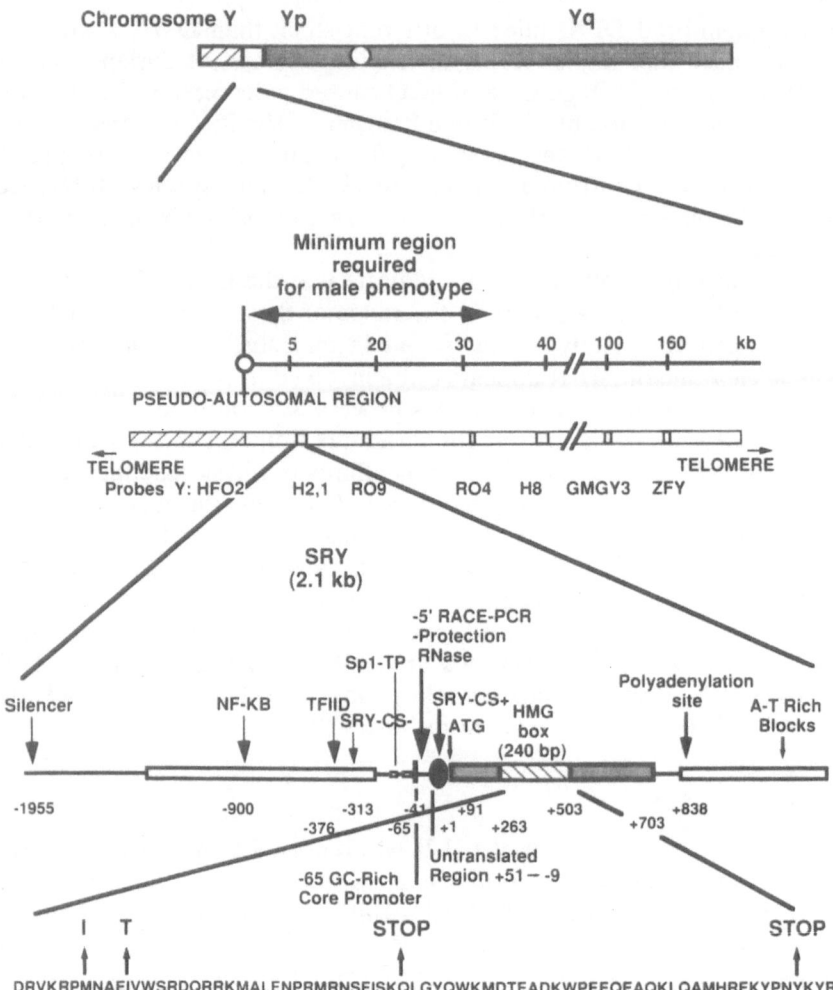

FIGURE 20.2. Characteristics of the human *SRY* gene. Adapted from references 105 and 106.

The lowest species in the evolutionary scale in which LH-R-like molecules were detected is *Pseudomonas maltophili*. In this bacterium, which by the way produces an hCG-like protein, receptors capable of binding hCG, but not any other human hormone, were described as having a kd similar to that reported in mammals (101).

Sexual Differentiation

Some of the most exciting discoveries of the last few years have been the isolation, chromosomal mapping, and cloning of the genes *KIT* (SCF-R),

TABLE 20.2. Comparative chromosomal gene mapping
of the SCF-R (*KIT*), SCF (*MGF*), and *SRY* genes.

Species	KIT	MGF	SRY
Human	4q11–22	12q22–24	Yp11.2
Gorilla	3	10	Yp
Chimpanzee	?	10	Yp
Orangutan	3	9	Yp
Baboon	5	?	Y
Rhesus	?	11	Y
Capuchin monkey	2	10	Y
Owl monkey	1	10	Y
African green monkey	?	3	Y
House mouse	5E	10D1-2	YA1
Norway rat	14	?	Y
Chinese hamster	1	1	Y
Rabbit	?	16	Y
Cat	B1	B4	Y
Mink	6	9	Y
Domestic pig	8	5	Y
Domestic cattle	6	5	Y
Domestic sheep	?	3	Y
Other mammals	?	?	Y

Note: The above is described in references 7, 57, and 102–
104. It should be noted that many locations are probable
chromosomal locations inferred by conservation of certain
chromosomal regions across species.

MGF (SCF), and *SRY* (sex-determining region Y) (Fig. 20.2). For comparative mapping of these three genes, see Table 20.2. Since LH-R is
required for the appearance of Leydig cells, SCF and SCF-R regulate
LH-R metabolism, and SRY is required for the differentiation of testes
and Sertoli cells, we have attempted to integrate a model for the testis-
determining pathway. By doing this, we are proposing that the *testis-
determining factor* (TDF) is really a pathway under the control of the
SRY gene (Fig. 20.3).

   The *SRY* gene belongs to a family of regulatory DNA-binding proteins
that are highly conserved throughout evolution (110, 111). As a matter of
fact, its HMG box is similar to that of the gene coding for the Mc mating-
type protein in yeast (110). Furthermore, as a person of Gaelic descent,
Dr. Ann McLaren was quick to point out that in Gaelic, *Mc* means *son of*
(112). This odd coincidence is not the only one we observe regarding the
SRY gene. Within the peptide sequence coded for by the HMG box, one
can find a methionine-alanine-leucine-glutamate sequence. When using
one-letter amino acid codes, this sequence spells MALE. How does *SRY*
carry out its developmental action? It appears that its main function is to
insure the differentiation and development of Sertoli cells from cells of
the urogenital ridge (111, 113).

SEXUAL DIFFERENTIATION

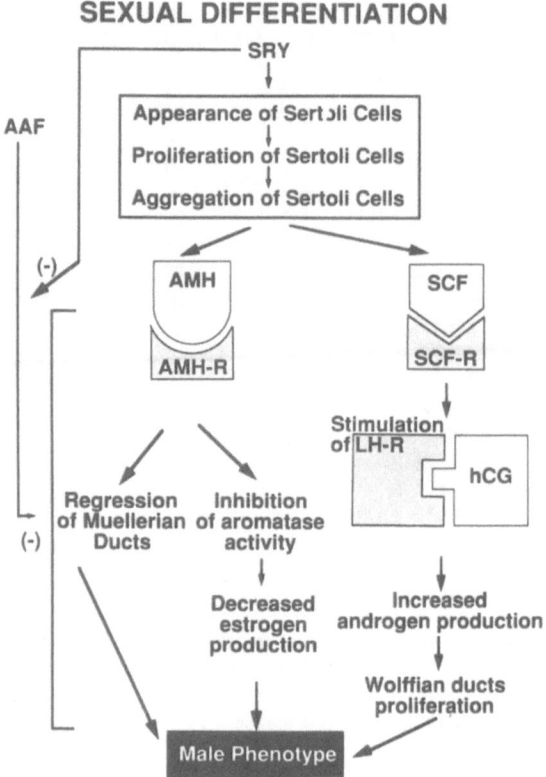

FIGURE 20.3. The testis-determining (pathway) factor (TDF). The flow diagram shows the events involved in the normal development of the male phenotype. Adapted from Vilain, McElreavey, Herskowitz, and Fellous (105), McLaren (107), Abbas, Bishop, and Fellous (108), and Lobaccaro and Sultan (109).

A second function for the *SRY* product has been proposed by Vilain et al. (105). They proposed the existence of a Z-factor(s) that would inhibit male-specific genes. We have renamed this factor(s) *antiandrogenic factor* (AAF) (Fig. 20.3). Thus, the presence of a normal copy of the *SRY* gene insures the presence of normal-functioning Sertoli cells. This results in the normal production of many factors, including SCF and the *antimuellerian hormone* (AMH) (114, 115). The SCF-R is present in Leydig cells, and the SCF/SCF-R interaction regulates LH-R metabolism (24, 116). The presence of normal-functioning LH-R results in Leydig cell differentiation and adequate androgen production (1, 3). Thus, under these circumstances, normal levels of AMH and androgens are available to the embryo. This obviously results in normal masculinization. However, it also becomes apparent that there are many major steps, some of which include other important developmental genes (i.e., steroid receptors) that if altered, are capable of interfering with the production of a normal male phenotype. Therefore, our proposal is for TDF to be a pathway rather than a single gene. This is in concordance with the action of other major developmental genes that exert their multiple actions by controlling complex signaling pathways rather than by having a direct effect on all the required steps of cell metabolism. Furthermore, the concept of the existence of a testis-determining pathway has already been advocated by others (117).

Concluding Remarks

In the two decades since the first description of LH-R, hundreds of articles have been published on the subject. However, even in this age of genetics and molecular biology, very little is known about the genetics of this molecule and its role in normal and abnormal genetically dependent events. It is our hope that this review, however incomplete, might help in initiating a better understanding of the relationship between LH-R levels, regulation, and genetics.

Acknowledgments. We thank Drs. Wesley G. Beamer, Terry A. Parkening, Thomas J. Collins, Theresa M. Siler-Khodr, John G.M. Shire, Harold G. Klemcke, Richard W. Steger, Robert D. Hilgers, and George L. Wolff for their past collaborations. We thank Ms. Lynette Bates, Sherie L. Hodges, Rita Meyers, Tim Began, and Cathy Henebry for their technical help. We thank the National Hormone and Pituitary Program and Dr. R. Canfield for the extrapure hCG used in the radioreceptor-assays. Studies performed in our laboratories were supported by NIH Grants HD-20033 and HD-20001 (A.B.), Grant Ma1080/1-2 from the Deutsche Forschungsgemeinschaft (A.M.), SIU-CRC Grant 16-91 (A.G.A.), the SIU-OB/GYN Departmental Research Fund (A.G.A.), and the ReproGen Research Fund (A.G.A).

References

1. Schwartz M, Imperato-McGinley J, Peterson RE, et al. Male pseudo-hermaphroditism secondary to an abnormality in Leydig cell differentiation. J Clin Endocrinol Metab 1981;53:123–7.
2. Pelliniemi LJ, Niemi M. Fine structure of the human foetal testis, I. The interstitial tissue. Z Zellforsch 1969;99:507–22.
3. Huhtaniemi I, Pelliniemi LJ. Fetal Leydig cells: cellular origin, morphology, life span, and special functional features. Proc Soc Exp Biol Med 1992;201: 125–40.
4. Vuotilainen R. Differentiation of the fetal gonad. Horm Res 1992;38(suppl 1):66–71.

5. Niemi M, Ikonen M. Steroid-3beta-ol-dehydrogenase activity in foetal Leydig's cells. Nature 1961;189:592–3.

6. Rousseau-Merck MF, Misrahi M, Atger M, Loosfelt H, Milgrom E, Berger R. Localisation of the human luteinizing hormone/choriogonadotropin receptor gene (LHCGR) to chromosome 2p21. Cytogenet Cell Genet 1990; 54:77–9.

7. O'Brien SJ, Graves JAM. Report of the committee on comparative gene mapping. Cytogenet Cell Genet 1991;58:1124–51.

8. Libert F, Vassart G, Parmentier M. Current developments in G-protein-coupled receptors. Curr Opin Cell Biol 1991;3:218–23.

9. Coumbarnous Y. Molecular basis of the specificity of binding of glycoprotein hormones to their receptors. Endocr Rev 1992;13:670–91.

10. Bohnet HG, Friesen HG. Effect of prolactin and growth hormone on prolactin and LH receptors in the dwarf mouse. J Reprod Fertil 1976;48:307–11.

11. Amador AG, Bartke A. Regulation of testicular human chorionic gonadotrophin binding in prolactin-deficient Snell dwarf mice. J Endocrinol 1982; 95:301–9.

12. Roux M, Bartke A, Dumont F, Dubois MP. Immunohistological study of the anterior pituitary gland—pars distalis and pars intermedia—in dwarf mice. Cell Tissue Res 1982;223:415–20.

13. Camper SA, Saunders TL, Katz RW, Reeves RH. The Pit-1 transcription factor gene is a candidate for the murine Snell dwarf mutation. Genomics 1990;8:586–90.

14. Li S, Crenshaw EB III, Rawson EJ, Simmons DM, Swanson LW, Rosenfeld MG. Dwarf locus mutants lacking three pituitary cell types result from mutations in the POU-domain gene pit-1. Nature 1990;347:528–33.

15. Amador AG, Bartke A, Beamer WG, Siler-Khodr TM. Autoregulation of testicular LH receptors in Ames dwarf (df/df) and pigmy (pg/pg) mice. Endocrinol Exp 1988;22:87–97.

16. Amador AG, Parkening TA, Beamer WG, Bartke A, Collins TJ. Testicular LH receptors and circulating hormone levels in three mouse models for inherited diseases (Tfm/y, lit/lit and hyt/hyt). Endocrinol Exp 1986;20:349–58.

17. Amador AG, Mayerhofer A, Bartke A. Mice with inherited or surgically induced hypothyroidism. Mouse News Lett 1988;81:87.

18. Amador A, Parkening T, Beamer W, Bartke A, Collins TJ. Autoregulation of testicular luteinizing hormone receptors in hypogonadal (hpg/hpg) mice. Biochem Genet 1984;22:395–401.

19. Huhtaniemi I. Gonadotrophin receptors: correlates with normal and pathological functions of the human ovary and testis. Clin Endocrinol Metab 1983;12:117–32.

20. Stallings MH, Matt KS, Amador A, et al. Regulation of testicular LH/hCG receptors in golden hamsters (Mesocricetus auratus) during development. J Reprod Fertil 1985;75:663–70.

21. Beamer WG, Cunliffe-Beamer TL, Shultz KL, Langley SH, Roderick TH. Juvenile spermatogonial depletion (jsd): a genetic defect of germ cell proliferation of male mice. Biol Reprod 1988;38:899–908.

22. Mizunuma M, Dohmae K, Tajima Y, Koshimizu U, Watanabe D, Nishimune Y. Loss of sperm in juvenile spermatogonial depletion (jsd) mutant mice is

ascribed to a defect of intratubular environment to support germ cell differentiation. J Cell Physiol 1992;150:188–93.

23. Amador A, Beamer WG, Bartke A. Evidence for a second locus (*jsd*) involved in regulating testicular LH receptor levels in mice. Genetics 1986; 113:s2.

24. Amador A, Parkening T, Beamer W, Bartke A, Collins TJ. Effects of alleles at the *W* locus on testicular luteinizing hormone receptors in adult mice. J Recept Res 1984;4:741–53.

25. Amador AG, Bartke A, Hodges SL. A new allele (*Sl^d*) involved in the regulation of testicular LH receptors and consequently, testosterone production in the mouse. Genome 1988;30(suppl 1):259.

26. Amador AG, Bartke A, Parkening TA, et al. Hormonal regulation of testicular human chorionic gonadotropin binding and steroidogenesis in adult mice with different forms of hereditary diabetes and obesity. Horm Res 1986;23:215–24.

27. Amador AG, Esquifino AI, Wolff GL, Mayerhofer A, Steger RW, Bartke A. Effects of a diabetogenic allele of the *a* locus on Leydig cell function in mice. In: Monduzzi, ed. Proc IV Int Cong Androl, S.p.A., Bologna, 1989: 1–4.

28. Amador AG, Mayerhofer A, Bartke A, et al. Further studies on Leydig cell function in viable yellow (A^vy/a) mice. Mouse Genome 1990;86: 257–8.

29. Amador A, Bartke A, Klemcke H, Dung HC. Pituitary and testicular function in lethargic mice. Indian J Com Anim Physiol 1983;1:34–8.

30. Wilkinson M. Abnormal gonadotrophin release from pituitaries of muscular dystrophic mice and hamsters. J Reprod Fertil 1984;71:463–6.

31. Amador AG, Mayerhofer A, Parkening TA, Collins TJ, Bartke A. Differential effects of two alleles of the *dy* locus on the pituitary-testicular axis of mice. Rev Esp Fisiol 1992;48:157–66.

32. Amador AG, Mayerhofer A, Bartke A, Steger RW. Transgenic mice expressing the gene for human growth hormone. Mouse News Lett 1989; 83:183–4.

33. Klemcke HG, Bartke A. Effect of chronic hyperprolactinemia in mice on plasma gonadotropin concentrations and testicular human chorionic gonadotropin binding sites. Endocrinology 1981;108:1763–8.

34. Lin T, Haskell H, Vinson N, Terracio L. Direct stimulatory effects of insulin-like growth factor on Leydig cell steroidogenesis in primary culture. Biochem Biophys Res Commun 1986a;137:950–6.

35. Lin T, Haskell H, Vinson N. Characterization of insulin and insulin-like growth factor I receptors of purified Leydig cells and their role in steroidogenesis in primary culture: a comparative study. Endocrinology 1986b;119: 1641–7.

36. Amador AG, Mayerhofer A, Bartke A, Steger RW, Wagner TE, Yun JS. Testicular function in transgenic mice expressing the gene for the bovine growth hormone. Mouse Genome 1990;86:257.

37. Amador AG, Bartke A, Mayerhofer A, Wagner TE, Yun JS. Testicular function in transgenic mice expressing the gene for the human growth hormone II (variant or placental growth hormone). Mouse Genome 1990; 87:99.

38. MacLeod J, Worsley I, Ray J, Friesen H, Liebhaber S, Cooke N. Human growth hormone-variant is a biologically active somatogen and lactogen. Endocrinology 1991;128:1298–302.

39. Amador A, Klemcke HG, Bartke A, Soares MJ, Siler-Khodr TM, Talamantes F. Effects of different numbers of ectopic pituitary transplants on regulation of testicular LH/hCG and prolactin receptors in the hamster (*Mesocricetus auratus*). J Reprod Fertil 1985;73:483–9.

40. Amador A, Bartke A, Mayerhofer A, Shire JGM, Wagner TE, Yun JS. Testicular function in transgenic mice expressing a gene construct for the bovine growth hormone that includes a PEPCK promoter. Mouse Genome 1991;89:264–5.

41. Amador A, Bartke A, Beamer W. Genetic variation in testicular LH receptors in the mouse. Endocr Res Comm 1982;9:79–88.

42. Amador AG, Bartke A, Parkening TA, Chamness GC. Genetic differences in androgen receptors and in autoregulation of testicular human chorionic gonadotropin binding sites in the mouse. J Recept Res 1983;3:495–512.

43. Amador AG, Parkening TA, Collins TJ, Bartke A, Silverman AY. Differences in testicular LH receptors and steroidogenesis between two stocks of outbred Swiss mice. Adv Contracept 1987;3:29–33.

44. Stalvey JRD, Payne AH. Luteinizing hormone receptors and testosterone production in whole testes and purified Leydig cells from the mouse: differences among inbred strains. Endocrinology 1983;112:1696–701.

45. Spearow JL, Bradford GE. Genetic variation in spontaneous ovulation rate and LH receptor induction in mice. J Reprod Fertil 1983;69:529–37.

46. Spearow JL. The mechanism of action of genes controlling reproduction. In: Land RB, Robinson DW, eds. Genetics of reproduction in sheep. London: Butterworths, 1985:203–15.

47. Iams SG, McMurtry JP, Wexler BC. Aldosterone, deoxycorticosterone, corticosterone, and prolactin changes during the lifespan of chronically and spontaneously hypertensive rats. Endocrinology 1979;104:1357–63.

48. Hodson CA, Louis TM, Burden HW, Iams SG. Serum gonadotropin and prolactin levels in young and old spontaneously hypertensive rats. IRCS Med Sci 1981;9:321.

49. Amador A, Steger RW, Bartke A, Johns A, Hayashi RH, Stallings MH. Pituitary and testicular function in spontaneously hypertensive rats. J Androl 1983;4:67–70.

50. Sharpe RM, McNeilly AS. The effect of induced hyperprolactinemia on Leydig cell function and LH-induced loss of LH receptors in the rat testis. Mol Cell Endocrinol 1979;16:19–27.

51. Amador AG, Esquifino AI, Bartke A, et al. Effects of diethylstilboestrol on testicular function and luteinizing hormone receptors. Rev Esp Fisiol 1989;45:245–54.

52. Thompson S, Huseby RA, Fox MA, Davis CL, Hunt RD. Spontaneous tumors in the Sprague-Dawley rat. J Natl Cancer Inst 1961;27:1037–57.

53. Jacobs BB, Huseby RA. Neoplasms occuring in aged Fisher rats, with special reference to testicular, uterine and thyroid tumors. J Natl Cancer Inst 1967;39:303–9.

54. Walsh PC. The endocrinology of testicular tumors. Recent Results Cancer Res 1979;60:196–201.

55. Amador A, Steger RW, Bartke A, et al. Testicular LH receptors during aging in Fisher 344 rats. J Androl 1985;6:61–4.
56. Sweeney C, Castrecane D, Doherty P, Bartke A. Effects of spontaneous Leydig cell tumors on testicular steroidogenesis. J Androl 1983;4:34.
57. Hedrich HJ. Mutant genes and polymorphic loci of the laboratory rat. In: Hedrich HJ, ed. Genetic monitoring of inbred strains of rats. Stuttgart, BDR: Gustav Fisher Verlag, 1990:289–409.
58. Amador A, Klemcke HG, Bartke A. Autoregulation of hCG binding in $H^{re}/+$ rats. J Androl 1984;5:19P.
59. Hakamata Y, Kikukawa K, Kame T, Suzuki K, Taya K, Sasamoto S. A new male hypogonadism mutant rat (hgn/hgn): concentrations of testosterone (T), luteinizing hormone (LH), and follicle-stimulating hormone (FSH) in the serum and the responsiveness of accesory sex organs to exogenous T, FSH, human chorionic gonadotropin, and luteinizing hormone-releasing hormone. Biol Reprod 1988;38:1145–53.
60. Amador AG, Siler-Khodr TM, Mayerhofer A, Steger RW, Bartke A. Intra-species differences in the regulation of the pituitary-Leydig cell axis in the rat. Genetics 1989;122:s32.
61. Barenton B, Hochereau-de Reviers MT, Perreau C. Breed differences in testicular histology and numbers of LH and FSH receptors in the lamb. IRCS Med Sci 1983;11:471.
62. Hochereau-de Reviers MT, Blanc MR, Colas G, Pelletier J. Parameters of male fertility and their genetic variation in sheep. In: Land RB, Robinson DW, eds. Genetics of reproduction in sheep. London: Butterworths, 1985: 301–14.
63. Finkel MS, Marks ES, Patterson RE, Speir EH, Steadman K, Keiser HR. Increased cardiac calcium channels in hamster cardiomyopathy. Am J Cardiol 1986;57:1205–6.
64. Wagner JA, Reynolds W, Weisman HF, Dudeck P, Weisfeldt ML. Calcium antagonist receptors in cardiomyopathic hamsters: selective increases in heart, muscle, brain. Science 1986;232:515–8.
65. Amador AG, Mayerhofer A, Bartke A, Chandrashekar V. Pituitary-testicular axis in cardiomyopathic Syrian hamsters. J Androl 1992;13:422–7.
66. Amador AG, Bartke A, Klemcke HG, Siler-Khodr TM, Stallings MH. Effects of photoperiod and hCG on the regulation of testicular LH/hCG receptors in Syrian hamsters (Mesocricetus auratus). J Reprod Fertil 1985; 74:693–700.
67. Amador AG, Bartke A. Actions of human chorionic gonadotropin on Leydig cell function in hamsters with photoperiod-induced testicular atrophy. Ann NY Acad Sci 1987;513:353–5.
68. Carayon P, Lefort G, Nisula B. Interaction of human chorionic gonadotropin and human luteinizing hormone with human thyroid membranes. Endocrinology 1980;106:1907–16.
69. Frazier AL, Robbins LS, Stork PJ, Sprengel R, Segaloff DL, Cone RD. Isolation of TSH and LH/CG receptor cDNAs from human thyroid: regulation by tissue specific splicing. Mol Endocrinol 1990;4:1264–76.
70. Sharpe RM, Wu FCW, Hargreave TB. Binding of human chorionic gonadotropin to testicular biopsy tissue from infertile men and the effect of prior treatment with human chorionic gonadotropin. J Endocrinol 1980;86:117–25.

71. Grizard G, Boucher D, Hermabessiere J, Grizard J. Testicular receptors of human chorionic gonadotrophin in adult men. Binding and degradation of the hormone. Acta Endocrinol (Copenh) 1982;101:293–300.
72. Leinonen P, Bolton N, Vihko R. Human testicular LH receptors: correlations with circulating gonadotrophins and testicular steroid secretion. Int J Androl 1982;5:145–57.
73. Namiki M, Kitamura M, Miyake O, et al. Reduction of testicular human chorionic gonadotropin receptors by human chorionic gonadotropin in infertile men. Arch Androl 1988;20:45–50.
74. Namiki M, Kitamura M, Nonomura N, et al. Reduction of testicular human chorionic gonadotropin receptors by human chorionic gonadotropin in vivo and in vitro. Horm Res 1988;29:156–61.
75. Namiki M, Kitamura M, Nonomura N, et al. Testicular gonadotropin receptors in men with hypogonadotropic hypogonadism. Arch Androl 1988; 20:81–5.
76. Berthezene F, Forest MG, Grimaud JA, Claustrat B, Monex R. Leydig cell agenesis: a cause of male pseudohermaphroditism. N Engl J Med 1976;295: 969–72.
77. Brown DM, Markland C, Dehner LP. Leydig cell hypoplasia: a cause of male pseudohermaphroditism. J Clin Endocrinol Metab 1978;46:1–7.
78. Wu RHK, Rosenfeld R, Fukushima D. Hypogonadism and Leydig cell hypoplasia unresponsive to human luteinizing hormone (hLH). Am J Med Sci 1984;287:23–5.
79. Nistal M, Jimenez F, Paniagua R. Sertoli cell types in the Sertoli-cell-only syndrome: relationships between Sertoli cell morphology and aetiology. Histopathology 1990;16:173–80.
80. Hammar M, Berg AA. Impaired Leydig cell function in vitro in testicular tissue from human males with "Sertoli cell only" syndrome. Andrologia 1985;17:37–41.
81. Berman MI, Sairam MR. Studies on primate gonadotropin receptors: comparison of follitropin and lutropin receptors in human testis. Can J Biochem Cell Biol 1983;61:561–8.
82. Huhtaniemi I, Bolton N, Leinonen P, Kontturi M, Vihko R. Testicular luteinizing hormone receptor content and in vitro stimulation of cyclic adenosine 3',5'-monophosphate and steroid production: a comparison between man and rat. J Clin Endocrinol Metab 1982;55:882–9.
83. Wahlstroem T, Huhtaniemi I, Hovatta O, Seppala M. Localization of luteinizing hormone, follicle-stimulating hormone, prolactin, and their receptors in human and rat testis using immunohistochemistry and radioreceptorassay. J Clin Endocrinol Metab 1983;57:825–30.
84. Sinha Hikkim AP, Amador AG, Bartke A, Russell LD. Structure/function relationship in active and inactive hamster Leydig cells: a correlative morphometric and endocrine study. Endocrinology 1989;125:1844–56.
85. Zaidi P, Wickings EJ, Arslan M, Nieschlag E. Characterization and comparison of testicular LH/hCG receptors of rhesus monkeys (*Macaca mulatta*) and green monkeys (*Cercopithecus aethiops*). Am J Primatol 1982;2:285–90.
86. Tahka KM, Rajaniemi H. Photoperiodic modulation of testicular LH receptors in the bank vole (*Clethrionomys glareolus*). J Reprod Fertil 1985;75: 513–9.

87. Amador A, Bartke A, Gust C, Nequin L, Borer K. Effects of short photo-period (SPP) on plasma prolactin (PRL) and testicular function in Siberian hamsters (*Phodopus sungorus*). Biol Reprod 1986;34(suppl 1):183.
88. Amador AG, Harper MJK, Bartke A, Hogan MP, Norris CJ. Interspecies differences in testicular LH receptors and in vitro testosterone production among rodents. Rev Esp Fisiol 1986;42:389–94.
89. Amador AG, Mayerhofer A, Bartke A. Interspecies differences in the effects of hCG on testicular function among rodents. Rev Esp Fisiol 1990; 46:197–204.
90. Bartke A, Schanbacher BD, Amador AG, Klemcke HG, Chandrashekar V. Effects of immunoneutralization of luteinizing hormone (LH)-releasing hormone on testicular prolactin and LH receptors in the golden hamster and on LH receptors in the Djungarian hamster. Endocrinology 1987;121: 2027–34.
91. West AP, Cooke BA. Regulation of the truncation of luteinizing hormone receptors at the plasma membrane is different in rat and mouse Leydig cells. Endocrinology 1991;128:363–70.
92. Sundby A, Torjesen P, Hansson V. Particulate and solubilized LH receptor in the pig and bull testis. Arch Androl 1982;9:233–44.
93. Mather JP, Saez JM, Haour F. Primary cultures of Leydig cells from rat, mouse and pig: advantages of porcine cells for the study of gonadotropin regulation of Leydig cell function. Steroids 1981;38:35–44.
94. Stewart F, Allen WR. Biological function and receptor binding activities of equine chorionic gonadotrophins. J Reprod Fertil 1981;62:527–36.
95. Kikuchi M, Ishii S. Radioiodination of chicken luteinizing hormone without affecting receptor binding potency. Biol Reprod 1989;41:1047–54.
96. Schlaghecke R. Binding of 125-J to rainbow trout (*Salmo gairdneri*) testis in vitro. Gen Comp Endocrinol 1983;49:261–9.
97. Le Ga F, Breton B, Bougoussa M. Gonadotropic hormone (GtH) receptors in the testis of the trout *Salmo gairdneri*; in vitro studies. Fish Physiol Biochem 1988;5:209–17.
98. Yan L, Swanson P, Dickhoff WW. Binding of gonadotropins (GTH I and GTH II) to coho salmon gonadal membrane preparations. J Exp Zool 1991;258:221–30.
99. Takada K, Kubokawa K, Ishii S. Specific gonadotropin binding sites in the bullfrog testis. Gen Comp Endocrinol 1986;61:302–12.
100. Kubokawa K, Ishii S. Receptors for native gonadotropins in amphibian liver. Gen Comp Endocrinol 1987;68:260–70.
101. Carrell DT, Odell WD. A bacterial binding site which binds human chorionic gonadotropin but not human luteinizing hormone. Endocr Res 1992;18:51–8.
102. Johansson M, Ellegren H, Marklund L, et al. The gene for dominant white color in the pig is closely linked to *ALB* and *PDGFRA* on chromosome 8. Genomics 1992;14:965–9.
103. Lyon MF, Kirby MC. Mouse chromosome atlas. Mouse Genome 1992;90: 22–44.
104. O'Brien SJ, Womack JE, Lyons LA, Moore KJ, Jenkins NA, Copeland NG. Anchored reference loci for comparative genome mapping in mammals. Nature Genet 1993;3:103–12.

105. Vilain E, McElreavey K, Herskowitz I, Fellous M. La determination du sexe: faits et nouveaux concepts. M/S Med Sci 1992;9:i–vii.
106. Su H, Lau Y-FC. Identification of the transcriptional unit, structural organization, and promoter sequence of the human sex-determing region Y (*SRY*) gene, using a reverse genetic approach. Am J Hum Genet 1993;52: 24–38.
107. McLaren A. Sex determination in mammals. Trends Genet 1988;4:153–7.
108. Abbas N, Bishop C, Fellous M. Genetique du sexe. La Recherche 1989; 213:1036–46.
109. Lobaccaro J-M, Sultan C. La differentiation sexuelle normale: genetique et endocrinologie moleculaires. CR Soc Biol 1992;186:314–31.
110. Sinclair AH, Berta P, Palmer MS, et al. A gene from the human sex-determining region encodes a protein with homology to a conserved DNA-binding motif. Nature 1990;346:240–4.
111. Gubbay J, Collignon J, Koopman P, et al. A gene mapping to the sex-determining region of the mouse Y chromosome is a member of a novel family of embryonically expressed genes. Nature 1990;346:245–50.
112. McLaren A. What makes a man a man? Nature 1990;346:216–7.
113. Koopman P, Muensterberg A, Capel B, Vivian N, Lovell-Badge R. Expression of a candidate sex-determining gene during mouse testis differentiation. Nature 1990;348:450–2.
114. Motro B, van der Kooy D, Rossant J, Reith A, Bernstein A. Contiguous patterns of c-*kit* and *steel* expression: analysis of mutations at the *W* and *Sl* loci. Development 1991;113:1207–21.
115. Josso N. Anti-muellerian hormone and Sertoli cell function. Horm Res 1992;38(suppl 2):72–6.
116. Yoshinaga K, Nishikawa S, Ogawa M, et al. Role of c-*kit* in mouse spermatogenesis: identification of spermatogonia a specific site of c-*kit* expression and function. Development 1991;113:689–99.
117. Hawkins JR. Genetic determinants of testis development in normal and abnormal individuals. Horm Res 1992;38(suppl 2):62–5.

Part IV

Regulation of Germ Cells,
Differentiation, and
Gene Expression

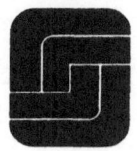

21

Development of Postnatal Gonocytes In Vivo and In Vitro

Joanne M. Orth and Michael P. McGuinness

Sperm production in the mature testis is the result of complex processes that have been the focus of intense study in recent years, resulting in an increased understanding of the mechanisms underlying spermatogenesis. However, far less information is available concerning the early postnatal development of the testis when a number of events occur that are critical for the later onset of spermatogenesis, including the resumption of germ cell mitosis and the arrival of some of these cells at the basement membrane.

Until recently, our appreciation of how spermatogenic cells develop during this period had been based solely on several important, but limited observations. For example, in fetal rats primordial germ cells apparently proliferate from day 12 (1) through day 16, after which they enter a period of mitotic quiescence that extends into the immediate postnatal period (2). By day 4 after birth, however, Clermont and Perey (3) observed large mitotic figures attributed to germ cells, or gonocytes, thus signaling a resumption of gonocyte division. Similarly, once migrating germ cells of fetal rats reach the gonadal ridge, they become nonmotile and are completely enveloped by Sertoli cells, with which they maintain close and exclusive contact during the ensuing prenatal and early postnatal period. However, by day 5 after birth, some gonocytes were seen with cytoplasmic processes directed toward the periphery of the seminiferous cord (4–6), and at or near the same time, some cells had apparently come into contact with the surrounding basement membrane (7). Notably, this change in gonocyte position, or *relocation*, anticipates compartmentalization of the epithelium that occurs about 10 days later (8). In addition, relocation seems to insure the survival of gonocytes since germ cells remaining in the center of the cords subsequently degenerate (9).

Thus, in response to unknown stimuli, two important events take place in the neonatal testis that precede spermatogenesis and are likely to be necessary for its normal onset. First, mitotically quiescent gonocytes reinitiate cell division shortly after birth. In addition, at or near the same time, many germ cells assume a new, peripheral location in the seminiferous cord, where they are likely to come into contact with matrix factors in the basement membrane. Recently, we have been using various approaches both in situ and in gonocyte-Sertoli cell cocultures to study these critical events, to probe their relationship to each other, to characterize the underlying cellular mechanisms, and to identify the factor(s) regulating them. These approaches are outlined in this chapter.

Development of Postnatal Gonocytes In Situ

Resumption of Proliferation

Although previous observations suggested that gonocytes resume mitosis after birth, no information was available detailing precisely when this occurs in rats, whether all cells act in concert, and whether division of these cells depends on extratesticular factors and/or is related to their position in the seminiferous cord. For these reasons we used an autoradiographic approach to assess the ability of gonocytes in situ to incorporate labeled thymidine with increasing postnatal age or with increasing time of incubation in serum- and hormone-free organ culture. Further, we asked whether gonocytes that begin to divide do so before or after relocating to the basement membrane.

DNA replication by gonocytes in seminiferous cords was detected and subsequently quantified in the autoradiographs (i) of testes from pups aged 1–5 days (day of birth = day 1) injected with ^{3}H-thymidine prior to sacrifice or (ii) of organ-cultured testis segments obtained on day 1, incubated for 1–4 days in vitro, and exposed to ^{3}H-thymidine for the final 4 h of incubation (10). The outcome of these analyses is summarized in Table 21.1. We found no proliferative gonocytes in vivo on either postnatal day 1 or 2; however, in testes from 3-day-old pups, approximately 1 in 10 cells incorporated ^{3}H-thymidine, and on each of the subsequent 2 days, statistically equivalent numbers of gonocytes were labeled. Thus, for 2 days after birth, all gonocytes remain quiescent; some then resume mitosis on the next day, day 3. In addition, not all gonocytes begin dividing at the same time; rather, additional cells are recruited to divide on at least the next 2 days of postnatal life.

When we carried out a parallel analysis of gonocytes in organ-cultured testes, we observed a pattern that was remarkably similar to that seen in vivo (Table 21.1). Again, when testes were cultured on the day of birth, no cells divided during the subsequent 2 days in vitro. However, on each of the next 2 days, some cells incorporated ^{3}H-thymidine and, hence, had begun to divide. The observation that gonocytes resume mitosis after 2 days in culture with no added hormones, and the like, makes it unlikely that any extratesticular factors, such as gonadotropins, have a significant

TABLE 21.1. Percentage of gonocytes labeled (±SEM) in autoradiographs of testes from rat pups given ^{3}H-thymidine in vivo on days 1–5 ($n = 3$/age) or in testes that were organ cultured on the day of birth for 1–4 days and exposed to ^{3}H-thymidine 4 h before fixation.

In vivo				Organ culture		
Age (days)	No. of cells scored	Gonocytes labeled (%)		Days	No. of cells scored	Gonocytes labeled (%)
1	1084	0		1	976	0
2	883	0		2	877	0
3	1072	10.6 (4.7)[a]		3	694	4.3 (2.3)[b]
4	774	7.5 (1.3)[a]		4	593	15.2 (0.6)[b]
5	950	13.5 (3.2)[a]				

[a] N.S.
[b] $P < 0.01$.
Source: Data modified from Orth and Boehm (11).

role in stimulating gonocytes to resume proliferation. Rather, our findings point to paracrine factors as possible regulators of gonocyte division after birth.

Relocation of Gonocytes to the Basement Membrane

Although it was clear from earlier reports that at least some postnatal gonocytes undergo a change in position from the central region of the seminiferous cord to its periphery (4–6), this potentially important phenomenon had remained unstudied and, hence, poorly understood. For our studies we first determined when gonocytes of rats begin relocating to the basement membrane and whether extratesticular influences were involved (10). On the day of birth, all gonocytes in vivo appear round and are separated from the basement membrane by portions of Sertoli cell cytoplasm (Fig. 21.1a). However, 3 days later on day 4, we first detected cells with peripherally oriented processes and a few cells (6.9%) in apparent contact with the basement membrane (Figs. 21.1b and 21.1c), an observation we subsequently confirmed by electron microscopy. In addition, gonocytes in testes that we had organ cultured on day 1 were first seen on the basement membrane on day 4 in vitro, essentially mimicking the time course observed in vivo and strongly suggesting that as for proliferation, relocation of these cells begins without the influence of extratesticular factors.

Relationship Between Onset of Proliferation and Relocation

Our findings suggested that at least some gonocytes divide before relocation begins since gonocytes in S-phase were first seen on day 3, one

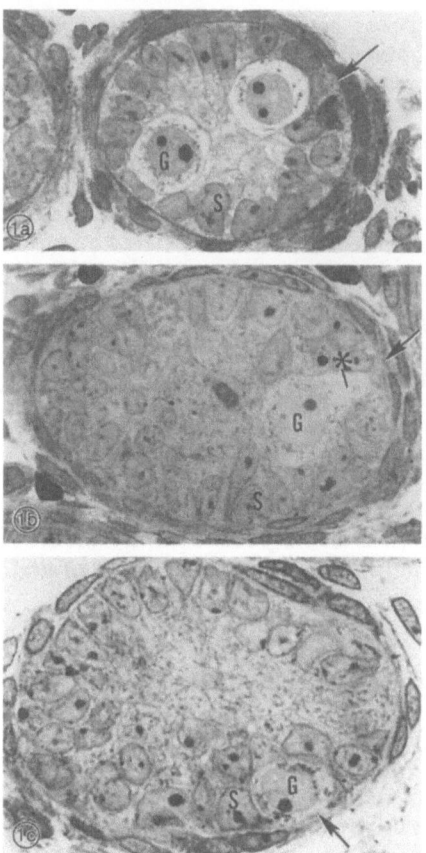

FIGURE 21.1. Light micrographs of seminiferous cords from testes of rat pups on days 1 (*a*) and 4 (*b*, *c*) postpartum. On the day of birth (*a*), all gonocytes (G) are round, and the great majority are separated from the basement membrane (arrows) by portions of Sertoli cells (S). In contrast, on day 4 some gonocytes display prominent cellular processes (asterisk in *b*) directed towards the basement membrane, while other cells at this age have come into contact with the basement membrane, as in (*c*). Reprinted with permission from McGuinness and Orth (10), © Wiley-Liss, a division of John Wiley & Sons, Inc., 1992.

day before any cells were detected on the basement membrane. This observation raised the possibility that mitosis might provide the impetus for relocation of the daughter cells to the periphery of the cord. For this reason we asked (i) whether the first gonocytes to divide subsequently move to the basement membrane and (ii) whether any gonocytes relocate without having first divided (10).

To answer the first question, we labeled the first group of dividing cells and determined their position 1–2 days later. The ^{3}H-thymidine was added to organ cultures of 2-day-old testes at the start of culture, before any cells had begun dividing; 24 h later some of these testes were fixed, while others were placed into unlabeled medium for an additional 24 h and then fixed. Autoradiographic analysis confirmed that although some gonocytes became labeled during the first 24 h, as expected, all cells remained central in the cords during that time. However, 24 h after removal of isotope and 48 h after the start of the culture period, many of the gonocytes on the basement membrane had labeled nuclei (Fig. 21.2A). Thus, at least some of the first cells to divide relocate during the next day.

To address the second issue, we labeled essentially *all* gonocytes dividing through the third postnatal day and then determined whether any relocated cells were unlabeled. Testes were organ cultured on the morning of day 2 for 48 h with ^{3}H-thymidine constantly present in the medium. Tissue was again fixed and processed for autoradiography after 24 or 48 h of incubation, and peripheral gonocytes were examined for the presence of labeled nuclei. We found that while many gonocytes on the basement membrane were labeled, many others were not (Fig. 21.2B). Because it is unlikely that any significant number of gonocytes dividing during the culture period would remain unlabeled under these conditions, we conclude that some gonocytes relocate without having first resumed mitosis. Thus, although both division and relocation occur during the same period of development, some cells relocate without having first divided, and vice versa. This suggests that the onset of gonocyte mitosis and the movement of gonocytes to the basement membrane occur independently of each other and thus may not share a common regulatory mechanism.

Development of Postnatal Gonocytes in Sertoli Cell-Gonocyte Cocultures

Characterization of Cocultures

Cultures containing primarily gonocytes and Sertoli cells were prepared from either 1-day-old or 5-day-old pups by sequential digestion of minced testes in hyaluronidase-collagenase and then in collagenase alone (11), followed by treatment in dissociation buffer to obtain a suspension of single cells (12). When representative aliquots were analyzed for cell size and the presence of alkaline phosphatase activity, less than 5% of the cells were judged to be peritubular cells; approximately 10% were gonocytes, and the remainder were Sertoli cells (11). Cells were plated on Matrigel-coated dishes in serum-free medium containing standard supplements, lactate, transferrin, and retinol. Sertoli cells in these cultures adhere rapidly (20–30 min) to the matrix, as do occasional peritubular cells, while gonocytes attach within a short time to the Sertoli cells.

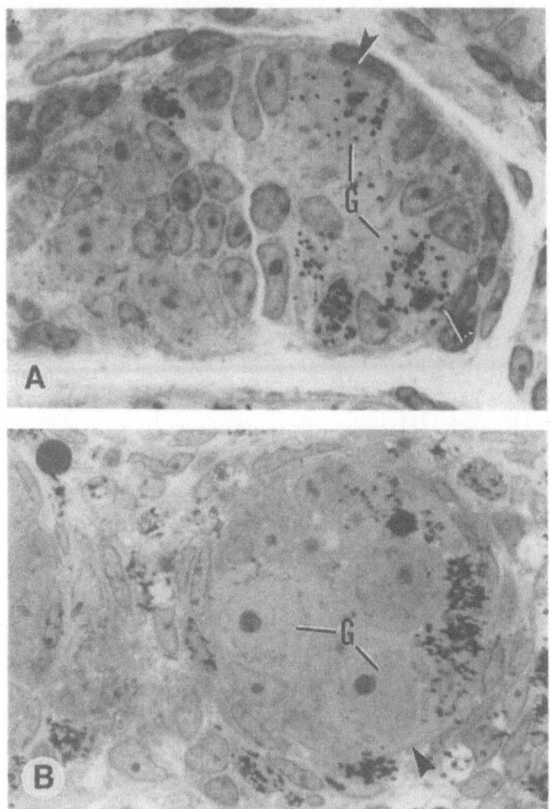

FIGURE 21.2. Autoradiographs of testis placed on organ culture on postnatal day 2 when no gonocytes are proliferative. When tissue exposed to ^{3}H-thymidine for day 1 in vitro was maintained an additional day without label (*A*), some labeled gonocytes (as at G) were at the basement membrane (arrowheads), indicating that some gonocytes divide before relocating. Other tissue was maintained in the presence of isotope for the entire culture period (*B*). In these samples many gonocytes that had moved to the periphery of the cord were labeled; however, some unlabeled cells (G) were also at the basement membrane (arrowhead), suggesting that not all gonocytes divide before relocating. Reprinted with permission from McGuinness and Orth (10), © Wiley-Liss, a division of John Wiley & Sons, Inc., 1992.

Initially, all of the attached gonocytes are uniformly round. However, within several hours some gonocytes isolated from 5-day-old pups develop an obvious, somewhat blunted cellular process, resulting in a morphology we have termed *elongated* (11). Ultrastructural analysis revealed that cells in these cultures assume relationships that are remarkably like those seen in situ shortly after birth and that processes of

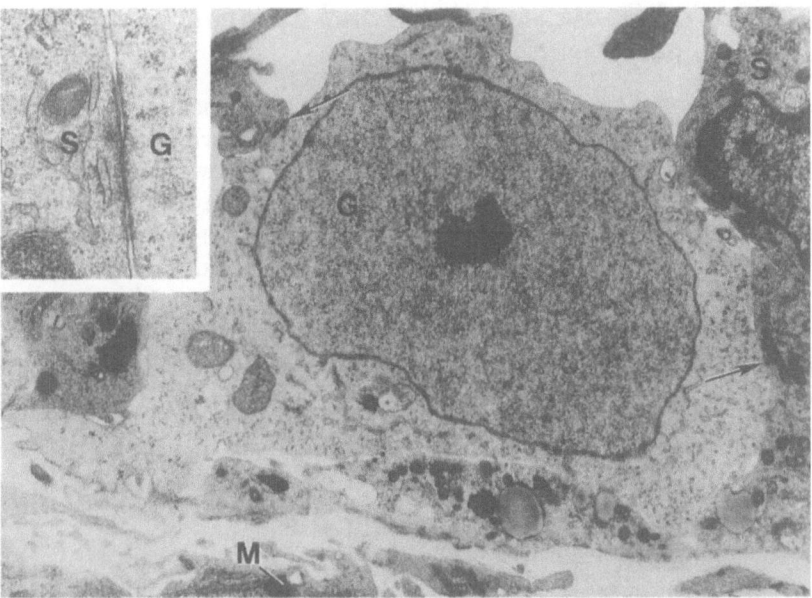

FIGURE 21.3. An electron micrograph showing a portion of a gonocyte-Sertoli cell coculture sectioned in a plane perpendicular to the underlying surface. A gonocyte (G) is surrounded by several Sertoli cells (S) and has extended a process between the latter towards the substrate. Numerous apparent adhesion plaques (arrows) are located between the gonocyte and Sertoli cells, and morphologically recognizable gap junctions (inset) can also be located between the two cell types. Peritubular myoid cells (M) can also be occasionally found beneath both the Sertoli cells and the gonocytes. Adapted with permission from Orth and Boehm (11), © The Endocrine Society, 1990.

gonocytes frequently pass between Sertoli cells to make contact with underlying Matrigel and/or peritubular cells (Fig. 21.3). Moreover, we also found that many gonocytes in the cocultures are structurally and functionally coupled to adjacent Sertoli cells. First, morphologically recognizable gap and adhesive junctions were detected between some of these cells. Also, when the transfer of a gap junction permeant probe, *Lucifer yellow* (LY), was studied in cultures simultaneously bead loaded with both LY and a nonpermeant probe, rhodamine dextran, we verified direct transfer of LY from Sertoli cells to both round and elongated gonocytes (11). Thus, gonocyte-Sertoli cell cultures provide an excellent system with which to explore structural and functional relationships between the two cell types and in which to study the mechanisms, factors, and the like, that are important for gonocyte development after birth.

Initiation of Gonocyte Mitosis and Elongation in Coculture

When cells from 5- to 6-day-old pups were placed in coculture on Matrigel, we noted the rapid appearance (1–2 h) of gonocytes with processes and of gonocytes that incorporated ^{3}H-thymidine (12). However, this was not the case when cells were prepared from pups on the day of birth. Although some Sertoli cells became labeled in ^{3}H-thymidine-treated cultures shortly after plating, no proliferative gonocytes were detected until the third day of culture, at which time a very few incorporated ^{3}H-thymidine. Similarly, in cultures prepared on day 1, all gonocytes remained round until day 2, when about 10% developed processes, or elongated (Fig. 21.4). This time

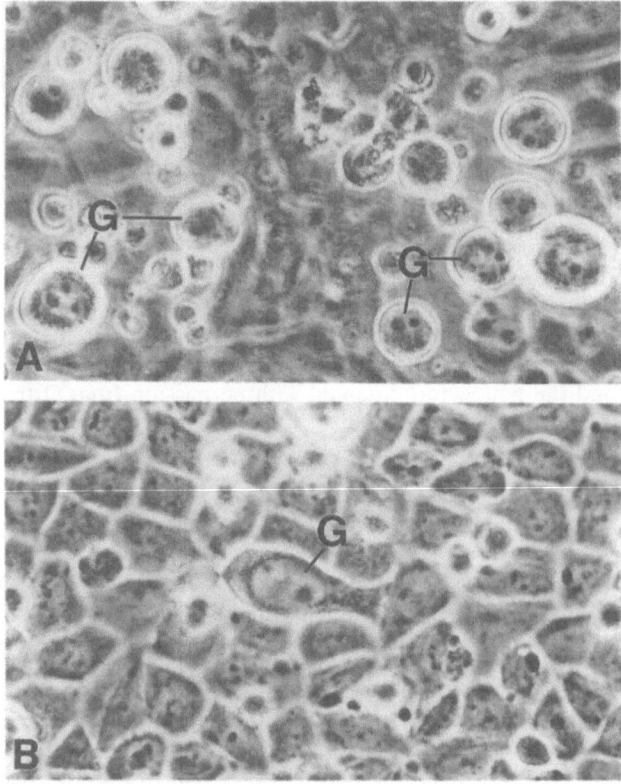

FIGURE 21.4. A gonocyte-Sertoli cell coculture established on the day of birth and viewed with phase contrast optics on either day 1 (*A*) or day 4 (*B*) in vitro. Shortly after plating, gonocytes (G) adhere to the underlying Sertoli cells, but remain uniformly round as in (*A*). However, by day 2 of culture, some gonocytes develop cellular processes directed toward either Sertoli cells, as in (*B*), or toward areas of bare Matrigel.

frame is very similar to that we observed both in vivo and in organ culture, suggesting that gonocytes from newborns undergo maturational processes in coculture that parallel those occurring in situ and that any stimulus for these processes is provided by elements within the culture, such as Sertoli (and possibly peritubular) cells.

Because whole cells could be readily observed in cocultures, we used this system to study division and elongation in the same population of cells. First, we determined by simple inspection of autoradiographs that a large majority of those gonocytes that incorporated ^{3}H-thymidine were round; however, occasional cells with obvious processes were also labeled, suggesting that gonocyte elongation is not incompatible with entry of these cells into S-phase. Next, we exposed some cultures of 5- to 6-day-old gonocytes to cytosine arabinoside to eliminate cells preparing to divide, an effect confirmed by ^{3}H-thymidine autoradiography, and then scored drug-treated gonocytes as elongated or nonelongated. Our findings indicated that in spite of the absence of proliferative gonocytes in the population, many of the remaining cells developed processes. Though somewhat preliminary, these observations lend additional support to our previous evidence obtained in situ and described above that suggests that the elongation of gonocytes is not a sequel to division, but rather an independent event under separate control.

Role of Migration in Gonocyte Relocation

The fact that gonocytes change their position within the seminiferous cord shortly after birth was reported some time ago (5) when early postnatal development of the testis was first examined with electron microscopy. Since that time, no information regarding the nature of this relocation or the underlying mechanism has become available. However, in our studies of various aspects of testicular development, we detected peripherally oriented processes on some gonocytes shortly after birth and noted that the general ultrastructure of these processes was consistent with that of pseudopods, as described for such migratory cells as macrophages and fibroblasts.

Since the ability of primordial germ cells, the ancestors of postnatal gonocytes, to migrate into and populate the gonadal ridge is well recognized, we hypothesized that some gonocytes might regain the ability to move actively after birth in order to relocate to the basement membrane. Such a scenario would require (i) that gonocytes adhere to Sertoli cells while they move since the surface of adjacent Sertoli cells would of necessity provide the only available "substrate" during this movement and (ii) that a gonocyte process, upon making contact with the basement membrane, would be able to attach to that matrix to facilitate relocation of the entire cell to the new position.

Adhesion of Gonocytes to Sertoli Cells and the Basement Membrane

Recently, we have gathered evidence indicating that postnatal gonocytes do attach to both Sertoli cells and extracellular matrix. First, observations on gonocytes in coculture indicate that under those conditions gonocytes attach rapidly to Sertoli cells (11). Also, we detected morphological evidence of apparent adhesion plaques between Sertoli cells and gonocytes both in coculture and in vivo (11). Moreover, by using a protocol involving processing of neonatal testes in hypertonic fixative, we have demonstrated that plasma membranes of gonocytes adhere tightly to those of Sertoli cells at many sites, in spite of shrinkage forces that cause obvious separation of the cells in other areas (Fig. 21.5). In addition, some of the adherent gonocytes seen after these hypertonic incubations had a peripherally directed process. Also, although we have not tested whether gonocyte processes attach to the basement membrane in vivo, we have gathered evidence that these cells, when removed from cocultures by washing and then replated onto bare Matrigel, develop processes that attach to the substrate (Fig. 21.6). Thus, it is likely that postnatal gonocytes adhere both to Sertoli cells and via their processes to matrix factors, thus providing them with the attachments necessary to allow them to move actively to the basement membrane.

Cellular Mechanisms in Gonocyte Relocation: Studies In Vivo

In examining the possibility that gonocytes resume migration to relocate after birth, we also asked whether cellular mechanisms mediating their

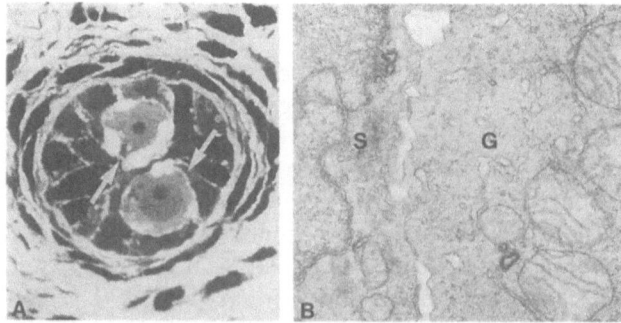

FIGURE 21.5. Tissue from a 5-day-old pup fixed in hypertonic buffer and viewed with either light (A) or electron (B) microscopy. Note areas (arrows) where gonocytes (G) are adherent to Sertoli cells (S) in spite of the shrinkage forces that cause obvious separation of the cells in other areas.

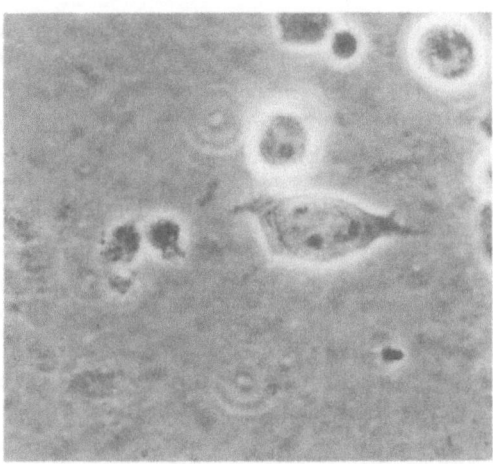

FIGURE 21.6. Phase contrast micrograph of an individual gonocyte from a 5-day-old pup removed several hours earlier from a coculture by washing and then replated onto bare Matrigel. Note the presence of several processes on this cell that have attached directly to the matrix substrate.

relocation in vivo and their development of processes in vitro are compatible with those described for motile cells. It is well recognized that the Golgi complex and cytoskeleton play a central role in promoting migration. For example, in macrophages and fibroblasts (13, 14) the Golgi complex responds rapidly to a chemotactic stimulus by assuming a location in the perinuclear cytoplasm facing the anticipated direction of movement, perhaps to provide a source of new membrane for a growing pseudopod.

Moreover, cytoskeletal elements are important for migration, as evidenced, for example, by the observation that motile cells cease to move in the presence of microtubule-disrupting drugs (15). This may indicate that microtubules emanating from the *microtubule-organizing center* (MTOC), which is closely related to the Golgi complex, direct membrane vesicles from that area to the leading edge of a moving cell. Thus, to determine whether relocating gonocytes utilize mechanisms similar to those of migrating cells, we explored the involvement of the Golgi complex and cytoskeleton in the behavior of postnatal gonocytes.

Development of the Golgi Complex

Because the Golgi complex is likely to respond rapidly when a cell is stimulated to migrate (14), we first compared the appearance of this organelle in gonocytes from newborn pups with that of the Golgi complex in cells from older pups, when some of these cells have begun to relocate.

For this we visualized the Golgi complex at the light microscope level with *thiamine pyrophosphatase* (TPPase) cytochemistry (16) and used conventional transmission electron microscopy to examine its ultrastructure.

When gonocytes from 1- and 5-day-old pups were reacted for TPPase and compared, we noted a striking difference in the outcome (Fig. 21.7). Although an easily detected area of reaction product was seen in virtually all Sertoli cells at both ages, we rarely encountered any recognizable product in gonocytes from 1-day-old pups. In contrast, on day 5 many gonocytes contained a large, localized area of cytoplasm filled with reaction product.

Similarly, when we examined and compared the ultrastructure of the Golgi complex in gonocytes on days 1 and 5, we also noted a striking difference. In the cells of newborns, the Golgi complex was difficult to locate and, when encountered, invariably consisted of a few dispersed stacks, each containing only 3 or 4 cisternae (Fig. 21.8A). However, by day 5 many gonocytes had developed a Golgi complex that was strikingly more extensive, with several large stacks and associated vesicles occupying a relatively large area of perinuclear cytoplasm (Fig. 21.8B). Thus, an apparent enlargement and/or reorganization of the Golgi complex occurs in gonocytes in vivo that correlates temporally with relocation of some cells to the basement membrane.

Orientation of the Golgi Complex

Since substantial evidence indicates that the Golgi complex faces the pseudopod in migrating cells, it would be of obvious value to localize this organelle in relocating gonocytes relative to the position of a cellular process and the direction of apparent movement. However, because our analyses in vivo were of necessity carried out on sectioned tissue, it was not feasible to relate the appearance or position of the Golgi complex in each cell to the presence or location of a cellular process on that cell. Nevertheless, with light microscopy we could identify a population of gonocytes in these sections that was likely to include those in the act of relocating. These cells, with and without processes in the plane of section, were measurably closer to the basement membrane than to the center of the cord and were termed *noncentral*.

When we visually divided the nucleus and perinuclear region of each of these cells into wedge-shaped thirds and scored the position of the TPPase-reacted Golgi complex in each cell relative to the 3 areas, we found that significantly more cells ($P < 0.01$) had a Golgi complex in that third facing the basement membrane than in either of the other 2 positions (44.4% vs. 26.2% and 29.4%). Moreover, this nonrandom distribution did not appear to be spurious since in cells identified as *central*—that is, equally separated from the basement membrane around their circumferences and, hence, presumably nonrelocating—the inci-

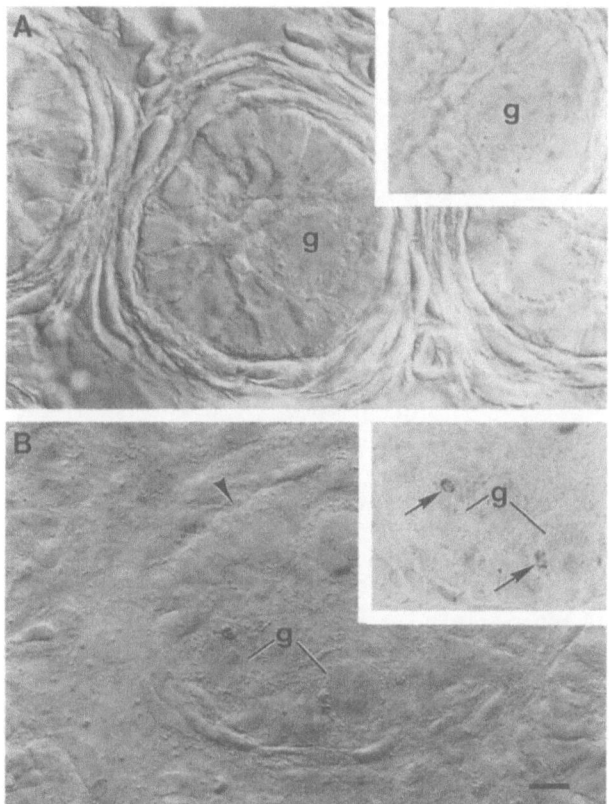

FIGURE 21.7. Companion *differential interference contrast* (DIC) and bright-field (insets) views of frozen, TPPase-reacted sections from testes of 1-day-old (*A*) and 5-day-old (*B*) rats. At birth all gonocytes (g) are round and well separated from the basement membrane by portions of Sertoli cells, as in (*A*). In addition, virtually no reaction product was seen in any of these cells (inset in *A*). However, by 5 days of age, many gonocytes were apparently en route to the basement membrane or already in contact with it (arrowhead in *B*), and a large area of TPPase reaction product was found in many gonocytes at this age (arrows, inset in *B*). Reprinted with permission from McGuinness and Orth (17).

dence of the Golgi complex in each of the randomly oriented thirds of the cells was verified as uniform (35.8%, 31.8%, and 32.4%). These in vivo observations, along with those indicating that the Golgi complex undergoes dramatic changes in some cells between 1 and 5 days after birth, suggest that postnatal gonocytes may utilize mechanisms similar to those of motile cells in order to reach the basement membrane of the seminiferous cord.

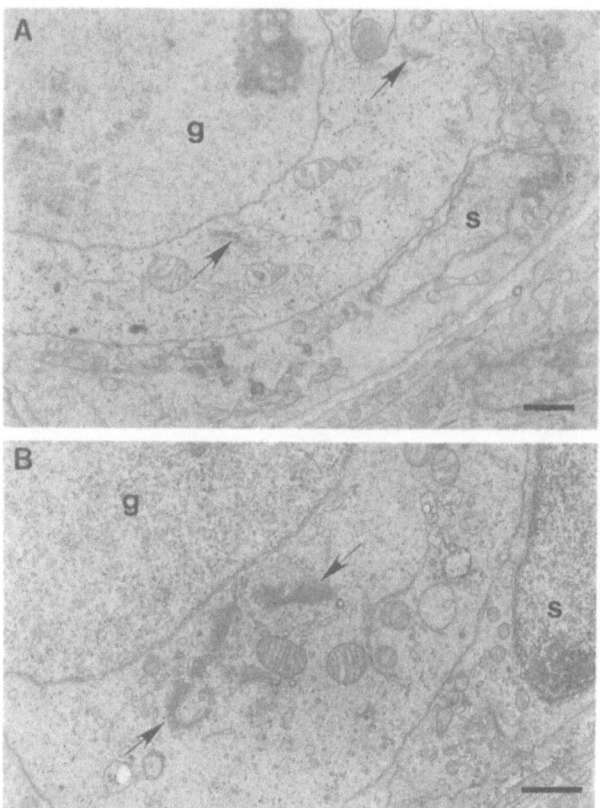

FIGURE 21.8. Electron microscopic views of portions of seminiferous cords from 1-day-old (*A*) and 5-day-old (*B*) rats. At birth the Golgi complex of gonocytes (g) typically consists of a few scattered stacks, each composed of 2 or 3 short cisternae and occasional nearby vesicles (arrows in *A*). However, a few days later the Golgi complex of many gonocytes changes dramatically (between arrows in *B*) and consists of a number of larger stacks occupying a single perinuclear position, with each stack made up of numerous cisternae and associated with many cytoplasmic vesicles. Reprinted with permission from McGuinness and Orth (17).

Cellular Mechanisms in Gonocyte Relocation: Studies in Coculture

Although the results described above suggested that gonocytes may move actively to the basement membrane, it was difficult to obtain direct evidence for this possibility in vivo. For this reason we extended our studies to include gonocyte-Sertoli cell cocultures (17). Notably, we had already documented that the development of gonocytes in coculture parallels that in vivo to a remarkable extent, strongly suggesting that cellular behavior in this system was likely to be similar to that in vivo. In addition, with this approach we could ask the same questions already addressed in vivo, with the added benefit of being able to visualize and study whole cells. Finally, by examining cells in coculture, we could test directly whether function of the Golgi complex and/or cytoskeleton is required for gonocytes to develop processes.

Development and Orientation of the Golgi Complex

For our studies in vitro, we visualized the Golgi complex by incubating cells in NBD-ceramide, a fluorescent lipid probe that preferentially partitions into Golgi membranes (18). Cultures were established on the day of birth and maintained for 1–5 days. On days 1 and 2, we detected only small areas of punctate fluorescence scattered in the perinuclear cytoplasm of all gonocytes (Figs. 21.9A and 21.9B), in contrast to an easily recognized area of fluorescence in most Sertoli cells. However, by day 3 some gonocytes displayed a remarkably different pattern from that seen in these cells on day 1, with bright fluorescence localized to a relatively large, well-defined area beside the nucleus (Figs. 21.9C and 21.9D). Similar observations were also recorded for gonocytes during the next 2 days of culture. Thus, a well-defined Golgi complex develops in some gonocytes after 2 days in vitro, when these germ cells first gain the ability to develop processes.

When we correlated the type of Golgi-associated fluorescence with the morphology of individual gonocytes, we found that a large, well-defined Golgi complex was present in many elongated cells, but also occurred in some nonelongated cells. From these studies we concluded that enlargement and/or reorganization of this organelle may be an early event preceding formation of a process, as is likely to be the case for motile cells responding to a chemotactic stimulus (14).

Because the outcome of our studies in vivo suggested that the Golgi complex was preferentially oriented toward the basement membrane in noncentral gonocytes, we carried out a parallel analysis on elongated cells in culture in which the position of the Golgi complex could be related to that of a cellular process. The nucleus and perinuclear cytoplasm of cells with a single, blunt process were divided visually into thirds, and the position of the Golgi complex was scored relative to the process. We found that the Golgi complex is located in the perinuclear cytoplasm facing the process in 71.7% of elongated cells, compared to an incidence in each of the other areas of 14.6% and 13.8% (17).

Notably, when we subjected nonelongated cells to a similar analysis, we found that the distribution of the Golgi complex relative to an arbitrarily chosen, fixed location in these cells was random. Thus, we concluded that the Golgi complex is preferentially oriented towards the cellular process of elongated gonocytes in coculture and is therefore likely

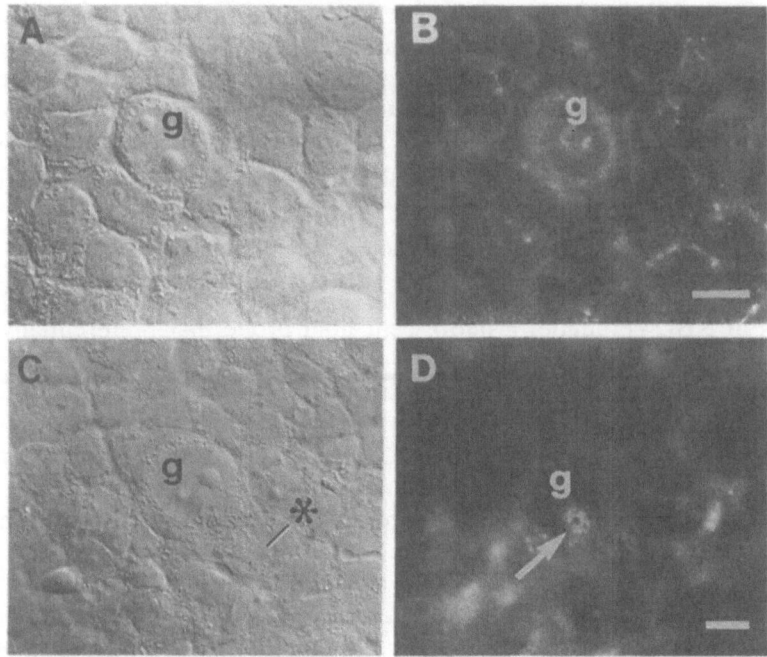

FIGURE 21.9. Corresponding DIC and fluorescent views of cocultures established on the day of birth, maintained for either 1 day (*A, B*) or 3 days (*C, D*), and then fixed and exposed to NBD-ceramide to visualize the Golgi complex. After 1 day in vitro, all gonocytes (g) remained round, as in (*A*), and NBD-ceramide appeared as only a few small spots of punctate fluorescence and some diffuse cytoplasmic fluorescence (*B*). However, by day 3 some gonocytes developed processes (asterisk in *C*). Many of these elongated cells, as well as some round gonocytes, had an apparently enlarged Golgi complex compared to that seen 2 days earlier, with bright and distinct fluorescence occupying a relatively large region of cytoplasm (arrow in *D*). Reprinted with permission from McGuinness and Orth (17).

to be of central importance in its formation and/or maintenance. In subsequent studies (see below), we tested directly whether disruption of the Golgi complex prevents elongation of gonocytes in coculture.

Cytoskeleton of Gonocytes in Coculture

To evaluate the cytoskeleton of gonocytes qualitatively, we used immunocytochemical approaches to visualize tubulin and actin in gonocytes in vitro. First, we incubated gonocytes isolated from 1- or 5-day-old pups with visual probes specific for *filamentous* (F) or *globular* (G) actin and noted a striking difference between cells of the two ages. In the gonocytes of newborns, virtually all actin was unpolymerized, or G-actin (Figs.

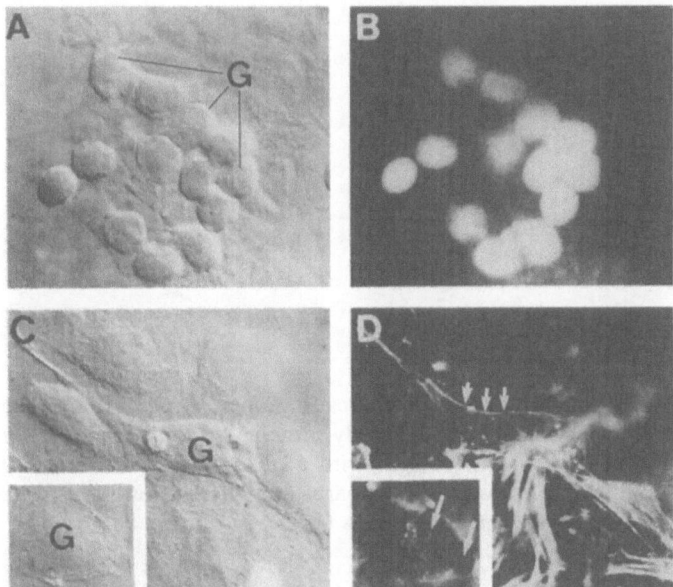

FIGURE 21.10. Companion DIC and fluorescent views of cocultured gonocytes (G) isolated from newborn pups (A, B) and exposed to DNAase I to demonstrate G-actin (B), and from 5-day-old pups (C, D) exposed to rhodamine-phallicidin to demonstrate F-actin (D). Gonocytes from newborns contain abundant unpolymerized G-actin, but little F-actin (not shown). In contrast, cells from 5-day-old pups usually contain an obvious cortical rim of F-actin (arrows in D), as well as F-actin localized to apparent focal adhesions (arrows in inset in D) on underlying Sertoli cells or Matrigel. In addition, gonocytes isolated from older pups generally contain little if any G-actin (not shown).

21.10A and 21.10B), with little F-actin detectable in any of these cells. In contrast, nearby Sertoli cells contained abundant actin filaments, but little G-actin. However, when gonocytes from 5-day-old pups were similarly studied, most of these cells, including many with processes, possessed an obvious cortical rim of F-actin, as well as punctate areas positive for F-actin that we interpreted as likely to be focal adhesions on underlying Sertoli cells (Figs. 21.10C and 21.10D). Although these observations are preliminary, they suggest that the actin cytoskeleton of gonocytes undergoes significant reorganization in the first few days after birth, a change that may be related to a renewed ability of postnatal gonocytes to move actively.

Substantial evidence from the literature implicates microtubules in facilitating pseudopod production in motile cells (13–15). For this reason we also examined the pattern of tubulin immunofluorescence in cocultured gonocytes, with the initial aim of localizing the MTOC in elongated

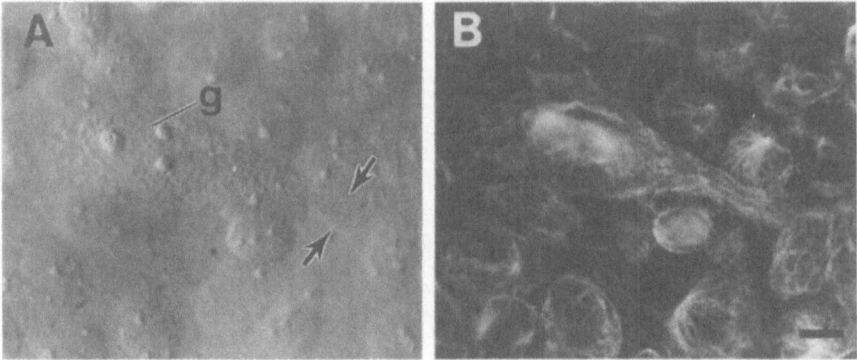

FIGURE 21.11. Immunofluorescence localization of α-tubulin in gonocyte-Sertoli cell cocultures. Gonocytes (g) invariably contained a distinct and elaborate microtubular network, and in those cells with a process (between arrows in *A*), this network could be seen to extend linearly into the process itself. Adapted with permission from McGuinness and Orth (17).

cells relative to the cellular process. We found that in general, gonocytes from 5-day-old pups displayed a very elaborate microtubular network, with fluorescence that was usually brighter and more distinct than that of most Sertoli cells (Fig. 21.11). Moreover, although a MTOC was recognizable in many of the latter, the fluorescence pattern in most gonocytes was so extensive that we were unable to discern an obvious MTOC in most of these cells. Thus, in subsequent studies we used a direct approach to test whether microtubules are involved in gonocyte elongation by exposing cultures to several agents that interfere with microtubular function in vitro, as described below.

Effect of Golgi Complex- and Microtubule-Specific Drugs on Gonocyte Elongation

To determine whether an intact Golgi complex is required for gonocytes to develop processes in coculture, we tested the ability of these cells to elongate in the presence of brefeldin A or monensin, two drugs that disrupt the Golgi complex via different mechanisms (19, 20), and then reevaluated the behavior of the same cultures 24 h after removal of the drug and replacement of standard medium. In addition, some cultures were exposed to NBD-ceramide or subjected to immunofluorescence to visualize the Golgi complex and tubulin, respectively, in order to view directly the effects of the drug on these organelles.

First, we found that (i) drug, but not vehicle, caused a fully reversible disruption of the Golgi complex and that (ii) the microtubular network was intact in drug-treated cells in spite of the effects on the Golgi

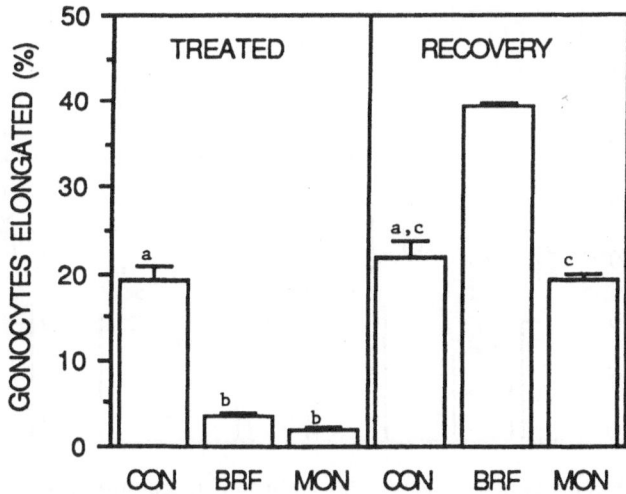

FIGURE 21.12. Effect of either brefeldin A (BRF) or monensin (MON) on the ability of gonocytes to develop processes. Cocultures were prepared from 5-day-old pups and exposed to BRF, MON, vehicle, or medium alone (CON) for 3h postplating. At that time some chambers were fixed, while others in each group were rinsed with drug- or vehicle-free medium and maintained an additional 24h, followed by fixation. In each group at each time, 1500 gonocytes were viewed and scored as elongated or nonelongated. Percentages of gonocytes elongated in vehicle-treated chambers were identical to those in CON chambers at either time (not shown). (a, b, c = not significant; all others = $P < 0.01$.) Reprinted with permission from McGuinness and Orth (17).

complex. Next, when we quantified development of a process by gonocytes in cultures containing brefeldin A or monensin (Fig. 21.12), we found only 3.5% and 1.8%, respectively, of the cells had elongated compared to 19.2% for vehicle-treated controls. In addition, when we removed the drugs and incubated the cultures in standard medium for an additional day, both drug-treated groups fully recovered the ability to elongate. Thus, an intact, functional Golgi complex appears to be an important requirement for development of processes by postnatal gonocytes in vitro.

To determine whether functional microtubules are required for gonocyte elongation, we incubated cocultures in either nocodazole or taxol to depolymerize or stabilize microtubules, respectively, and tested the ability of gonocytes in these cultures to elongate. Again, we also viewed the Golgi complex and microtubules directly in these cultures, as above, and maintained some treated cultures in drug-free medium for an additional 24h to evaluate the ability of the cells to recover. From these studies we found that (i) microtubules were mostly absent and the Golgi

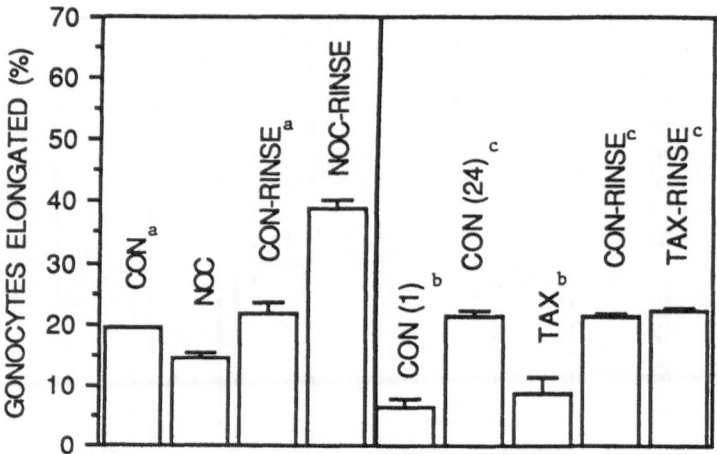

FIGURE 21.13. Percentage of gonocytes elongated in cultures exposed to either nocodazole (NOC) or taxol (TAX) for 24 h compared to that in cultures exposed to medium alone for the same length of time (CON). For both groups the percentage of gonocytes elongated was not altered by the presence of either vehicle alone (not shown). For taxol gonocyte elongation was compared in chambers treated with taxol for 24 h to that seen both 1-h postplating (CON [1]), the time of taxol addition, and 24-h postplating (CON [24]), when taxol-treated chambers were fixed. We also tested the ability of cells to recover from either drug by incubating other drug-treated chambers in drug-free medium for an additional 24 h (NOC-RINSE and TAX-RINSE) and comparing the percentages of gonocytes elongated in these cultures to those in time-matched chambers incubated in medium alone (CON-RINSE). (a, b, c = not significant; all others = $P < 0.01$). Reprinted with permission from McGuinness and Orth (17).

complex dispersed in nocodazole-treated cultures, but both organelles returned to normal after drug removal; (ii) in the taxol-treated cells, stabilized microtubules appeared as well-defined, linear structures and the Golgi complex was apparently intact; and (iii) treatment with nocodazole or taxol either significantly reduced (nocodazole) or virtually abolished (taxol) the ability of gonocytes to elongate, and these effects were wholly reversible (Fig. 21.13).

We concluded from these studies that microtubules have an important role in gonocyte maturation since their dispersal is incompatible with development of processes by these cells, at least in vitro. Moreover, it appears that microtubules that are intact but stabilized by taxol are also unable to support gonocyte elongation. Interestingly, taxol-stabilized microtubules have been shown by others to be incapable of participating in the movement of various vesicular organelles (21), some of which may play a role in inserting new membrane into a forming pseudopod.

Thus, taken together, our findings from studies with drug-treated co-cultures suggest that functional interaction between the Golgi complex and the cytoskeleton of gonocytes is an essential component of the mechanism underlying their elongation. Although the Golgi complex is apparently unaffected by taxol treatment, it is not by itself sufficient to support formation of processes by cells whose microtubules have been stabilized. Similarly, although both brefeldin A and monensin appear to leave microtubules intact, they are not sufficient to allow gonocytes without an intact Golgi complex to elongate. Thus, the Golgi complex and the microtubular network of postnatal gonocytes seem to comprise a functional unit that is essential for formation of processes on these cells, a conclusion that provides strong support for the hypothesis that postnatal gonocytes regain the ability to migrate shortly after birth.

Regulation of Development of Postnatal Gonocytes

Findings from our studies on gonocytes in organ culture and in coculture with Sertoli cells indicate that newborn gonocytes develop the ability to divide and to form processes and apparently migrate without extratesticular factors, as already described. Thus, our work in progress is in large part focused on identifying and characterizing paracrine, intratesticular elements that regulate gonocyte maturation. Findings to date suggest that matrix components of the basement membrane, and possible secreted peptide factors, are likely to modify gonocyte function in the neonatal testis, as summarized below.

Role of Laminin in Gonocyte Division and Elongation in Coculture

To determine whether the underlying matrix influences gonocyte development in vitro, we isolated cells from 5- to 6-day-old pups and cocultured them on various substrates, including Matrigel, pure fibronectin, pure laminin, or polylysine-treated plastic (12). In addition, other cells were cultured on similar substrates containing antibodies against laminin or fibronectin, with the same antiserum present in the medium. When the ability of gonocytes to elongate under each of these conditions was evaluated, the data indicated that the greatest percentage of cells produced processes when cultured on laminin alone, even more than when cultured under standard conditions on Matrigel (Table 21.2), while few cells elongated on plastic, and none elongated on fibronectin. Similarly, when we measured gonocyte proliferation on Matrigel, laminin, or plastic, we also observed enhanced mitosis in laminin-containing chambers compared to plastic, although in this case more cells divided on Matrigel than on

TABLE 21.2. Effect of extracellular matrix on elongation (±SEM) of 5- to 6-day-old gonocytes in vitro.

	Gonocytes elongated (%)	Total N		Gonocytes elongated (%)	Total N
PL—no matrix	5.2 (1.8)[a,c]	1405	PL + anti-LM	2.9 (0.8)[c,e]	900
Matrigel (MG)	19.7 (0.7)[b]	1812	MG + anti-LM	11.0 (0.7)[d]	912
Fibronectin (FN)	13.2 (1.4)[d]	905	FN + anti-LM	0[a,e]	900
Laminin (LM)	24.0 (1.8)[b]	1805	LM + anti-FN	30.0 (1.0)	909

[a,b] $P < 0.05$.
[c,d,e] N.S.
All others: $P < 0.01$.

TABLE 21.3. Effect of extracellular matrix on proliferation (±SEM) of 5- to 6-day-old gonocytes in vitro.

	Gonocytes labeled (%)	Total N
PL—no matrix	8.4 (0.7)[a]	901
Matrigel	24.7 (0.5)[a]	907
Laminin	16.5 (2.9)[a]	903

[a] $P < 0.01$.

laminin (Table 21.3). Thus, the presence of a matrix substrate is critical in allowing gonocytes to elongate or to divide.

Moreover, laminin seems to play a central role in promoting formation of processes by these cells, a conclusion supported by substantial evidence from other systems in which laminin has been shown to stimulate process formation and eventual movement of migrating cells (22, 23). In addition, our pilot observation that gonocytes can themselves attach to Matrigel (Fig. 21.6) suggests that laminin may affect them directly. However, we cannot at present rule out the possibility that the effect of laminin on these cells is an indirect one resulting from its influence on Sertoli or peritubular cells. Nevertheless, regardless of the exact mechanism, our data strongly suggest that cell-matrix interactions are of central importance for postnatal maturation of the gonocyte population.

Summary

In the testis of the neonatal rat pup, gonocytes take the first steps in a complex, long-ranging process that will ultimately lead at maturity to the production of viable sperm. During the first few days after birth, these cells once again begin dividing, and at about the same time, some of them move peripherally to contact and apparently attach to the basement membrane. Results of our studies in vivo and in vitro have led to the conclusion that while some temporal overlap may occur, cells that resume dividing do so independently of their position in the cord, and, similarly, cells that move peripherally do so with or without having first divided. We also found that behavior of postnatal gonocytes in vivo and in vitro conforms to that described for migratory cells moving in response to a directional signal, leading us to conclude that at least some of these cells regain the ability to move actively after birth. Moreover, gonocytes adhere avidly to adjacent Sertoli cells, between which they pass as they approach and eventually contact the basement membrane.

The impetus for resumption of both mitosis and motility by the post-natal gonocyte population originates from within the testis, most likely from Sertoli cells and possibly peritubular cells. In addition, the makeup of the basement membrane, determined at least in adults by both Sertoli and peritubular cells (24), is important in promoting gonocyte division and relocation; in particular, laminin is a likely requirement for these maturational events. Taken together, these observations, albeit incomplete, serve to highlight the complexity and overall importance of the neonatal period in insuring subsequent normal development and successful spermatogenic function of the mammalian testis.

References

1. Hilscher B, Hilscher W, Delbruck G, Lerouge-Benard B. Autoradiographische bestimmung der S-phasen-dauer der gonocyten bei der wistarratte durch einfachund doppelmarkierung. Z Zellforsch 1972;125:229–51.
2. Orth JM. Proliferation of Sertoli cells in fetal and postnatal rats: a quantitative autoradiographic study. Anat Rec 1982;203:485–92.
3. Clermont Y, Perey B. Quantitative study of the cell population of the seminiferous tubules in immature rats. Am J Anat 1957;100:241–66.
4. Huckins C. Changes in gonocytes at the time of initiation of spermatogenesis in the rat. Anat Rec 1963;145:243.
5. Novi AM, Saba P. An electron microscopic study of the development of rat testis in the first 10 postnatal days. Z Zellforsch 1968;86:313–26.
6. Clark JM, Eddy EM. Fine structural observations on the origin and associations of primordial germ cells of the mouse. Dev Biol 1975;47:136–55.
7. Beaumont HM, Mandl AM. A quantitative study of primordial germ cells in the male rat. J Embryol Exp Morph 1963;11:715–40.
8. Vitale R, Fawcett DW, Dym M. The normal development of the blood-testis barrier and the effects of clomiphene and estrogen treatment. Anat Rec 1973;176:333–44.
9. Roosen-Runge EC, Leik J. Gonocyte degeneration in the postnatal male rat. Am J Anat 1968;122:275–300.
10. McGuinness MP, Orth JM. Reinitiation of gonocyte mitosis and movement of gonocytes to the basement membrane in testes of newborn rats in vivo and in vitro. Anat Rec 1992;233:527–37.
11. Orth JM, Boehm R. Functional coupling of neonatal rat Sertoli cells and gonocytes in coculture. Endocrinology 1990;127:2812–20.

12. Orth JM, McGuinness MP. Neonatal gonocytes cocultured with Sertoli cells on a laminin-containing matrix resume mitosis and elongate. Endocrinology 1991;129:1119–21.
13. Nemere I, Kupfer A, Singer SJ. Reorientation of the Golgi apparatus and the microtubule-organizing center inside macrophages subjected to a chemotactic gradient. Cell Motil Cytoskeleton 1985;5:17–29.
14. Kupfer A, Louvard D, Singer SJ. Polarization of the Golgi apparatus and the microtubule-organizing center in cultured fibroblasts at the edge of an experimental wound. Proc Natl Acad Sci USA 1982;79:2603–7.
15. Goldman RD. The role of three cytoplasmic fibers in BHK-21 cell motility, 1. Microtubules and the effects of colchicine. J Cell Biol 1971;51:752–62.
16. Novikoff AB, Goldfischer S. Nucleosidediphosphatase activity in the Golgi apparatus and its usefulness for cytological studies. Proc Natl Acad Sci USA 1961;47:802–10.
17. McGuinness MP, Orth JM. Gonocytes of male rats resume migratory activity postnatally. Eur J Cell Biol 1992;59:196–210.
18. Pagano RE, Sepansik MA, Martin OC. Molecular trapping of a fluorescent ceramide analogue at the Golgi apparatus of fixed cells: interaction with endogenous lipids provides a trans-Golgi marker for both light and electron microscopy. J Cell Biol 1989;109:2067–79.
19. Donaldson JG, Lippincott-Schwartz J, Bloom GS, Kreis TE, Klausner RD. Dissociation of a 110-kD peripheral membrane protein from the Golgi apparatus is an early event in Brefeldin A action. J Cell Biol 1990;111:2295–306.
20. Tartakoff A, Vassalli P. Comparative studies of intracellular transport of secretory proteins. J Cell Biol 1978;79:694–707.
21. Herman B, Albertini DF. A time-lapse video image intensification analysis of cytoplasmic organelle movements during endosome translocation. J Cell Biol 1984;98:565–76.
22. Wujek J, Haleem-Smith H, Yamada Y, Lipsky R, Lan Y, Freese E. Evidence that the B2 chain of laminin is responsible for the neurite outgrowth promoting activity of astrocyte extracellular matrix. Dev Brain Res 1990;55:237–47.
23. Davis L, Ogle R, Little C. Embryonic heart mesenchymal cell migration in laminin. Dev Biol 1989;133:37–43.
24. Skinner M, Tung P, Fritz I. Cooperativity between Sertoli cells and testicular peritubular cells in the production and deposition of extracellular matrix components. Cell Biol 1985;100:1941–7.

22

Role of Retinoids in Spermatogonial Proliferation and Differentiation and the Meiotic Prophase

DIRK G. DE ROOIJ, ANS M.M. VAN PELT, HENK JG VAN DE KANT,
PAUL T. VAN DER SAAG, ANTOINE H.F.M. PETERS,
CHRISTA HEYTING, AND PETER DE BOER

A detailed study was made of the events taking place during the early onset of *vitamin A deficiency* (VAD). Pregnant Wistar rats (18–20 days p.c.) were fed a VAD diet or a normal balanced diet. Newborn male rats received the same diet. Body growth of the animals on the VAD diet was found to slow down after the rats were 40 days old. Cell counts revealed that at a very early stage, between 44 and 46 days, soon after the slowing down of the body growth, the division of A_1- into A_2-spermatogonia no longer occurred, indicating an arrest at this stage of the spermatogenic process. All other types of differentiating spermatogonia were found to develop normally into spermatocytes except for the *intermediate* (In) spermatogonia, some of which were seen to degenerate. As a result of the inhibition of the formation of the A_2-spermatogonia, a maturation depletion was found in which cell numbers decreased to those normally present in stage VIII. This depletion could be calculated to lead to a complete depletion at about day 58.

At day 46 the transition of the *preleptotene* (preL) spermatocytes was already found to be delayed from stage IX to stage X. From day 51 onwards the pachytene spermatocytes in early stage I were morphologically abnormal and were observed to degenerate in late stage I. Immunofluorescent immunohistochemistry with antibodies against a 30- to 33-kd protein of the axial and lateral filaments of the synaptonemal complex and a 125-kd protein of the transverse proteins revealed that the formation of abnormal synaptonemal complexes was severely hampered.

The first results of an in situ hybridization experiment indicated that the A-spermatogonia in the VAD testis possess the nuclear *retinoic acid receptors* (RAR) α and γ. Hence, the effect of vitamin A/retinoic acid replacement on spermatogonial differentiation and proliferation may be a direct one.

It is concluded that at least in the early stages, the effect of VAD on spermatogenesis is very specific. The division of A_1- into A_2-spermatogonia ceases to occur, either because the undifferentiated spermatogonia are no longer able to differentiate into A_1-spermatogonia or because the A_1-spermatogonia are blocked in G2 phase. Almost simultaneously, the start of the meiotic prophase becomes delayed, and the formation of the synaptonemal complexes becomes hampered. Apparently, both the undifferentiated spermatogonia and the early spermatocytes need specific, *retinoic acid* (RA)-dependent factors. These factors are not likely to be common cell cycle-related proteins, as normal S-phase and mitosis can still take place during the onset of VAD and even in the ultimate VAD situation.

Vitamin A is essential for the maintenance of the spermatogenic process (1–4). VAD causes extensive loss of germ cells, and, eventually, only Sertoli cells, spermatogonia, and some spermatocytes are left in the seminiferous tubules (5, 6). The damage caused by VAD is reversible, as replacement of vitamin A in the diet restores normal spermatogenesis (7, 8). In recent years it has been established that during VAD the production of A_2-spermatogonia becomes arrested, while there is also an arrest at the preL spermatocyte level (9–12). The administration of either vitamin A (9–12) or RA (13) causes a massive production of A_2-spermatogonia throughout the testis that synchronously develop along the spermatogenic line. In this way synchronized testes of rats and mice can be obtained in which a restricted number of epithelial stages are present, enabling the study of epithelial stage-related processes in rats and mice (14–20).

In this chapter the effect of VAD on spermatogonia and preL spermatocytes is reviewed, and new data on the specificity of the effect of VAD on spermatogonia and (pre)leptotene spermatocytes are described. Furthermore, some preliminary results are presented relating to the question of whether or not the action of vitamin A/RA on spermatogonia is a direct action or an indirect one via Sertoli cells.

Initial Effect of VAD on Spermatogonial Numbers in Relation to Effects on Body Weight

A number of studies have been performed on the numbers and behavior of spermatogonia and of preL spermatocytes in the ultimate situation of severe VAD (9–12). Still, the earliest events during the onset of the shortage of retinoids may give a better insight into the precise moments during the spermatogenic process at which these molecules play an es-

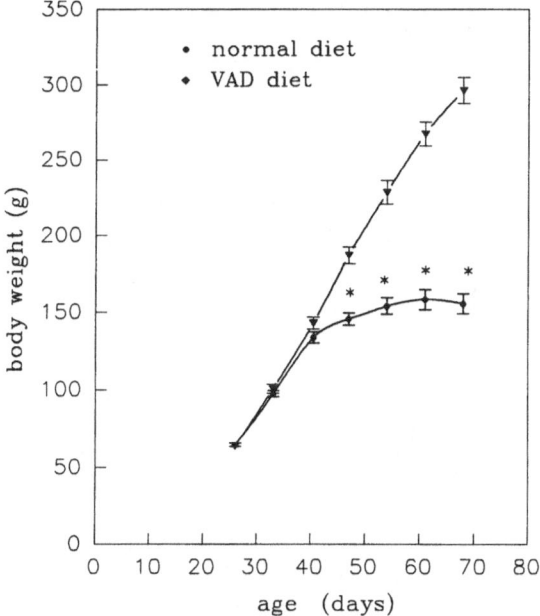

FIGURE 22.1. Growth curve of rats between 26 and 70 days of age put on a normal or VAD diet. Pregnant Wistar rats (18–20 days p.c.) were fed a VAD diet (Teklad Trucking, Madison, WI) or a normal balanced diet (Hope Farms, Woerden, The Netherlands). Newborn male rats received the same diet. From day 26 onwards these rats were weighed with weekly intervals.

sential role. Therefore, we have now studied spermatogonial numbers at the time when body growth begins to slow down.

Body growth was followed in rats receiving either a normal or a VAD diet. No significant differences were seen between the weights of the control rats and those on a VAD diet up to day 40 (Fig. 22.1). However, after day 40 there was a rather abrupt slowing down of the rate of growth of the rats given a VAD diet. Finally, after day 61 the animals started to lose weight and began to show the severe signs of the ultimate VAD condition.

At 44 days of age, increasing numbers of spermatogonia were seen from epithelial stage VIII onwards, each wave of divisions of the differentiating-type spermatogonia (A_1–A_4, In, and B) rendering a significant increase in cell numbers (Fig. 22.2). Although cell counts in age-matched controls have not been performed as yet, it was clear that at this time spermatogonial numbers in the rats given a VAD diet were at least close to normal.

The numbers of A-spermatogonia in stage VIII did not significantly change from days 44 to 51 (Fig. 22.2A). In contrast, in stage X, in which

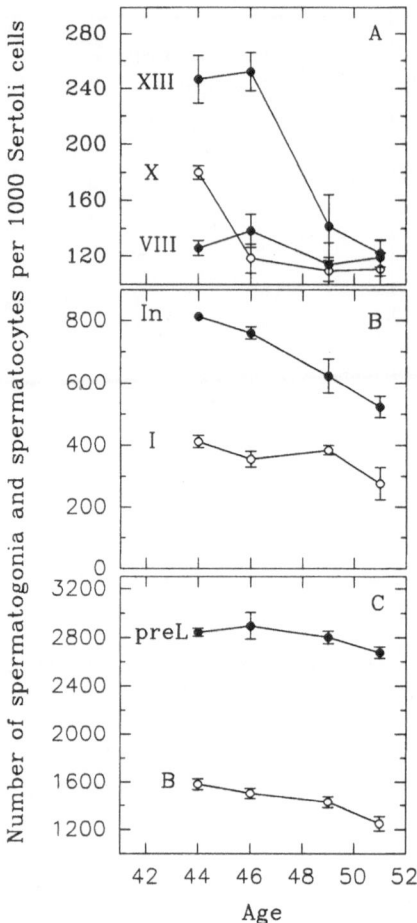

FIGURE 22.2. Numbers of spermatogonia and preL spermatocytes per 1000 Sertoli cells at 44, 46, 49, and 51 days of age in testes of rats fed a VAD diet (mean ± SEM; $n = 4$). Testes were fixed in Bouin's fluid and embedded in glycol methacrylate Technovit 7100 (Kulzer and Co., GmbH, Wehrheim, FRG). Five-micrometer sections were made and stained with periodic acid-Schiff and Gill's hematoxylin No. 3 (Polysciences Inc., Warrington, USA). In each animal and for each epithelial stage, counts were made in 20 tubular sections comprising 350–900 Sertoli cells (Sertoli cell numbers depending on the tubular shrinkage caused by the VAD). Data were analyzed with the t-test modified according to Bonferroni, as described in reference 21. A shows the total numbers of A-spermatogonia in stages VIII, X, and XIII; B shows the total numbers of A-spermatogonia in early stage I and the numbers of In spermatogonia in stages II–III; and C shows the numbers of B-spermatogonia in stage V and the numbers of preL spermatocytes in stage VIII.

the population of A-spermatogonia consists mainly of A_2-spermatogonia, a sharp, significant, decrease was seen at day 46 ($P < 0.01$) to the level seen in stage VIII. Thereafter, the numbers of A-spermatogonia in stage X remained constant at this level. In stage XIII, in which the population of A-spermatogonia consists mainly of A_3-spermatogonia, spermatogonial numbers at day 46 were similar to those at day 44. However, a sharp decrease was seen at day 49 ($P < 0.05$), although in one animal the number of A-spermatogonia was only slightly less than at day 46. At day 51 the number of A-spermatogonia in stage XIII was similar to that in stage VIII.

In whole mounts of seminiferous tubules, it was checked whether the decrease in the numbers of A-spermatogonia in stage X was due to a lack of the wave of divisions of A_1- into A_2-spermatogonia or to enhanced degeneration of these cells. From day 46 onwards no waves of mitotic spermatogonia could be found in stage IX. Hence, the numbers of A-spermatogonia in stage X decreased because of the disappearance of the divisions of the A_1-spermatogonia.

In early stage I the population of A-spermatogonia consists mainly of A_4-spermatogonia. No significant changes in the numbers of A-spermatogonia in stage I were observed from 44 to 51 days (Fig. 22.2B). However, at 51 days spermatogonial numbers in stage I in 3 animals were low and normal in the fourth, indicating that at that time cell numbers were starting to decrease. The numbers of In spermatogonia in stages II and III were found to decrease gradually, the decrease being significant for the first time at day 49 ($P < 0.05$). The numbers of B-spermatogonia were significantly decreased only at day 51 ($P < 0.01$) (Fig. 22.2C). No significant changes in the numbers of preL spermatocytes were seen during the period studied.

These results show that very early during the onset of VAD, spermatogonial proliferation and differentiation become affected. At day 46 it already became evident that the formation of A_2-spermatogonia had been inhibited. As, at that time, the subsequent depletion already completely covered stage X, it can be inferred that the actual onset of the effect had taken place some days earlier. Furthermore, a slight effect on the numbers of In spermatogonia was seen that occurred earlier than could be expected from the depletion originating from the arrest in the formation of A_2-spermatogonia. Also, some degenerating In spermatogonia were observed. Apparently, the In spermatogonia also need retinoids for their development, although a shortage in retinoids does not cause such a drastic effect as on the formation of A_2-spermatogonia.

In conclusion, at the onset of the growth retardation of rats on a VAD diet, there is already a complete arrest of the formation of A_2-spermatogonia, indicating that relatively high levels of retinoids are needed for this process to occur. Also, some In spermatogonia were found to degenerate during the period of 46–51 days, suggesting that these cells, too, need some retinoids for their development. However, no effects of the increasing state of VAD on the other types of spermatogonia (A_2, A_3, A_4, and B) were observed, the numbers of spermatogonia remaining at the control level as long as they did not decrease as a result of the maturation depletion caused by the block in A_2-formation and the degeneration of some In spermatogonia.

Initiation of the Meiotic Prophase During the Onset of VAD

At 44 days of age, the preL spermatocytes were seen clearly to enter leptotene stage in early epithelial stage IX (Fig. 22.3). However, from 46 days onward the transition to leptotene stage was delayed and occurred during stage X. Furthermore, at 51 and 54 days, the pachytene spermatocytes in early stage I were abnormal, as the condensed chromosomes had a more vague appearance than in the normal situation. Subsequently, in late stage I all of the pachytene spermatocytes were found to degenerate. At 61 and 65 days when, due to massive degeneration of spermatids, epithelial stages can no longer be recognized—and in addition to the presence of some preL spermatocytes—spermatocytes were also seen with the appearance of leptotene stage up to the abnormal pachytene type of cells seen in early stage I at 51 and 54 days (Fig. 22.4).

Thus, early during the onset of VAD, while no effects can still be seen on spermatids and late spermatocytes, the preL spermatocytes have difficulties in entering the meiotic prophase. Nevertheless, from day 46 up to day 54, the preL spermatocytes were always seen to aquire a leptotene-like appearance during epithelial stage X despite the fact that less and less vitamin A was available during this period. This led us to wonder whether these spermatocytes really did start the meiotic prophase or if, perhaps, some abortive chromosome condensation took place.

To investigate whether or not the preL spermatocytes during the onset of VAD still started a normal, although possibly delayed, meiotic prophase, fluorescent immunohistochemistry was performed (22) using antibodies directed against proteins of the synaptonemal complex. A monoclonal antibody was used that recognizes a 30- to 33-kd protein that is a constituent of the axial and lateral elements of the synaptonemal complexes (23). A polyclonal antibody against the 125-kd protein (SCP1) that detects the transversal filaments of the synaptonemal complex (24), indicative for meiotic pairing, was also employed. The use of these antibodies permits the unequivocal determination of the various stages of the meiotic prophase. Frozen sections were used from rats at ages 44, 49, and 54 days.

The preliminary observations in the VAD testes indicate, first of all, that the preL spermatocytes did start the meiotic prophase, as proteins of

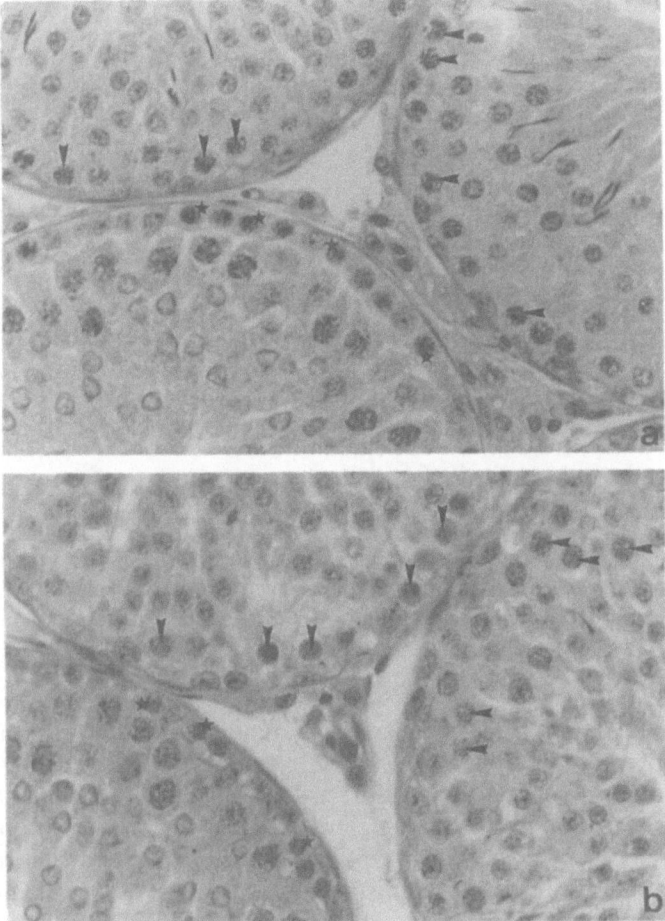

FIGURE 22.3. Testis section of a rat at day 44 showing normal pachytene spermatocytes (arrowheads) in epithelial stage I in 2 tubular cross sections and leptotene spermatocytes in epithelial stage IX (asterisks) (*a*). *b:* Tubular cross sections at day 54 show pachytene-like spermatocytes (horizontal arrowheads) with a vague appearance in early stage I and degenerating spermatocytes (vertical arrowheads) in late stage I. Furthermore, preL spermatocytes are shown in stage IX (asterisks).

the synaptonemal complex were found in early spermatocytes during the whole period studied; that is, up to 54 days (Fig. 22.5). Second, in the normal testis the initiation of the assembly of the axial elements during leptotene can be seen to coincide with the initiation of diplotene in the same tubular cross sections. Hence, normally, tubular cross sections showing diplotene spermatocytes also contain leptotene spermatocytes.

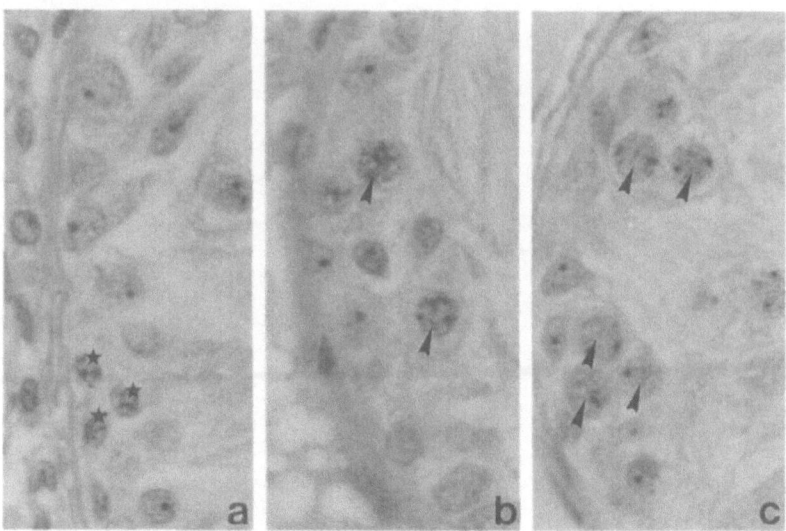

FIGURE 22.4. Tubular cross sections at day 65 during severe VAD. Shown are preL spermatocytes (asterisks) (*a*), leptotene/zygotene-like spermatocytes (arrowheads) (*b*), and pachytene-like spermatocytes (arrowheads) (*c*).

However, at days 49 and 54, early diplotene spermatocytes were found without spermatocytes staining with these antibodies, indicating a delay in the initiation of meiosis. Third, at day 49 enlarged zygotene spermatocytes were found in which meiotic pairing was limited to only a few homologs. Fourth, also from day 49 onwards, pachytene spermatocytes were encountered in which the distance between the lateral elements was visible through the fluorescent label, indicating an incorrect spacing of these lateral elements (Fig. 22.5). Fifth, at day 49, but more prominently at day 54, synaptonemal complexes were encountered that showed distinct interruptions. Sixth, also at day 54 some pachytene spermatocytes were found in which the signal for the transverse filaments had disappeared from the synaptonemal complex (Fig. 22.5). At this stage it is not yet clear whether the abnormal pachytene spermatocytes in early epithelial stage I, as identified by routine microscopy, are in fact delayed zygotene spermatocytes with a sharp reduction in pairing initiation sites.

In conclusion, VAD seems to affect both the visible signs of the onset of meiosis and the proper course of its events. The onset is delayed, and the progression of zygotene development is hindered. The marker proteins that we have studied did become expressed, but the assembly of a proper synaptonemal complex was hampered. In at least some spermatocytes that succeeded in the formation of a complete synaptonemal complex, later effects were seen as specified by an incorrect spacing of the

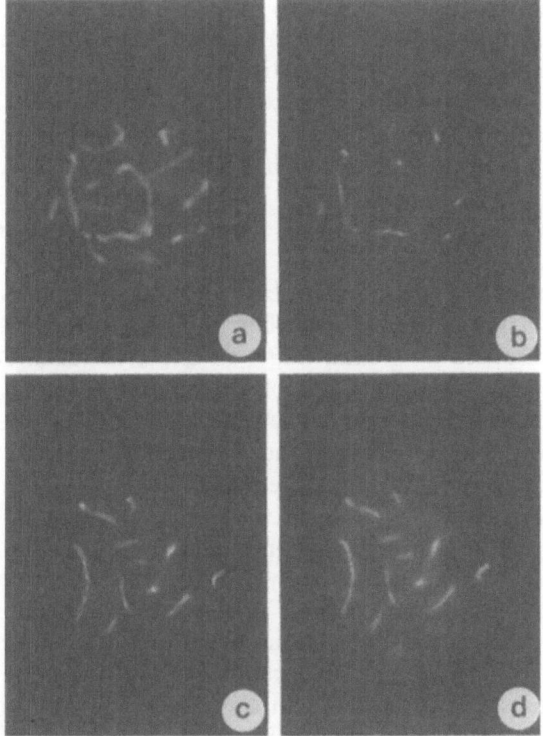

FIGURE 22.5. Pachytene spermatocytes at 54 days of vitamin A depletion with an apparently normal appearance of synaptonemal complexes as shown by an anti-30- to anti-33-kd SC monoclonal antibody fluorescent labeling (*a*), but with a strongly diminished signal for the 125-kd transverse filament proteins (*b*). Furthermore, pachytene spermatocytes are shown at 49 days of vitamin A depletion with synaptonemal complexes that show a visible separation of the lateral elements (*c*). Nevertheless, the signal for the transverse filaments is present (*d*).

lateral elements. Whether the precocious loss of the signal for the transverse filaments is the result of the latter observation or can occur independently is as yet not clear.

Cellular Localization of RAR mRNAs in the Testis

Since vitamin A/RA clearly has an effect on spermatogonial proliferation and differentiation (13), the question arises whether or not this effect is direct or indirect via Sertoli cells. One way to answer this question is to study the occurrence of nuclear receptors for RA in spermatogonia and Sertoli cells. Recently, the expression in the testis of the mRNAs for the receptors RARα (25–28), RARβ (25, 28), and RARγ (28), as well as *retinoic X receptor* α (RXRα) (28) was studied. All these receptors were found to be expressed in the testis, especially after vitamin A replacement. Also, in two studies the expression of some RARs in particular testicular cell types was studied. Kim and Griswold (25) reported the expression of a 1.8-kb and a 4.7-kb RARα transcript exclusively in pachytene spermatocytes and a 2.7-kb RARα transcript and a 2.9-kb RARβ transcript in Sertoli cells. They found no RARβ expression in spermatocytes or spermatids. Eskild et al. (27) found a 4-kb and a 7-kb RARα transcript expressed uniquely in round spermatids.

To date, it is not known whether any of the RAR subtypes are expressed in A-spermatogonia. Using in situ hybridization we now have done some preliminary studies of the hybridization of RAR mRNAs in the Sertoli cells and the A-spermatogonia present in the vitamin A-deficient mouse testis 6 h after administration of RA, which is during the time that the induction of the synchronized differentiation and proliferation of A-spermatogonia takes place.

We used mouse tissue for hybridization with the murine RAR riboprobes because extensive homology was expected to be necessary for RNA-RNA in situ hybridization. Cpb-N mice were made vitamin A deficient as described earlier (16). After attaining vitamin A deficiency, the mice received an intraperitoneal injection of 1-mg RA (Sigma, St. Louis, MO). The RA was stored in ethanol in a concentration of 50 mg/mL. Before injection and for each animal, 20 μL of the RA solution was mixed with 80-μL sesame oil. Animals were sacrificed 6 h after the injection because shortly after replacement of RA in the rat, there is an increased expression of all RAR subtypes (28). The probes used for in situ hybridization were derived from cDNA fragments of mRARα (148 bp in the C-D region), mRARβ (181 bp in the F region), and mRARγ (247 bp in the E-F region), all subcloned in pBluescript SK(−). The ^{35}S-labeled sense and antisense probes were prepared according to Wilkinson and Green (29). Whole testes of vitamin A-deficient mice treated with RA were fixed in 4% paraformaldehyde and embedded in paraffin. Sections of 3 μm were made and transferred to TESPA-coated slides (TESPA = 3-aminopropyltriethoxysilane, Sigma, St. Louis, MO).

The in situ hybridization was carried out according to the procedure of Wilkinson and Green (29). The hybridization mix contained 50% formamide, 0.3 M NaCl, 20 mM Tris-HCl, 5 mM EDTA (pH 8.0), 10% dextran sulphate, 1x Denhardt's solution, and 10 mM Na-phosphate (pH 8.0). The concentration of the probes was 2×10^5 cpm/μL. Hybridization was carried out overnight at 55°C, and the slides were washed under high stringency. The slides were dipped in NTB-2 Kodak emulsion (1:1 dilution) (Kodak, Rochester, NY), exposed for 23 days at 4°C, developed with D-19 Kodak developer, and stained with haemaluin (Mayer)-eosine (alcoholic). The number of silver grains was determined above 100 A-spermatogonia and in at least 25 areas of 325 μm² above the Sertoli cell cytoplasm in testis sections labeled with either the sense or the antisense probe of the various RAR subtypes.

Localization of RARα mRNA

In the testis of the vitamin A-deficient mouse 6 h after injection of RA, Sertoli cells with their cytoplasm throughout the tubular cross sections and A-spermatogonia at the basement membrane were observed. Within the seminiferous tubules a clear hybridization of RARα was found. The number of grains was determined per 325 μm² above the cytoplasm of the Sertoli cells in sections labeled with the antisense or the sense probe for RARα. The hybridization found for RARα with the antisense probe in the Sertoli cells was significantly higher than that after labeling with the sense probe, indicating a specific expression of RARα in the Sertoli cells (Table 22.1). Determining the number of silver grains above A-spermatogonia in sections of the same animal labeled with the sense or antisense probe for RARα also revealed significant hybridization in the A-spermatogonia (Table 22.2).

Localization of RARβ mRNA

In the RA-replaced mouse testis, hybridization with the RARβ mRNA was found in the seminiferous tubules. However, the density of the label

TABLE 22.1. Number of grains per 325 μm² above the cytoplasm of the Sertoli cells after hybridization with the antisense or sense RNA probe for nuclear RARα, RARβ, and RARγ.

	Sense (mean ± SD)	Antisense (mean ± SD)
RARα	2.8 ± 1.9	7.2 ± 3.4*
RARβ	1.3 ± 1.1	4.5 ± 2.0*
RARγ	3.8 ± 2.6	3.1 ± 2.1

* Significantly different from sense probe: $P < 0.01$ ($n > 25$).

TABLE 22.2. Number of silver grains above A-spermatogonia after hybridization with the antisense or sense RNA probe for nuclear RARα, RARβ, and RARγ.

	Sense (mean ± SD)	Antisense (mean ± SD)
RARα	1.4 ± 1.5	4.9 ± 4.0*
RARβ	0.4 ± 0.7	0.8 ± 1.0
RARγ	0.9 ± 1.0	2.6 ± 2.0*

* Significantly different from sense probe: $P < 0.01$ ($n > 100$).

was low. Hybridization was found primarily on the luminal site in the cytoplasm of Sertoli cells and was found to be significantly above background using hybridization with the sense probe for RARβ (Table 22.1). Also, the number of grains above A-spermatogonia was slightly, but not significantly, greater than the background level (Table 22.2).

Localization of RARγ mRNA

Hybridization of RARγ mRNA was also found in the seminiferous tubules. The density of the grains found above Sertoli cells was not significantly different from the background level when comparing the hybridization of the antisense RNA with that of the sense probe of RARγ (Table 22.1). However, within the A-spermatogonia significant hybridization was found (Table 22.2).

Conclusions

In the seminiferous epithelium of the VAD mouse 6 h after injection of RA, RARα mRNA was found in Sertoli cells and A-spermatogonia. RARβ mRNA was found in Sertoli cells, but no significant amount of this mRNA was found in A-spermatogonia. The expression of RARα and RARβ observed in Sertoli cells is in agreement with the results of other studies using Northern blot analysis, in which the expression of RARα mRNA (25, 27) and RARβ mRNA was found (25). In Sertoli cells we found no RARγ mRNA. However, there was a clear expression of RARγ mRNA in A-spermatogonia.

These results indicate that A-spermatogonia express RARα and RARγ mRNA. However, in this in situ hybridization study, we used ^{35}S-labeled probes. Although the scatter of the signal of this radioisotope is less than 1-cell in diameter, silver grains caused by the labeled probe in the border of 1 cell, to some extent may be located above the adjacent cell. In the situation of the testis, the label found in the A-spermatogonia could be the scatter of the signal in the adjacent Sertoli cells. This could be the case for the hybridization of RARβ in A-spermatogonia. However, it could not explain the relatively large difference between the hybridization after labeling with the antisense and sense probe for RARα in both the A-spermatogonia and in the Sertoli cells and the fact that no significant RARγ mRNA hybridization was found in the Sertoli cells while clearly present in A-spermatogonia. Nevertheless, the expression of mRNAs for the different RAR subtypes in A-spermatogonia will be studied in further detail by in situ hybridization with ^{3}H-labeled probes and/or the use of nonradioactive probes. Northern blot analysis is not possible since A-spermatogonia cannot be purified as yet from the adult testis.

The expression of at least two of the RARs in A-spermatogonia indicates that the action of RA on the induction of differentiation and proliferation of these cells in the vitamin A-deficient testis could be direct. This is in accordance with our previous finding that besides vitamin A, RA is also able to fully support the spermatogenic process (13). However, it also remains possible that RA-induced gene expression in Sertoli cells plays an additional role, as Sertoli cells are known to produce many growth factors (recently reviewed in 30).

Discussion

The present results have shown that at the first signs of VAD in the rats, at about 44–46 days of age, the formation of A_2-spermatogonia already becomes completely inhibited. Apparently, a relatively high concentration of retinoids is necessary for this division to take place. The A_2-, A_3-, A_4-, In, and B-spermatogonia already present at day 44 were found to develop normally and to divide normally into the subsequent generations of differentiating spermatogonia up to at least day 51, despite further decreasing levels of retinoids. However, some In spermatogonia were found to degenerate, suggesting that to a certain extent this type of cell is also dependent on vitamin A.

In whole mounts of seminiferous tubules, it was found that the wave of divisions of A_1- into A_2-spermatogonia is lacking from day 46 onwards and that the decrease in the numbers of A-spermatogonia in stage X is not caused by degeneration of these cells. During the period between 44 and 51 days, the numbers of A-spermatogonia in stages X and XIII both decreased to those in stage VIII. Hence, the A-spermatogonia in stage VIII do not divide and also do not degenerate from day 46 onwards.

As a consequence of the block in the formation of A_2-spermatogonia, a maturation depletion takes place during which, in more and more stages, the number of spermatogonia decreases to the level present in stage VIII. It can be calculated that the depletion of the spermatogonia will become complete at about day 58, coinciding with the onset of the severe signs of VAD.

It is intriguing that the A-spermatogonia in stage VIII are the only spermatogonia really affected by the onset of the VAD. All other spermatogonia proliferate to some extent and differentiate normally into preL spermatocytes. This suggests that vitamin A/RA has a very specific effect on the development of the A-spermatogonia in stage VIII and that this effect does not likely consist of the inhibition of the expression of a normal cell cycle-related gene that is active in all proliferating germ cells.

As described by a number of authors (9–12), in the ultimate VAD condition in the testis, only A-spermatogonia and a relatively low number of preL spermatocytes are left. The precise nature of the A-spermatogonia in the VAD testis is still a matter of debate. In the normal seminiferous epithelium, the undifferentiated spermatogonia proliferate until about stages III–IV, when most of them become arrested in G1 phase. During the period of proliferation, A_{al}-spermatogonia are formed, most of which differentiate into A_1-spermatogonia that will enter S-phase in stage VIII and subsequently divide into A_2-spermatogonia (reviewed in 9, 31).

According to Van Pelt and De Rooij (9, 11), the A-spermatogonia in the severe VAD state are comparable with the A-spermatogonia normally present in stage VII; that is, a mixture of largely quiescent A_s-, A_{pr}-, and A_{al}-spermatogonia. In this model RA is necessary to induce the differentiation of the A_{al}-spermatogonia into A_1-spermatogonia that subsequently start their cell cycle to divide into A_2-spermatogonia. In contrast, according to Griswold et al. (10) and Ismail et al. (12), in the VAD testis the arrest of spermatogenesis is in the G2 phase of the A_1-spermatogonia, these spermatogonia being unable to start their mitosis into A_2-spermatogonia. The difference in opinion arose from the fact that one group found an increased number of spermatogonial mitoses shortly after the administration of vitamin A (10, 12), and the other did not (9, 11). More extensive studies of the labeling index and the mitotic index of the A-spermatogonia in the VAD testis before and after administration of RA are presently being carried out in an attempt to clarify this issue. Nevertheless, the VAD testis offers the unique possibility to study either the induction of the differentiation of the undifferentiated spermatogonia into A_1-spermatogonia or the genes involved in a G2 arrest mechanism specific for type A_1-spermatogonia.

Also early during the VAD, when no effect can be seen as yet on late spermatocytes and on spermatids, preL spermatocytes were found to delay the onset of leptotene stage. However, this delay was only temporary because as long as stages could be identified, leptotene stage was found to begin during epithelial stage X. Also, during severe VAD, at days 61 and 65, leptotene spermatocytes were also seen. Hence, it can be concluded that for the initiation of the meiotic prophase, RA is not required. However, after the delayed onset of the meiotic prophase in these spermatocytes, problems arose with the assembly of a normal synaptonemal complex, and these spermatocytes were seen to degenerate in epithelial stage I. In the normal epithelium these spermatocytes would be in early pachytene stage. Apparently, in the VAD situation the (pre)leptotene spermatocytes are missing one or more RA-dependent factors vital for a proper meiotic prophase.

It has been established that spermatocytes contain the nuclear receptor RARα. Hence, the effect of RA on early spermatocytes may be a direct one. However, an as-yet-unsolved question is whether the effect of vitamin A/RA on A-spermatogonia is a direct one or an indirect one via Sertoli cells. It has been well established now that Sertoli cells possess nuclear receptors for RA, RARα, and RARβ, and these cells will contain target genes for RA. The present results indicate that the A-spermatogonia present in the VAD testis contain nuclear receptors

RARα and RARγ. Hence, A-spermatogonia may be directly affected by RA as well. Studies are now being carried out to find possible target genes for RA in these spermatogonia.

In conclusion, the present results suggest that both A-spermatogonia and (pre)leptotene spermatocytes need very specific, RA-dependent factors to enable them to start the series of divisions of the differentiating spermatogonia and to traverse a normal meiotic prophase, respectively. These factors are not likely to be common cell cycle-related proteins, as normal S-phase and mitosis can still take place during the onset of VAD and even in the ultimate VAD situation. The genes encoding for these specific factors will now be the target for our future studies.

Acknowledgment. This study was supported in part by the Dutch Science Foundation (NWO) through GB-MW (Medical Sciences).

References

1. Wolbach BS, Howe PR. Tissue changes following deprivation of fat-soluble A vitamin. J Exp Med 1925;42:753–77.
2. Mason KE. Differences in testes injury and repair after vitamin A deficiency, vitamin E deficiency and inanition. Am J Anat 1933;52:153–239.
3. Howell JMcC, Thompson JN, Pitt GAJ. Histology of the lesions produced in the reproductive tract of animals fed a diet deficient in vitamin A alcohol but containing vitamin A acid, I. The male rat. J Reprod Fertil 1963;5:159–67.
4. Thompson JN, Howell JMcC, Pitt GAJ. Vitamin A and reproduction in rats. Proc R Soc Lond [Biol] 1964;159:510–35.
5. Mitranond V, Sobhon P, Tosukhowong P, Chindaduangrat W. Cytological changes in the testes of vitamin A-deficient rats, I. Quantitation of germinal cells in the seminiferous tubules. Acta Anat 1979;103:159–68.
6. Unni E, Rao MRS, Ganguly J. Histological and ultrastructural studies on the effect of vitamin A depletion and subsequent repletion with vitamin A on germ cells and sertoli cells in rat testis. Indian J Exp Biol 1983;21:180–92.
7. Huang HFS, Hembree WC. Spermatogenic response to vitamin A in vitamin A deficient rats. Biol Reprod 1979;21:891–904.
8. Huang HFS, Durenfurth I, Hembree WC. Endocrine changes associated with germ cell loss during vitamin A deficiency and vitamin A-induced recovery of spermatogenesis. Endocrinology 1983;112:1163–71.
9. De Rooij DG, Van Dissel-Emiliani FMF, Van Pelt AMM. Regulation of spermatogonial proliferation. In: Ewing LL, Robaire B, eds. Regulation of testicular function: signaling molecules and cell-cell communication. Ann NY Acad Sci 1989;564:140–53.
10. Griswold MD, Bishop PD, Kim KH, Ping R, Siiteri JE, Morales C. Function of vitamin A in normal and synchronized seminiferous tubules. In: Ewing LL, Robaire B, eds. Regulation of testicular function: signaling molecules and cell-cell communication. Ann NY Acad Sci 1989;564:154–72.

11. Van Pelt AMM, De Rooij DG. The origin of the synchronization of the seminiferous epithelium in vitamin A-deficient rats after vitamin A replacement. Biol Reprod 1990;42:677–82.
12. Ismail N, Morales C, Clermont Y. Role of spermatogonia in the stage-synchronization of the seminiferous epithelium in vitamin-A-deficient rats. Am J Anat 1990;188:57–63.
13. Van Pelt AMM, De Rooij DG. Retinoic acid is able to reinitiate spermatogenesis in vitamin A-deficient rats and high, replicate doses support the full development of spermatogenic cells. Endocrinology 1991;128:697–704.
14. Morales C, Griswold MD. Retinol-induced stage synchronization in seminiferous tubules of the rat. Endocrinology 1987;121:432–4.
15. Bartlett JMS, Weinbauer GF, Nieschlag E. Quantitative analysis of germ cell numbers and relation to intratesticular testosterone following vitamin A-induced synchronization of spermatogenesis in the rat. J Endocrinol 1989;123: 403–12.
16. Van Pelt AMM, De Rooij DG. Synchronization of the seminiferous epithelium after vitamin A replacement in vitamin A-deficient mice. Biol Reprod 1990;43:363–7.
17. Huang HFS, Marshall GR, Nieschlag E. Enrichment of the stages of the seminiferous epithelium in vitamin A-replaced-vitamin A-deficient rats. J Reprod Fertil 1990;88:51–60.
18. Van Beek MEAB, Meistrich ML. A method for quantifying synchrony in testes of rats treated with vitamin A deprivation and readministration. Biol Reprod 1990;42:424–31.
19. Bartlett JMS, Weinbauer GF, Nieschlag E. Stability of spermatogenic synchronization achieved by depletion and restoration of vitamin A in rats. Biol Reprod 1990;42:603–12.
20. Van Beek MEAB, Meistrich ML. Stage-synchronized seminiferous epithelium in rats after manipulation of retinol levels. Biol Reprod 1991;45:235–44.
21. Fleis JL. In: Barnett V, Bradley RA, Hunter JS, et al., eds. The design and analysis of clinical experiments. New York: Wyley HY, 1986;104–5.
22. Heyting C, Dietrich AJJ. Meiotic chromosome preparation and protein labelling. Methods Cell Biol 1991;35:177–202.
23. Heyting C, Dettmers RJ, Dietrich AJJ, Redeker EJW, Vink ACG. Two major components of synaptonemal complexes are specific for meiotic prophase nuclei. Chromosoma 1988;96:325–32.
24. Meuwissen RLJ, Offenberg HH, Dietrich AJJ, Riesewijk A, Van Iersel M, Heyting C. A coiled-coil related protein specific for synapsed regions of meiotic prophase chromosomes. EMBO J 1992;11:5091–100.
25. Kim KH, Griswold MD. The regulation of retinoic acid receptor messenger RNA levels during spermatogenesis. Mol Endocrinol 1990;4:1679–88.
26. Haq RU, Pfahl M, Chytil F. Retinoic acid affects the expression of nuclear retinoic acid receptors in tissues of retinol-deficient rats. Proc Natl Acad Sci USA 1991;88:8272–6.
27. Eskild W, Ree AH, Levy FO, Jahnsen T, Hansson V. Cellular localization of messenger RNAs for retinoic acid receptor-α, cellular retinol-binding protein, and cellular retinoic acid-binding protein in rat testis—evidence for germ cell-specific mRNAs. Biol Reprod 1991;44:53–61.

28. Van Pelt AMM, Van den Brink CE, De Rooij DG, Van der Saag PT. Effects of retinoids on retinoic acid receptor mRNA levels in the vitamin A-deficient rat testis. Endocrinology 1992;131:344–50.
29. Wilkinson DG, Green J. In situ hybridization and the three-dimensional reconstruction of serial sections. In: Copp AJ, Cockroft DL, eds. Post-implantation mammalian embryos: a practical approach. IRL Press, 1990: 155–71.
30. Jégou B. The Sertoli cell. Baillieres Clin Endocrinol Metab 1992;6:273–311.
31. De Rooij DG. Proliferation and differentiation of undifferentiated spermatogonia in the mammalian testis. In: Potten CS, ed. Stem cells: their identification and characterization. Edinburgh: Churchill Livingstone, 1983: 89–117.

23

Role and Regulation of Spermatogenic Cell-Specific Gene Expression: Enzymes of Glycolysis

E.M. Eddy, J.E. Welch, C. Mori, K.D. Fulcher, and D.A. O'Brien

Many of the unique structural and functional features of spermatogenic cells result from the expression of genes producing male germ cell-specific mRNAs (1). There are four features of genes that are expressed during spermatogenesis that may contribute to this outcome. The first is that some genes are transcribed only in spermatogenic cells, and there are no similar genes expressed in somatic cells; for example, protamines (2, 3). The second is that some genes are transcribed only in spermatogenic cells, but are cognates of similar genes expressed in somatic cells. The cognate genes encode proteins with comparable sequences in spermatogenic cells and somatic cells that are assumed to have like functions; for example, cytochrome c_T of spermatogenic cells and cytochrome c_S of somatic cells (4, 5). The third is that some genes are transcribed both in spermatogenic cells and somatic cells, but the mRNAs produced in spermatogenic cells differ in size or sequence from those transcribed in somatic cells. These differences arise during transcription and/or processing of the nascent transcript into mRNA; for example, angiotensin-converting enzyme (6). The fourth is that the expression of many genes in spermatogenic cells is developmentally regulated, with transcription beginning at a discrete time during either mitotic, meiotic, or postmeiotic phases of spermatogenesis (7, 8).

We assume that the cell- and stage-specific patterns of expression of these genes are regulated by mechanisms intrinsic to spermatogenic cells, with cell type-specific combinations of transcription factors binding to gene-specific regulatory elements to promote or suppress transcription. However, it is likely that extrinsic cues are necessary to modulate these intrinsic mechanisms and thereby enable the successful progression of spermatogenesis. We are currently identifying genes expressed in spermatogenic cells, characterizing their developmental patterns of expression, and probing the intrinsic mechanisms regulating expression of these genes in male germ cells. The longer-term goal is to determine how extrinsic mechanisms modulate spermatogenic cell-specific expression of specific genes. There is a critical need to understand how cues coming from somatic cells of the testis or other germ cells influence transcription in spermatogenic cells. Among the genes that we are studying for these purposes are those encoding the glycolytic enzymes glyceraldehyde-3-phosphate dehydrogenase and hexokinase.

Energy production in most cells requires metabolism of glucose to pyruvate by the enzymes in the glycolytic pathway (Fig. 23.1). Although some ATP is generated by this process, the major role of glycolysis is to fuel the Krebs cycle and oxidative phosphorylation of the respiratory chain where most ATP is made. Although glycolysis is a highly conserved metabolic process, several enzymes in the glycolytic pathway appear to have spermatogenic cell-specific isozymes. Cognate genes for glycolytic enzymes that are currently known to be expressed only in spermatogenic cells are those for phosphoglycerate kinase-2 (9, 10) and glyceraldehyde 3-phosphate dehydrogenase-s (11, 12). Furthermore, we have recently identified cDNAs representing three hexokinase mRNAs that are unique to spermatogenic cells and appear to be products of another cognate gene (13).

In addition, several other enzymes in the glycolytic pathway of spermatogenic cells are reported to have unique structural or functional properties based on enzymology and protein chemistry studies. These are phosphoglucose isomerase (14), aldolase (15), phosphoglycerate mutase (16), and enolase (17). We have hypothesized that some of these glycolytic enzymes are also products of cognate genes expressed only in spermatogenic cells. Furthermore, two other enzymes involved in energy metabolism are encoded by genes expressed only in spermatogenic cells. These are lactate dehydrogenase-C (18, 19) and cytochrome c_T (4).

Glyceraldehyde-3-Phosphate Dehydrogenase

Glyceraldehyde-3-phosphate dehydrogenase (Gapd, EC1.2.2.12) removes a hydrogen and adds a phosphate to glyceraldehyde-3-phosphate to form 1,3-diphosphoglycerate during glycolysis (Fig. 23.1); during this process it converts NAD^+ to NADH. To study this enzyme, we isolated a cDNA clone encoding Gapd from a mouse spermatogenic cell expression library (12). By Northern blotting this cDNA was found to hybridize with a 1.5-kb mRNA first present on day 20 in mouse testes (12) and on day 29 in rat testes (unpublished observations). These ages correspond to the beginning of round spermatid development in juvenile males in these species. The cDNA also hybridized with a 1.5-kb mRNA present in round sper-

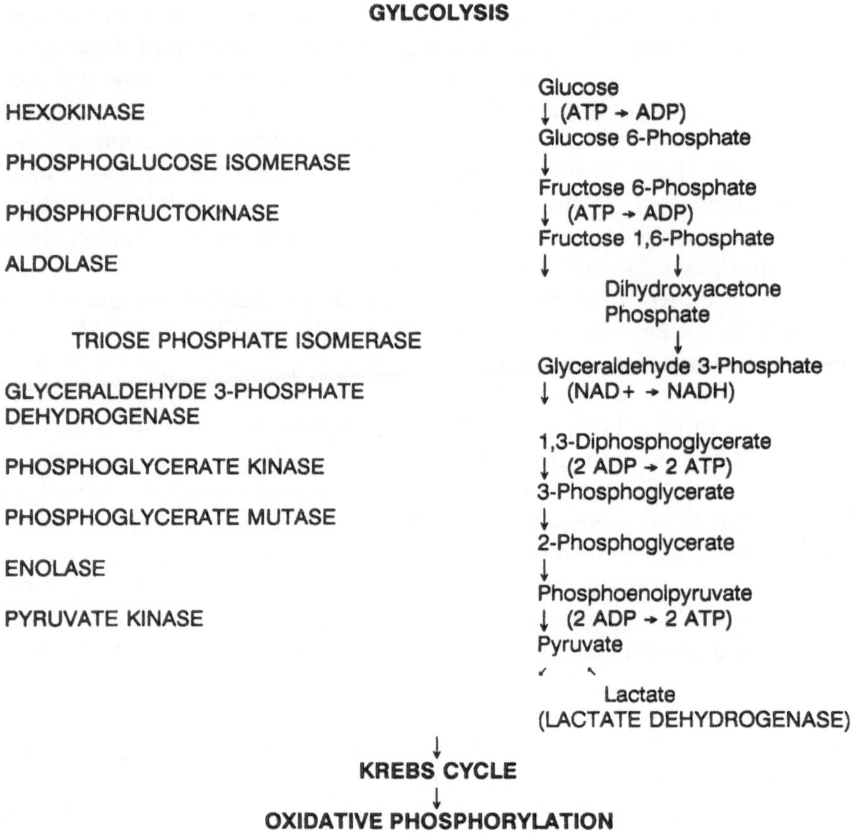

GYLCOLYSIS

Glucose

HEXOKINASE $\downarrow$ (ATP → ADP)

Glucose 6-Phosphate

PHOSPHOGLUCOSE ISOMERASE $\downarrow$

Fructose 6-Phosphate

PHOSPHOFRUCTOKINASE $\downarrow$ (ATP → ADP)

Fructose 1,6-Phosphate

ALDOLASE $\downarrow$ $\downarrow$

Dihydroxyacetone
Phosphate

TRIOSE PHOSPHATE ISOMERASE $\downarrow$

Glyceraldehyde 3-Phosphate

GLYCERALDEHYDE 3-PHOSPHATE $\downarrow$ (NAD+ → NADH)
DEHYDROGENASE

1,3-Diphosphoglycerate

PHOSPHOGLYCERATE KINASE $\downarrow$ (2 ADP → 2 ATP)

3-Phosphoglycerate

PHOSPHOGLYCERATE MUTASE $\downarrow$

2-Phosphoglycerate

ENOLASE $\downarrow$

Phosphoenolpyruvate

PYRUVATE KINASE $\downarrow$ (2 ADP → 2 ATP)

Pyruvate

Lactate
(LACTATE DEHYDROGENASE)

$\downarrow$

KREBS CYCLE

$\downarrow$

OXIDATIVE PHOSPHORYLATION

FIGURE 23.1. Pathway of glycolysis. Cognate genes for phosphoglycerate kinase-2 (*Pgk-2*) and *Gapd-s* have previously been reported to be expressed only in spermatogenic cells, as described in references 9–12. There may also be a gene for a spermatogenic cell-specific hexokinase (*Hk1-s*), as decribed in reference 13. In addition, earlier studies reported that unique isozymes may be present in sperm or spermatogenic cells for phosphoglucose isomerase (described in reference 14), aldolase (described in reference 15), phosphoglycerate mutase (described in reference 16), and enolase (described in reference 17). Genes for 2 other enzymes involved in energy metabolism, lactate dehydrogenase-c (*Ldh-c*) and cytochrome c_T, are also expressed only in spermatogenic cells, as described in references 4, 18, and 19.

matids and condensing spermatids, but not with mRNA from pachytene spermatocytes isolated from mouse testes or with mRNA from 16 different mouse somatic tissues (12).

Analysis of the nucleotide sequence of this and overlapping cDNAs indicated a greater than 70% predicted amino acid sequence similarity with a protein encoded by a mouse somatic Gapd cDNA (20). It also contained the appropriate amino acids in the locations necessary to form the NAD^+-binding pocket and thioester substrate-binding site (12). A rat somatic cell Gapd cDNA provided by Dr. P. Fort (21) was found to hybridize with a 1.3-kb mRNA in 16 somatic tissues and the testis of mouse and also to hybridize weakly with a 1.5-kb mRNA in mouse testis. However, the first 105 amino acids of the predicted mouse spermatogenic cell Gapd protein sequence had no similarity to any known protein. The cDNA appeared to represent the product of a *Gapd* gene expressed only in spermatogenic cells, which we have termed *Gapd-s* (12). Subsequent Southern blotting studies have verified that *Gapd-s* is a single-copy gene in the mouse and distinct from the *Gapd* gene expressed in somatic tissues (unpublished observations).

The Northern blots of mRNA from mouse testes of different ages and of mRNA from isolated spermatogenic cells suggested that the *Gapd-s* gene is first expressed during the postmeiotic phase of spermatogenesis. This was examined more critically by in situ hybridization on sections of mouse testes with an antisense cRNA probe of *Gapd-s*. These studies demonstrated that *Gapd-s* transcription begins during the early cap phase of spermiogenesis in steps 4–6 of round spermatid development (11). They also demonstrated that expression of the *Gapd-s* gene occurs at the same stage of spermatogenesis in juvenile and adult mice. An additional finding was that the *Gapd-s* mRNA is abruptly degraded between steps 15 and 16 of spermiogenesis (11).

Hexokinase

Hexokinase (EC2.7.1.1) is the initial enzyme of glycolysis and utilizes an ATP to convert glucose to glucose 6-phosphate (Fig. 23.1). Earlier studies suggested that there is a sperm-type hexokinase activity (22–24). This isozyme was not seen in testes of 10- or 15-day-old rats, but was detected at 20 days and thereafter by starch gel electrophoresis (24). To determine if a spermatogenic cell-specific gene for hexokinase is expressed in spermatogenic cells, we screened a mouse spermatogenic cell cDNA library with a mouse hepatoma cell line hexokinase (*Hk1*) cDNA (25). A spermatogenic cell hexokinase cDNA was isolated that had 99% identity to the *Hk1* cDNA over most of its length, but contained a unique sequence at the 5' end (13). We refer to this clone as *Hk1-sa*. Northern blot analysis with a probe derived from the unique 5' region of *Hk1-sa* demonstrated hybridization with a 4.1-kb mRNA present in testis, but not in other mouse tissues. However, a probe derived from the region common to *Hk1* and *Hk1-sa* hybridized with a 4.1-kb mRNA that was detected in most somatic tissues examined and was abundant in the testis (13).

The developmental expression of the spermatogenic cell hexokinase mRNA was determined by Northern blotting with the probe derived from the unique 5' region of *Hk1-sa*. A 4.1-kb mRNA was first present in low abundance in testes of 16-day-old juvenile mice and in higher abundance in 22-day-old mice (13). In addition, a 4.1-kb mRNA was present in low abundance in juvenile rat testes at day 23 and at higher abundance at day 29 (unpublished observations). For both species the appearance of the lower-abundance mRNA corresponds to the age when pachytene spermatocytes are the most advanced spermatogenic cells, and the appearance of higher-abundance mRNA corresponds to the age when round spermatids become numerous. These findings were confirmed by in situ hybridization (13).

Because the original cDNA did not represent a full-length mRNA sequence, a mouse spermatogenic cell library was screened with the probe from the unique 5' region. In addition to a clone containing more of the *Hk1-sa* 5' sequence, two other closely related cDNAs were isolated (13). Like *Hk1-sa*, these two clones (referred to as *Hk1-sb* and *Hk1-sc*) were nearly identical to the *Hk1* cDNA except at the 5' end. Part of the sequence in this region was the same for the three spermatogenic cell hexokinase clones, but each clone also had a unique sequence at its extreme 5' end. In addition, *Hk1-sb* contained a specific internal sequence that was not present in *Hk1*, *Hk1-sa*, or *Hk1-sc*. Using probes derived from the extreme 5' unique sequences of *Hk1-sa* and *Hk1-sb*, it was determined by Northern blotting that the probe specific for *Hk1-sa* hybridized with a relatively low abundance mRNA that first appeared in pachytene spermatocytes, while the probe specific for *Hk1-sb* hybridized with an mRNA first present in higher abundance in spermatids (13).

Discussion

It is likely that unique mRNAs are present in each of the different cell types of the body. As in male germ cells, some of these may come from genes expressed only in one cell type, some may result from cell type-specific transcriptional processes, and some may be developmentally regulated. However, male germ cells appear to be unusual in the frequency with which these events occur (1, 7, 8, 26). One of the paradoxes of spermatogenesis is that it results in a cell with a highly differentiated phenotype, but with totipotent developmental capacity. Perhaps it is the requirement of meeting these diverse needs that results in the high degree of unique gene expression in spermatogenic cells.

Gapd appears to have a key role in the regulation of glycolysis in round spermatids and spermatozoa. Although spermatids contain the glycolytic pathway enzymes (27, 28) and can metabolize glucose to lactate (27, 29), the addition of even low amounts of glucose to isolated rat spermatids causes a rapid depletion of ATP stores and a buildup in concentration of fructose 1,6-phosphate (30, 31). This has been interpreted to indicate that glycolysis is blocked by an enzyme downstream of phosphofructokinase (Fig. 23.1) in the pathway (30). The concentration of intermediates after Gapd did not increase in these studies, suggesting that this enzyme is inhibited and limits glycolysis in round spermatids (27, 30, 32). The depletion of ATP that occurs upon addition of glucose may result from 2 ATPs being used to metabolize glucose to glyceraldehyde 3-phosphate. It is not until later in glycolysis that 4 new ATPs are generated.

Other studies have suggested how Gapd is regulated in spermatids and how ATP levels are maintained in these cells in the absence of effective glycolysis. Gapd activity is inhibited in cell-free extracts of spermatids by ATP and other adenine nucleotides (27). In addition, ATP-generating pathways are inhibited in cells that have a high energy charge (a high ratio of ATP to other adenosine nucleotides) (33, 34). Even in the absence of effective glycolysis, hamster spermatids appear able to maintain a high energy charge by utilizing lactate (35), and lactate is the preferred energy substrate for rat spermatids (31, 36–38). The glucose concentration within the seminiferous epithelium is apparently low, while the lactate concentration is about the same as that in plasma (39), conditions that would favor the utilization of lactate by spermatids as their main energy substrate. A major role for the spermatogenic cell-specific lactate de-hydrogenase-C isozyme (18, 19) may be to convert lactate to pyruvate for utilization in the citric acid cycle and thereby to maintain ATP levels in spermatids.

Although glycolysis is inhibited in spermatids, it is active in sper-matozoa. In vitro fertilization studies indicate that glucose is required in the medium for sperm to carry out capacitation, the acrosome reaction, hypermotility, and fertilization (40–42). These processes do not occur if lactate, pyruvate, or fructose is substituted for glucose as the energy substrate. Fertilization is not blocked by oligomycin (41), an inhibitor, or oxidative phosphorylation, suggesting that the glycolytic pathway serves a key role in sperm energy production. Reproductive toxicology studies have indicated that a metabolite of α-chlorohydrin can cause infertility by inhibiting Gapd activity in sperm (43, 44).

Taken together, these studies suggest that Gapd regulates glycolysis in postmeiotic germ cells by being switched off in spermatids and switched on in sperm. Although adenine nucleotides may be responsible for in-hibiting Gapd activity in spermatids, how the enzyme becomes activated in sperm is unknown. It is probably not due to a drop in ATP because addition of glucose to spermatids in vitro depletes ATP levels without enabling glycolysis. However, if the adenine nucleotide that inhibits Gapd binds with high affinity, ATP levels may be depleted beyond recovery before Gapd can become active. Another possibility is that modification of the Gapd in sperm by phosphorylation or some other posttranslational process allows it to become active. An additional possibility is that multiple processes are involved in regulating glycolysis in male germ cells, perhaps with additional enzymes in the glycolytic pathway also being inhibited in spermatids.

The occurrence of a common sequence near the 5' end of *Hk1-sa*, *Hk1-sb*, and *Hk1-sc* suggests that they are products of a common gene. Preliminary Southern blotting results are consistent with this (unpublished observations), supporting the suggestion that they are also the products of a hexokinase cognate gene expressed only in spermatogenic cells. In addition, there appears to be developmentally regulated use of alternative start sites and alternative splicing events during expression of this gene to produce different *Hk1-s* mRNAs during distinct phases of spermatogenesis. Additional studies will be necessary to confirm these observations, but the regulation of expression of the *Hk1-s* gene in spermatogenic cells appears to be a complex process.

The putative *Hk1-s* gene is similar to other members of the hexokinase gene family. There are four hexokinase isozymes reported in mammals, and they vary in their tissue distribution and kinetic properties. The type 1, 2, and 3 isozymes consist of single polypeptide chains of ~100 kd. These enzymes are inhibited by an elevated concentration of their product, *glucose 6-phosphate* (Glc-6-P). The type 4 isozyme, usually referred to as *glucokinase*, is a single polypeptide of ~50 kd and is insensitive to Glc-6-P (45).

Genes have been cloned for each of these four hexokinases. The three spermatogenic cell-specific hexokinase cDNAs (*Hk1-sa*, *Hk1-sb*, and *Hk1-sc*) show considerable similarity to the somatic *Hk1* sequence except at the 5' end. Sequence comparisons between mouse, rat, bovine, and human *Hk1* sequences indicate that the 15 amino acids at the N-terminal end are identical (46). This sequence corresponds to a hydrophobic domain believed to bind to porin in the outer mitochondrial membrane (47, 48). *Porin* is a pore-forming protein involved in the movement of adenine nucleotides across the outer mitochondrial membrane. However, the *Hk1-s* cDNAs lack the porin-binding domain, and this region is occupied instead by the spermatogenic cell-specific sequence.

It is unknown what properties the common spermatogenic cell-specific sequence or the unique 5' terminal sequences confer on the proteins encoded by the *Hk1-sa*, *Hk1-sb*, and *Hk1-sc* mRNAs. However, the N-terminal region of *Gapd-s* is also quite different from that present in *Gapd* (12). The N-terminal regions of the cognate spermatogenic and somatic cell proteins do not include the portions likely to be responsible for enzymatic activity. Although the *Hk1-s* and *Gapd-s* N-terminal sequences do not show obvious similarities, they might serve similar purposes. It has been reported that glycolytic enzymes of mammalian sperm appear to associate as a complex with insoluble components of the cell

(49). Others have noted that glycolytic enzymes cofractionate with sperm tail components (50) or are difficult to extract from sperm (51). Aldolase was found to associate with sperm components by noncovalent, ionic, and hydrophobic interactions and not by disulfide or covalent bonds (15).

These observations have led to the suggestion that glycolytic enzymes may function as multienzyme complexes in sperm (15, 49). We speculate that the unique N-terminal sequences of *Gapd-s*, *Hk1-s*, and perhaps those of other spermatogenic cell-specific glycolytic enzymes, may be involved in binding to other sperm components or in forming multienzyme complexes.

Acknowledgments. This work was supported in part by NIH Grants HD-26485 to D.A.O. and P30-HD-18968 to the Laboratories for Reproductive Biology, University of North Carolina at Chapel Hill.

References

1. Eddy EM, Welch JE, O'Brien DA. Gene expression during spermatogenesis. In: de Kretser DM, ed. The molecular biology of the male reproductive system. Orlando, FL: Academic Press, 1993.
2. Hecht NB. Regulation of haploid expressed genes in male germ cells. J Reprod Fertil 1990;88:679–93.
3. Oliva R, Dixon GH. Vertebrate protamine genes and the histone-to-protamine replacement reaction. In: Cohn ME, Moldave K, eds. Progress in nucleic acid research and molecular biology; vol 40. New York: Academic Press, 1991: 25–94.
4. Virbasius JV, Scarpulla RC. Structure and expression of rodent genes encoding the testis-specific cytochrome c: differences in gene structure and evolution between somatic and testicular variants. J Biol Chem 1988;263: 6791–6.
5. Hake LE, Hecht NB. Utilization of an alternative transcription initiation site of somatic cytochrome c in the mouse produces a testis-specific cytochrome c mRNA. J Biol Chem 1993;268:4788–97.
6. Howard TE, Shai S-Y, Langford KG, Martin BM, Bernstein KE. Transcription of testicular angiotensin-converting enzyme (ACE) is initiated within the 12th intron of the somatic ACE gene. Mol Cell Biol 1990;10:4294–302.
7. Eddy EM, O'Brien DA, Welch JE. Mammalian sperm development in vivo and in vitro. In: Wassarman PM, ed. Elements of mammalian fertilization; vol I. Boca Raton, FL: CRC Press, 1991:1–28.
8. Wolgemuth DJ, Watrin F. List of cloned mouse genes with unique expression patterns during spermatogenesis. Mammalian Genome 1991;1:283–8.
9. McCarrey JR, Thomas K. Human testis-specific PGK gene lacks introns and possesses characteristics of a processed gene. Nature 1987;326:501–5.
10. Boer PH, Adra CN, Lau Y-F, McBurney MW. The testis-specific phosphoglycerate kinase gene Pgk-2 is a recruited retroposon. Mol Cell Biol 1987;7:3107–12.

11. Mori C, Welch JE, Sakai Y, Eddy EM. In situ localization of spermatogenic cell-specific glyceraldehyde 3-phosphate dehydrogenase (Gapd-s) messenger ribonucleic acid in mice. Biol Reprod 1992;46:859–68.
12. Welch JE, Schatte EC, O'Brien DA, Eddy EM. Expression of a glyceraldehyde 3-phosphate dehydrogenase gene specific to mouse spermatogenic cells. Biol Reprod 1992;46:869–78.
13. Mori C, Welch JE, Fulcher KD, O'Brien DA, Eddy EM. Unique hexokinase mRNAs lacking the porin-binding domain are developmentally expressed in mouse spermatogenic cells. Biol Reprod 1993.
14. Buehr M, McLaren A. An electrophoretically detectable modification of glucosephosphate isomerase in mouse spermatozoa. J Reprod Fertil 1981; 63:169–73.
15. Gillis BA, Tamblyn TA. Association of bovine sperm aldolase with sperm subcellular components. Biol Reprod 1984;31:25–35.
16. Fundele R, Winking H, Illmensee K, Jagerbauer E-M. Developmental activation of phosphoglycerate mutase-2 in the testis of the mouse. Dev Biol 1987;124:562–6.
17. Edwards YH, Grootegoed JA. A sperm-specific enolase. J Reprod Fertil 1983;68:305–10.
18. Millan JL, Driscoll CE, LeVan KM, Goldberg E. Epitopes of human testis-specific lactate dehydrogenase deduced from a cDNA sequence. Proc Natl Acad Sci USA 1987;84:5311–5.
19. Sakai I, Sharief FS, Li S-L. Molecular cloning and nucleotide sequence of the cDNA for sperm-specific lactate dehydrogenase-C from mouse. Biochem J 1987;242:619–22.
20. Sabath DE, Broome HE, Prystowsky MB. Glyceraldehyde-3-phosphate dehydrogenase mRNA is a major interleukin 2-induced transcript in a cloned T-helper lymphocyte. Gene 1990;91:185–91.
21. Fort P, Marty L, Piechaczyk M, et al. Various rat adult tissues express only one minor mRNA species from the glyceraldehyde-3-dehydrogenase multigenic family. Nucleic Acids Res 1985;13:1431–42.
22. Katzen HM. The multiple forms of mammalian hexokinase and their significance to the action of insulin. In: Weber G, ed. Advances in enzyme regulation; vol 5. New York: Pergamon Press, 1967:335–56.
23. Katzen HM, Solerman DD, Cirillo VJ. Tissue distribution and physiological significance of multiple forms of hexokinase. Ann NY Acad Sci 1968;151: 351–8.
24. Sosa A, Altamirano E, Hernandez P, Rosado A. Developmental pattern of rat testis hexokinase. Life Sci 1972;11:499–510.
25. Arora KK, Fanciulli M, Pedersen PL. Glucose phosphorylation in tumor cells: cloning, sequencing, and overexpression in active form of a full length cDNA encoding a mitochondrial bindable form of hexokinase. J Biol Chem 1990;265:6481–8.
26. Erickson RP. Molecular genetics of mammalian spermatogenesis. In: Gwatkin RBL, ed. Genes in mammalian reproduction. New York: Wiley-Liss, 1993: 1–26.
27. Nakamura M, Fujiwara A, Yasamasu I, Okinaga S, Arai K. Regulation of glucose metabolism by adenine nucleotides in round spermatids from rat testes. J Biol Chem 1982;257:13945–50.

28. Tvermyr SM, Froysa A, Jutte NHPM, Hansson V. Glucose metabolism in rat germ cells: mechanism of action of gossypol. Ann NY Acad Sci 1984;438: 543–5.
29. Mita M, Hall PF. Metabolism of round spermatids from rats: lactate as preferred substrate. Biol Reprod 1982;26:445–55.
30. Nakamura M, Okinaga S, Arai K. Studies of metabolism of round spermatids: glucose as unfavorable substrate. Biol Reprod 1986;35:927–35.
31. Grootegoed JA, Jansen R, van der Molen HJ. Effect of glucose on ATP dephosphorylation in rat spermatids. J Reprod Fertil 1986;77:99–107.
32. Nakamura M, Okinaga S, Arai K. Metabolism of round spermatids: evidence that lactate is preferred substrate. Am J Physiol 1984;247:E234–42.
33. Atkinson DE, Chapman AG. The adenylate energy charge in the study of enzymes in vitro. Methods Enzymol 1979;55:229–35.
34. Malhotra OP, Bernhard SA. Noncovalent modulation by ATP of the acyl transfer from acyl-glyceraldehyde-3-phosphate dehydrogenase to phosphate. Biochemistry 1989;28:124–8.
35. Mackenbach P, Den Boer PJ, de Jong JW, Grootegoed JA. Effects of glucose and adenosine in the ATP content of hamster spermatids. Reprod Fertil Dev 1990;2:145–52.
36. Jutte NHPM, Jansen R, Grootegoed JA, Rommerts FFG, van der Molen HJ. FSH stimulation of the production of pyruvate and lactate by rat Sertoli cells may be involved in hormonal regulation of spermatogenesis. J Reprod Fertil 1983;68:219–26.
37. Mita M, Hall P. Metabolism of round spermatids from rats: lactate as the preferred substrate. Biol Reprod 1982;26:445–55.
38. Nakamura M, Hino A, Yasumasu I, Kato J. Stimulation of protein synthesis in round spermatids from rat testes by lactate. J Biochem (Tokyo) 1981; 89:1309–15.
39. Setchell BP, Scott TW, Voglmayr JK, Waites GMH. Characteristics of testicular spermatozoa and the fluid which transports them into the epididymis. Biol Reprod 1969;1:40–66.
40. Hoppe PC. Glucose requirement for mouse sperm capacitation in vitro. Biol Reprod 1976;15:39–45.
41. Fraser LR, Quinn PJ. A glycolytic product is obligatory for initiation of the sperm acrosome reaction and whiplash motility required for fertilization in the mouse. J Reprod Fertil 1981;61:25–35.
42. Cooper TG. The onset and maintenance of hyperactivated motility of spermatozoa in the mouse. Gamete Res 1984;9:55–74.
43. Jones AR. The antifertility actions of α-chlorohydrin in the male. Life Sci 1978;23:1625–46.
44. Stevenson D, Jones AR. Production of (S)-3-chloroactaldehyde from (S)-α-chlorohydrin by boar spermatozoa and the inhibition of glyceraldehyde-3-phosphate dehydrogenase in vitro. J Reprod Fertil 1985;74:157–65.
45. Wilson JE. Regulation of mammalian hexokinase activity. In: Beitlher R, ed. Regulation of carbohydrate metabolism. Boca Raton, FL: CRC Press, 1984: 45–85.
46. Griffin LD, Gelb BD, Wheeler DA, Davision D, Adams V, McCabe ERB. Mammalian hexokinase 1: evolutionary conservation and structure to function analysis. Genomics 1991;11:1014–24.

47. Adams V, Griffin L, Towbin J, Gelb B, Worley K, McCabe ERB. Porin interaction with hexokinase and glycerol kinase: metabolic microcompartmentation at the outer mitochondrial membrane. Biochem Med Metab Biol 1991;45:271–91.
48. Smith AD, Wilson JE. Disposition of mitochondrially bound hexokinase at the membrane surface, deduced from reactivity with monoclonal antibodies recognizing epitopes of defined location. Arch Biochem Biophys 1991;287: 359–66.
49. Storey BT, Kayne FJ. Energy metabolism of spermatozoa, VII. Interactions between lactate, pyruvate and malate as oxidative substrates for rabbit sperm mitochondria. Biol Reprod 1978;18:527–36.
50. Mohri H, Mohri T, Ernster L. Isolation and enzymatic properties of the midpiece of bull spermatozoa. Exp Cell Res 1965;38:217–46.
51. Harrison RAP. Glycolytic enzymes in mammalian spermatozoa: activities and stabilities of hexokinase and phosphofructokinase in various fractions from sperm homogenates. Biochem J 1971;124:741–50.

24

Regulation of the Cyclic Nucleotide-Dependent Pathway in Seminiferous Tubule Cells

Marco Conti, Carla Boitani, Cristina D'Alessandris,
Saveria Iona, Lucia Monaco, Anna Rita Morena,
Claudio Sette, Elena Vicini, Gaetano Frajese,
and Mario Stefanini

It is well established that the development and maturation of the male gamete is dependent on germ cell interaction with somatic cells (1–3). Thus, in the protected environment of the seminiferous tubules, germ cells and somatic cells are continuously exchanging the signals necessary to integrate their functions. In the last 15 years, there has been a major effort to elucidate the biochemical nature of the signals mediating these cell-to-cell interactions (3–5). One concept that has emerged from these studies is that the mechanisms of signal transduction must be operating in seminiferous tubule cells. These mechanisms are necessary for somatic and germ cells to receive and elaborate extracellular signals (6) and to translate them into changes in cell function. Activation of signal transduction pathways must also play an important role in mediating another cell-cell interaction, that between the spermatozoon and the oocyte (7, 8).

A large body of work supports the idea that the cyclic nucleotide-mediated signal transduction is one of the pathways involved in the differentiation and maturation of somatic Sertoli cells (3, 6, 9). This pathway mediates most of the effects of the gonadotropin FSH. Conversely, little is known about the role of this signal transduction pathway in the differentiation of male germ cells (6), although in the female cAMP has been implicated in the control of oocyte meiosis (10). Receptors (11) and G-proteins (12, 13), components of the cyclic nucleotide-dependent signal transduction, have now been extensively characterized.

Receptors and G-proteins modulate the activity of *effectors*, enzymes that regulate second-messenger levels. One of these effectors is the adenylate cyclase that catalyzes the synthesis of cAMP (14). The *phos-*

phodiesterases (PDEs) (15), enzymes that degrade and inactivate cAMP, are also to be considered effectors since hormonal regulation of these enzymes has been amply documented (16). The presence of adenylate cyclase and PDEs implies that steady state cAMP intracellular levels are the result of a balance between the rate of synthesis and the rate of degradation (17). Therefore, modulation of the rate of degradation is as effective as regulation of synthesis to control cAMP levels. Experiments transfecting a cDNA coding for a PDE into hormone-responsive cells have provided conclusive proof of this pivotal role of PDEs. Using this approach, we have demonstrated that even small changes in the rate of degradation have a major impact on intracellular cAMP levels (18).

Until recently, our understanding of the cAMP-degrading machinery had major gaps because of the astounding number and complexity of the PDE forms present in the cell. On the basis of the most recent data on structural similarities, kinetic properties, and regulation, PDEs have been classified into five families (19): type I, *calmodulin-regulated PDEs* (CaM-PDEs); type II, *cGMP-stimulated PDEs* (cGS-PDEs); type III, *cGMP-inhibited PDEs* (cGI-PDEs); type IV, *cAMP-specific PDEs* (cAMP-PDEs); and type V, *cGMP-specific PDEs* (cGMP-PDEs). The cloning of the different PDE genes has shown that each family is in turn composed of several members so that at least 25 different PDE forms have been identified (19).

In this chapter we summarize our studies on the expression and regulation of the cAMP-degrading machinery in seminiferous tubule germ cells and Sertoli cells. We also discuss our hypotheses on the role of this pathway in somatic and germ cell function and summarize the data available in support of these ideas.

Only Some Components of the cAMP-Dependent Pathway Are Present in Developing Germ Cells

Scant information on the presence of receptors linked to cAMP regulation in germ cells is available (6). Biochemical studies using radioactive ligands have provided inconclusive results on the presence of binding. However, homologous screening of germ cell cDNA libraries indicates that mRNAs that encode receptors belonging to the family of the 7 transmembrane domains are present in germ cells (20, 21). The natures of these receptors, their physiological ligands, and the G-protein with which they are interacting are unknown. Inspection of the deduced amino acid sequences encoded in these cDNAs suggests that one is similar to the receptors expressed in the olfactory system that serve to recognize olfactory cues (20). The other receptor shares homologies with the receptors that belong to the adenosine receptor family (21).

In contrast to the lack of data on receptors acting through cAMP, it is well established that all the other components of the cAMP-dependent pathway are expressed in germ cells (6). Adenylate cyclase activity is present in the particulate and soluble fractions of spermatids (22–25). These soluble and particulate forms are, however, anomalous in that they are not activated by cholera toxin, forskolin, or GTP analogs (23–25). These findings support the hypothesis that these adenylate cyclases do not interact with a G-protein.

Several studies have also shown that inhibitory Gi1, Gi2, Gi3, and Go proteins are expressed during spermatogenesis (26–28); however, the Gs protein could not be detected in differentiating and mature germ cells (26–28). These findings indicate that the activation of cAMP synthesis is unique in germ cells and cannot occur via a receptor-Gs coupling as in somatic cells. Of interest is the observation that in *Dictyostelium*, an adenylate cyclase not regulated by G-proteins is expressed at the time of sporulation (29), a process similar to gamete production. This apparent absence of the transduction system necessary to stimulate adenylate cyclase in germ cells opens the possibility that cAMP is constitutively produced at a constant level and that its regulation is exerted at the level of degradation by PDEs. This would be similar to what occurs in the retina. In this tissue a light-activated receptor regulates the cyclic nucleotide degradation via regulation of a PDE (30). As discussed below, several different PDE forms are present in germ cells at different stages of development, and their expression appears to be regulated in conjunction with differentiation. It is then possible that these are the enzymes controlling cAMP levels during germ cell development.

Phosphodiesterase Expression in Somatic and Germ Cells of the Testis: Cell-Specific PDE Gene Expression

Early studies from our laboratory have shown that two major forms of cAMP-degrading activity are expressed in the testis (31–33). These activities have been separated by ion exchange chromatography of extracts from total testes and from enriched somatic or germ cell populations (31–33). Using this chromatographic analysis a Ca^{++}-calmodulin-regulated phosphodiesterase (type I, CaM-PDE) could be distinguished from a cAMP-specific phosphodiesterase (type IV, cAMP-PDE). The two forms have been extensively characterized in terms of kinetics characteristics, sensitivity to different inhibitors, and substrate specificity (31–33). With the development of new tools to study these PDE forms, it has become clear that the two peaks of PDE activity are not homogeneous. The molecular cloning of the cDNAs coding for these PDEs has greatly helped in understanding the basis of this heterogeneity (34–41).

Complementary DNAs for PDEs have been isolated and characterized from bovine and rat tissues and indicate that at least two genes encoding CaM-PDEs are present (38–41). These genes encode similar proteins of different molecular masses (61-kd and 63-kd CaM-PDEs). The data available indicate that these forms are regulated by Ca^{++}, calmodulin, and phosphorylation (15, 42). Possible differences in RNA splicing and in the regulation of the two proteins by phosphorylation have also been reported (19, 38, 39). Northern blot or RNAse protection analysis of testis mRNA demonstrated that only the 61-kd mRNA is present in this organ (38–41). However, it is still to be determined unequivocally whether the CaM-PDE forms expressed in somatic and germ cells of the testis derive from a single or multiple genes. Data from our laboratory and others indicate that the germ cell CaM-PDE has physicochemical properties different from those of somatic cells, at least in the mouse (33, 43). This suggests that a third gene encoding a CaM-PDE may be present in the mammalian genome.

The family of genes encoding the cAMP-PDE is even more complex. The available cloning data indicate that four cAMP-PDE genes are present in rodents and expressed in the testis and other organs (34–37). These have been termed *ratPDE1*, *ratPDE2*, *ratPDE3*, and *ratPDE4*. Recently, human cDNAs derived from four genes have been characterized, demonstrating that the presence of multiple genes is not a peculiarity of the rat (44, 45) . These genes encode proteins that are homologous to the *Drosophila melanogaster* dunce PDE (46). In *Drosophila* disruption of the dunce gene produces infertility and impaired central nervous system function. The sterility observed in the female is due to a lesion affecting both the somatic cells and the egg (47, 48). It is also intriguing that a locus (*sam*) whose disruption produces sperm immotility has been mapped within the dunce gene (49).

To determine whether all four genes were expressed in all seminiferous tubule cells, several experimental approaches have been used. Separation of somatic and germ cells of the seminiferous tubules and identification of the forms expressed demonstrated that distinct cAMP-PDE genes are expressed in somatic and germ cells (34). Germ cells express transcripts derived from the ratPDE1 and ratPDE2 genes, while Sertoli cells express transcripts corresponding to ratPDE3 and ratPDE4. This cell-specific gene expression is also consistent with the observation of changes in transcript levels after in utero irradiation. The selective loss of germ cells with irradiation is accompanied by the disappearance of ratPDE1 and ratPDE2 transcripts from the testis (34).

Regulation of Phosphodiesterase Expression During Germ Cell Differentiation

Northern analysis of mRNA derived from pachytene spermatocytes, round spermatids, and elongating spermatids confirmed the expression of ratPDE1 and ratPDE2 and demonstrated a differential regulation of the two PDE mRNAs during spermatogenesis (50). Northern blot analysis with a ratPDE1 probe showed that 4.0- and 2.8-kb mRNAs were found predominantly in pachytene spermatocytes (50). The low-molecular weight transcript of 2.8 kb was still present in low amounts in round spermatids, but no transcripts were detected in the elongating spermatids/condensing body fraction, even after longer exposure of the blot (50). Probing of similar blots with a ratPDE2 cDNA detected a 4.0-kb transcript in pachytene spermatocytes (50). The expression of ratPDE2 transcripts occurred mostly in round spermatids where five transcripts of 4.0, 3.5, 3.1, 2.8, and 2.4 kb were evident. A 2.8-kb transcript was retained in elongating spermatids/residual bodies. The above data are in agreement with developmental data showing that ratPDE1 mRNA appears in the testis at an earlier age than the ratPDE2 mRNA (50). This is again consistent with the expression at different stages of spermatogenesis.

The developmental regulation of the two cAMP-PDE genes expressed in germ cells has been recently confirmed by in situ hybridization of the seminiferous tubule. In fact, the mRNA derived from the ratPDE1 gene is exclusively localized in late pachytene spermatocytes (Boitani et al., manuscript in preparation). Although more rigorous studies are required to confirm this finding, it was also noted that ratPDE1 mRNA was mostly localized in the nuclei of these cells. A similar nuclear localization has been reported for the RNA coding for CREB, a transacting factor that mediates cAMP-regulated gene transcription (51).

Similar conclusions on the differential expression of the two genes could be reached by studying the PDE protein expression during spermatogenesis. Separation of the two ratPDE1 and ratPDE2 forms was achieved by HPLC ion exchange chromatography of extracts from pachytene spermatocytes and round spermatids (50). The ratPDE1 form was found present in pachytene spermatocytes, decreased markedly in round spermatids, and became undetectable in elongating spermatids. Conversely, the ratPDE2 protein was maximally expressed in round spermatids and, although at a reduced level, was still present in elongating spermatids (50). Since spermatozoa also express a cAMP-PDE (32), it is likely that this is the product of the ratPDE2 gene expressed earlier during spermiogenesis.

Role of the Cyclic Nucleotide-Dependent Pathway in the Spermatozoon

The above data indicate that the expression of PDEs is regulated during spermatogenesis, as has been shown for other components of the cyclic nucleotide-dependent pathway (6). This is an indication that this pathway must have a role during germ cell differentiation. However, demonstrating the exact function of cAMP during spermatogenesis remains a major challenge, mostly because spermatogenesis cannot be reproduced in vitro.

A possibility that needs to be considered is that the major function of cAMP is to regulate spermatozoon final maturation. Consistent with this view is the well-established notion that cAMP plays an important role in regulation of the spermatozoon motility (52). Changes in cAMP levels in the spermatozoon during its transit in the epididymis are associated with the acquisition of motility (52, 53). Furthermore, it is well established that manipulation of cAMP levels in the mature gamete using PDE inhibitors causes or increases flagellar movements (52, 54, 55). A regulation of cAMP levels has also been implicated in the modulation of acrosomal reaction (52, 56). Thus, the most important functions involved in the fertilizing ability of the spermatozoon appear to be under the control of cyclic nucleotides.

However, little is known about how cAMP levels are regulated in the spermatozoon. As for the immature germ cells, spermatozoon adenylate cyclase does not interact with a G-protein, nor has a Gs protein been detected by biochemical or immunological methods in the male gamete (26). Conversely, on the basis of our developmental data, both a CaM-PDE and a cAMP-PDE are present in spermatozoa (32, 57). How these PDEs are regulated in the maturing gamete is still unclear. It has been reported that the PDE activity decreases during sperm transit in the epididymis (55). Hoskins and collaborators have proposed that this PDE regulation is the mechanism by which cAMP is increased in the spermatozoon during its transit in the epididymis (55). This increase is in turn responsible for activation of motility. Interestingly, our laboratory and others have shown that PDEs are mostly bound to particulate structures of the mature spermatozoon (32, 57). The available data show that the regulatory subunit of the cAMP-dependent protein kinase is a major component of the flagellum and is tightly bound to flagellar structures (52, 58). Wasco et al. have shown that the CaM-PDE present in spermatozoa is also tightly bound to flagellar components (57). If this PDE localization is confirmed, it would be a further indication of the important role of these enzymes in the control of cAMP levels and sperm motility.

Regulation of cAMP-PDEs in the Sertoli Cell

Because the Sertoli cell response to FSH undergoes major changes during testis development and during the seminiferous tubule cycle, our efforts have focused on understanding the biochemical mechanisms causing these changes in FSH sensitivity during the life span of the Sertoli cell. Studies conducted by our laboratory over the last five years indicate that several modes of regulation affect PDE expression in these cells and that these regulations are responsible for the changes in FSH sensitivity. Changes in cAMP-PDE expression and activity have been observed in vitro during acute FSH stimulation, in vivo during testicular development, and, more recently, during the seminiferous tubule cycle. Regardless of the role of these enzymes in the function of the Sertoli cell, the gene encoding the cAMP-PDEs can be used as a paradigm of gonadotropin-regulated, development-dependent gene regulation in the Sertoli cell.

Acute Regulation of PDEs by FSH In Vitro and In Vivo

Immature Sertoli cells respond to FSH with a transient increase in cAMP levels. Under the most widely used culture conditions (defined DMEM medium, 32°C), cAMP levels reach a maximum in about 15–30 min after the addition of a saturating concentration of hormone (100–500 ng/mL of ovine FSH) (59, 60). They remain at a plateau for an additional 20–30 min and decrease steadily in the following 3 h. In the continuous presence of hormone, cAMP levels remain slightly elevated over basal levels for several days. The cause of this transient increase in intracellular cAMP is dependent on a complex series of events set in motion by FSH binding to the Sertoli cell (60–62). These events are also responsible for the onset of desensitization of this and other cells.

Diffusion of cAMP outside the Sertoli cell contributes to the decrease in intracellular cAMP, but this would be by itself ineffective if major changes in synthesis and degradation of the second messenger were not occurring at the same time. Uncoupling of the receptor from G-proteins and adenylate cyclase causes a decrease in the cAMP rate of synthesis in spite of the continuous presence of hormone (60–62). This uncoupling also prevents the cells from responding to a second challenge with FSH (cell desensitization). The decrease in cAMP synthesis during this phase of response to FSH is associated with an increase in cAMP degradation resulting from de novo synthesis of PDE isoenzymes (62).

This FSH-dependent mechanism of PDE regulation has been extensively investigated in our laboratory. Two cAMP-PDE genes, ratPDE3 and ratPDE4, are expressed in the Sertoli cell (35, 63). FSH stimulation of these cells induces an increase in the rate of transcription of the ratPDE3 gene and an increase in ratPDE3 and ratPDE4 mRNA levels (63). The stimulation of the PDE activity is therefore the result of an increase in PDE protein accumulation (63). That this increase has a major role in reducing intracellular cAMP concentration and in producing desensitization is indicated by a number of studies. Phosphodiesterase inhibitors, for instance, are able to delay or block, albeit incompletely, the reduction in cAMP levels (62, 64). They also reverse the state of desensitization of the Sertoli cell (62, 64). Furthermore, transfection of a cDNA encoding the FSH-stimulated PDE in a heterologous hormone-responsive system, the MA-10 cells, demonstrates that a dramatic reduction in cAMP levels follows expression of the PDE (18). These cells that overexpress the ratPDE3 PDE have a phenotype of desensitized cells because response to hCG is markedly reduced. Therefore, regulation of cAMP synthesis and degradation are both responsible for the overall changes in FSH responsiveness in vitro.

Some puzzling differences are observed when the FSH stimulation of the ratPDE3 and ratPDE4 gene is compared. Although ratPDE4 mRNA is increased by the FSH treatment, no clear increase in the rate of ratPDE4 gene transcription has been observed (63). On the contrary, as mentioned before, FSH produces more than a 10-fold increase in the ratPDE3 rate of transcription. This difference in transcription activation indicates that mRNA stabilization may be the primary mechanism for the increase in ratPDE4 mRNA levels in the Sertoli cell.

Furthermore, by using selective antibodies against the ratPDE3 and ratPDE4 proteins, it was found that most of the increase in PDE activity measured in the cell extract is associated with an increase in ratPDE3 protein (63). Little ratPDE4 protein could be immunoprecipitated from the soluble fraction of the Sertoli cell or detected by Western analysis with the available antibodies (Iona et al., manuscript in preparation). A possible explanation of this is that the intrinsic translatability of the ratPDE4 mRNA might be much lower than that of ratPDE3 mRNA. In support of this is the finding that the 5' untranslated end of the ratPDE4 mRNA is characterized by the presence of short open reading frames (Monaco et al., manuscript submitted). This feature has been associated with a low rate of translation into protein (65). It is also possible that the ratPDE4 protein may turn over very rapidly because it is unstable. Alternatively, the ratPDE4 protein might be anchored to the particulate fraction of the cell and might not be completely recovered in the soluble extract. Regardless of the cause of the above-described differences, these data indicate that the two PDEs are differentially regulated in the Sertoli cell.

A similar regulation of ratPDE3 mRNA has been observed in vivo. Injection of FSH in immature rats causes an increase in ratPDE3 mRNA and an increase in PDE protein and activity (63, 66). This confirms that the observed phenomenon is physiologically relevant and not an artifactual result of the culture conditions.

cAMP-PDE Regulation During Testicular Maturation

A reduction in FSH responsiveness of the Sertoli cell has been observed during testicular development (67). As determined for the acute desensitization, this reduction in responsiveness is most likely caused by a combined reduction in synthesis and increases in degradation of cAMP. Several laboratories have reported a reduction in adenylate cyclase responsiveness of the Sertoli cell during testis development (68, 69). Means and collaborators were the first to observe that an increase in PDE activity is associated with Sertoli cell maturation (70). Our studies have demonstrated that this increase is restricted to an increase in the cAMP-PDE activity expressed in this cell. Preliminary data suggest that this increase is mostly due to an increase in ratPDE3 mRNA. Available antibodies will serve to determine whether an increase in mRNA is followed by an increase in protein levels.

This increase in cAMP-PDE activity and its impact on cell responsiveness to FSH have also been studied by a different approach. Recently, inhibitors specific for the cAMP PDEs have become available (71, 72). For instance, the antidepressant rolipram inhibits the cAMP-PDEs at 1000-fold-lower concentrations than those required to inhibit the activity of the CaM-PDE. Therefore, it is now possible to study changes in cAMP-PDE activity in the intact cell by measuring the rolipram effect on cAMP levels. Using this approach, cAMP levels have been measured in the maturing Sertoli cell following FSH stimulation in the absence or presence of rolipram (D'Alessandris et al., manuscript in preparation). Rolipram has minimal effects on the response of the immature Sertoli cells, but markedly potentiates the FSH-dependent cAMP accumulation in Sertoli cells from 30- or 40-day-old rats. This is unlike what is observed with the nonselective PDE inhibitor IBMX that potentiates the cAMP responses at all ages studied (67). These data, therefore, provide an additional proof that a cAMP-PDE becomes activated in the Sertoli cell during testicular development and is responsible, at least in part, for the age-dependent reduction in FSH responsiveness.

PDE Regulation During the Seminiferous Tubule Cycle

A fluctuation in Sertoli cell responsiveness to FSH has been observed in segments of the seminiferous tubules separated by the transillumination method (73). Thus, Sertoli cells at stages XIII–IV of the seminiferous tubule cycle respond to FSH with an increase in cAMP, while little FSH stimulation is observed in Sertoli cells at stages VII–VIII of the cycle (73). Since these fluctuations take place in the absence of changes in circulating FSH, it is believed that association with germ cells is responsible for this phenomenon (73, 74). A change in receptor number could be the initial cause of the increased sensitivity of the cells to FSH since higher levels of FSH receptors and mRNA have been observed in stages XIII–II (73, 75). However, it has been reported that PDE activity in the seminiferous tubule is also fluctuating and is maximal at stages VII and VIII, when the response is reduced (73). As discussed above, somatic and germ cells of the seminiferous tubules express multiple PDE forms, and activity measured in total seminiferous tubule homogenates may not reflect a change occurring in the Sertoli cell.

For this reason, we have reevaluated the expression of the cAMP-PDEs in the Sertoli cell using the new tools that we have developed to distinguish between different PDE forms. Interestingly, the mRNA levels for ratPDE3, which is expressed exclusively in Sertoli cells, undergo marked fluctuations during the seminiferous tubule cycle. Using in situ hybridization to monitor ratPDE3 mRNA abundance, we have shown that mRNA levels are maximal in stages II–VI and minimal in stages VIII–IX of the cycle (Boitani et al., manuscript in preparation). This is an initial indication that the ratPDE3 gene expression, as do other regulatory proteins present in the Sertoli cell, undergoes cyclical variations during the seminiferous cycle. It also confirms that the series of events initially observed in vitro are also occurring in vivo. Thus, as the Sertoli cell becomes progressively more sensitive to FSH during stages XIII–IV of the seminiferous tubule cycle, an increase in intracellular cAMP is produced. This in turn causes an increase in the transcription of the ratPDE3 and a consequent increase in the ratPDE3 mRNA. It remains to be determined whether this increase in mRNA is followed by an increase in ratPDE3 protein and activity.

The reason why high PDE mRNA does not coincide with the stage of minimal responsiveness of the Sertoli cell might depend on several possibilities. As previously mentioned, the increase in total seminiferous tubule PDE might reflect the activity present in germ cells. It is also possible that unlike what is observed in vitro, ratPDE3 mRNA translation is subject to regulation and is delayed until later stages. Alternatively, the protein might be subjected to posttranslational modification that serves to increase the activity in stages VII and VIII. Use of the monoclonal antibodies that specifically recognize ratPDE3 should help us to determine the exact succession of these events. It is also possible that the decrease in Sertoli cell response observed at stages VII and VIII is the result of more stable, long-lasting changes in gene expression. According to this view, the induction of a cAMP-PDE in the Sertoli cell occurs earlier in the cycle and is only an initial event that triggers the inactivation of the cAMP-dependent pathway. This is followed by a progressive reduction in expression of all the components of the cAMP-dependent machinery that culminates with the FSH insensitivity observed in stages VII and VIII.

Conclusion and Perspective

Our data demonstrate that expression of the cAMP-degrading machinery is finely regulated in both somatic and germ cells of the seminiferous tubule. The presence of the components of the cAMP-dependent signal transduction in germ cells may reflect the assembly of a machinery that will be utilized later during sperm maturation. Understanding how this machinery is assembled and how it functions in the mature gamete will provide insight into the mechanism of activation of motility. It will also provide powerful pharmacological tools to manipulate spermatozoon motility in vitro. We have also shown that FSH is a primary regulator of cAMP degradation. This regulation has a major impact on the sensitivity of the Sertoli cell to FSH. Since similar regulations also occur in vivo during testis maturation and the seminiferous tubule cycle, we will be able to use the molecular and immunological tools that we have developed to understand the role of this regulation in the Sertoli cell. The cAMP-PDE genes could then be used as a paradigm for the mechanisms that regulate gene expression during the seminiferous tubule cycle.

Acknowledgments. The work described was supported by NIH Grant HD-20788 and grants from the Italian Ministry of Education and the University of Rome and from the Consiglio Nazionale delle Ricerche. Partial support from the Instituto Pasteur, Fondazione Cenci Bolognetti, is also acknowledged.

References

1. Steinberger A, Steinberger E. Spermatogenic function of the testis. In Greep RO, Astwood EB, eds. Handbook of physiology. Baltimore, MD: Williams and Wilkins, 1975:1–19.
2. Russell LD. Sertoli-germ cell interrelations: a review. Gamete Res 1980;3: 179–202.
3. Stefanini M, Conti M, Geremia R, Ziparo E. Regulation of mammalian spermatogenesis. In: Metz CB, Monroy A, eds. Biology of fertilization. New York: Academic Press, 1980:59–102.
4. Ritzen EM. Chemical messengers between Sertoli cells and neighboring cells. J Steroid Biochem 1983;49:499–504.
5. Skinner MK. Cell-cell interactions in the testis. Endocr Rev 1991;12:45–77.
6. Conti M, Monaco L. Modulatory mechanisms of the hormonal response of testicular cells. In: Stefanini M, Conti M, Geremia R, Ziparo E, eds. Molecular and cellular endocrinology of the testis. Amsterdam: Excerpta Medica, 1986:89–100.
7. Garbers DL. The regulation of sperm function by the egg. In: Schatten H, Schatten G, eds. The molecular biology of fertilization. New York: Academic Press, 1989:3–16.
8. Wasserman PM. Early events in mammalian fertilization. Annu Rev Cell Biol 1987;3:109–42.
9. Means AR, Dedman JR, Tash JS, Tindall DJ, van Sickle M, Welsh MJ. Regulation of the testis Sertoli cell by follicle stimulating hormone. Annu Rev Physiol 1980;42:59–69.
10. Maller JL. Interaction of steroids with the cyclic nucleotide system in amphibian oocytes. Adv Cyclic Nucl Res 1983;15:295–336.
11. O'Dowd BF, Lefkowitz RJ, Caron MG. Structure of the adrenergic and related receptors. Annu Rev Neurosci 1989;12:67–83.
12. Gilman AG. G proteins: transducers of receptor-generated signals. Annu Rev Biochem 1987;56:615–49.

13. Spiegel AM. Signal transduction by guanine nucleotide binding proteins. Mol Cell Endocrinol 1987;49:1–16.
14. Krupinski J, Coussen F, Bakalyar HA, et al. Adenylyl cyclase amino acid sequence: possible channel- or transporter-like structure. Science 1989;244: 1558–64.
15. Beavo JA. Multiple isozymes of cyclic nucleotide phosphodiesterase. Adv Second Messenger Phosphoprotein Res 1988;22:1–38.
16. Conti M, Jin SLC, Monaco L, Repaske D, Swinnen JV. Hormonal regulation of cyclic nucleotide phosphodiesterases. Endocr Rev 1991;12:218–34.
17. Barber R, Clark RB, Kelly LA, Butcher RW. A model of desensitization in intact cells. Adv Cyclic Nucl Res 1978;9:507–16.
18. Swinnen JV, D'Souza B, Conti M, Ascoli M. Modulation of MA-10 Leydig tumor cell responsiveness by genetic manipulation of a cAMP-phosphodiesterase. J Biol Chem 1991;266:14383–9.
19. Beavo JA, Reifsnyder DH. Primary sequence of cyclic nucleotide phosphodiesterase isozymes and design of selective inhibitors. TIPS 1990;11: 150–4.
20. Parmentier M, Libert F, Schurmans S, et al. Expression of members of the putative olfactory receptor gene family in mammalian germ cells. Nature 1992;3:453–5.
21. Meyerhof W, Muller-Brechlin R, Richter D. Molecular cloning of a novel putative G-protein coupled receptor expressed during rat spermiogenesis. FEBS Lett 1991;284:155–60.
22. Braun T. Evidence for multiple, cell-specific, distinctive adenylate cyclase systems in rat testis. In: Dufau ML, Means AR, eds. Hormone binding and target cell activation in the testis. 1974:243–64.
23. Neer EJ. Physical and functional properties of adenylate cyclase from mature rat testis. J Biol Chem 1978;253:5808–12.
24. Adamo S, Conti M, Geremia R, Monesi V. Particulate and soluble adenylate cyclase activities of mouse male germ cells. Biochem Biophys Res Commun 1980;97:607–13.
25. Gordeladze JO, Hansson V. Purification and kinetic properties of the soluble, Mn^{++} dependent adenylate cyclase of rat testis. Mol Cell Endocrinol 1981; 23:125–36.
26. Kopf GS, Woolkalis MJ, Gerton GL. Evidence for a guanine nucleotide-binding regulatory protein in invertebrate and mammalian sperm: identification by islet-activating protein-catalyzed ADP-ribosylation and immunochemical methods. J Biol Chem 1986;261:7327–31.
27. Paulsen RH, Paulssen EJ, Gordeladze JO, Hansson V, Haugen TB. Cell-specific expression of guanine nucleotide-binding proteins in rat testicular cells. Biol Reprod 1991;45:566–71.
28. Karnik NS, Newman S, Kopf GS, Gerton GL. Developmental expression of G protein a-subunits in mouse spermatogenic cells: evidence that Ga1 is associated with the developing acrosome. Dev Biol 1992;152:393–402.
29. Pitt GS, Milona N, Borleis J, Lin KC, Reed RR, Devreotes PN. Structurally distinct and stage-specific adenyl cyclase genes play different roles in dictyostelium development. Cell 1992;69:305–15.
30. Hurley JB. Molecular properties of the cGMP cascade of vertebrate photoreceptors. Annu Rev Physiol 1987;49:793–812.

31. Geremia R, Rossi P, Pezzotti R, Conti M. Cyclic nucleotide phosphodiesterase in developing rat testis: identification of somatic and germ-cell forms. Mol Cell Endocrinol 1982;28:37–53.
32. Rossi P, Pezzotti R, Conti M, Geremia R. Cyclic nucleotide phosphodiesterases in somatic and germ cells of mouse seminiferous tubules. J Reprod Fertil 1985;74:317–27.
33. Geremia R, Rossi P, Mocini D, Pezzotti R, Conti M. Characterization of calmodulin-dependent high affinity cyclic AMP and cGMP phosphodiesterase from male mouse germ cells. Biochem J 1984;217:693–700.
34. Swinnen JV, Joseph DR, Conti M. Molecular cloning of rat homologues of the *Drosophila melanogaster* dunce cAMP phosphodiesterase: evidence for a family of genes. Proc Natl Acad Sci USA 1989;86:5325–9.
35. Swinnen JV, Joseph DR, Conti M. The mRNA encoding a high affinity cAMP phosphodiesterase is regulated by hormones and cAMP. Proc Natl Acad Sci USA 1989;86:8197–201.
36. Davis RL, Takaysasu H, Eberwine M, Myres J. Cloning and characterization of mammalian homologs of the *Drosophila* dunce[+] gene. Proc Natl Acad Sci USA 1989;86:3604–8.
37. Colicelli J, Birchmeier C, Michaeli T, O'Neill K, Riggs M, Wigler M. Isolation and characterization of a mammalian gene encoding a high-affinity cAMP phosphodiesterase. Proc Natl Acad Sci USA 1989;86:3599–603.
38. Bentley KJ, Kadlecek A, Sherbert CH, et al. Molecular cloning of cDNA encoding a 63-kDa calmodulin-stimulated phosphodiesterase from bovine brain. J Biol Chem 1992;267:18676–82.
39. Repaske DS, Swinnen JV, Jin CS, Van Wyk JJ, Conti M. A polymerase chain reaction strategy to identify and clone cyclic nucleotide phosphodiesterases: molecular cloning of a 63 kDa calmodulin-dependent phosphodiesterase. J Biol Chem 1992;267:18683–8.
40. Polli JW, Kincaid RL. Molecular cloning of DNA encoding a calmodulin-dependent phosphodiesterase enriched in striatum. Proc Natl Acad Sci USA 1992;89:11079–83.
41. Sonnenburg WK, Seger D, Beavo JA. Molecular cloning of a cDNA encoding the "61-kDa" calmodulin-stimulated cyclic nucleotide phosphodiesterase: tissue-specific expression of structurally related isoforms. J Biol Chem 1993; 268:645–52.
42. Sharma RK, Wang JH. Differential regulation of bovine brain calmodulin-dependent cyclic nucleotide phosphodiesterase isoenzymes by cyclic AMP-dependent protein kinase and calmodulin-dependent phosphatase. Proc Natl Acad Sci USA 1985;82:2603–7.
43. Rossi P, Giorgi M, Geremia R, Kinkaid R. Testis-specific calmodulin-dependent phosphodiesterase. J Biol Chem 1988;263:15521–7.
44. Livi GP, Kmetz P, McHale MM, et al. Cloning and expression of a cDNA for a human low-Km rolipram-sensitive cyclic AMP phosphodiesterase. Mol Cell Biol 1990;10:2678–86.
45. McLaughlin MM, Cieslinski LB, Burman M, Torphy TJ, Livi GP. A low-Km, rolipram-sensitive, cAMP-specific phosphodiesterase from human brain. Cloning and expression of cDNA, biochemical characterization of recombinant protein, and tissue distribution of mRNA. J Biol Chem 1993;268: 6470–6.

46. Chen CN, Denome S, Davis RL. Molecular analysis of cDNA clones and the corresponding genomic coding sequences of the *Drosophila* dunce[+] gene, the structural gene for cAMP phosphodiesterase. Proc Natl Acad Sci USA 1986;83:9313–7.

47. Davis RL, Kiger JA. A partial characterization of the cyclic nucleotide phosphodiesterase of *Drosophila melanogaster*. Arch Biochem Biophys 1980; 203:412–21.

48. Bellen HJ, Gregori BK, Olsson CL, Kiger JA. Two *Drosophila* learning mutants, dunce and rutabaga, provide evidence of a maternal role for cAMP on embryogenesis. Dev Biol 1987;121:432–44.

49. Salz HK, Davis RL, Kiger JA. Genetic analysis of chromosome 3D4: the dunce and sperm-amotile genes of *Drosophila melanogaster*. Genetics 1982; 100:587–96.

50. Welch JE, Swinnen JV, O'Brien DA, Eddy EM, Conti M. Unique adenosine 3′,5′ cyclic monophosphate phosphodiesterase messenger ribonucleic acids in rat spermatogenic cells: evidence for differential gene expression during spermatogenesis. Biol Reprod 1992;46:1027–33.

51. Waeber G, Meyer TE, LeSieur M, Hermann HL, Gerard N, Habener JF. Developmental stage-specific expression of cyclic adenosine 3′-5′-monophosphate response element-binding protein CREB during spermatogenesis involves alternate exon splicing. Mol Endocrinol 1991;5:1418–30.

52. Tash JS, Means AR. Cyclic adenosine 3′5′ monophosphate, calcium and protein phosphorylation in flagellar motility. Biol Reprod 1983;28:75–104.

53. Austin CR. Sperm maturation in the male and female genital tracts. In: Metz CB, Monroy A, eds. Biology of fertilization. New York: Academic Press, 1985;121–55.

54. Garbers DL, Lust WD, First NL, Lardy HA. Effects of phosphodiesterase inhibitors and cyclic nucleotides on sperm respiration and motility. Biochemistry 1971;10:1825–31.

55. Hoskins DD, Hall ML, Munsterman D. Induction of motility in immature spermatozoa by cyclic AMP phosphodiesterase inhibitors and seminal plasma. Biol Reprod 1975;13:168–76.

56. Fraser LR. Mechanisms controlling mammalian fertilisation. Oxf Rev Reprod Biol 1984;6:174–225.

57. Wasco WM, Orr GA. Function of calmodulin in mammalian sperm: presence of a calmodulin-dependent cyclic nucleotide phosphodiesterase associated with demembranated rat caudal epididymal sperm. Biochem Biophys Res Commun 1984;118:636–42.

58. Horowitz JA, Toeg H, Orr GA. Characterization and localization of cAMP-dependent protein kinase in rat caudal epididymal sperm. J Biol Chem 1984;259:832–8.

59. Hall SH, Joseph DR, French FS, Conti M. Follicle stimulating hormone induces transient expression of the proto-oncogene c-*fos* in primary Sertoli cell cultures. Mol Endocrinol 1988;2:55–61.

60. Verhoeven G, Cailleau J, de Moor P. Desensitization of cultured rat Sertoli cells by follicle-stimulating hormone and by L-isoproterenol. Mol Cell Endocrinol 1980;20:113–26.

61. Attramadal H, Le Gac F, Jahnsen T, Hansson V. β-adrenergic regulation of Sertoli cell adenylyl cyclase: desensitization by homologous hormone. Mol Cell Endocrinol 1984;34:1–6.

62. Conti M, Toscano MV, Petrelli L, Geremia R, Stefanini M. Involvement of phosphodiesterase in the refractoriness of the Sertoli cell. Endocrinology 1983;113:1845–53.
63. Swinnen JV, Tsikalas K, Conti M. Properties and hormonal regulation of two structurally related cAMP-phosphodiesterases from the rat Sertoli cell. J Biol Chem 1991;266:18370–7.
64. Conti M, Monaco L, Geremia R, Stefanini M. Effect of phosphodiesterase inhibitors on Sertoli cell refractoriness: reversal of the impaired androgen aromatization. Endocrinology 1986;118:901–8.
65. Kozac M. An analysis of vertebrate mRNA sequences: intimations of translational control. J Cell Biol 1991;115:887–903.
66. Conti M, Toscano MV, Geremia R, Stefanini M. Follicle-stimulating hormone regulates in vivo testicular phosphodiesterase. Mol Cell Endocrinol 1983; 29:73–89.
67. Steinberger A, Hintz M, Heindel JJ. Changes in cAMP responses to FSH in isolated rat Sertoli cells during sexual maturation. Biol Reprod 1978;19: 566–72.
68. Van Sickle M, Oberwetter JM, Birnbaumer L, Means AR. Developmental changes in the hormonal regulation of rat testis Sertoli cell adenylyl cyclase. Endocrinology 1981;109:1270–80.
69. Heindel JJ, Clark RB. Loss of hormonal stimulation of Sertoli cell adenylate cyclase with testis maturation. Ann NY Acad Sci 1982;383:460–1.
70. Fakunding JL, Tindall DJ, Dedman JR, Mena CR, Means AR. Biochemical actions of follicle-stimulating hormone in the Sertoli cell of the rat testis. Endocrinology 1976;98:392–402.
71. Weishaar RE, Cain ME, Bristol JA. A new generation of phosphodiesterase inhibitors: multiple molecular forms of phosphodiesterases and the potential for drug selectivity. J Med Chem 1985;28:537–45.
72. Nicholson CD, Challis RAJ, Shahid M. Differential modulation of tissue function and therapeutic potential of selective inhibitors of cyclic nucleotide phosphodiesterase isoenzymes. Trends Pharmacol Sci 1991;12:19–24.
73. Parvinen M. Regulation of the seminiferous epithelium. Endocr Rev 1982; 3:404–17.
74. Kangasniemi M, Kaipia A, Toppari J, Mali P, Huhtaniemi I, Parvinen M. Cellular regulation of basal and FSH-stimulated cyclic AMP production in irradiated rat testes. Anat Rec 1990;227:32–6.
75. Heckert LL, Griswold MD. Expression of follicle-stimulating hormone receptor mRNA in rat testes and Sertoli cells. Mol Endocrinol 1991;5:670–7.

25

Testis-Specific Proprotein Convertase 4: Gene Structure, Optional Exons, and mRNA Isoforms

Majambu Mbikay, Marie-Laure Raffin-Sanson, Haidy Tadros, Francine Sirois, Nabil G. Seidah, and Michel Chrétien

Proprotein convertase 4 (PC4) is a member of the recently discovered family of eukaryotic serine endoproteinases characterized by their structural similarity to bacterial subtilisin and to the yeast Kex2 gene product, as well as by their cleavage specificity after pairs of basic residues (1–8). This family also includes furin, PC1/PC3, PC2, PACE4, and PC5/PC6. Unlike the other convertases that have wide, albeit distinct, tissue distributions (1–4, 7, 8), PC4 expression is restricted to testicular germ cells (5, 6, 10).

Northern blot analysis of testicular RNA has revealed PC4 mRNA forms of 3.5 and 2.8 kb in mouse and rat and a 1.9-kb form in rat (6). The amino acid sequence deduced from the 2.5-kb cDNA (form A) cloned from rat testis (6) differs in its carboxyl region from that deduced from its mouse homolog (5) due to frame-shifting base insertions (or deletions) in their respective mRNAs. Moreover, when the polymerase chain reaction was applied to total cDNA reverse transcribed from testis mRNA in order to amplify this terminal region, two additional DNA products (B and C) representing truncated forms of PC4 were identified in both species (6). To determine whether these various forms of PC4 mRNA are generated by alternate splicing of the primary transcript of the PC4 gene, the cloning and the structural characterization of this gene in mouse (mPC4) was undertaken.

Materials and Methods

Standard Molecular Biology Techniques

Routine molecular biology techniques were conducted essentially as described by Sambrook et al. (11) or Ausubel et al. (12).

Cloning and Sequencing of mPC4 Gene

A mouse cosmid library was purchased from Stratagene (La Jolla, CA). It was screened for PC4 sequences by hybridization using rat cDNA as probe (6). The DNA from positive clones was purified, digested with restriction enzymes, and analyzed by Southern blot. Hybridizing fragments were subcloned into plasmid vectors and sequenced.

In Vitro Transcription and Primer Extension

Drosophila embryo nuclear extracts produced by Promega (Madison, WI) were used for in vitro transcription assay. The assay was conducted according to instructions by the manufacturer using as template a plasmid carrying a 5.5-kb PC4 gene fragment that contained 3.9 kb of the upstream region and the structural domain down to the second intron. Primer extension was performed using a 5' [^{32}P]-labeled antisense oligodeoxyribonucleotide derived from the first exon.

Amplification of the 5' End of mPC4 mRNA

The cDNA was reverse transcribed from total RNA from mouse testis using an oligo(dT) primer. Two PC4-specific primers derived from the 5' region of the corresponding mRNA were used in the polymerase chain reaction. The amplified DNA fragments were cloned and sequenced.

Ribonuclease Protection Assay

The assay was performed as recently described by Saccomanno et al. (13). Briefly, polyadenylated RNA from mouse testis was incubated overnight at 42°C with a uniformly labeled [^{32}P]-cRNA probe. Unprotected RNA was digested with ribonuclease T2, and protected fragments were resolved by electrophoresis in a sequencing gel and revealed by autoradiography.

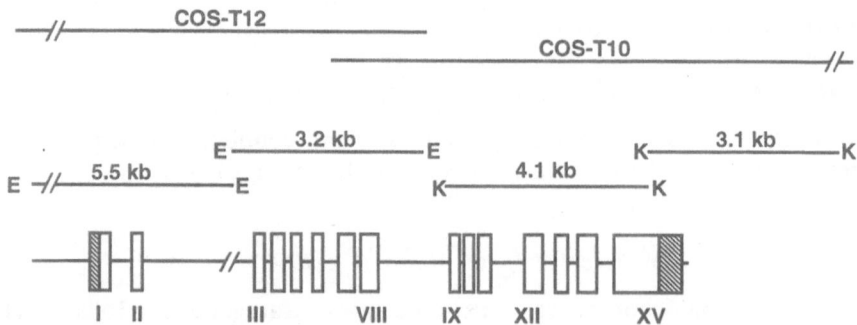

FIGURE 25.1. Schematic representation of mPC4 gene structure. The structure was derived from the sequence of *Eco*RI (E) and *Kpn*I (K) fragments of 2 overlapping cosmid clones, Cos-T10 and Cos-T12. Boxes represent exons. Non-coding segments of the exons are shaded.

Results

Exon-Intron Organization of the mPC4 Gene

The entire structure of the mPC4 gene was deduced from 2 overlapping cosmid clones and 4 plasmid subclones (Fig. 25.1) Excluding the up-stream region, the gene is about 9.5 kb long and contains 15 exons and 14 introns. All exons except exon 15 are shorter than 0.3 kb, and all introns except intron 2 are less than 1 kb long (Table 25.1). The positions of the introns relative to the mRNA and protein sequences are shown in Figure 25.2. The 3 active residues of the catalytic triad, Asp^{129}, His^{170}, and Ser^{344}, are each located in a distinct exon, as are Asn^{271} that influences substrate binding and the conserved $ArgArgGlyAsp^{478}$ sequence that may mediate membrane attachment (2, 4).

Signals for Transcription Initiation and Regulation

Figure 25.3A shows the sequence of the upstream 1-kb region of the mPC4 gene as well as the first exon. The upstream region lacks a can-onical TATAA motif, but contains 7 Sp1 binding sites (GGGCGG motifs) and 3 CTF binding sites (CCAAT box). The PC4 RNA was transcribed in vitro from a 3.9-kb fragment of the gene containing these signals to determine the sites of transcription initiation. Analysis of the transcript by primer extension revealed multiple bands. Relative to the sequencing reaction ladder obtained from the gene fragment using the same primer, the two most intense bands migrated at the same positions as the A and C located 27 and 67 *nucleotides* (nts) before the initiator codon, respectively (Fig. 25.3B). These bases probably represent the major initiation sites in this transcription system. Other, weaker bands were detectable further

TABLE 25.1. Exon-intron organization of mPC4 gene.

Exon no.	Exon size (bp)	Sequences around splice sites		
		5' splice site	(Intron, bp)	3' splice site
1	207	GGA CAG gttggt Gly Gln[34]	(219)	tcccag ATC TTC Ile Phe
2	105	CCC AAG gtgaga Pro Lys[69]	(>1300)	ctgcag GTG CGC Val Arg
3	93	TAC ATG gtgagc Tyr Met[100]	(81)	tcctag AAC AAG Asn Lys
4	129	AAT TAT gtagtt Asn Tyr[143]	(100)	ctgcag GAC CCT Asp Pro
5	81	ACC CG gtgagc Asn Ar[168]	nd	atacag A CAT g His
6	85	GGA G gtacca Gly G[199]	(175)	gcgcag GC GTG ly Val
7	173	ACA AAG gtgaac Thr Lys[256]	(89)	acccag GGC CGC Gly Arg
8	213	CAG ATC gtgagc Gln Ile[327]	(~920)	tcccag GTC ACC Val Thr
9	101	GCC AA gtaggt Ala As[361]	(73)	ccccag C CCG n Pro
10	104	CAA G gtgcgg Gln V[396]	(71)	gcgcag TG AGC al Ser
11	118	CCC AC gtgaga Pro Th[435]	(306)	caccag C CCA r Pro
12	186	ATC AG gtgtga Leu Ar[497]	(141)	ccacag A CCG g Pro
13	125	ACA G gtgcca Thr G[539]	(154)	gcccag GA ACT ly Thr
14	129	CAG G gtgagt Gln A[586]	(127)	atgcag AA AGT sp Ser
15	659	AAATAAAA Poly(A)[+] signal		

Note: All exons and most introns were fully sequenced. The sizes of some introns were approximated by PCR using flanking exonic primers. (nd = not determined.)

upstream, about 200 nts before the ATG (not shown). Potential regulatory elements recognizable in the promoter region include 2 imperfect cAMP-responsive elements and 1 glucocorticoid-responsive element (Fig. 25.3A).

Size Variation of the First Exon in PC4 Gene

While attempting to amplify the 5' end of rat and mouse PC4 mRNA by PCR on total testicular cDNA in order to verify its sequence, three DNA products were obtained. After sequencing, one of these products turned out to represent the previously described form A of PC4 mRNA (5, 6). The other two were novel forms, D and E, that lacked 29 and 115 nts

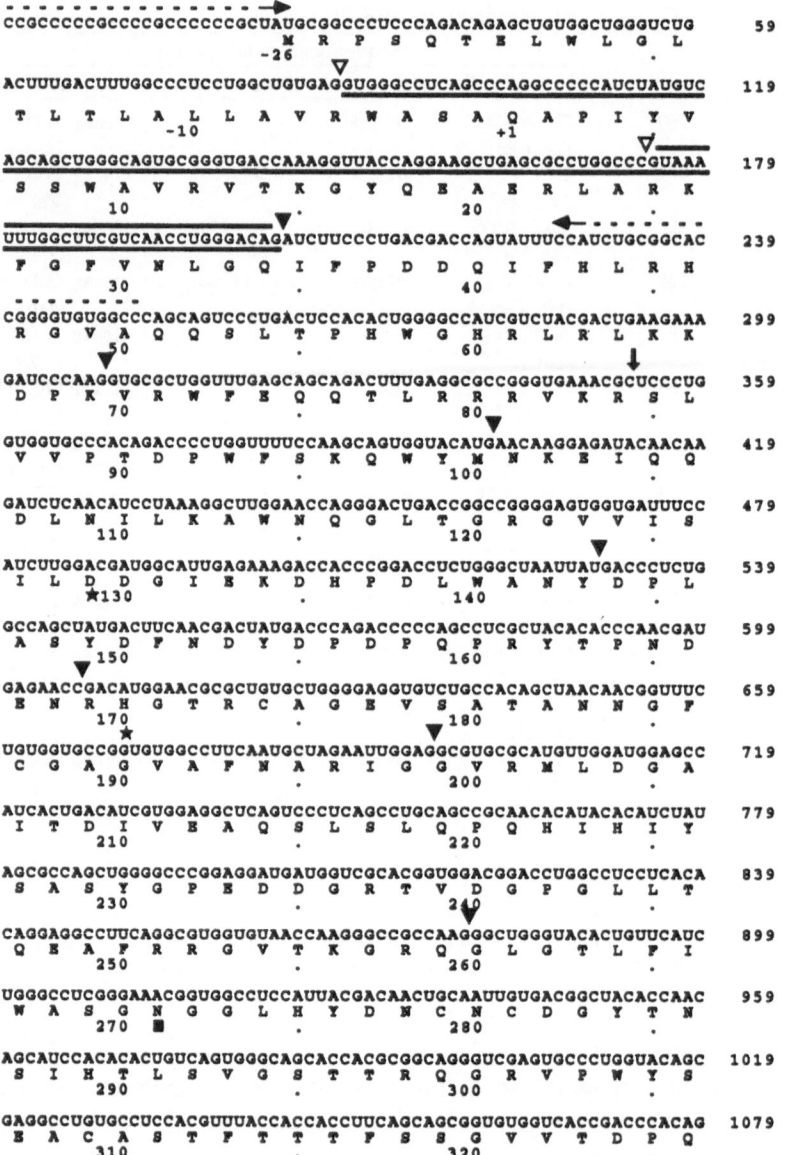

FIGURE 25.2. Sequence of mPC4 mRNA and protein. The positions where introns are located in the gene are indicated with filled arrowheads. Alternate donor splice sites are indicated with open arrowheads, and the alternate acceptor splice site is indicated with an open diamond. The optional exonic segments at the extremities of the mRNA are illustrated with overlines for isoforms D (5') and B (3') and with underlines for isoforms E (5') and C (3'). The positions of the PCR

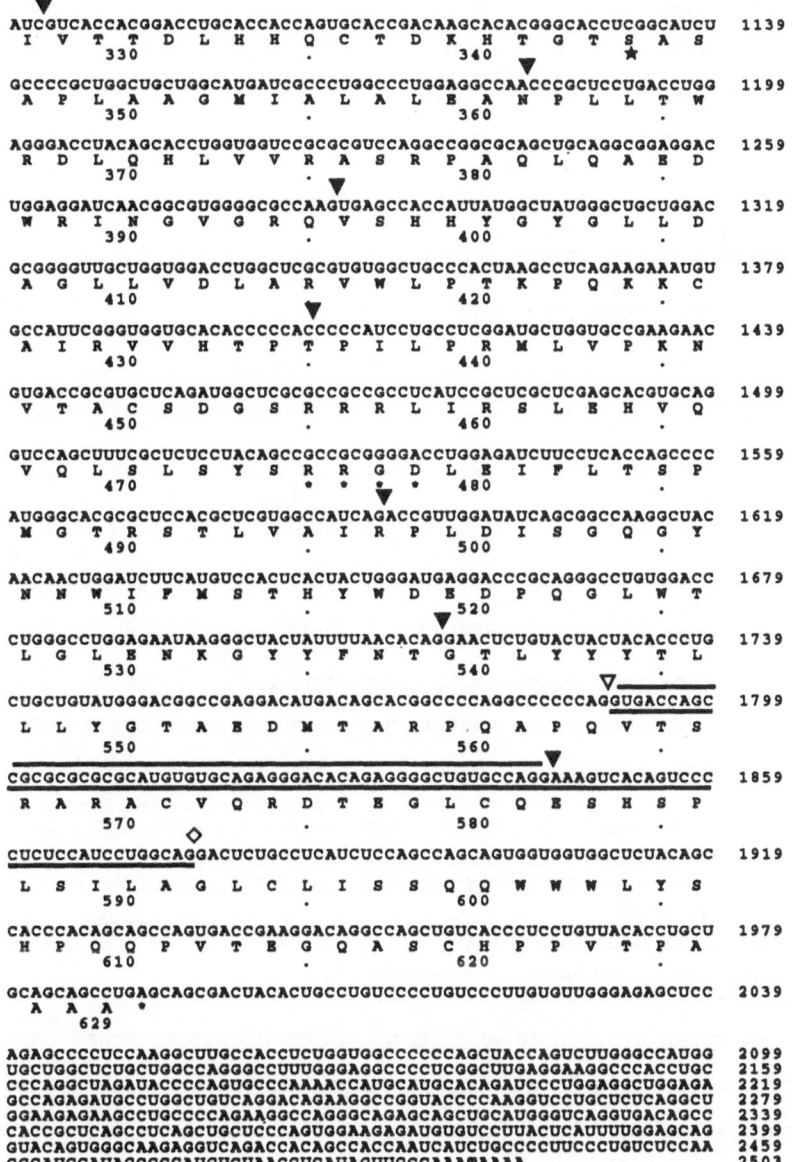

```
▼
AUCGUCACCACGGACCUGCACCACCAGUGCACCGACAAGCACACGGGCACCUCGGCAUCU  1139
I  V  T  T  D  L  H  H  Q  C  T  D  K  H  T  G  T  S  A  S
      330                      340              ★

GCCCCGCUGGCUGCUGGCAUGAUCGCCCUGGCCCUGGAGGCCAACCCGCUCCUGACCUGG  1199
A  P  L  A  A  G  M  I  A  L  A  L  E  A  N  P  L  L  T  W
      350                      360
                                              ▼
AGGGACCUACAGCACCUGGUGGUCCGCGCGUCCAGGCCGGCGCAGCUGCAGGCGGAGGAC  1259
R  D  L  Q  H  L  V  V  R  A  S  R  P  A  Q  L  Q  A  E  D
      370                      380
                      ▼
UGGAGGAUCAACGGCGUGGGGCGCCAAGUGAGCCACCAUUAUGGCUAUGGGCUGCUGGAC  1319
W  R  I  N  G  V  G  R  Q  V  S  H  H  Y  G  Y  G  L  L  D
      390                      400

GCGGGGUUGCUGGUGGACCUGGCUCGCGUGUGGCUGCCCACUAAGCCUCAGAAGAAAUGU  1379
A  G  L  L  V  D  L  A  R  V  W  L  P  T  K  P  Q  K  K  C
      410                      420
                ▼
GCCAUUCGGGUGGUGCACACCCCCACCCCCAUCCUGCCUCGGAUGCUGGUGCCGAAGAAC  1439
A  I  R  V  V  H  T  P  T  P  I  L  P  R  M  L  V  P  K  N
      430                      440

GUGACCGCGUGCUCAGAUGGCUCGCGCCGCCGCCUCAUCCGCUCGCUCGAGCACGUGCAG  1499
V  T  A  C  S  D  G  S  R  R  R  L  I  R  S  L  E  H  V  Q
      450                      460

GUCCAGCUUUCGCUCUCCUACAGCCGCCGCGGGGACCUGGAGAGAUCUUCCUCACCAGCCCC  1559
V  Q  L  S  L  S  Y  S  R  R  G  D  L  E  I  F  L  T  S  P
      470          •  •  ▼  •  480

AUGGGCACGCGCUCCACGCUCGUGGCCAUCAGACCGUUGGAUAUCAGCGGCCAAGGCUAC  1619
M  G  T  R  S  T  L  V  A  I  R  P  L  D  I  S  G  Q  G  Y
      490                      500

AACAACUGGAUCUUCAUGUCCACUCACUACUGGGAUGAGGACCCGCAGGGCCUGUGGACC  1679
N  N  W  I  F  M  S  T  H  Y  W  D  E  D  P  Q  G  L  W  T
      510                      520
                             ▼
CUGGGCCUGGAGAAUAAAGGGCUACUAUUUUAAACACAGGAACUCUGUACUACUACACCCUG  1739
L  G  L  E  N  K  G  Y  Y  F  N  T  G  T  L  Y  Y  Y  T  L
      530                      540    ▽
```
```
CUGCUGUAUGGGACGGCCGAGGACAUGACAGCACGGCCCCAGGCCCCCCAGGUGACCAGC  1799
L  L  Y  G  T  A  E  D  M  T  A  R  P  Q  A  P  Q  V  T  S
      550                      560
                                                  ▼
CGCGCGCGCGCAUGUGUGCAGAGGGACACAGAGGGGCUGUGCCAGGAAAGUCACAGUCCC  1859
R  A  R  A  C  V  Q  R  D  T  E  G  L  C  Q  E  S  H  S  P
      570                      580
         ◇
CUCUCCAUCCUGGCAGGACUCUGCCUCAUCUCCAGCCAGCAGUGGUGGUGGCUCUACAGC  1919
L  S  I  L  A  G  L  C  L  I  S  S  Q  Q  W  W  W  L  Y  S
      590                      600

CACCCACAGCAGCCAGUGACCGAAGGACAGGCCAGCUGUCACCCUCCUGUUACACCUGCU  1979
H  P  Q  Q  P  V  T  E  G  Q  A  S  C  H  P  P  V  T  P  A
      610                      620

GCAGCAGCCUGAGCAGCGACUACACUGCCUGUCCCCUGUCCCUUGUGUUGGGAGAGCUCC  2039
A  A  A  •
      629
```
```
AGAGCCCCUCCAAGGCUUGCCACCUCUGGUGGCCCCCCAGCUACCAGUCUUGGGCCAUGG  2099
UGCUGGCUCUGCUGGCCAGGGCCUUUGGGAGGCCCUCGGCUUGAGGAAGGCCCACCUGC  2159
CCCAGGCUAGAUACCCCAGUGCCCAAAACCAUGCAUGCACAGAUCCCUGGAGGCUGGAGA  2219
GCCAGAGAUGCCUGGCUGUCAGGACAGAAGGCCGGUACCCCAAGGUCCUGCUCUCAGGCU  2279
GGAAGAGAAGCCUGCCCCAGAAGGCCAGGGCAGAGCAGCUGCAUGGGUCAGGUGACAGCC  2339
CACCGCUCAGCCUCAGCUGCUCCCAGUGGAAGAGAUGUGUCCUUACUCAUUUUUGGAGCAG  2399
GUACAGUGGGCAAGAGGUCAGACCACAGCCACCAAUCAUCUGCCCCUUCCCUGUCUCCAA  2459
GCCAUCCAUAGCCCCAUGUCUAACCUCAUAGUUGCCAAAUAAAA  2503
```

FIGURE 25.2. (*Continued*)
primers that served to identify the 5′ variations are illustrated with thin dashed arrows. The vertical black arrow points to the zymogen processing site. Codon +100 encodes the first in-frame Met in mRNA isoforms D and E. The catalytic residues are indicated with stars, the Asp[271] with a black square, and the RRDG with asterisks.

A

```
      -965   cccatgacat gagtcctttc cagtgatggg acccaggacc tcatggtgcc cgttgggaac

      -905   caagtgacat gcacccctct gtcaaaggct cacccgcggc gagataccgc tgcttctcta

      -845   ctccggtcct tgcttcgagc gccccgagcc gttcggtggc cacgttcttc tgttccagaa

      -785   aacgttcgaa gcgctggccg agaccgtcca tggcagcggc gggcgcgggg caggccactg
                                                                         Sp1
      -725   aagcacgccc gcaacctcgc gcaccggcgc cgccgagagc gcactttcac caccgaccca
              Sp1              Sp1
      -665   ccgcccgccc ccgccccgc cccagtgctt tcggccccgg accctggcct ccgccccgcc
                         CTF
      -605   ccccgctctg gccaatggga gggcaggcgg acaaaggagg tgggtactgc cccgtctccc
                                                      Sp1
      -545   cggccgcatg gccctctggc cagggtgagg gccgggcgg cagcgaggag tcccgagggt

      -485   gcggatcccc aggttgatgt gtatgcgtgt tattgtcaag agggtgtggc acgatagtga
              CTF                       CREB               CREB
      -425   ttggacaagg ggtagattat tgacggagag cgattgacgt gggtggagct agttgaggaa
                 Sp1                                   CTF
      -365   cagatgggcg gagcctggta aagatcagcg ctgattggac agaccagaaa ggaggcggga
                                                                         Sp1
      -305   ctacactgga aaaacgaagc tgatcacgaa ggaggggggcg tggagaaaag tgggcggggc

      -245   caaataaagg ttgaaggcac ctgttacctt gaacctgctc ctaagggcag atggaggggt
                 GRE
      -185   tgtccttgta taggtggaga aagtgaggtt cccgaaggct taggtcccta tgaatgtgtc

      -125   actcctagtg gaatattccc acccctccca ggtgggggag gggtggggga ggaaggaagc

       -65   gcaaggcccc ttgggagttg acagcttcgc gttgctatgg tgtccaggct gggagtgctt
                1                                   MetArgPr oSerGlnThr GluLeuTrpL
        -5   cctgaAGCCC CGCCCCCGCC CCGCCCCCCG CTATGCGGCC CTCCCAGACA GAGCTGTGGC
              euGlyLeuTh rLeuThrLeu AlaLeuLeuA laValArgTr pAlaSerAla GlnAlaProI
       +56   TGGGTCTGAC TTTGACTTTG GCCCTCCTGG CTGTGAGGTG GGCCTCAGCC CAGGCCCCCA
              LeTrpValSe rSerTrpAla ValArgValT hrLysGlyTy rGlnGluAla GluArgLeuA
      +116   TCTATGTCAG CAGCTGGGCA GTGCGGGTGA CCAAAGGTTA CCAGGAAGCT GAGCGCCTGG
              laArgLysPh eGlyPheVal AsnLeuGlyG ln
      +176   CCCGTAAATT TGGCTTCGTC AACCTGGGAC AGgtgggtgg gacgtggtgt gtgggtctgg

      +236   acaacctcag atctggacaa atgggagggt gtccccaaga cctcccccaaa tatgtatctg

      +296   tgtcccctc catgttcaat ttgggattct gtggctgtca tatgacccct ggctgaggtg

      +356   tataaagtcg tggagatcct atcaaaggag cagggtccca gctatgcccc tgttgtccgt
                            IlePhePro AspAspGlnT yrPheHisLe uArgHisArg GlyValAlaG
      +416   cgcattccca gATCTTCCCT GACAACCAGT ATTTCCATCT GCGGCACCGG GGTGTGGCCC
              lnGlnSerLe uThrProHis TrpGlyHisA rgLeuArgLe uLysLysAsp ProLys
      +476   AGCAGTCCCT GACTCCACAC TGGGGCCATC CGCTACGACT GAAGAAAGAT CCCAAGgtga

      +536   gagtgctc gtggatgctgt gtcccaacca gtagccttct gtcaccgttcc tgctatgagt
```

FIGURE 25.3. Sequence of the immediate upstream region of mPC4 gene (*A*). Numbering is relative to the major transcription site (no +1) as determined by primer extension. Potential signals for transcription initiation and regulation are

B GATC

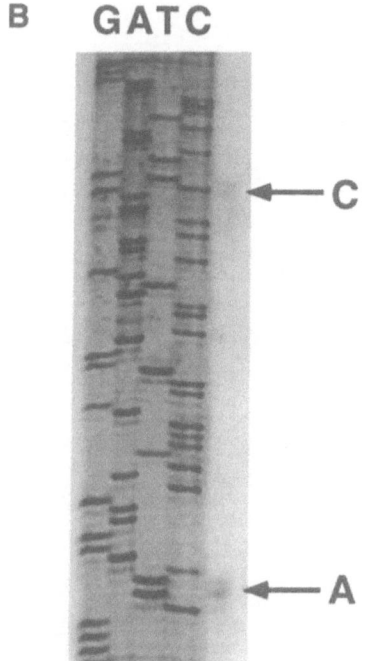

FIGURE 25.3 (*Continued*)
underlined and the interacting factors are named above (e.g., Sp1). *B:* Shown is
an autoradiogram of primer extension of PC4 gene transcripts. The position of
the primer in the first exon is illustrated with a thin line in A.

of the 3'-most region of the first exon, respectively (Fig. 25.1). These
deletions place the initiator ATG of PC4-A out of the reading frame with
the rest of the coding sequence. The first in-frame ATG would be the one
specified by codon 126 that is located past the sequence encoding the
signal peptide and the prodomain of PC4-A.

To confirm the existence and the relative abundance of these alternate
forms of PC4 mRNA, ribonuclease protection assay was performed using
total or polyadenylated RNA from mouse testis and a cRNA probe
representing the genomic fragment −336 to +248 (Fig. 25.3A). Total
RNA from liver was used as negative control. A complex pattern of
bands specifically protected by mouse testis RNA was obtained (Fig.
25.4). The two most intense bands of about 204 and 207 nts in size would
be the expected size of exon 1-A if transcription initiated around the
major site identified in Figure 25.3A. The 175-nts band may represent
exon 1-D that lacks the 29 terminal nucleotides of exon 1-A. The next
most intense band is about 90 nts long and may correspond to exon 1-E
that consists of the first 95 nts of exon 1-A.

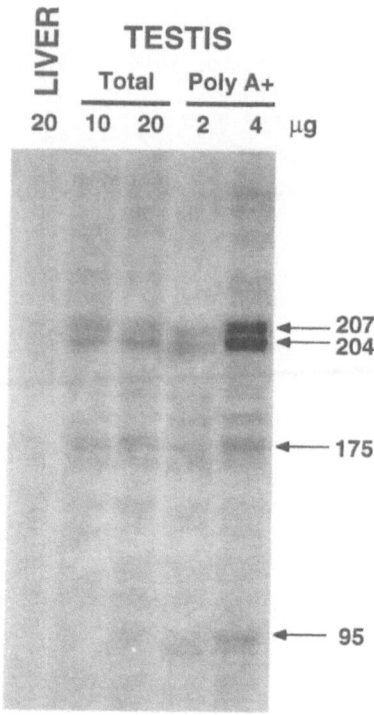

FIGURE 25.4. Autoradiogram of protected fragments from a ribonuclease protection assay. The probe extended from −336 to +248 (see Fig. 25.3A). The sizes of the major fragments were determined in reference to a ladder of products of sequencing reactions (not shown). Note the presence of minor bands of larger size that may be products of transcription initiation at upstream sites.

Discussion

We have elucidated the intron-exon organization of the mPC4 gene. The size of the structural segment of the gene (~9.5 kb) is just about 4-fold longer than that of its mRNA (2.5 kb). This size is comparable to that of the *human furin gene* ([hfur] ~10 kb) (14), but is nearly 15-fold shorter than that of the PC2 gene (~140 kb) in humans (hPC2) (15) and mouse (mPC2) (our unpublished observations) that carries very long introns. Both the mPC4 and the hfur genes contain 14 introns, whereas the hPC2 gene has 13. The first 13 introns of mPC4 and hfur genes interrupt the coding sequences at identical positions (not shown). Twelve out of the 13 introns of the mPC2 gene are also at the same positions. These observations support the proposition that these genes derive from a common ancestral gene. They also point to more evolutionary closeness between

PC4 and furin than between PC4 and PC2. This closeness is also suggested by the degree of homology between these proteins: Within the first 600 residues, there is 74% sequence similarity between hfur and mPC4, 64% between mPC4 and mPC2, and 63% between hfur and mPC2. The 3 residues of the catalytic triad are each contained within a distinct exon in accordance with the concept that exons define functional domains, which might be shuffled around the genome to create new genes and new functions (16). However, in view of the high sequence conservation in this family of proteins, it is more likely that their genes evolved by duplications and translocations rather than by exon shuffling.

The upstream region of the gene for mPC4 lacks a TATAA box, but like hPC2 and mPC2 genes (16, and our unpublished observations), it carries multiple GGGCGG motifs characteristic of many housekeeping genes. Transcription appears to be initiated at multiple sites, as determined in an in vitro transcription assay driven by *Drosophila* embryo nuclear extracts (Fig. 25.3B), suggesting that the 5' untranslated region of PC4 mRNA could be heterogeneous in size. Further heterogeneity affecting the coding sequence was revealed by the sequence of 2 PCR fragments derived from the 5' end of mPC4 cDNA that lacked segments of exon 1-A (Fig. 25.2). A ribonuclease protection assay applied to mouse testicular RNA indicated that mPC4-A is the predominant isoform (Fig. 25.4). Less-labeled fragments of sizes compatible with the alternate isoforms D and E were also detectable. Other minor bands revealed by this assay may represent fragments from transcripts initiated at different sites and differentially spliced.

Intriguingly, translation of mPC4-D and mPC4-E mRNA would result in truncated proteins if started at the initiator ATG of mPC4-A mRNA. Codon 126 at the end of exon 3 is the first ATG in frame with the rest of the coding sequence. The protein encoded from this codon will lack the signal peptide and the prodomains. Whether such a protein will be located in the cytosol, whether it will negotiate the secretory pathway due to internal signals, or whether it will properly fold and become active remains to be determined. All other proprotein convertases so far identified are secreted or located in secretory compartments. If a cytosolic location and an enzymatic activity for the translation product of mPC4-D and mPC4-E mRNA are confirmed, this will extend the action of this convertase outside of the secretory pathway, on substrates belonging to the cytosol, to the nucleus, or to mitochondria.

Three carboxyl-end isoforms of mPC4 resulting from insertion/deletion in the mRNA sequence have previously been reported (6). In combination with the three 5'-end isoforms described here, the possible existence of at least 9 mPC4 mRNA isoforms can be predicted. Knowing the structure of the gene, one can explain the production of all these isoforms by the utilization of alternate splicing sites, as illustrated in Figure 25.2.

The functional significance of these multiple isoforms of mPC4 will be assessed when more is known about the biosynthesis, regulation, and biological function of the various enzymes and their natural substrates. Testicular cells are known to produce alternate forms of many mRNAs, including those of proenkephalin, proopiomelanocortin, pim-1, and c-*fos* oncoproteins (reviewed in 17). The propensity of these cells to utilize transcription initiation and splice sites not recognized in other cells could be, among other possibilities, an indication of the absence of a refined proofreading mechanism or an indication of an excess of splicing factors that relieves competition among splice sites, thus allowing the indiscriminate recognition of cryptic ones. It could also represent a posttranscriptional mechanism of regulation or inactivation of the functions encoded by these mRNAs. The PC4 gene, with its multiplicity of transcripts, could represent an interesting paradigm for evaluating these hypotheses.

Acknowledgments. This work was supported by grants from the Medical Research Council of Canada and the J.A. de Sève Foundation.

References

1. Barr PJ. Mammalian subtilisins: the long-sought dibasic processing endonucleases. Cell 1991;66:1–3.
2. Seidah NG, Chrétien M. Proprotein and prohormone convertases of the subtilisin family: recent developments and future perspectives. Trends Endocrinol Metabol 1992;3:133–40.
3. Steiner DF, Smeekens SP, Ohagi S, Chan SJ. The new enzymology of precursor processing endoproteases. J Biol Chem 1992;267:23435–8.
4. Seidah NG, Day R, Marcinkiewicz M, Chrétien M. Mammalian paired basic amino acid convertases of prohormones and proproteins. Ann NY Acad Sci 1993.
5. Nakayama K, Kim WS, Torii S, et al. Identification of the fourth member of the mammalian endoprotease family homologous to the yeast Kex2 protease: its testis specific expression. J Biol Chem 1992;267:5897–900.
6. Seidah NG, Day R, Hamelin J, Gaspar A, Collard MW, Chrétien M. Testicular expression of PC4 in the rat: molecular diversity of a novel germ cell-specific Kex2/subtilisin-like proprotein convertase. Mol Endocrinol 1992; 6:1559–70.
7. Nakagawa T, Hosaka M, Torii S, Wanatabe T, Murakami K, Nakayama K. Identification and functional expression of a new member of the mammalian Kex2-like processing endoprotease family: its striking similarity to PACE4. J Biochem 1992;113:132–5.
8. Lusson J, Vieau D, Hamelin J, Day R, Chrétien M, Seidah NG. cDNA structure of the mouse and rat subtilisin/kexin-like PC5: a novel candidate pro-protein convertase expressed in endocrine and non-endocrine cells. Proc Natl Acad Sci USA (in press).

 9. Day R, Schafer MKH, Watson SJ, Chrétien M, Seidah NG. Distribution and regulation of prohormone convertases PC1 and PC2 in the rat pituitary. Mol Endocrinol 1992;6:485–7.
10. Torii S, Yamagishi T, Murakami K, Nakayama K. Localization of Kex-like processing endonucleases, furin and PC4, within mouse testis by in situ hybridization. FEBS Lett 1993;316:12–6.
11. Ausubel FM, Brent R, Kingston RE, et al. Current protocols in molecular biology. New York: Wiley and Sons, 1987.
12. Sambrook J, Maniatis T, Fritsch EF. Molecular cloning: a laboratory manual. 2nd ed. Cold Spring Harbor, NY: Cold Spring Harbor Laboratory, 1989.
13. Saccomanno CL, Bordonaro M, Chen JS, Nordstrom JL. A faster ribonuclease protection assay. Biotechniques 1992;13:846–50.
14. Barr PJ, Mason OB, Landsberg KE, Wong PA, Kiefer MC, Brake AJ. cDNA and gene structure of a human subtilisin-like protease with cleavage specificity for paired basic amino acids residues. DNA Cell Biol 1991; 10:319–28.
15. Ohagi S, Lamendola J, Lebeau MM, et al. Identification and analysis of the gene encoding human PC2, a prohormone convertase expressed in neuroendocrine tissues. Proc Natl Acad Sci USA 1992;89:4977–81.
16. Gilbert W. Why genes in pieces? Nature 1978;251:501.
17. Wolgemuth DJ, Watrin F. List of cloned mouse genes with unique expression patterns during spermatogenesis. Mammalian Genome 1991;1:283–8.

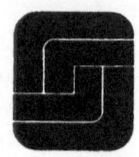

26

Neonatal Propylthiouracil Treatment as a Model System for Studying Factors Controlling Testis Growth and Sperm Production

PAUL S. COOKE, REX A. HESS, JOHN D. KIRBY, DAVID BUNICK, AND MATTHEW P. HARDY

Work from our laboratories and others has demonstrated that transient neonatal hypothyroidism results in increased adult testis size and sperm production. This initially paradoxical finding contrasted with previous results showing that early hypothyroidism inhibited testicular development. This suggested that initial inhibition was superseded by changes in the testis or elsewhere, leading to unprecedented increases when the animal was allowed to return to a euthyroid state. This system therefore provided a unique and powerful new tool to explore the complex developmental interrelationship between thyroid hormone and testis.

This chapter is devoted to a review and consideration of research using this system, which involves the administration of the reversible goitrogen *propylthiouracil* (PTU) during early postnatal development. The development of this system and the early results obtained with it were reviewed previously (1) and will not be discussed in detail. Instead, the emphasis here is on more recent data that have revealed at least a portion of the mechanism of this effect and have extended our understanding of the gamut of changes induced during development and adulthood by early PTU treatment.

The unique feature of this system is that hypothyroidism is limited to the period of early development, with subsequent return to a euthyroid state. This allows the permanent changes induced by early hypothyroidism to be examined at later ages after metabolic and other temporary changes resulting from hypothyroidism have been reversed. In this system male rats are made hypothyroid by adding PTU to the mother's drinking water from birth until day 25 and then giving no further treatment (1, 2). The PTU ingested by the mother passes through the milk, causing severe hypothyroidism in the pups (3). The concentration of PTU used in all experiments discussed here was 0.1%. However, recent results demonstrated that PTU concentrations as low as 0.006%, 15-fold lower than the original 0.1% dose, could maximally increase testis size and sperm production at 90 days of age while minimizing side effects (4). Neonatal treatment with methimazole, another commonly used goitrogen, also increased adult testis size and sperm production, although its potency was only half that of PTU (4).

Pup growth is markedly suppressed by 0.1% PTU (1). The treated rats grow rapidly after cessation of PTU administration, but their weights always remain 15%–25% lower than normal (1, 2). During PTU treatment *thyroxine* (T4) levels and *triiodothyronine* (T3) levels are decreased, while *thyroid stimulating hormone* (TSH) levels are elevated. Levels of all these hormones return to normal by about day 45–50 (3).

Testis weights in PTU-treated rats are increased by over 80% compared to controls at 160 days of age (1, 2). The increased testis size is accompanied by increased sperm production. *Daily sperm production* (DSP) is increased by about 140% in PTU-treated rats at 160 days of age (1, 5). Testosterone levels in adult PTU-treated rats are normal despite the increases in testis size and sperm production (2).

Endocrine Effects of PTU Treatment During Development and Adulthood

The gonadotropins (FSH and LH) and testicular hormones (testosterone and inhibin/activin) are the major regulators of testicular growth and maturation. It was therefore important to measure endocrine profiles during development and adulthood to determine if increased testis growth and sperm production were correlated with changes in these hormones. In male rats given PTU neonatally, serum FSH and LH levels were reduced 40%–50% during treatment (3, 6). This decrease persisted through adult life (3). Interestingly, these animals failed to show the prepubertal FSH peak typical of male rats. In golden hamsters, which also show an increase in adult testis size and sperm production following neonatal PTU treatment, similar decreases in gonadotropins were also observed (Kirby et al., unpublished observations).

In PTU-treated rats the pubertal rise in testosterone is delayed by about 15 days, but then this hormone increases rapidly starting at day 60 and reaches normal adult levels that are subsequently maintained. Immunoreactive inhibin α levels are initially suppressed (3, 6), rise after PTU treatment, and are about 2-fold greater than normal during adulthood (3).

The increased testis size despite chronically reduced gonadotropin levels is paradoxical, as are the normal testosterone levels despite reduced

LH. However, these results indicate that the increased testis growth does not reflect transient or permanent increases in gonadotropins or testosterone (3).

Early Postnatal Life as Critical Period for PTU Effects

An important step towards elucidating the mechanism of the PTU effect was to determine when this drug was effective during development in producing eventual increases in adult testis weight and sperm production. If such a critical window could be identified, this would suggest that PTU acted by altering a developmental process occurring during this time.

The first week of postnatal life was identified as the critical period for PTU effects (7). PTU treatment starting at birth or day 4 produced increases in adult size and DSP proportional to the length of treatment. Conversely, treatment starting on day 8 postnatal or later was ineffective (7). Meisami et al. (8) subsequently confirmed this early postnatal period as critical for the PTU effect, although these authors found a slightly broader window of PTU effectiveness.

Sertoli cell proliferation is high at birth, decreases steadily during neonatal life, then ceases at 3 weeks of age when the adult Sertoli population is established. Since Sertoli cell number appears to be the major regulator of sperm production and there is a temporal correspondence between Sertoli cell proliferation and the critical period for PTU treatment, we postulated that increased testis size and sperm production in treated rats resulted from increased Sertoli cell proliferation (7).

PTU Prolongs Neonatal Proliferation and Increases Adult Numbers of Sertoli Cells

If PTU acted by prolonging Sertoli cell proliferation, it was necessary to determine how the normal pattern of Sertoli cell proliferation was altered and to verify that adult Sertoli cell numbers were increased. Work by van Haaster et al. (6) and our laboratories (9) established that PTU acts by lengthening the period of neonatal Sertoli cell proliferation in the rat, thus, in effect, slowing the normal decline in the rate of Sertoli cell mitogenesis during neonatal life. Subsequent work in the mouse, where PTU also increases testis size and DSP, confirmed that Sertoli cell proliferation was extended after PTU treatment (Joyce and Cooke, unpublished observations). This increased neonatal Sertoli cell proliferation also results in increased adult populations of these cells. Van Haaster et al. (8) reported an 84% increase in Sertoli cell numbers in testes from 36-day-old PTU-treated rats, while Hess et al. (10) found a larger increase (157%) in adult PTU-treated rats.

The present data suggest that thyroid hormones normally inhibit mitogenesis of Sertoli cells, and reduced thyroid hormone levels therefore lead to increased Sertoli cell proliferation. Sertoli cells express thyroid hormone receptors, and thyroid hormones promote maturation of these cells during development (reviewed in 11), indicating that thyroid effects on Sertoli cells may be predominately direct through these cells' receptors for thyroid hormones.

PTU Alters mRNA Levels for Important Sertoli Cell Proteins

The morphological and proliferative data indicate a delayed maturation and prolonged expansion of Sertoli cells in the PTU-treated animal. A molecular study is in progress to better understand functional changes in Sertoli cells induced by transient neonatal hypothyroidism. Preliminary results obtained with Northern blot analysis of testes from treated and control rats (from birth to day 90) indicate differences in degree and time period of expression of a number of Sertoli cell genes. Genes associated with the early mitotic state of the cell, *mullerian inhibiting substance* (MIS) and c-*erb*Aα (thyroid hormone receptor) show increased early expression. The MIS and c-*erb*Aα mRNA levels remain very high until 40 days of age, whereas in control animals these mRNAs drop to low levels by 20 days of age (Bunick et al., unpublished observations). This prolonged expression of both MIS and c-*erb*Aα, normally linked to Sertoli cell mitogenic activity, correlates very well with earlier observations on the increased period of mitogenesis in treated rats.

The expression of Sertoli cell mRNAs that are associated with differentiated cell function generally shows either a delay of 20 days in peak expression or follows the expression pattern of the control with generally higher peak levels of transcript. The transcripts for *sulfated glycoprotein* (SGP2), *androgen binding protein* (ABP), and inhibin βB fall into the former category, while SGP1, transferrin, and inhibin α fall into the latter.

In conclusion, the mRNA profiles for MIS and c-*erb*Aα strongly support the morphological data that in treated rats Sertoli cells are dividing well beyond their normal time frame; whether these two proteins are involved in the extended proliferation of these cells or simply reflect this proliferative state is presently unclear. The dual-expression pattern of mRNAs associated with differentiated Sertoli cell function suggests that more than one mechanism is functioning to regulate the expression of these genes. One mechanism could easily involve the association with proliferating germ cells, while the other, possibly a hormonal or paracrine signal, could be independent of germ cell proliferation.

PTU Effects on Leydig Cells

The increases in testis size (2) and the interstitial volume (10) in PTU-treated rats suggest that Leydig cell numbers might also be increased. Conversely, the normal adult levels of testosterone in these rats indicate that Leydig cell number and/or function may not be affected, despite dramatic increases in Sertoli and germ cells and suppression of LH. We recently determined that Leydig cell numbers in adult PTU-treated rats were 69% greater than controls, but the average volume of these cells was reduced 20% (12). This volume decrease was suggestive of reduced steroidogenic potential, which was confirmed by the finding that LH-stimulated testosterone production in purified adult Leydig cells from treated rats was reduced by 55%. This reduction could be partially explained by the 50% decrease in LH receptors in PTU-treated Leydig cells. However, this was not the complete mechanism of the lowered steroidogenic potential because an even larger reduction in testosterone production (73%) occurred in the presence of substrate-saturating amounts of 22(R)hydroxycholesterol (12). Therefore, decreased activity of one or more steroidogenic enzymes is involved in reduced testosterone production by adult Leydig cells from PTU-treated rats.

These results indicate that neonatal PTU increases adult Leydig cell numbers, but that peripheral testosterone concentration is unchanged due to diminished steroidogenic potential in these cells. Leydig cells do not express thyroid hormone receptors. Therefore, increased Leydig cell numbers in treated rats may be coupled to the increased numbers of Sertoli (8–10), germ (10), and peritubular cells (Joyce, Cooke, unpublished observations) and mediated by paracrine factors produced by those cells.

PTU Effects on Testicular Morphology During Development and Adulthood

The sharp contrast between the inhibitory effects of goitrogen treatment on early testis development (6, 11) and the eventual stimulatory effects of early PTU treatment on the adult testis (1, 2, 5) indicates that the period following the cessation of PTU treatment must be one of pronounced change, as maturational deficits induced by goitrogen treatment are reversed, and the testis manifests unique changes that will accompany its increased adult testis size and sperm production. We recently examined the histology and patterns of cell proliferation in testes of treated rats from birth to day 90 to gain insight into the changes induced by PTU treatment.

In treated rats seminiferous tubule lumen formation was delayed 15 days compared to controls. Seminiferous tubule diameter was initially decreased by PTU administration, but began increasing rapidly by day 40, equaling and then exceeding that of controls by days 65 and 90, respectively (Cooke, Hess, unpublished observations). Gonocyte proliferation and differentiation were similar in treated and control testes. Germ cell degeneration, high during PTU treatment, declined rapidly following PTU treatment and by day 50 equaled controls. The differentiation of prespermatogonia into type A spermatogonia was delayed in treated testes; however, prespermatogonia of treated rats proliferated rapidly during this period (Cooke, Hess, unpublished observations).

In summary, maturational deficits resulting from hypothyroidism are quickly reversed after PTU treatment is stopped. Continued neonatal proliferation of undifferentiated prespermatogonia, in concert with increased Sertoli cell proliferation, may be critical for the expansion of spermatogenic clonal units and increased adult sperm production.

At 90 days of age, seminiferous tubules and interstitium in treated testes appeared histologically normal and were proportionally increased (10). There were significant increases in mean seminiferous tubule diameter (11%), tubule length (44%), and tubule volume (60%) in treated testes. In addition to increased Sertoli cell numbers in treated rats, the numbers of leptotene spermatocytes and round spermatids were doubled (10). The seminiferous epithelium occupied a greater percentage of the tubular area (at the expense of the tubular lumen), presumably as a result of increased height, and the number of Sertoli cells/unit area along the basement membrane was increased. Thus, the seminiferous tubule in the testis of the PTU-treated rat was found to be more compact, with Sertoli cells closer together and germ cells pushed toward the lumen. This compaction caused an overall increase in tubular area and epithelial height and may account for the 25% increase in the efficiency of spermatogenesis (DSP/g testis) in treated rats (5).

Possible Mechanism for the PTU Effect on Testis

In conclusion, we propose the following mechanism for the PTU-induced increases in testis size and sperm production. Thyroid hormones normally act directly on Sertoli cells during neonatal life to cause them to gradually mature and change from the mitogenic cells characteristic of the early postnatal period to the nonmitogenic, secretory cells characteristic of mature animals. This effect may involve modulation of responsiveness to FSH, the major mitogenic factor for Sertoli cells that is also critical for initiation of secretory activity at maturation. In PTU-treated animals, a reduction in the level of thyroid hormones slows the normal transition of Sertoli cells from their mitogenic to their mature secretory state, resulting in an extended mitogenic period and large increases in adult populations of these cells. This increased Sertoli cell population may then stimulate increased proliferation of prespermatogonia, leading to increased numbers of stem cell spermatogonia. The increased Sertoli cell population may also stimulate increased Leydig and peritubular cell proliferation, most likely through the paracrine actions of growth factors.

The stimulatory effects of early hypothyroidism on populations of Sertoli, Leydig, germ, and peritubular cells are initially masked by the immaturity of Sertoli cells in hypothyroid animals. This immaturity results in retarded seminiferous tubule lumen opening and the inability of these Sertoli cells to support advanced stages of germ cell development, which leads to the germ cell degeneration observed in these animals. The reestablishment of euthyroidism allows Sertoli cells to fully mature; the increased Sertoli cell population then interacts with the increased spermatogonial population to produce full spermatogenesis, leading to morphological changes and increases in testis size and sperm production characteristic of the adult PTU-treated animal. There are obviously many important gaps in our knowledge of how the PTU effect works, and this proposed model will undoubtedly require revision as knowledge in this area increases. However, the PTU system and other work related to thyroid effects on the testis should provide insight into the role of these hormones in developing testis and insight into the broader questions of how testis growth and sperm production are regulated.

Acknowledgments. This work was supported by NIH Grants HD-29376 (P. Cooke), ES-05214 (R. Hess), DK-45821 (J. Kirby), and HD-285963 (D. Bunick); USDA Grant 91-37203-6890 (J. Kirby); and a grant from the Population Council (M. Hardy).

References

1. Cooke PS. Thyroid hormones and testis development: a model system for increasing testis growth and sperm production. Ann NY Acad Sci 1991; 637:122–32.
2. Cooke PS, Meisami E. Early postnatal hypothyroidism causes increased adult size of testis and other reproductive organs but does not increase testosterone levels. Endocrinology 1991;129:237–43.
3. Kirby JD, Jetton AE, Cooke PS, et al. Developmental hormonal profiles accompanying the neonatal hypothyroidism induced increases in adult testis size and sperm production in the rat. Endocrinology 1992;131:559–65.
4. Cooke PS, Kirby JD, Porcelli J. Increased testis growth and sperm production in adult rats following transient neonatal goitrogen treatment: optimization of the propylthiouracil dose and effects of methimazole. J Reprod Fertil 1993.
5. Cooke PS, Hess RA, Porcelli J, Meisami E. Increased sperm production in adult rats following transient neonatal hypothyroidism. Endocrinology 1991;129:244–8.

6. Van Haaster LH, De Jong FH, Docter R, De Rooij DG. The effect of hypothyroidism on Sertoli cell proliferation and differentiation and hormone levels during testicular development in the rat. Endocrinology 1992;131: 1574–6.
7. Cooke PS, Porcelli J, Hess RA. Induction of increased testis growth and sperm production in the adult rat by neonatal administration of the goitrogen propylthiouracil (PTU): the critical period. Biol Reprod 1992;46:146–52.
8. Meisami E, Sendera TJ, Clay LB. Paradoxical hypertrophy and plasticity of the testis in rats recovering from early thyroid deficiency: a growth study including effects of age and duration of hypothyroidism. J Endocrinol 1992;135:495–505.
9. Cooke PS, Hess RA. The pattern of Sertoli cell proliferation is altered in the neonatal hypothyroidism model for increased testis size. J Androl 1992; 13(suppl):47.
10. Hess RA, Cooke PS, Bunick D, Kirby JD. Adult testicular enlargement induced by neonatal hypothyroidism is accompanied by increased Sertoli and germ cell numbers. Endocrinology 1993.
11. Francavilla S, Cordeschi G, Properzi G, et al. Effect of thyroid hormone on the pre- and post-natal development of the rat testis. J Endocrinol 1991;129:35–42.
12. Hardy MP, Kirby JD, Hess RA, Cooke PS. Leydig cells increase their numbers but decline in steroidogenic function in the adult rat after neonatal hypothyroidism. Endocrinology 1993.

Part V

Recent Developments in the Studies
of Male Reproductive Functions

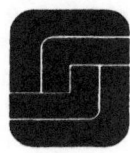

27

Molecular Biology of Human Androgen Insensitivity Syndrome

TERRY R. BROWN, YING-TAI CHANG, PAOLO GHIRRI,
CLAUDE J. MIGEON, KOICHI MURONO, PATRICIA SCHERER,
AND ZHIFENG ZHOU

Testosterone (T) is synthesized and secreted by testicular Leydig cells. High levels of T must be maintained within the seminiferous tubules of the testes to promote and maintain the process of spermatogenesis, whereas relatively lower concentrations of T are available to peripheral tissues from the general circulation. Androgen-dependent target tissues concentrate androgens by means of the specific, high-affinity binding of T and its potent 5α-reduced metabolite, *dihydrotestosterone* (DHT), to intracellular *androgen receptors* (AR). The intranuclear binding of the androgen receptor-steroid complex to DNA leads to the regulation of androgen-dependent gene transcription. Early aspects of male sex differentiation and development during the first trimester of fetal life are dependent on the normal biological function of the AR in tissues of the internal male reproductive tract and the external genitalia.

In karyotypically normal 46,XY subjects, incomplete masculinization of the fetal external genitalia may result from insensitivity to androgens due to abnormalities of the AR (1). Indeed, variable phenotypic expression has been observed, ranging from a normal-appearing female phenotype in the complete form of *androgen insensitivity* (AIS) to male genitalia with hypospadias in the less severe partial forms of AIS. The heterogeneity in phenotypic expression of AIS is due to a variety of AR defects, some of which are detectable by biochemical characterization of steroid binding in cultured genital skin fibroblasts of affected subjects. For example, AR binding in cells from subjects with *complete androgen insensitivity* (CAIS) may be undetectable in the so-called *receptor-negative* (AR−) form, whereas quantitatively normal AR binding, termed the *receptor-positive* (AR+) form, may be present for others. In the partial form of AIS, AR binding either may be deficient (AR±) or may be quantitatively normal (AR+). Therefore, it should be readily apparent that the measurement of

411

AR binding activity in cells does not necessarily predict the phenotypic presentation of AIS and that defects in the biological activity of AR may be independent of quantitative measures of steroid receptor number.

The cloning and sequencing of human cDNA encoding the AR have presented the opportunity to pursue molecular analyses of the genetic defects in the AR gene in affected subjects with AIS. The heterogeneity of naturally occurring mutations thus far identified in the AR gene is representative of the presence of an often recognizable, yet highly variable, phenotype in subjects with AIS. Mutations in the AR provide a base upon which we can better understand the structure and function of the receptor in normal physiology and disease.

Structure and Function of the AR

The human AR is a member of the ligand-dependent superfamily of nuclear transcription factors that includes receptors for estrogen, glucocorticoid, mineralocorticoid, progesterone, retinoic acid, thyroid hormone, and vitamin D (2, 3). The amino acid sequence of the DNA and steroid binding domains of AR share a particularly high degree of conservation with the progesterone, glucocorticoid, and mineralocorticoid receptors (2). The earlier-described X-chromosomal linkage of the AR gene (4) and the synthesis of consensus oligonucleotides to the extraordinarily conserved region of the DNA binding domain common to these receptors led to the identification and isolation of a putative AR genomic DNA clone from a human X-chromosome-specific DNA library by oligonucleotide hybridization screening methods (5). The isolation of an AR-specific genomic DNA clone led directly to the identification and characterization of full-length human AR cDNA (2, 3).

The human AR gene locus is proximal to the centromere on the long arm of the X-chromosome in the region q11-12 (6). The human AR gene encodes a protein of 919 amino acids with a molecular weight of ~110,000 (5) (Fig. 27.1). Molecular and biochemical studies of human AR function revealed the presence of the three major conserved domains common to each of the aforementioned classes of nuclear receptor (7–9). The aminoterminus exhibits the greatest degree of amino acid sequence divergence among these receptors and is highly variable in length (2), but it is essential for receptor-mediated transcriptional regulation of hormone-responsive genes (7–9). In the human AR this region includes amino acids 1–537 and is encoded by the first and largest of the 8 exons of the gene (10). Another interesting feature of this transcriptional activation domain of the human AR is the homopolymeric amino acid repeats of proline, glutamine, and glycine, with the latter two polymeric regions being of variable length and, thus, polymorphic within the normal population.

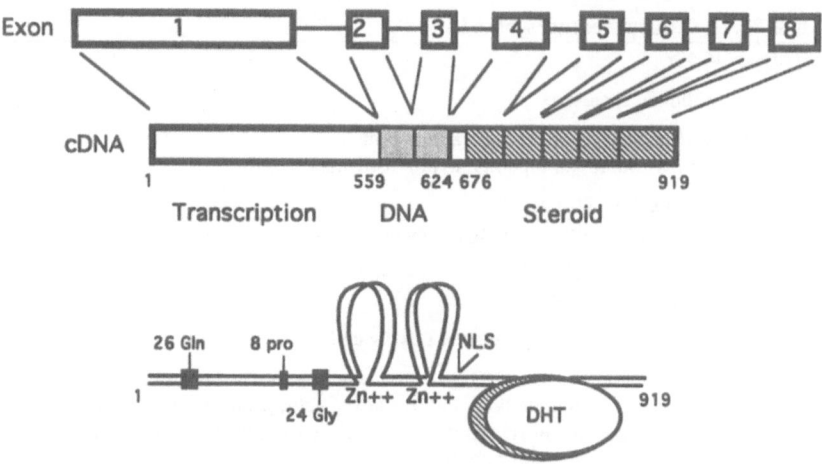

FIGURE 27.1. Schematic representation of the AR gene structure showing the presence of 8 exons (top); the AR cDNA (cDNA) with the 3 functional regions encoding the N-terminal transcriptional activation (amino acids 1–559), the DNA binding (amino acids 559–624, shaded area), and steroid binding (amino acids 676–919, diagonally hatched area) domains (middle); and the AR protein with relative positions shown for the N-terminal polymeric amino acids, the 2 zinc fingers required for DNA binding, the nuclear localization signal sequence (NLS), and steroid (5α-dihydrotestosterone [DHT]) binding (bottom).

The most highly conserved region in the AR is the central, cysteine-rich DNA binding domain between amino acids 559 and 624 that forms the 2 *zinc fingers* critical for specific nucleotide recognition and binding to DNA. Exons 2 and 3 of the AR gene encode the first and second zinc fingers, respectively. An amino acid sequence, RLKLLKGN, directly adjoining the DNA binding domain from residues 629–636, specifies a common signal for nuclear translocation of the AR (8). The steroid binding domain of AR is contained within the carboxyterminus, amino acid residues 676–919, encoded within exons 4–8. Common to the family of steroid receptors, additional functional domains for interaction with the 90-kd heat-shock protein and receptor dimerization, as well as receptor phosphorylation, may also play a role in human AR action (8, 9).

Complete Androgen Insensitivity

Subjects with a 46,XY karyotype but female external genitalia and a short vaginal pouch internally are characteristic of CAIS (1) (Table 27.1). Despite normal male levels of T produced by testicular Leydig cells

414   T.R. Brown et al.

TABLE 27.1. Clinical features of AIS.

Karyotype: 46,XY

Inheritance: X-linked recessive

Genitalia: *complete:* female with blind vaginal pouch
            partial: variable, ambiguous, hypoplastic male

Wolffian ducts: *complete:* absent
                partial: rudimentary, hypoplastic, normal

Mullerian ducts: *complete:* absent
                 partial: absent

Gonads: testes, azoospermia, hyperplastic Leydig cells

Habitus: *complete:* scant or absent pubic and axillary hair; normal female breast
         development and female body habitus at puberty; primary amenorrhea
         partial: decreased or normal pubic and axillary hair, beard growth, and body hair;
         gynecomastia common at puberty

Hormone profile: elevated plasma LH and T; elevated estradiol (for men); normal or
                 elevated FSH

Cell/molecular basis: end-organ insensitivity to androgenic and metabolic effects of T/DHT;
                      mutation in AR gene causing defect in expression of receptor or its
                      biological function

during fetal life, the Wolffian duct derivatives remain hypoplastic due to the AIS. Mullerian ducts regress in the fetus in response to the secretion of mullerian inhibiting substance from Sertoli cells. The gonads are testes located in the abdominal or inguinal region characterized postpubertally by hyperplastic Leydig cells and seminiferous tubules containing immature Sertoli cells and primitive germ cells that have not progressed beyond the spermatogonia stage. At puberty, normal female secondary sex characteristics appear accompanied by breast development in response to estrogens formed by aromatization of T, but pubic and axillary hair are usually sparse. In early infancy and postpubertally, the hallmark endocrine findings in AIS are serum T-concentrations within or above the normal male range and elevated LH levels.

Receptor-Negative Form of CAIS

The earlier biochemical findings of absent AR binding activity in many subjects with CAIS led to the presumption that genetic mutations in the AR gene were responsible for this abnormality. Following the isolation and characterization of cDNA fragments encoding portions of the human AR, we elected to concentrate our efforts on those subjects with the AR− form of CAIS in an attempt to identify gross AR gene mutations in the form of major deletions, insertions, or rearrangements.

Gene Deletions

Genomic DNA was isolated from peripheral blood lymphocytes of 6 unrelated 46,XY phenotypic female subjects known to have undetectable AR binding activity in cultured genital skin fibroblasts. Southern blots of these DNA samples were hybridized with human AR cDNA probes (Fig. 27.2). In one affected subject the cDNA fragments designated hAR1 and hAR2 failed to hybridize to the expected DNA fragments formed by digestion with several different restriction endonucleases, whereas a third cDNA fragment, hAR3, hybridized as expected in all subjects (11). The steroid binding domain of the human AR is encoded by the cDNA region included in hAR1 and hAR2, suggesting that a partial deletion in the AR gene was responsible for the absence of androgen binding in this particular subject. Analysis of other family members confirmed the deletion and its association with the AR− form of CAIS (Fig. 27.3).

The deletion of exons 4–8 encoding the entire steroid binding domain of AR has since been confirmed by amplification of genomic DNA using the *polymerase chain reaction* (PCR) (Fig. 27.2). Hybridization of hAR cDNA on Northern blots of poly(A)$^+$ RNA extracted from genital skin fibroblasts of one member of this family failed to detect either the full-length 10-kb AR mRNA or smaller mRNA species (12), suggesting that

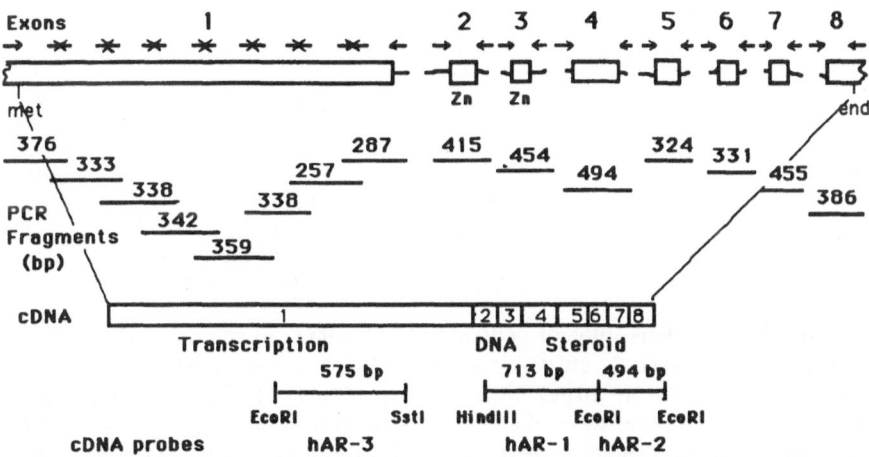

FIGURE 27.2. Strategy for molecular analyses of the AR gene structure. Shown are the exon structure (top) of the AR gene and the positions of oligonucleotide primers used for amplification by the PCR of DNA fragments covering the entire cDNA sequence (middle) encoding the functional domains of the AR protein. The cDNA fragments hAR1, hAR2, and hAR3 (bottom) were used as hybridization probes of AR gene structure on Southern blots and AR gene transcription on Northern blots.

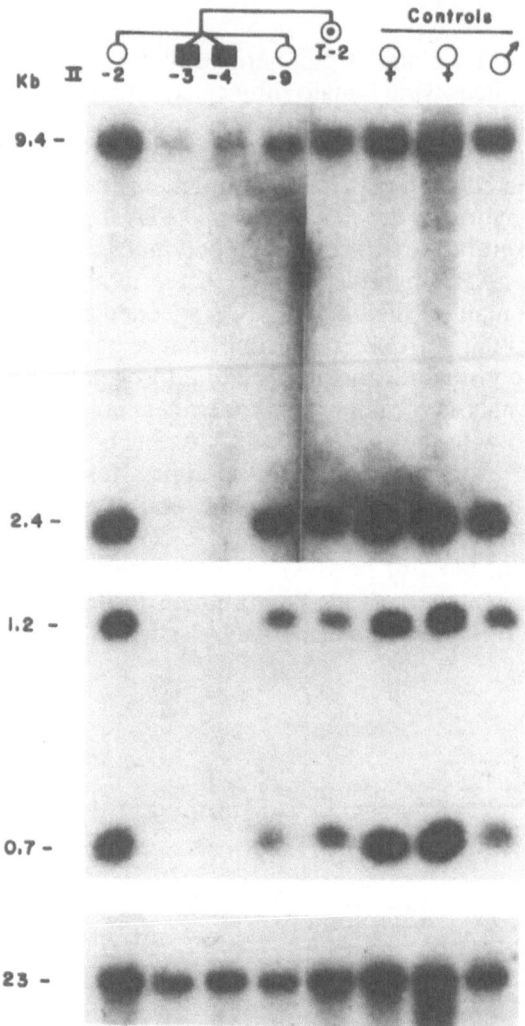

FIGURE 27.3. Southern blot of genomic DNA from 5 members of the pedigree with 2 subjects affected with CAIS (top, left) and 3 control subjects (2 females and 1 male). DNA (~10 µg) was digested with *Eco*RI, size-fractionated by agarose gel electrophoresis, and analyzed by hybridization with ^{32}P-labeled hAR1 (top panel), hAR2 (middle panel), and hAR3 (bottom panel) cDNA fragments. Reprinted from Brown, Lubahn, Wilson, Joseph, French, and Migeon (11).

the partial AR gene deletion resulted in an extremely labile, truncated transcript that was processed inappropriately.

Complete deletion of the androgen receptor gene represents the null phenotype of CAIS and has been reported in 2 unrelated subjects (13,

14). In addition, we have identified another family in which 2 affected siblings have a partial AR gene deletion of exons 3–8 encoding the second zinc finger of the DNA binding domain and the entire carboxy-terminal steroid binding domain (15). We are presently investigating a family with a deletion of exon 2 encoding the first zinc finger of the DNA binding domain in 4 members from 2 generations of a family with the AR– form of CAIS (unpublished observation). In one family with CAIS, a complex recombination event centered around intron 5 of the human AR gene is postulated to have given rise to a 5-kb deletion surrounding exon 5 in 2 affected siblings, whereas the deletion of exons 6 and 7 occurred in their affected, genotypic 46,XY "aunt" (16).

Whereas each of the aforementioned deletions occurred in subjects with absent AR steroid binding, deletion of exon 3 encoding the second zinc finger of the DNA binding domain occurred in a subject with the AR+ form of CAIS (17). The deletion of exon 3 is unique and allows the use of an alternative splice junction to produce a truncated AR protein lacking 39 amino acids that binds androgen with normal affinity and is localized intracellularly to the nucleus, but fails to bind a representative androgen response element within DNA and does not activate androgen-sensitive reporter gene transcription. Furthermore, we have identified a deletion in exon 1 of a subject with partial AIS, suggesting the possibility that initiation of transcription/translation of the AR might occur downstream from the normal start site, resulting in the synthesis of a truncated protein with partial activity (unpublished observation).

In general, however, gross alterations in the AR gene structure caused by deletions (Fig. 27.4), insertions, or rearrangements are not a common mechanism leading to AIS, even among those subjects with the AR– form. This became evident in our early studies using Southern blots (11)

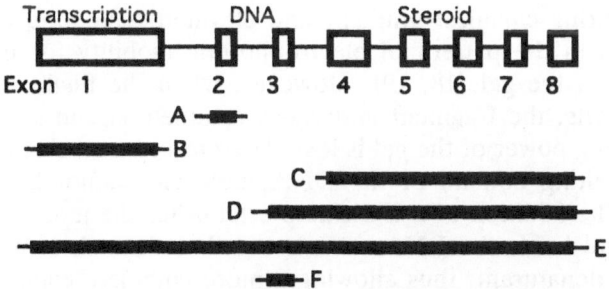

Figure 27.4. Deletions occurring in the AR gene of subjects with AIS. A schematic diagram of the AR gene is shown (top), and the gene segments that are deleted are indicated by the solid bars below. The 5 families, A–E, have been studied in our laboratory, whereas the deletion in family F was described in reference 17. All the deletions occur in families with CAIS, except for family B, which has PAIS.

to analyze the genomic organization of the AR gene and Northern blots (12) to demonstrate the transcription of a normal-sized AR mRNA in many subjects with CAIS.

Single Nucleotide Substitutions

In our laboratory we have applied a rapid genetic screening technique, termed *denaturing gradient gel electrophoresis* (DGGE), for the detection of human AR gene mutations at the level of sensitivity involving a single nucleotide (12). Exons of the human AR gene are amplified as DNA fragments of several hundred bp from genomic DNA by PCR and subjected to analysis by DGGE (Fig. 27.5, top). The oligonucleotide primers for PCR (Fig. 27.2) were designed so as to anneal to intron regions adjoining the 5' and 3' splice junctions of exons 2–8, whereas primers for the much larger exon 1 were designed to cover several overlapping regions (10). Moreover, the AR-specific 5' oligonucleotide primer was synthesized to contain an additional 40-bp GC clamp to improve the sensitivity and resolution of the DGGE method.

The DGGE method relies on the melting property of DNA under denaturing conditions created by the combination of temperature, urea, and formamide (18, 19). The AR-specific DNA fragments are electrophoresed in polyacrylamide gels containing an ascending gradient of denaturant formed by urea and formamide (100% denaturant = 7 M urea/40% urea) at a constant, but elevated (60°C) temperature. Regions in the DNA, termed *melting domains*, undergo a strand separation to produce partially denatured molecules with decreased electrophoretic mobility. As these DNA molecules continue to move slowly into higher concentrations of denaturant, additional melting domains undergo strand separation. Most importantly, single-nucleotide base changes in any of these domains will alter their melting characteristics, as determined by the denaturing chemical gradient and elevated temperature, leading to differences in the pattern of electrophoretic mobility for unique DNA fragments in the gel (18, 19). However, when the final or most stable domain melts, the fragment undergoes complete strand separation, and the resolving power of the gel is lost. Therefore, base substitutions in the highest melting domain of the DNA molecule cannot be detected by DGGE. The addition of a GC clamp establishes the most stable melting domain and holds the DNA strands together through a higher concentration of denaturant, thus allowing a more complete comparison of the various melting domains and, hence, of the specific nucleotide sequence within the amplified DNA fragment (20).

While this method may not detect all mutations, its simplicity and the absence of a requirement for use of radioactive nucleotide incorporation into the PCR-amplified DNA fragment make it attractive as an initial genetic screening procedure (21). DNA fragments exhibiting altered

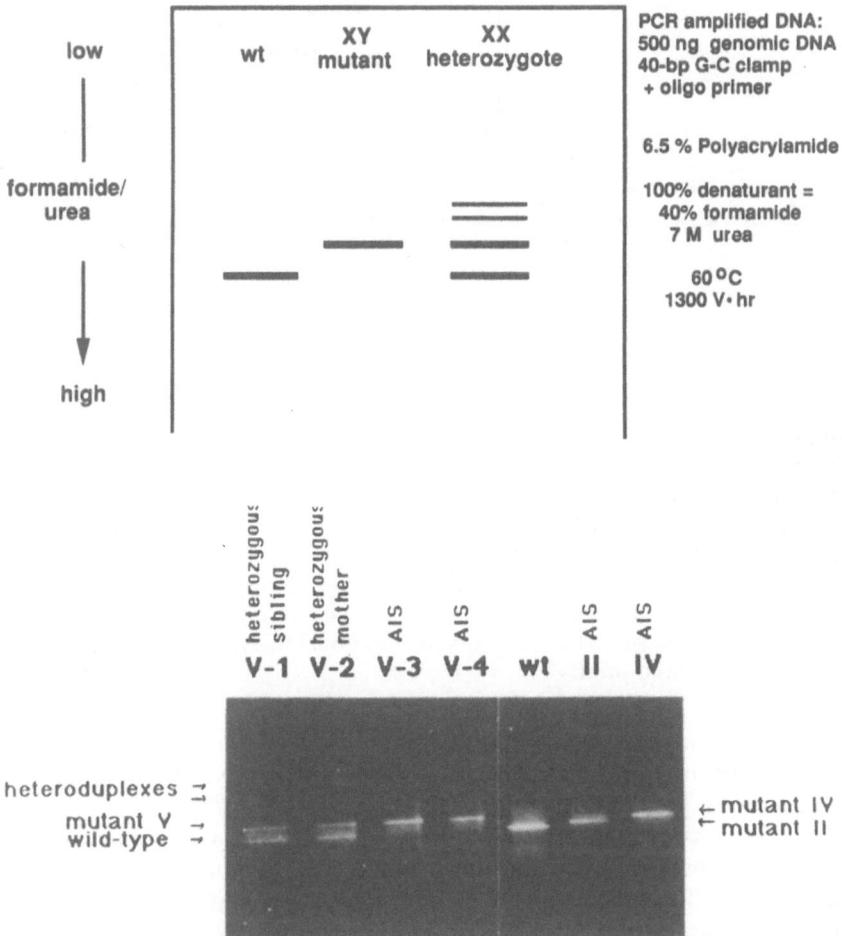

FIGURE 27.5. Schematic of DGGE (top) and a representative gel (bottom) show-ing the migration of PCR-amplified DNA fragments from exon 5 of the AR gene from individual members of 3 different families (II, IV, and V) with subjects affected with AIS. Note the retarded mobility of DNA fragments in the affected AIS subjects containing point mutations within exon 5 and the characteristic pattern of 4 different ethidium bromide-stained bands in heterozygotes. The normal (wt) fragment is used as a control.

mobility relative to control samples (Fig. 27.5, bottom) are subjected to nucleotide sequence analysis (22, 23). The resolving power of the DGGE method can also be increased when normal and mutant amplified AR exon DNA fragments are mixed, denatured, and allowed to reanneal prior to electrophoresis, thus forming DNA heteroduplexes containing nucleotide base mismatches (21). The DGGE method has allowed us to

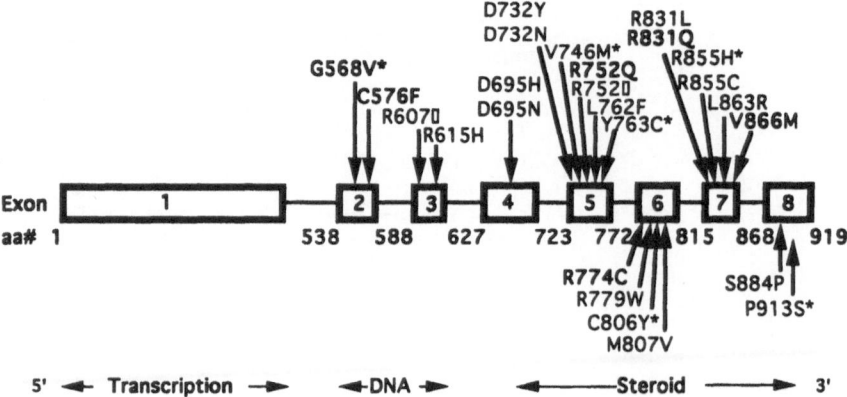

FIGURE 27.6. Diagram of point mutations in the human AR gene of subjects with AIS. The numerical position of the amino acid residues affected is flanked on the left by the normal amino acid and on the right by the amino acid encoded by the nucleotide substitution. The single-letter amino acid code is used, and the open boxes represent termination codons. The presence of an asterisk denotes the phenotypic presence of partial AIS in an affected subject; all other mutations occur in subjects with completely female external genitalia, indicative of CAIS. Those mutations presented in the text with additional relevant supporting data are indicated by the bold print.

detect a multitude of single nucleotide base mutations in the human AR gene coding region, as described below.

The majority of subjects with the AR– form of CAIS have single-nucleotide substitutions within the coding region of the human AR gene that result in alternative splicing of AR mRNA, premature termination of AR gene transcription, or missense mutations (Fig. 27.6). In one subject with all coding exons present and without sequence alterations within the AR open reading frame, a single-nucleotide substitution was detected at the consensus splice donor site located at the 3' terminus of exon 4 (24). Inactivation of the consensus splice site resulted in alternative use of a cryptic splice donor site within exon 4 and transcription of a truncated AR mRNA that encodes a protein with a 41-amino acid deletion (residues 682–722) lacking steroid binding and being transcriptionally inactive.

Some mutations, termed *nonsense mutations*, create premature termination codons for transcription of AR mRNA (Fig. 27.6). This leads to truncated mRNAs that are more labile or that are translated into AR proteins lacking the functional domains encoded by the normal downstream codons. Nonsense mutations have been identified in exons 1, 3, 4, 5, 6, 7, and 8 of subjects with CAIS (25, 26). Cultured fibroblasts from most of these subjects are reported to lack androgen binding, and when the mutated AR cDNA has been expressed in transient transfection assays, both androgen binding and transcriptional activation of reporter genes have been undetectable (25). However, a recent report demonstrated that an aminoterminal truncation of the AR due to a premature termination signal in exon 1, but with downstream initiation, resulted in a diminished amount of qualitatively abnormal AR in a subject with CAIS (27). These data suggest that the N-terminal domain of the AR is of critical importance for in vivo function.

Taken together, these data confirm deletion mapping studies performed with the human AR, as well as with the other steroid receptors, and demonstrate the essential requirement for steroid binding that is predicated on an intact carboxyterminal domain (7–9). For example, the occurrence of a premature termination codon in exon 8 that normally encodes amino acid residue 883 (26) suggests that most, if not all, of the carboxyterminal steroid binding domain must be intact for ligand binding and transcriptional activation to occur. That CAIS with female phenotype occurs in these subjects, even when the premature termination codon is located 3' of the DNA binding domain, also implies that constitutive transcriptional activity of these putatively truncated receptor proteins does not occur in vivo. These findings suggest that although constitutive transcriptional activation of reporter genes by in vitro-expressed carboxyterminal truncated ARs can occur (8, 9), the ligand binding domain normally acts to repress transcription when either the ligand is absent or its high-affinity binding is prevented by a mutation within this domain.

The majority of subjects with the AR− form of CAIS have missense mutations that cause amino acid substitutions within the steroid binding domain of the human AR (25, 26). The results from our laboratory exemplify the heterogeneity of such mutations. Missense mutations in exons 4, 5, 6, and 7 are the molecular basis for undetectable ligand binding among numerous subjects affected with AIS. In DNA from 2 subjects with undetectable AR binding in cultured genital skin fibroblasts, we identified missense mutations in exon 6 and in exon 7 of the human AR gene, respectively (12). For one subject (R774C), amino acid residue 774 was changed from arginine (CGC), R, to cysteine (TGC), C; and in the other subject (R831Q), a glutamine (CGA), Q, was substituted for arginine (CAA), R, at amino acid residue 831. Mutagenesis of human AR cDNA at the respective nucleotide positions within an expression vector and transfection of this plasmid into a mammalian cell line resulted in the synthesis of an AR protein that was detectable by immunoblotting but that failed to bind either DHT or the synthetic androgen methyltrienolone (R1881) (Fig. 27.7). The mutant R774C or R831Q AR expression plasmids were cotransfected with a second plasmid containing an androgen-responsive *mouse mammary tumor virus* (MMTV) promoter driving *chloramphenicol acetyltransferase* (CAT) reporter gene transcription. CAT activity was absent in the extracts from transfected cells in the presence or absence of androgen (12).

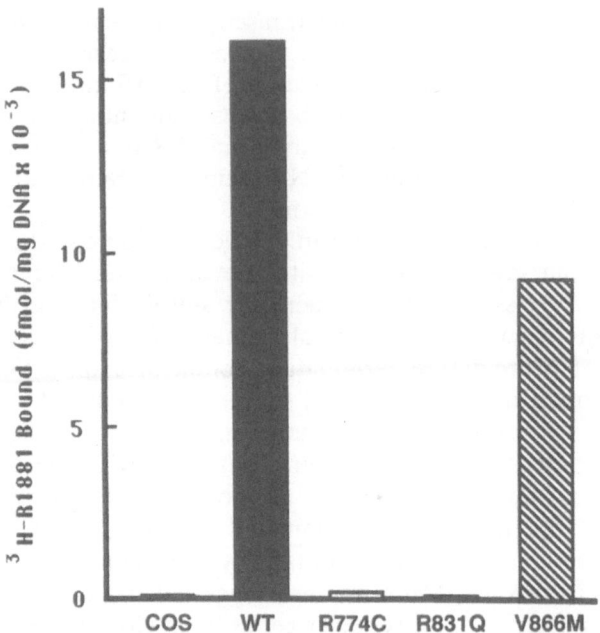

FIGURE 27.7. Androgen binding in COS monkey kidney cells transfected with an AR cDNA expression plasmid. Cells were transfected with plasmid DNA lacking the AR cDNA insert (COS) or containing the normal (wt) or mutant (R774C, R831Q, or V866M) AR cDNA insert. Specific AR binding was quantitated in total cell extracts following incubation with a saturating concentration (5 nM) of the synthetic androgen methyltrienolone (^{3}H-R1881). (R = arginine; C = cysteine; Q = glutamine; V = valine; M = methionine.)

We have identified other missense mutations in exon 4 (D695N), exon 5 (D732Y, D732N, R752Q, and Y762F), exon 6 (M807V), and exon 7 (R855C and L863R). Each of these subjects has a complete female phenotype, and the identified mutations are the only ones detected within the AR open reading frame of each subject. The AR binding in COS monkey kidney cells transfected with an AR expression vector containing mutations D732N (GAC, aspartic acid → AAC, asparagine), R752Q (CGA, arginine → CAA, glutamine), and L762F (CTC, leucine → TTC, phenylalanine) is negligible. The mutation R752Q that is present in the AR gene from 2 independent families investigated in our laboratory is the same mutation that is responsible for AIS in the Tfm rat model (28).

In summary, a heterogeneous group of human AR missense mutations in exons 4, 5, 6, and 7 encoding the steroid binding domain of the human AR is related to the AR− form of CAIS. Although missense mutations have been identified within exon 8, which encodes the furthest carboxyterminal region of the steroid binding domain, only partial forms of AIS are related to these amino acid substitutions.

Receptor-Positive Form of CAIS

The AR binding in cultured genital skin fibroblasts of some subjects with the CAIS syndrome may be quantitatively normal, thus adding to the heterogeneity of the abnormality (29, 30). This property of AR binding immediately suggests the normality of the steroid binding domain of the receptor and also suggests that the defect lies within a separate functional domain, such as that for DNA binding.

Several mutations within the DNA binding domain of the human AR have been identified (17, 25, 26, 31). As mentioned earlier, one such mutation represents a rather surprising finding: the deletion of exon 3 that encodes the second zinc finger required for receptor binding to DNA (17). Other subjects with the AR+ form of CAIS have single-nucleotide substitutions in exons 2 and 3. Within the first zinc finger encoded by exon 2, a missense mutation occurs at the base of the loop, resulting in the substitution of a phenylalanine residue for a cysteine (C576F) residue that in turn leads to the disruption of the essential coordinate binding of zinc to form the DNA binding finger motif (31). This mutation prevents an expressed AR fusion protein from binding to an androgen response element DNA sequence, as analyzed by an electrophoretic mobility shift assay (Fig. 27.8). Another missense mutation (R615H) occurs immediately 3' to the second zinc finger coding region within exon 3.

By contrast, other subjects with CAIS may have quantitatively normal AR binding, but are distinguished by functional abnormalities in the AR. Qualitative defects in the AR, such as increased thermolability, accelerated ligand dissociation, and impaired up-regulation, have been demonstrated (30, 32–34). We have extensively studied one pedigree with 3 affected 46,XY subjects with the AR+ form of CAIS (30). Assays of AR binding in cultured genital skin fibroblasts showed a decreased binding affinity for DHT, thermolability of the receptor, and an accelerated rate of dissociation for DHT. Sequence analysis of the entire coding region of the AR gene revealed a single-nucleotide substitution (G → A) within exon 7 (10). This same mutation was present in each of the 3 affected siblings, and the mother was heterozygous for G- and A-nucleotides at the same position. Transfection of the mutant AR cDNA that encoded the methionine replacement of valine (V866M) revealed the expression of a receptor protein that bound androgen (Fig. 27.7), but that had a decreased affinity for DHT (normal = 0.38 nM vs. V866M = 2.46 nM) and a significantly reduced ability to activate reporter gene transcription in cotransfection studies (12). The mutant receptor was almost completely inactive in transcriptional activity assays in the phys-

protein   AR-wt-GST   AR-C576F-GST
                        1   2   3   4

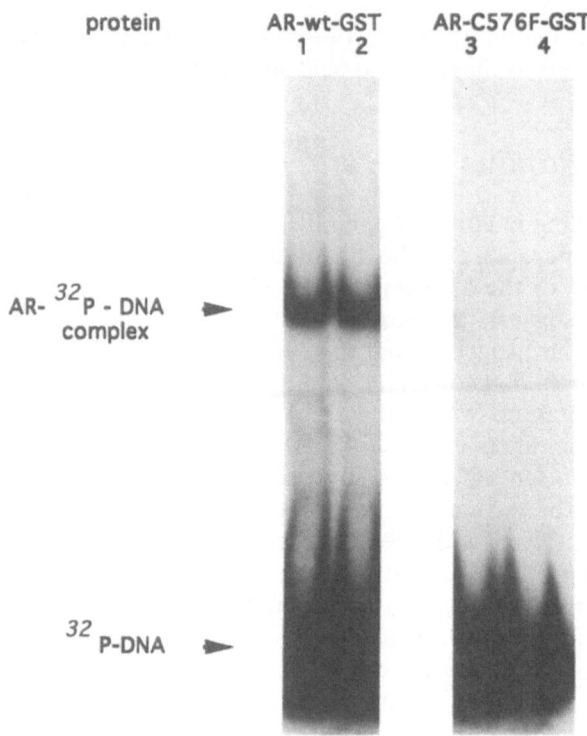

AR- ^{32}P - DNA   ➤
complex

^{32} P-DNA   ➤

FIGURE 27.8. Polyacrylamide gel electrophoretic mobility shift assay of a consensus glucocorticoid/androgen DNA response element binding to a fusion protein containing the normal or mutated DNA binding domain of the AR. A double-stranded DNA fragment representing a GRE/ARE response element, 5′-AGAACAcagTGTTCT-3′, was radiolabeled with ^{32}P-ATP and incubated with 100 (lanes 1 and 3) or 50 (lanes 2 and 4) ng of a purified glutathione S-transferase/androgan receptor fusion (GST-AR) protein purified from *E. coli* following expression in the pGEX-2T plasmid vector. The AR portion of the fusion protein included amino acid residues 559–645 from the normal (wt) or mutant (C576F) AR. The formation of a protein-DNA complex is indicated by the upward shift of the radiolabeled DNA appearing as a band on the autoradiogram.

iologic range of androgen concentrations (0.1 nM) when compared to the normal receptor (Fig. 27.9).

Another subject with a completely female phenotype and a valine for methionine amino acid substitution, M807V (ATG → GTG), may also fall into the AR+ category of CAIS. Preliminary studies have shown the presence of AR binding activity in COS cells transfected with this mutant AR expression vector and the stimulation of CAT reporter gene activity at high concentrations (10 nM) of R1881.

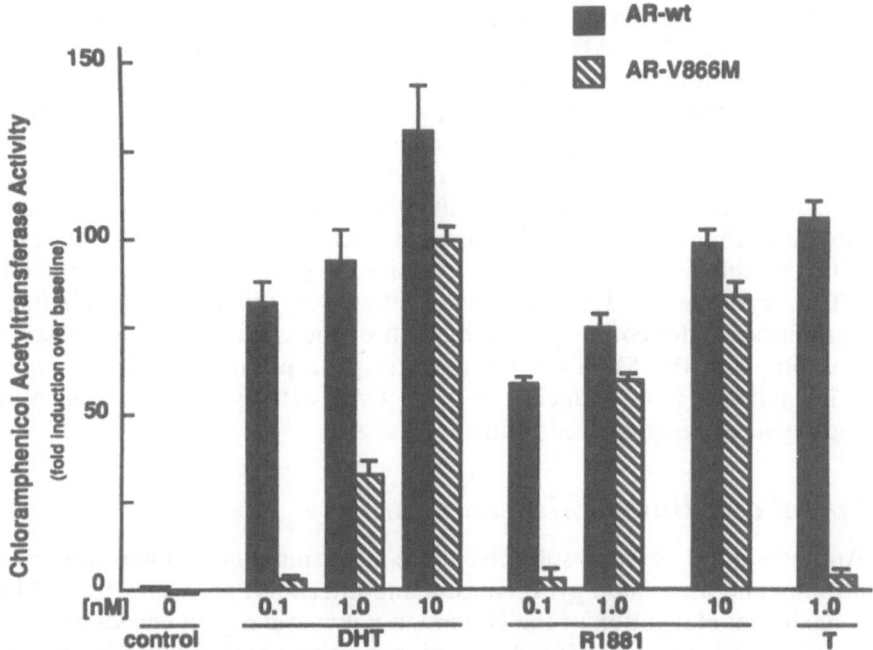

FIGURE 27.9. Concentration-dependent transcriptional activation of CAT by androgens for normal and mutant AR. The CV-1 monkey kidney cells were cotransfected with the normal (wt, solid bars) or mutant (V866M, diagonally hatched bars) AR expression vector and the reporter gene plasmid vector containing the MMTV promoter-CAT construct. Acetylation of chloramphenicol was quantitated in extracts from cells incubated in the absence or presence of different steroids at various concentrations. (DHT = 5α-dihydrotestosterone; R1881 = methyltrienolone; T = testosterone.) Modified with permission from Brown, Lubahn, Wilson, French, Migeon, and Corden (12), © The Endocrine Society, 1990.

Partial AIS

Varying degrees of sexual ambiguity are present in other subjects with an endocrine profile consistent with the diagnosis of AIS (1). Most likely, those subjects described earlier by Reifenstein, Gilbert-Dreyfus, Lubs, and Rosewater and associates form part of the phenotypic continuum related to insensitivity to androgen. In the *partial* form of AIS (PAIS), AR binding may be either quantitatively normal (AR+) or partially deficient (AR±) (35).

In our laboratory we have identified missense mutations within both the DNA binding and steroid binding domains that are associated with the phenotype of PAIS. Single-nucleotide substitutions within exon 2 (G568V) of the DNA binding domain and exons 5 (V746M and Y763C),

6 (C806Y), 7 (R855H), and 8 (P913S) of the steroid binding domain have been detected. The AR binding is quantitatively normal in COS cells transfected with the mutant AR expression plasmid G568V (G<u>G</u>G, glycine → G<u>T</u>T, valine), but CAT reporter gene activity is significantly reduced in cotransfection experiments that measure the biologic function of the mutant AR to stimulate androgen-regulated gene transcription. In comparison to the aforementioned mutation in the DNA binding domain, mutations in the steroid binding domain, such as V746M (<u>G</u>TG, valine → <u>A</u>TG, methionine), Y763C (<u>C</u>TC, tyrosine → <u>T</u>TC, cysteine), and C806Y (T<u>G</u>C, cysteine → T<u>A</u>C, tyrosine), effectively diminish AR binding. Furthermore, the cotransfection of each of these mutant AR expression plasmids with the MMTV-CAT reporter gene plasmid results in a partially, but not fully, inducible level of CAT activity in the presence of androgen, when compared to the normal AR.

Spinal and Bulbar Muscular Atrophy

Another group of male subjects, those with spinal and bulbar muscular atrophy and associated gynecomastia and infertility, are characterized by a significant elongation of the glutamine polymeric region encoded within exon 1 of the AR gene (36). This region is polymorphic within the normal population and varies in length from approximately 17 to 26 glutamine residues. However, in male subjects with spinal and bulbar muscular atrophy, the number of glutamine residues is increased to between 40 and 52. The exact role of the homopolymeric region in determining the biologic function of the human AR is unknown at present, but may represent target cell-specific alterations of AR activity in neuronal cells.

Summary

Cloning of the human AR cDNA has made it possible to establish the causal relationship of mutations in the AR gene with the AIS syndromes. Upon inspection, one is immediately impressed by the diverse nature of the mutations involved in the complete and partial forms of AIS and the heterogeneous distribution of these mutations throughout the coding region of the AR gene. Because of the large number and diverse array of these naturally occurring mutations and their associated clinical phenotypes, there is great potential for understanding the structure-function relationships of the AR from the in vitro expression of the mutant receptors in various cell lines.

Whereas studies to date have focused on the identification of mutations and their correlation with steroid binding and transactivation of reporter genes in heterologous cell lines, future studies will be directed toward understanding the interaction of the functional domains of the

AR, AR binding to specific androgen response elements of DNA, AR dimerization, AR phosphorylation, AR interaction with heat-shock protein(s), the nuclear localization of AR, and the role of accessory factors that direct cell- and temporal-specific regulation of gene transactivation by AR. All represent future areas for intensive investigation using the naturally occurring mutations of AR, found in AIS subjects with a clinically identifiable phenotype, as the basis for understanding mechanisms of hormone action.

Acknowledgments. We thank our many clinical collaborators for providing the human subject materials necessary to conduct our research efforts. This work was supported by NIH Research Grants RO1-DK-43147 and DK-00180 and NIH Training Grant T32-DK-07116. Dr. Chang was supported in part by an Eli Lilly-Lawson Wilkins Society Fellowship, Dr. Paolo Ghirri was supported by a fellowship from the CNR, Italy, and Dr. Murono was supported by a stipend from the Asahikawa Medical College, Japan.

References

1. Brown TR, Migeon CJ. Androgen insensitivity syndromes: paradox of phenotypic feminization with male genotype and normal testicular androgen secretion. In: Cohen MP, Foa PP, eds. Hormone resistance and other endocrine paradoxes. New York: Springer-Verlag, 1987:157–203.
2. Lubahn DB, Joseph DR, Sar M, et al. The human androgen receptor: complementary deoxyribonucleic acid cloning, sequence analysis, and gene expression in prostate. Mol Endocrinol 1988;2:1265–75.
3. Chang C, Kokontis J, Liao S. Structural analysis of complementary DNA and amino acid sequences of human and rat androgen receptors. Proc Natl Acad Sci USA 1988;85:7211–5.
4. Migeon BR, Brown TR, Axelman J, Rothwell SW, Migeon CJ. Studies on the locus for androgen receptor localization on the human X-chromosome and evidence for homology with the Tfm locus in the mouse. Proc Natl Acad Sci USA 1981;78:6339–43.
5. Lubahn DB, Joseph DR, Sullivan PM, Willard HF, French FS, Wilson EM. Cloning of human androgen receptor complementary DNA and localization of the X-chromosome. Science 1988;240:327–30.
6. Brown CJ, Gross SJ, Lubahn DB, et al. Androgen receptor locus on the human X-chromosome: regional localization of Xq11-12 and description of a DNA polymorphism. Am J Hum Genet 1989;44:264–9.
7. Rundlett SE, Wu S-P, Miesfield RL. Functional characterization of the androgen receptor confirms that the molecular basis of androgen action is transcriptional regulation. Mol Endocrinol 1990;4:708–14.
8. Simental JA, Sar M, Lane MV, French FS, Wilson EM. Transcriptional activation and nuclear targeting signals of the human androgen receptor. J Biol Chem 1991;266:510–8.

9. Jenster G, van der Korput HAGM, van Vroonhoven C, van der Kwast TH, Trapman J, Brinkman AO. Domains of the human androgen receptor involved in steroid binding, transcriptional activation and subcellular localization. Mol Endocrinol 1991;5:1396–404.

10. Lubahn DB, Brown TR, Simental JA, et al. Sequence of the intron/exon junctions of the coding region of the human androgen receptor gene and identification of a point mutation in a family with complete androgen insensitivity. Proc Natl Acad Sci USA 1989;86:9534–8.

11. Brown TR, Lubahn DB, Wilson EM, Joseph DR, French FS, Migeon CJ. Deletion of the steroid-binding domain of the human androgen receptor gene in one family with complete androgen insensitivity syndrome: evidence for further genetic heterogeneity in this syndrome. Proc Natl Acad Sci USA 1988;85:8151–5.

12. Brown TR, Lubahn DB, Wilson EM, French FS, Migeon CJ, Corden JL. Functional characterization of naturally occurring mutant androgen receptors from patients with complete androgen insensitivity. Mol Endocrinol 1990; 4:1759–72.

13. Trifiro M, Gottlieb B, Pinsky L, et al. The 56/58 kDa androgen-binding protein in male genital skin fibroblasts with a deleted androgen receptor gene. Mol Cell Endocrinol 1991;75:37–47.

14. Quigley CA, Friedman KJ, Johnson A, et al. Complete deletion of the androgen receptor gene: definition of the null phenotype of the androgen insensitivity syndrome and determination of carrier status. J Clin Endocrinol Metab 1992;74:927–33.

15. French FS, Lubahn DB, Brown TR, et al. Molecular basis of androgen insensitivity. Recent Prog Horm Res 1990;46:1–42.

16. MacLean HE, Chu S, Warne GL, Zajac JD. Related individuals with different androgen receptor gene deletions. J Clin Invest 1993;91:1123–8.

17. Quigley CA, Evans BA, Simental JA, et al. Complete androgen insensitivity due to deletion of exon C of the androgen receptor gene highlights the functional importance of the second zinc finger of the androgen receptor in vivo. Mol Endocrinol 1992;6:1103–12.

18. Fischer S, Lerman L. DNA fragments differing by single-base-pair substitutions are separated in denaturing gradient gel: correspondence with melting theory. Proc Natl Acad Sci USA 1983;80:1579–83.

19. Myers RM, Lerman LS, Fischer SG, Maniatis T. Modification of the melting properties of duplex DNA by denaturing gradient gel electrophoresis. Nucleic Acids Res 1985;13:3111–29.

20. Sheffield V, Cox DR, Lerman LS, Myers RM. Attachment of a GC-clamp to genomic DNA fragments by the polymerase chain reaction results in improved detection of single base changes. Proc Natl Acad Sci USA 1989; 86:232–6.

21. Theophilius BDM, Latham T, Gabrowski GA, Smith FI. Comparison of RNase A, a chemical cleavage and GC-clamped denaturing gradient gel electrophoresis for the detection of mutations in exon 9 of the human acid β-glucosidase gene. Nucleic Acids Res 1989;19:7707–22.

22. Traystman MD, Higuchi M, Kasper CK, Antonarakis SE, Kazazian HH Jr. Use of denaturing gradient gel electrophoresis to detect point mutations in the factor VIII gene. Genomics 1990;6:293–301.

23. Wong C, Dowling CD, Saiki RK, Higuchi RG, Erlich HA, Kazazian HH Jr. Characterization of β-thalassemia mutations using direct genomic sequencing of amplified single copy DNA. Nature 1987;330:384–6.

24. Ris-Stalpers C, Kuiper GGJM, Faber PW, et al. Aberrant splicing of androgen receptor mRNA results in the synthesis of a nonfunctional receptor protein in a patient with androgen insensitivity. Proc Natl Acad Sci USA 1990; 87:7866–70.

25. McPhaul MJ, Marcelli M, Tilley WD, Griffin JE, Wilson JD. Androgen resistance caused by mutations in the androgen receptor gene. FASEB J 1991;5:2910–5.

26. Pinsky L, Trifiro M, Kaufman M, et al. Androgen resistance due to mutation of the androgen receptor. Clin Invest Med 1992;15:65–81.

27. Zoppi S, Wilson CM, Harbison MD, et al. Complete testicular feminization caused by an amino-terminal truncation of the androgen receptor with downstream initiation. J Clin Invest 1993;91:1105–12.

28. Yarbrough WG, Quarmby VE, Simental JA, et al. A single base mutation in the androgen receptor gene causes androgen insensitivity in the testicular feminized rat. J Biol Chem 1990;265:8893–900.

29. Amrhein JA, Meyer WJ III, Jones HW Jr, Migeon CJ. Androgen insensitivity in man: evidence for genetic heterogeneity. Proc Natl Acad Sci USA 1976;73:891–4.

30. Brown TR, Maes M, Rothwell SW, Migeon CJ. Human complete androgen insensitivity with normal dihydrotestosterone receptor binding capacity in cultured genital skin fibroblasts: evidence for a qualitative abnormality of the receptor. J Clin Endocrinol Metab 1982;55:61–9.

31. Chang YT, Migeon CJ, Brown TR. Human androgen insensitivity syndrome due to androgen receptor gene point mutations in subjects with normal androgen receptor levels but impaired biological activity [Abstract 28]. 73rd annu meet Endocr Soc, 1991.

32. Griffin JE. Testicular feminization associated with a thermolabile androgen receptor in cultured human fibroblasts. J Clin Invest 1979;64:1624–31.

33. Kaufman M, Pinsky L, Feder-Hollander R. Defective up-regulation of the androgen receptor in human androgen insensitivity. Nature 1981;293:735–7.

34. Griffin JE, Durrant JL. Qualitative defects in families with androgen resistance: failure of stabilization of the fibroblast cytosol androgen receptor. J Clin Endocrinol Metab 1982;55:465–74.

35. Amrhein JA, Klingensmith G, Walsh PC, McKusick VA, Migeon CJ. Partial androgen insensitivity: the Reifenstein syndrome revisited. N Engl J Med 1977;297:350–6.

36. La Spada AR, Wilson EM, Lubahn DB, Harding AE, Fischbeck KH. Androgen receptor gene mutations in X-linked spinal and bulbar muscular atrophy. Nature 1991;352:77–9.

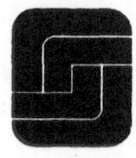

28

Congenital Absence of the Vas Deferens and the Cystic Fibrosis Transmembrane Conductance Regulator Gene: Evidence for a Common Genetic Background

Pasquale Patrizio and Ricardo H. Asch

Bilateral *congenital absence of the vas deferens* (CAVD) with obstructive azoospermia is a form of male sterility found in otherwise normal men, with an estimated incidence of 1% of all the cases of male infertility, thus affecting about 25,000–35,000 men in the United States alone (1). For these patients, once considered to be irreversibly sterile, the only option for fathering their own progeny is the combined use of *microsurgical epididymal sperm aspiration* (MESA) and *in vitro fertilization* (IVF) (2, 3). Spermatozoa trapped in the blind-ending portions of the epididymis are surgically retrieved and used to inseminate in vitro oocytes harvested from the wives of these patients. The success of this approach has been documented by reports of pregnancies and births (4).

However, despite the simplicity of the diagnostic procedure, easily recognizable by careful scrotal examination, and the possibility of offering treatment, the etiology of this peculiar male reproductive disorder remained unknown until recently (1, 5, 6). This chapter focuses on the evidence that isolated CAVD and *cystic fibrosis* (CF) share a common genetic background, with CAVD representing a milder or variant form of the same disease. Furthermore, new data showing a very close link between particular CF mutations and the fertilizing capacity of epididymal sperm is also presented.

Etiology of CAVD

Two factors contributed to the understanding of this anomaly: The first was the observation that CAVD is almost invariably present in males with

CF disease (7–9), and the second was the recent identification, on the long arm of chromosome 7, of the gene responsible for CF (10, 11). This gene, designated *cystic fibrosis transmembrane conductance regulator* (CFTR), codes for a transmembrane protein consisting of 1480 amino acids that appears to regulate the transmembrane transport of chloride ions in epithelial cells (12, 13).

The availability of assays for detecting CFTR mutations led us to investigate the hypothesis that the infertile population of men with CAVD might in reality have the same genetic defects as patients with cystic fibrosis. To fully evaluate this genetic hypothesis, screening for CFTR mutations also included the offspring generated with MESA and IVF and the parents of men with CAVD, representing three generations of subjects.

A total of 66 patients with CAVD, their wives, 32 of their parents, and 13 of their offspring were evaluated for 12 known mutations in the CFTR gene. None of the patients had signs or symptoms of CF disease (normal chest X ray, no malabsorption, normal liver function tests, and only a few had sweat chloride tests that were borderline high), nor had they relatives or siblings with CF disease. In all, testicular biopsies showed the presence of normal spermatogenesis (14), and the serum levels of FSH and testosterone were within the normal range.

Genomic DNA was obtained from peripheral lymphocytes and amplified by PCR methods; the resulting products were analyzed for 12 mutations in the CFTR gene, accounting for almost 90% of those identified. Exons 4, 10, 11, 20, and 21 were amplified, and then by using *allele-specific oligonucleotide* (ASO) probe analysis, the following mutations were tested: *Delta F508* (DF508), G542X, G551D, R553X, W1282X, N1303K, *Delta I507* (DI507), 1717G-A, R560T, S549N, R117H, and 621+1. Of the 66 patients with isolated CAVD tested, 42 (64%) were positive for at least one CFTR mutation; of the 49 wives (controls), 2 (4%) were found to be a CFTR carrier, which matched the incidence of carrier frequency for CF mutations in the general population (4%)(15). The complete details of CFTR screening are summarized in Table 28.1.

Five patients were found to be compound heterozygotes, 4 with genotypes DF508/R117H and 1 with R553X/R117H. Since these genotypes are also found in individuals with full-blown CF disease, these findings alone strongly suggest that isolated CAVD is a mild or atypical form of CF

TABLE 28.1. Incidence of CFTR gene mutations in 3 generations.

Patients	42/66 = 64%
Parents	13/32 = 41%
Offspring	3/13 = 23%
Wives	2/49 = 4%

TABLE 28.2. Total CFTR mutations detected in patients with CAVD.

N	Mutation	Incidence
22	DF508	19.6% (26/132)*
10	W1282X	7.5% (10/132)
1	N1303K	0.7% (1/132)
1	G542X	0.7% (1/132)
1	1717G-A	0.7% (1/132)
1	R553X	—
1	R553X/R117H	1.5% (2/132)
1	R117H/R117H	—
4	DF508/R117H	5.3% (7/132)

*Indicates 26 DF508 alleles among the 132 chromosomes tested.

TABLE 28.3. Three-generation allele distribution in compound heterozygotes and homozygote.

Case	Patient allele	Paternal allele	Maternal allele	Wife allele	Offspring Allele	Phenotype
1.	DF508/R117H	DF508	R117H	Neg	(g) DF508	Normal
2.	DF508/R117H*	R117H	DF508	R117H	?	?
3.	R117H/R117H	R117H	R117H	Neg	—	—
4.	DF508/R117H	N/A	N/A	Neg	(b) DF508	Normal

*Normal sweat chloride tests.
N/A = data not available; b = boy; g = girl.

disease. More to the point, 1 patient was found to be homozygous for R117H/R117H, and yet, except for chronic sinusitis he did not show any symptom of CF disease (normal sweat chloride test and normal pancreatic function). The complete results of the CFTR mutations detected and their corresponding incidence are reported in Table 28.2. In 22 cases the CFTR mutation identified was DF508, a 3-bp deletion. This mutation, so called because a residue of phenylalanine in position 508 of the CFTR protein is lost, is the most common (75% of the cases in North American Caucasians) and lethal of the known mutations for cystic fibrosis. In 10 patients, mostly of Jewish Askenazi ethnicity, the mutation identified was W1282X, a nonsense mutation coding for the production of truncated polypeptides. The next most frequent mutation was the R117H found in 5 subjects. This is a missense mutation of the transmembrane region of the CFTR protein and is usually found in patients having a very mild form of CF (16). The remaining nonsense mutations (1717G-A, R553X, and G542X) are all very rare. Table 28.3 details the distributions of the mutations in the three generations of subjects examined.

TABLE 28.4. Comparisons of mutation frequency between CF and CAVD patients.

Mutation	CF patients	CAVD patients
R117H	0.3% (3/812)*	5.3% (7/132)
W1282X	1.6% (9/578)	7.5% (10/132)
DF508	73.0% (593/812)	19.6% (26/132)

* Denotes 3 R117H alleles among 812 CF alleles.

All of the 13 offspring have normal phenotype, including the 3 who were found to be carriers of DF508. Interestingly, 13 parents of patients with CAVD were found to be carriers for CF mutations, and two points deserve particular comment. First, every time a mutation was identified in the genomic DNA of the parents, that same mutation was inherited by their son with CAVD without the appearance of new mutations. Second, 6 of the parents were fathers, and despite being CF carriers, all have normal vas deferens and are fertile.

Since the prevalence of a single identifiable CF mutation among the patients with CAVD is very high, it could be hypothesized that heterozygosity for CF mutations (carrier) is sufficient to cause isolated CAVD (17, 18). However, when the results of the genetic screening are analyzed for each component of the family pedigree (father-patient with CAVD offspring), it becomes clear that the presence of a mutation on one chromosome only is not sufficient to give a phenotypic expression. Therefore, since male parents and male offspring found to be carriers all have normal vas, it can be thought that men with CAVD found to be only carriers are in reality compound or double heterozygotes, with the CF mutation on the opposite chromosome either not yet identified or simply not included in the panel of the mutations tested in the original study. Similarly, patients testing negative to the screening (24/66) must carry 2 mutations not identified.

Table 28.4 compares mutation frequency between CF and CAVD patients. The frequency distribution of the CF mutations among the chromosomes of the CAVD patients is markedly different from that seen in patients affected by CF disease. In fact, less common mutations (R117H, W1282X, and R553X) are seen in greater frequency in our cohort of patients.

Particularly interesting is the observation that the 5 patients identified as compound heterozygotes and the 1 homozygote all have R117H mutation opposite either to DF508 (4 cases), or to R553X (1 case), or to R117H. All had normal sweat chloride tests and isolated CAVD as the only phenotypic expression of CF disease. Since R117H is a mild CF mutation, it appears that when it is opposed to the most common and lethal DF508, it may moderate the more severe phenotype.

In conclusion, the discovery of CFTR mutations in infertile men with isolated CAVD serves to define CAVD as a variant and mild form of CF that exclusively affects the development of the mesonephric duct and that is inherited in autosomic recessive fashion in sons of parents that are carriers of different CF mutations. Screening for CF mutations can be proposed as an additional diagnostic criterion during the workup of azoospermic males.

As a final point, it must be stressed that from now on, each couple interested in MESA and IVF needs genetic testing for CF and counseling. The risk of producing offspring with CF is extremely small if the wives are negative, while the counseling becomes more difficult in cases where wives are CF carriers (case 2 from Table 28.3). In such instances we cannot assess the risk for a future child to have CF or if male, only isolated CAVD.

Effect of CFTR Gene on IVF Results

One of the limiting factors for the successful treatment of male sterility due to CAVD is the low (<20%) and extremely unpredictable rate of IVF with their epididymal sperm. Also, it has been a common observation in patients that had multiple attempts of MESA and IVF that the fertilization capacity of their epididymal sperm can be predicted according to the results of their first procedure. In other words, if a patient had fertilization and embryos at their first sperm aspiration procedure, they also had fertilization in subsequent attempts. On the other hand, if there was failure of fertilization at the first attempt, such failure was repeated at each new attempt.

Many studies have been performed in order to clarify these patterns and to identify a possible parameter that could have a predictive value for the fertilizing capacity of epididymal sperm, but none so far has proved to be conclusive (19–23). The identification of isolated CAVD as a mild form of CF disease has opened a new avenue for the interpretation of the IVF results. In fact, by analyzing the fertilization and pregnancy rates according to the type and severity of the CF mutations, a striking pattern emerged (24).

Table 28.5 presents the IVF results according to the three groups of patients identified: group I ($n = 21$) formed by patients with DF508 mutation only; group II ($n = 18$) formed by patients with other various mutations, including the compound heterozygotes; and group III ($n = 21$) formed by patients with no detectable CF mutations. In group I the fertilization rate was 7%; in group II it was 21%; and in group III it was 23%. The difference between groups I and III was statistically significant ($P < 0.000001$). The pregnancy rates per patient among the three groups were 10%, 39%, and 48%, respectively, with a statistically significant

TABLE 28.5. Overall results of fertilization and pregnancy rate according to CFTR mutations.

	Group I (DF508) (n = 21)	Group II (other) (n = 18)	Group III (neg) (n = 21)
No. eggs	475	362	414
No. embryos	32	74	96
Fertilization rate (%)	7*	21	23*
No. with fertilization	10	13	13
Pregnancy	2 (10%)**	7 (39%)	10 (48%)**

$*P < 0.000001$; $**P < 0.01$.

difference between groups I and III ($P < 0.01$). Patients that were found to be compound heterozygotes with R117H (a very mild mutation) had a fertilization rate of 27%, thus leading us to believe that R117H may moderate not only the phenotypic expression of the DF508, but also the fertilizing ability of their sperm.

Twelve patients had repetitive MESA attempts, of which 5 were in group I (DF508), 3 were in group II (other CF mutations), and 4 were in group III (no CF mutations). Of those in group I, 3 consistently failed to fertilize (one with 4 attempts, one with 3, and one with 2), while the remaining consistently fertilized. In group II one patient had 2 attempts but never fertilized, while the other 2 patients (4 and 2 attempts, respectively) always did. Of the 4 patients in group III, one never fertilized, while the other 3 (all with 2 attempts) always did.

The explanation for these observations can be derived from a recent work by Trezise and Buckwald (25) where by means of in situ hybridization, the presence of CFTR messenger mRNA was demonstrated in rat pancreas, salivary glands, and testes. Specifically in the testis it was demonstrated that CFTR expression is maximal in the round spermatids at stages VII and VIII.

Our clinical data can be reconciled with the concept that the expression of normal CFTR activity during spermatogenesis must be crucial. A defective CFTR activity present in CF patients and in patients with isolated CAVD could account for biochemical disturbances (alteration of the Cl⁻ channel or lack of transport of alternative substrates) in the mechanisms necessary for sperm morphogenesis and maturation. Men with DF508 might have the highest degree of abnormal CFTR activity; therefore, their epididymal sperm, regardless from which segment of the epididymis it is retrieved, will be biochemically defective and impaired in the fertilization function.

In conclusion, men with DF508 mutation not opposed to a known mild mutation need to be informed about the high likelihood of a poor IVF outcome. It will be also interesting to see if micromanipulation methodologies may achieve fertilization in this specific group of patients and in those who showed failure to fertilize with routine IVF. Finally, once additional CF mutations are identified, it will be possible to identify with more accuracy which CF genotype is associated with the best or the poorest IVF outcome. These data have opened a new avenue for researching the causes of low or failed IVF of epididymal sperm and highlight a previously unrecognized function of the CFTR gene activity during spermatogenesis.

References

1. Patrizio P, Asch RH. Epididymal sperm in assisted reproduction. Ann Acad Med Singapore 1992;21:533–7.
2. Silber SJ, Ord T, Borrero C, Balmaceda JP, Asch RH. New treatment for infertility due to congenital absence of vas (letter). Lancet 1987;2:850.
3. Silber SJ, Ord T, Balmaceda JP, Patrizio P, Asch RH. Congenital absence of vas deferens: study on the fertilizing capacity of human epididymal sperm. N Engl J Med 1990;323:1788–92.
4. Patrizio P, Silber SJ, Ord T, Balmaceda JP, Asch RH. Two births after microsurgical sperm aspiration in congenital absence of vas deferens (letter). Lancet 1988;2:1364.
5. Anguiano A, Oates RH, Amos JA, et al. Congenital bilateral absence of the vas deferens: a primarily genital form of cystic fibrosis. J Am Med Assoc 1992;67:1794–7.
6. Patrizio P, Asch RH, Handelin B, Silber SJ. Aetiology of congenital absence of the vas deferens: genetic study of three generations. Hum Reprod 1993; 8:215–20.
7. Kaplan E, Shwachman H, Perlmutter AD, Rule A, Khaw KT, Holsclaw DS. Reproductive failure in males with cystic fibrosis. N Engl J Med 1968;279:65–9.
8. Holsclaw DS, Perlmutter AD, Jockin H, Shwachman H. Genital abnormalities in male patients with cystic fibrosis. J Urol 1971;106:568–74.
9. Taussig LM, Lobeck CC, Di Sant'Agnese PA, Ackerman DR, Kattwinkel J. Fertility in males with cystic fibrosis. N Engl J Med 1972;287:586–9.
10. Rommens JM, Iannuzzi MC, Kerem B, et al. Identification of the cystic fibrosis gene: chromosome walking and jumping. Science 1989;245:1059–65.
11. Riordan JR, Rommens JM, Kerem BS, et al. Identification of the cystic fibrosis gene: cloning and characterization of complementary DNA. Science 1989;245:1066–73.
12. Widdicombe J, Welsh MJ, Finkbeiner WE. Cystic fibrosis decreases the apical membrane chloride permeability of monolayers cultured from cells of trachea epithelium. Proc Natl Acad Sci USA 1985;82:6167–71.
13. Quinton PM. Chloride impermeability in cystic fibrosis. Nature 1983;301: 421–2.
14. Silber SJ, Patrizio P, Asch RH. Quantitative evaluation of spermatogenesis by testicular histology in men with congenital absence of the vas deferens undergoing epididymal sperm aspiration. Hum Reprod 1990;5:89–93.
15. Boat TJ, Webb MJ, Beaudet AL. Cystic fibrosis. In: Scriver CR, Beaudet AL, Sly WJ, Valle D, eds. Metabolic basis of inherited diseases. New York: McGraw Hill, 1989;6:2649–80.

16. Dean M, White MB, Amos J, et al. Multiple mutations in highly conserved residues are found in mildly affected cystic fibrosis patients. Cell 1990;61: 863–70.
17. Dumur V, Gervais R, Rigot JM, et al. Abnormal distribution of CF DF-508 allele in azoospermic men with congenital aplasia of epididymis and vas deferens (letter). Lancet 1990;336:512.
18. Rigot JM, Lafitte JJ, Dumur V, et al. Cystic fibrosis and congenital absence of the vas deferens (letter). N Engl J Med 1991;325:64–5.
19. Davis RO, Overstreet JW, Asch RH, Ord T, Silber SJ. Movement characteristics of human epididymal sperm used for fertilization of human oocytes in vitro. Fertil Steril 1991;56:1128–35.
20. Ord T, Marello E, Patrizio P, Balmaceda JP, Silber SJ, Asch RH. The role of the laboratory in the handling of epididymal sperm for assisted reproductive technologies. Fertil Steril 1992;57:1103–6.
21. Asch RH, Silber SJ. Microsurgical epididymal sperm aspiration and assisted reproductive techniques. Ann NY Acad Sci 1992;626:101–10.
22. Patrizio P, Silber SJ, Ord T, Moretti-Rojas I, Asch RH. Relationship of epididymal sperm antibodies to their in vitro fertilization capacity in men with congenital absence of the vas deferens. Fertil Steril 1992;58:1006–10.
23. Rojas FJ, La AT, Ord T, et al. Penetration of zona-free hamster oocytes using human sperm aspirated from the epididymis of men with congenital absence of the vas deferens: comparison with human in vitro fertilization. Fertil Steril 1992;58:1000–5.
24. Patrizio P, Ord T, Silber SJ, Asch RH. Cystic fibrosis mutations impair the fertilization rate of epididymal sperm from men with congenital absence of the vas deferens. Hum Reprod 1993.
25. Trezise AEO, Buchwald M. In vitro cell-specific expression of the cystic fibrosis transmembrane conductance regulator. Nature 1991;353:434–7.

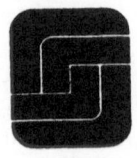

29

GnRH Neurons in an Interpretation of Kallmann's Syndrome

Marlene Schwanzel-Fukuda and Donald W. Pfaff

Since virtually all neurons in the mammalian brain are born adjacent to the linings of the cerebral ventricles and migrate a small distance to their final functional positions, it was surprising that the neurons that control vertebrate reproduction, those expressing *gonadotropin releasing hormone* (GnRH and LHRH), do not follow that rule. Rapidly gathering information about the origin and migration of GnRH neurons has recently been reviewed (1), and this chapter briefly sets some of those facts against a background of ideas about Kallmann's syndrome and about gene transfer into brain.

During a developmental study using immunocytochemistry applied to the central nervous system and olfactory apparatus of fetal mice, we discovered that GnRH immunoreactive neurons are first detected in the epithelium of the medial side of the olfactory pit at about 11 days of gestation (2). In this work a temporal series of immunocytochemical studies, spanning the time from 10 to 20 days of fetal life in mice, showed that by days 12 and 13, cords of GnRH cells migrate across the nasal septum toward the forebrain. After crossing the critical juncture between the olfactory apparatus and the forebrain, from 14 to 20 days of gestation in mice, GnRH neurons arrive in the basal forebrain, in and around the preoptic area.

In these observations and in subsequent experiments, a striking feature of the migrating GnRH neurons on the nasal septum was their tendency to cluster. Subsequently, a combination of electron microscopy with GnRH immunocytochemistry applied to those neurons residing in the olfactory epithelium showed an unusual distribution of immunoreactivity: It was most evident tightly surrounding the nuclear envelope. As GnRH neurons migrated, their shape was much different—they were elongated according to the axis of migration—and more of the GnRH immunoreactivity was seen distributed throughout the cytoplasm. At no time before or during migration did we observe immunoreactive secretory vesicles.

Thus, there is a striking absence of evidence that GnRH is secreted during migration in such a manner as to guide the movements of its own cells.

Is the GnRH gene actually being expressed in cells whose origin lies in the olfactory placode, or has the prohormone been absorbed from someplace else? Likewise, could these unusual cells be producing a product that is immunologically similar to GnRH, but is actually something different? We used riboprobes antisense to the GnRH message to detect cells expressing the GnRH gene; as controls, sense riboprobes were prepared and used in a similar fashion. In fetal mice of the appropriate ages, we found that cells near the olfactory placode actually produce GnRH mRNA (3). In these results as well, GnRH gene-expressing cells tended to be found in clusters.

Is GnRH merely one of many peptides that have a neuroendocrine function and whose progenitor cells make the migration from olfactory placode to brain? To answer this, we first replicated the GnRH neuronal migration in a new batch of experimental mouse tissue. Then, using tissue from the same animals, we performed immunocytochemistry for several other peptides of neuroendocrine importance. It was clear that at least among the peptides tested so far, GnRH neurons are unique in making the migration from the olfactory placode (3).

For example, antiserum against neuropeptide Y revealed no cells in or near the olfactory placode or on the nasal septum even though positive control tissue indicated that under our immunocytochemical conditions, the antiserum did recognize neuropeptide Y cells in the adult mouse brain. Likewise, somatostatin immunocytochemistry performed with tissue from the fetal mouse olfactory apparatus gave negative results even though the same antibody used as a positive control in adult mouse brain tissue revealed immunoreactive cells. Since we sampled only a few among the large number of peptides delivered to the anterior pituitary, to posterior pituitary, or those that influence releasing factor cells, it is entirely possible that another neuroendocrine cell type is also born in the olfactory epithelium. Also to be taken into account are those nonendocrine cells connected with the olfactory system proper that migrate towards the brain and that might have important influences on the related vomeronasal and nervus terminalis systems.

If, indeed, the migration of GnRH neurons occurs, then we should be able to rule out the alternative hypothesis that immunocytochemical studies, instead, were sensitive to a *wave of expression* in cells distributed along the proposed migration route, even though the cells themselves were actually stationary. This alternative hypothesis has been ruled out in two ways, both involving blockage of migration (also, see below). First, in mice the application of an antiserum to *neural cell adhesion molecule* (NCAM) can block migration if and only if it blocks the olfactory placode on the proper day of life in the fetal mouse (4). Second, in human tissue

Kallmann's syndrome can be occasioned by a blockage of GnRH neurons as they attempt to leave the nose and head for the brain. In both cases the interruption of an event in the nasal cavity prevents the normal appearance of GnRH neurons in the basal forebrain, those providing support for the very existence and direction of GnRH neuronal migration.

It remained a possibility that the origin of GnRH neurons in the olfactory pit might be a phenomenon unique to this rodent species instead of a phenomenon common to other vertebrates. Results from several laboratories indicate that the migration as discovered in mice applies to a large number of vertebrates. In our own studies results from guinea pigs, rats, chickens, and human tissue support the discovery in mice.

Kallmann's Syndrome

Since the initial mouse results concerned neurons important for the physiology of reproduction and since the migration phenomenon linked olfactory tissue with reproductive neuroendocrine tissue, there was a striking parallel to the pattern of deficiencies seen in *Kallmann's syndrome:* hypogonadotropic hypogonadism coupled with anosmia. Human GnRH deficiency syndromes have recently been reviewed (5). From our mouse results it appeared that in Kallmann's syndrome a GnRH deficiency could potentially be explained by a failure of migration of GnRH neurons and that migration failures in the olfactory system proper might lead to a failure of induction of the olfactory bulbs, thus explaining the anosmia.

Collaborating with medical geneticist David Bick (6), we found that even at a fetal age where matched human controls had GnRH-expressing neurons as expected in the preoptic area and hypothalamus, the Kallmann's tissue sample had no GnRH immunoreactive cells in the brain. Instead, in tissue from the Kallmann's individual that had massive X-chromosome damage, GnRH neurons were dammed up in the nose. These observations indicated clearly that the gene for GnRH itself was being expressed normally to produce an immunoreactive peptide, but that the GnRH neuronal migration had been interrupted.

It is important to note that even though this particular Kallmann's case had X-chromosome damage, by no means do all Kallmann's sufferers have such damage. Instead, any abnormality that interferes with GnRH neuronal migration—and this could encompass a panoply of mechanical and chemical alterations—could lead to Kallmann's symptoms. We hypothesize, therefore, that a multiplicity of conditions, sometimes quite separate from obvious chromosomal damage, could lead to partial reproductive failures as regards perturbations of the normal GnRH neuronal migration pattern. Further, it should be emphasized that this sort of reproductive failure need not be an all-or-none situation. Partial failures of migration might lead to abnormal patterns of GnRH secretion under circumstances where the synthesis or release capacities of the resultant neuroendocrine neuronal population are challenged. That is, a wide variety of environmental conditions—at the biological level—and a large number of synaptic inputs—at the neural circuitry level—are required for normal, biologically adaptive reproductive processes.

Environmentally, not only must steroid hormones properly influence neurons in limbic and hypothalamic structures, but adequate nutrition, adequate water and salt supplies, time of day, length of day, and appropriate temperature are also required, as well as the presence of nesting material, the presence of an adequate mating partner, and a relative absence of stress. When any of these environmental parameters are marginal, deficiencies in a suboptimal population of GnRH neurons whose migration was only partly disrupted might be revealed. At the neural circuitry level, a rich variety of transmitter and peptide inputs to GnRH neurons have been described by anatomical and pharmacological means (7, 8). A population of GnRH neurons whose migration was incomplete or poorly timed might have less than a full complement of neurotransmitter and neuropeptide influences. In turn, such a situation may lead to reproductive failure under circumstances that are less than optimal.

A gene absent or damaged in at least some cases of Kallmann's syndrome has been cloned (9, 10). That is, previously, a portion of Kallmann's cases, X-linked Kallmann's, had been localized in the Xp22.3 region near the position of a steroid sulfatase gene. Deletions and mutations of the candidate gene were strongly correlated with Kallmann's symptoms. Indeed, some features of the amino acid sequence of the gene make it reasonable to speculate that it is related to the neuronal migration essential for normal reproductive function. Not only is its structure consistent with an extracellular location, but fibronectin repeats present in the candidate Kallmann's gene are also found in many proteins related to cell adhesion. Thus, at least for X-linked Kallmann's, the first interpretation could be that in such patients a protein essential for normal GnRH neuronal migration was absent or damaged.

However, things do not seem to be that simple. First, looking at the basal expression of the Kallmann's gene protein product in the neural tissue did not yield neuroanatomical results whose temporal and spatial properties make it easy to explain how the absence of such expression would lead to a failure of GnRH neuronal migration (11, 12). Second, working in collaboration with Dr. Petit and her colleagues, we have preliminary results that have a paradoxical character. That is, performing immunocytochemistry with antibodies to peptides that represent part of the candidate Kallmann's gene, we have new results indicating that it is actually easier to see the immunoreactive product in brain tissue from the fetal Kallmann's tissue than it is from control human tissue. This does not fit the simplest interpretation given above. Nonspecific immunocyto-chemical staining alone cannot explain these results since the control human brain tissue of 3 individuals has been uniformly negative.

One possible insight to this paradox comes from the observation (13) that there is a homologous gene on the Y-chromosome at Yq11. The structure of the Y-chromosome pseudogene shows deletions and base substitutions, as well as a relation to the steroid sulfatase pseudogene on the Y-chromosome, consistent with a chromosomal rearrangement event onto the long arm of the Y-chromosome during primate evolution (13). The high degree of homology between the X- and the Y-Kallmann's gene sequences (86% to 98% for exons and 86% to 99% for introns) suggests that the peptide used to generate the antibody for the preliminary im-munocytochemical studies may be represented in the Y-chromosome pseudogene, and this indeed appears to be the case. Thus, an essential feature of the genetic pathology of Kallmann's syndrome (eventually yielding failure of neuronal migration) may not simply be the absence of the protein from the X-chromosome, but instead the presence of a protein product from the Y-chromosome pseudogene. While both this new work and the interpretations deriving therefrom are in their initial stages, it may not be too early to speculate that the simple absence of the candidate Kallmann's gene from the X-chromosome will not explain the disease and that, instead, the presence of a normal X-Kallmann's gene helps to suppress the pseudogene gene on the Y-chromosome.

In summary, there is not yet an overall, simple explanation of Kall-mann's. First, the X-linked Kallmann's represents only a fraction of all cases, and linkages to other chromosomes must all also be explored. Second, even for X-linked Kallmann's, it is not yet possible to reason directly from the absence of an X-chromosome gene to a failure of neuronal migration.

Mechanisms

It is highly unlikely that a single feature of the GnRH neuronal migration route explains the entire passage of these neurons from olfactory epi-thelium to basal forebrain. Among mechanical signals, those signals that exclude are the most obvious; that is, cells unable to go in a medial direction because of the septum nor in a lateral direction because of the surface of the nasal epithelium. Chemical signals for migration are cer-tainly more complicated. They can have either an attractive or propulsive force—NCAM may be one of these—but they also could cause a nerve cell to stop migrating—cytotactin could be one of these.

Using double-labeled immunocytochemistry for GnRH as well as for NCAM, we found that GnRH-expressing cells migrate into the brain along and within the medial parts of a scaffolding formed by the NCAM-immunoreactive axons of the vomeronasal and terminal nerves. Never have GnRH cells been seen on the migration route independent of this NCAM scaffolding (14); yet, we have never seen the same cell expressing both GnRH and NCAM. The question of where the NCAM comes from is under investigation in this laboratory. NCAM-immunoreactive cells that migrate out of the epithelium of the olfactory pit appear to be *pioneer cells* that form an aggregate between the olfactory apparatus and the developing forebrain. The central processes of the olfactory vomeronasal and terminal nerves grow into the cellular aggregate that in turn adheres to the tip of the rostral forebrain. The olfactory nerves, originating from the epithelium of the lateral olfactory pit, grow into the lateral parts of the aggregate and eventually form the olfactory nerve layer of the olfactory bulb. The vomeronasal and terminal nerves, as well as the GnRH neurons, originate from the epithelium of the medial part of the olfactory pit and form the migration route for GnRH neurons into the brain. By day E11.5 in the embryonic mouse, the GnRH neurons may be seen on this NCAM-immunoreactive migration route.

It appears that the pioneer cells may not be neurons. Our results so far show that the majority of cells in the NCAM-immunoreactive aggregate that joins olfactory apparatus to forebrain are not *neuron-specific enolase* (NSE)-positive cells (15). If they are glial cells, the type of glial cell and stage of maturity they would be have not been determined. NSE-immunoreactive cells in the aggregate are limited to the medial edge, and here, double-labeled immunocytochemistry indicates that the GnRH-immunoreactive cells are NSE positive. Thus, it appears that GnRH neurons traveling up the scaffolding traverse the medial side of the NCAM-positive cellular aggregate, using it as a bridge between the epithelium of the medial olfactory pit and the basal forebrain.

While the correlation between NCAM immunoreactivity and migrating GnRH neurons is visually dramatic, it remained to be shown that NCAM plays a causal role in the migration. The question was addressed by using a surgical approach and microinjection techniques to deliver small amounts of NCAM antibody into the area of the olfactory pit in 10-day-old embryonic mice (16). If and only if immunocytochemistry proved that the antibody actually penetrated the epithelium of the olfactory pit was the number of GnRH neurons in the epithelium and on the migration route along the nasal septum reduced to almost zero. This is highly unlikely to be a nonspecific disruption of neuronal movement since vehicle administration, control immunoglobulin administration, and purposeful mechanical disruption did not have the same effect. Moreover, when the anti-NCAM antibody was delivered but did not penetrate the epithelium of the olfactory pit, the same striking result was not seen.

Results from another study, preliminary and still ongoing, further indicate that the GnRH neuronal migration can survive morphologic perturbations. Injection of methotrexate, a folic acid (vitamin B) antagonist, into a pregnant mouse on day 9 of gestation results in a variety of midline frontal-nasal abnormalities due to its action on the vasculature. Despite these morphologic abnormalities of the olfactory placode and forebrain, it is possible to see GnRH neurons migrating along the expected route (Schwanzel-Fukuda, study in progress).

While the interruption of migration by the successful administration of an anti-NCAM antibody was the predicted result, another surprising effect also occurred. Not only were fewer GnRH cells seen on the migration route past the olfactory placode, but also fewer cells were seen expressing immunoreactive GnRH in the epithelium of the medial olfactory pit itself. On the one hand, this finding raises a complication: It seems possible that GnRH neurons are actually migrating, but because they are not expressing an immunoreactive product, we cannot see them. A second and more likely possibility is that NCAM is not only necessary for normal migration, but is also necessary for the differentiation of the olfactory epithelium cells to the stage of producing GnRH in the first place. Thus, this cell adhesion molecule may foster the differentiation of GnRH neurons as well as the migration.

Gene Transfer into Specific Regions of Adult Brain

Although genetic loss or damage associated with Kallmann's syndrome is distinct from the damage to the GnRH gene evident in hypogonadal mice, both raise the question of how to prove that specific molecular events are required for specific endocrine or behavioral events. After all, the entire body of cloning and sequencing work for genes of interest to endocrinologists and neurobiologists, as well as an entire generation of work on regulation and expression, provide only *correlational* evidence for the involvement of specific genes and specific functions. One infers that because elevation of expression of a particular gene is associated with heightened function, the two events are correlated, but this is not a certain conclusion. Therefore, in order to prove causality, active manipulation of gene expression is required. Transfection of naked DNA, even using lipofection techniques, can be done, but it would be desirable to have processes of gene administration that are more efficient and that last for a longer duration.

Along these lines, although the creation of transgenic mice has permitted an entirely new type of experiment in developmental biology, such studies are prohibitively difficult and expensive for work on adult brain function, particularly for species other than mice. Therefore, launching a new field of work based on the neurotropism of the herpes simplex virus and derived from the literature on defective viral vectors (17) was important for permitting the application of foreign genes to selected groups of neurons in the adult brain. In a defective viral vector, termed an *Amplicon*, using the CMV immediate early promoter with the β-galactosidase gene as a reporter, it was proven possible to express foreign genes in predetermined regions of the adult rat brain for the first time (18).

At least three purposes can be envisioned for this tool. The first and most obvious use is gene replacement therapy in which the gene administered in our defective viral vector replaces that which is missing from a mutant. Second, for antisense experimentation or therapy, the defective viral vector can be used to heighten the strength and, in particular, the duration of action of antisense sequences that would block translation of the mRNA for a specific gene. Finally and most subtly, the defective viral vector can be used for in vivo promoter analysis. A very large percentage of promoter analyses are done in arbitrarily chosen cell lines in which the factors that permit basal or regulated expression might be different from the cell type to which the conclusions will be applied. With the Amplicon we can use fragments of the promoter of any given gene of neurobiologic interest to drive the reporter β-galactosidase in order to find out which portions of the promoter are necessary for basal or regulated expression. Thus, we have shown that 2700 bases of the rat preproenkephalin promoter allows expression among cell types that would ordinarily express endogenous enkephalin and that successful expression can last as long as 2 months (19).

Using this molecular tool for the analysis of GnRH synthesis and function, we note that the hypogonadal mouse has deficient gonadal function caused by a large deletion in the distal half of the gene for the GnRH precursor (20). In an exciting experiment the introduction of an intact GnRH gene by transgenic procedures into these mutant mice resulted in a reversal of reproductive deficiencies (21). To analyze the exact features of the GnRH gene and the promoter that is necessary, we are using the defective viral vector technology mentioned above (Byrnes et al., unpublished data) and have prepared vectors containing the normal GnRH gene with or without a second cassette containing the β-galactosidase marker. In collaboration with Dr. Harry Charlton, these vectors are microinjected into the preoptic area and hypothalamus of hypogonadal mice.

The ability to yield transcription and expression of GnRH as detected by in situ hybridization and immunocytochemistry, respectively, is a proximal goal of these experiments, while the effects on reproductive physiological end points will be a long-term goal. Technically, avoiding killing the would-be GnRH neurons by excessive viral load is expected to be a major problem. On the other hand, the ability of relatively small numbers of GnRH neurons in the mouse to support gonadotropin release gives reason for optimism in this GnRH gene transfer project.

References

1. Schwanzel-Fukuda M, Jorgenson KL, Bergen H, Weesner G, Pfaff DW. Biology of normal LHRH neurons during and after their migration from olfactory placode. Endocr Rev 1992;13:623–34.
2. Schwanzel-Fukuda M, Pfaff DW. Origin of luteinizing hormone-releasing hormone neurons. Nature 1989;338:161–5.
3. Schwanzel-Fukuda M, Zheng LM, Bergen HT, Weesner GD, Pfaff DW. LHRH neurons: functions and development. Prog Brain Res 1992;93: 189–203.
4. Schwanzel-Fukuda M, et al. Disruption of LHRH neuronal migration by anti-NCAM antibody on olfactory placode. J Comp Neurol (submitted).
5. Crowley WF, Jameson JL. GnRH deficiency: perspectives from clinical investigation. Endocr Rev 1992;13:635.
6. Schwanzel-Fukuda M, Bick D, Pfaff DW. Luteinizing hormone-releasing hormone (LHRH)-expressing cells do not migrate normally in an inherited hypogonadal (Kallmann) syndrome. Mol Brain Res 1989;6:311–9.
7. Silverman A. The gonadotropin-releasing hormone (GnRH) neuronal systems: immunocytochemistry. In: Knobil E, Neill J, eds. The physiology of reproduction. New York: Raven Press, 1988:1283–304.
8. Weiner RI, Findell PR, Kordon C. Role of classic and peptide neuromediators in the neuroendocrine regulation of LH and prolactin. In: Knobil E, Neill J, eds. The physiology of reproduction. New York: Raven Press, 1988:1235–81.
9. Legouis R, Petit C, et al. The candidate gene for the X-linked Kallmann syndrome encodes a protein related to adhesion molecules. Cell 1991;67: 423–35.
10. Franco B, Guioli S, Pragliola A, et al. A gene deleted in Kallmann's syndrome shares homology with neural cell adhesion and axonal path-finding molecules. Nature 1991;353:529–36.
11. Legouis R, Ayer-LeLievre C, Leibovici M, Lapointe F, Petit C. Neuronal expression of the Kallmann gene during chicken development. PNAS. (submitted).
12. Rugarli EI, Lutz B, Kuratani C, Borsani G, Ballabio A, Eichele G. Expression pattern of the Kallmann's syndrome gene in the olfactory system indicates a role in neural target recognition mechanisms (submitted).
13. del Castillo I, Cohen-Salmon M, Blanchard S, Lutfalla G, Petit C. Structure of the gene responsible for the X-linked Kallmann syndrome and its homologous pseudogene on the Y chromosome. Nature (in press).
14. Schwanzel-Fukuda M, Abraham S, Crossin KL, Edelman GM, Pfaff DW. Immunocytochemical demonstration of neural cell adhesion molecule (NCAM) along the migration route of luteinizing hormone-releasing hormone (LHRH) neurons in mice. J Comp Neurol 1992;32:1–8.
15. Schwanzel-Fukuda M, Pfaff DW. Neuron specific enolase (NSE)-immunoreactivity is present in cells on the migration route of luteinizing hormone-releasing hormone (LHRH) neurons originating from the olfactory placode in mice. Soc Neurosci Abstr 1992;18:233.
16. Schwanzel-Fukuda M, et al. Disruption of LHRH neuronal migration by anti-NCAM antibody on olfactory placode. J Comp Neurol (submitted).

17. Kaplitt MG, Rabkin SD, Pfaff DW. Molecular alterations in nerve cells: direct manipulation and physiological mediation. Curr Top Neuroendocrinol 1993;11:169–91.
18. Kaplitt MG, Pfaus J, Rabkin S, Pfaff DW. Expression of a functional foreign gene in adult mammalian brain following in vivo transfer via a herpes simplex virus type 1 defective viral vector. Mol Cell Neurosci 1991;2:320–30.
19. Kaplitt MG, Rabkin S, Pfaff DW. In vivo promoter analysis in brain using a herpes simplex vector. PNAS (submitted).
20. Mason AJ, Hayflick JS, Zoeller RT, et al. A deletion truncating the gonadotropin-releasing hormone gene is responsible for hypogonadism in the hpg mouse. Science 1986;234:1366–71.
21. Mason AJ, Pitts SL, Nikolics K, et al. The hypogonadal mouse: reproductive functions restored by gene therapy. Science 1986;234:1372–8.

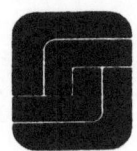

Author Index

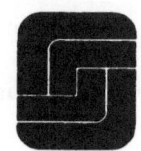

Subject Index